IEE ENERGY SERIES 1

SERIES EDITORS: M. BARAK AND PROFESSOR D.T. SWIFT-HOOK

Electrochemical Power Sources

PRIMARY & SECONDARY BATTERIES

Edited by

M. Barak, M.Sc., D.Phil., F.R.I.C., C.Eng., F.I.E.E.

Consultant

Burgess Hill, Sussex

England

PETER PEREGRINUS LTD.

on behalf of the

Institution of Electrical Engineers

Published by: The Institution of Electrical Engineers, London
and New York
Peter Peregrinus Ltd., Stevenage, UK, and New York

British Library Cataloguing in Publication Data

Electrochemical power sources.
- (Institution of Electrical Engineers.
IEE energy series; 1).
1. Electric batteries
2. Electric cells
I. Barak, M II. Series
621.35 TK2901 80-40412

ISBN 0-906048-26-5

Printed in England by A. Wheaton & Co., Ltd., Exeter

To Jo, but for whose inexhaustible patience, Chapter 4 would never have been completed.

Authors

BARAK, M.
Consultant
Baydon Cottage, 61 Folders Lane, Burgess Hill, Sussex

DICKINSON, T. (deceased)
Senior Lecturer in Physical Chemistry
Chemistry Department
The University
Newcastle upon Tyne

FALK, U.
Director, Technical Co-ordination
NIFE Jungner AB
Oskarshamn
Sweden

SUDWORTH, J.L.
Head of the Electrochemistry Section
R & D Division
British Rail Technical Centre
Derby

THIRSK, H.R.
Professor of Electrochemistry
Chemistry Department
The University
Newcastle upon Tyne

TYE, F.L.
General Manager
Group Technical Centre
Berec Group Ltd.
London

Contents

		Page
Authors		*iv*
Preface		*xii*
1	**Primary and secondary batteries: fuel cells and metal/air cells**	1
	Dr. M. Barak	
	1.1 Introduction	1
	1.1.1 General principles	1
	1.1.2 Common cells and batteries with aqueous electrolytes	4
	1.1.3 Properties of materials for anodes and cathodes	5
	1.1.4 Cells in the research and development stage or pilot plant production	7
	1.1.5 Historical notes	7
	1.1.5.1 Power sources and electrochemistry	7
	1.1.5.2 Gas polarisation and the Leclanché cell	11
	1.1.5.3 Origins of the lead/acid storage battery; the Planté process	12
	1.1.5.4 The Faure-Brush pasted plate	15
	1.1.5.5 Standard cells	15
	1.1.5.6 Alkaline batteries	16
	1.1.5.7 Fuel cells	16
	1.1.5.8 Metal/air (or oxygen) cells	18
	1.1.5.9 Metal/halogen cells	19
	1.1.5.10 High temperature batteries	20
	1.1.5.11 Cells with lithium anodes	20
	1.1.5.12 Cells with solid electrolytes at room temperatures	20
	References	21
2	**Definitions and basic principles**	23
	H.R. Thirsk	23
	2.1 Introduction	23
	2.2 Symbols	24
	2.3 Electrodes and the galvanic cell	24

2.4 Basic thermodynamics and the efficiency of cells 27
2.5 Efficiency 29
2.6 The metal electrode solution interface 30
2.7 Adsorption and kinetics 32
2.8 Electrochemical kinetics 33
2.9 Steady state procedures 34
2.10 Simple electron transfer 35
2.11 Rotating disc electrode 37
2.12 Non-steady state methods 39
2.13 Nucleation and growth: potential step and a.c. perturbations 39
2.14 A.C. perturbations 44
References 47

3 Primary batteries for civilian use 50
F.L. Tye
3.1 Introduction 50
3.2 Standardisation and nomenclature 51
3.2.1 Standardisation 51
3.2.2 Nomenclature 52
3.3 Terminals and dimensions 58
3.3.1 Polarity 58
3.3.2 Cylindrical batteries 58
3.3.3 Button batteries 61
3.3.4 Square cell batteries 63
3.3.5 Multi-cell batteries 63
3.3.6 Advantages of the IEC system 64
3.4 Construction and electrode structure 67
3.4.1 Cylindrical Leclanché and alkaline manganese batteries 67
3.4.2 Square cell Leclanché batteries 70
3.4.3 Neutral air-depolarised batteries 71
3.4.4 Flat cell Leclanché batteries 71
3.4.5 Comparison of electrode volumes 72
3.4.6 Button batteries: Mercury (II) oxide/zinc and Silver (I) oxide/zinc 75
3.4.7 Cylindrical mercury batteries 78
3.5 Capacity and electrical performance 78
3.5.1 Total capacity 78
3.5.2 Practical discharge capacities 79
3.5.3 IEC application tests 84
3.5.4 Discharge durations of Leclanché and alkaline manganese batteries on IEC application tests 88
3.5.5 Capacity yield of Leclanché batteries 91
3.5.6 Capacity yield of alkaline manganese batteries 95
3.5.7 Neutral air-depolarised batteries 96
3.5.8 Button batteries 97
3.6 Overall cell reactions 99
3.7 Open-circuit cell potentials 101

3.7.1 Heterogeneous and homogeneous reactions 101
3.7.2 Effect of composition on the potential of manganese oxyhydroxy electrodes 103
3.7.3 Calculated standard cell potentials 105
3.7.4 Equilibrium pH effects 108
3.7.5 Influence of compositional changes during discharge on open-circuit voltage 110
3.8 Cathode discharge mechanisms 113
3.8.1 Leclanché cells 113
3.8.2 Alkaline manganese cells 120
3.8.3 Mercury cells 122
3.8.4 Silver (Ag_2O) cells 123
3.9 Anode discharge mechanisms 125
3.9.1 Leclanché cells 125
3.9.2 Alkaline cells 128
Acknowledgments 136
References 137

4 Lead/acid storage batteries 151
M. Barak
4.1 Introduction 151
4.2 Classes of storage batteries 152
4.2.1 Automotive batteries 152
4.2.2 Traction batteries 154
4.2.3 Stationary batteries 156
4.2.4 Miscellaneous applications 156
4.3 Active materials and electrochemical reactions 156
4.3.1 Active materials 156
4.3.1.1 Positive electrode or 'cathode', lead dioxide (PbO_2) 156
4.3.1.2 Negative electrode or 'anode', spongy lead (Pb) 157
4.3.2 Electrolyte 157
4.3.2.1 Purity standards for sulphuric acid 157
4.3.2.2 Purity standards for water 161
4.3.3 Electrode potentials 162
4.3.4 Electrochemical reactions 163
4.4 Polarisation and kinetic aspects 166
4.4.1 Activation polarisation 167
4.4.2 Concentration polarisation 167
4.4.3 Resistance polarisation 169
4.4.4 Discharge-voltage characteristics 170
4.4.4.1 Effect of current density on capacity 170
4.4.4.2 Effect of temperature on capacity 173
4.4.4.3 Effect of plate thickness on the coefficient of use of the active materials 175
4.4.5 Characteristics during charge 177
4.4.6 Over-voltage and the Tafel expression 180

4.5 Oxides of lead 182
4.5.1 Lead monoxide PbO. Ratio Pb:O = 1:1 183
4.5.2 Red lead or minimum Pb_3O_4. Ratio Pb:O = 1:1·33 183
4.5.3 Lead dioxide PbO_2. Ratio PbO = 1:2 (nominally) 184
4.5.3.1 Stoichiometry 184
4.5.3.2 Polymorphism 185
4.5.4 Mechanism of the formation of lead sulphate during discharge 187
4.5.4.1 The lead anode 187
4.5.4.2 The PbO_2 cathode 188
4.5.4.3 Coup de fouet or Spannungsak 190
4.6 Manufacturing processes 191
4.6.1 The Planté process 191
4.6.2 The pasted plate 192
4.6.2.1 Design of die-cast grids for pasted plates 193
4.6.2.2 Wrought metal grids 197
4.6.2.3 Lead antimony alloys for grids 199
4.6.2.4 Alternative hardeners 200
4.6.2.5 Effect of additives to lead antimony alloys 202
4.6.2.6 Dispersion-strengthened lead (DSL) 206
4.6.2.7 Miscellaneous metals for grids 208
4.6.2.8 Plastics for grids 209
4.6.2.9 Alloys for maintenance-free batteries 210
4.6.2.10 The effect of antimony on the positive active material 216
4.6.3 Grid casting machines 220
4.6.4 Hand casting 221
4.6.5 Oxides for pastes: methods of manufacture 221
4.6.5.1 Grey oxide 221
4.6.5.2 The Barton or Linklater pot process 222
4.6.5.3 Fume oxide 223
4.6.6 Properties of oxides 223
4.6.6.1 Polymorphism 223
4.6.6.2 Particle size 223
4.6.6.3 Surface area 225
4.6.6.4 Porosity 226
4.6.6.5 Apparent density 227
4.6.6.6 Water (or acid) absorption 227
4.6.6.7 Purity standards 227
4.6 7 The paste-making process 228
4.6.7.1 Paste density 230
4.6.7.2 Additives to negative paste 230
4.6.7.3 Additives to positive paste 231
4.6.8 The pasting process 232
4.6.8.1 Machine pasting 232

4.6.8.2 Hand pasting 233
4.6.8.3 Chemical processes in setting and drying 233
4.6.9 Formation 242
4.6.9.1 Tank formation 242
4.6.9.2 Jar formation 243
4.6.10 Drying processes 243
4.6.10.1 Tank-formed plates 243
4.6.10.2 The dry-charging process 244
4.6.11 Separators and 'retainers' 245
4.6.12 Containers and covers 248
4.6.13 Assembly of cells and batteries 249
4.6.14 Weight analysis of typical SLI 12V batteries 250
4.5 Performance of SLI batteries 251
4.7.1 General characteristics 251
4.7.2 Specifications for SLI batteries 253
4.7.2.1 Cranking current in amps (CCA) 254
4.7.2.2 Reserve capacity 255
4.7.2.3 Capacity at the 20h rate at ambient temperatures – 23°C (77°F) 255
4.7.2.4 Miscellaneous tests 255
4.7.2.5 High rate performance at – 17·8°C 256
4.7.2.6 Effect of temperature on high rate performance 257
4.7.2.7 Charging equipment for SLI batteries 258
4.8 Batteries for electric traction 261
4.8.1 Flat-plate, glass wool assemblies 263
4.8.1.1 Alloys for grids 263
4.8.1.2 Positive and negative pastes 264
4.8.1.3 Separators, assembly and containers 264
4.8.1.4 Formation 265
4.8.2 Tubular plate assemblies 265
4.8.2.1 The Ironclad plate 265
4.8.2.2 New plastic materials for tubes 266
4.8.2.3 Alloys and spine-casting procedures 268
4.8.2.4 Tube-filling material 269
4.8.2.5 Cell assembly, separators and formation 270
4.9 Performance of traction cells and batteries 271
4.9.1 General aspects 271
4.9.2 Discharge voltage characteristics 273
4.9.3 Energy and power densities 275
4.9.4 Weight analyses of traction cells 278
4.9.5 Charging of traction batteries 278
4.9.5.1 Single-step chargers 279
4.9.5.2 Two-step chargers 281
4.9.5.3 Automatic control devices 282
4.10 Stationary batteries 282
4.10.1 Service and battery types 282
4.10.2 High performance Planté batteries (type HPP) 283

4.10.3 Cells with tubular and flat pasted plates 285
4.10.3.1 Telephone and telecommunications services 286
4.10.3.2 Static un-interruptible power systems (UPS) 288
4.11 Non-spill batteries for cordless appliances 289
4.11.1 Gelled electrolyte 290
4.11.2 Supported active material 290
4.11.2.1 Applications of batteries with supported active materials (SAM) 291
4.11.3 Developments in semi-sealed and sealed batteries 291
4.11.4 Gas-tight assemblies 293
4.11.4.1 Direct gas re-combination 295
4.11.5.2 Indirect gas re-combination 297
4.11.5 Practical semi-sealed cylindrical cells 300
4.12 Batteries for aircraft 306
4.13 Batteries for submarines 308
4.13.2 Standard types 308
4.13.2 Double-decker construction 312
Acknowledgments 313
References 314

5 Alkaline storage batteries 324
U. Falk
5.1 Introduction 324
5.2 Nickel/cadmium pocket type batteries 331
5.2.1 Reaction mechanisms 331
5.2.2 Manufacturing processes 333
5.2.3 Performance characteristics 339
5.2.4 Applications 346
5.3 Nickel/iron batteries 349
5.3.1 Reaction mechanisms 350
5.3.2 Manufacturing processes 350
5.3.3 Performance characteristics 353
5.3.4 Applications 357
5.4 Nickel/cadmium sintered plate batteries 358
5.4.1 Reaction mechanisms 358
5.4.2 Manufacturing processes 359
5.4.3 Performance characteristics 368
5.4.4 Applications 378
5.5 Silver/zinc and silver/cadmium batteries 380
5.5.1 Reaction mechanisms 380
5.5.2 Manufacturing processes 382
5.5.3 Performance characteristics 386
5.5.4 Applications 390
5.6 Nickel/zinc batteries 391
5.6.1 Reaction mechanisms 391
5.6.2 Manufacturing processes 392
5.6.3 Performance characteristics 394

5.6.4 Applications 397
References 398

6 **High temperature batteries** **403**
J.L. Sudworth
6.1 Introduction 403
6.2 Thermal management 404
6.3 Lithium/chlorine cells 407
6.4 Lithium-aluminium chlorine cells 410
6.4.1 The positive electrode 410
6.4.2 The negative electrode 411
6.4.3 Cell design and performance 412
6.5 Lithium/sulphur cells 412
6.6 Lithium alloy/metal sulphide cells 415
6.6.1 Lithium-aluminium/iron sulphide cells 415
6.6.1.1 Li-Al/FeS_2 cells 415
6.6.1.2 Li-Al/FeS cells 416
6.6.1.3 Cells assembled in the discharged state 417
6.6.1.4 Behaviour of Li-Al/FeS_2 cells on overcharge 419
6.6.2 Li-Si/FeS_2 cells 420
6.6.3 Status of lithium alloy/iron sulphide cells 421
6.7 Sodium/sulphur cell 421
6.7.1 Introduction 421
6.7.2 The sodium/sulphur cell with beta alumina electrolyte 423
6.7.2.1 The solid electrolyte 423
6.7.2.1.1 Properties of beta″ alumina 425
6.7.2.1.2 Preparation of beta alumina ceramics 428
6.7.2.2 Flat plate cells 429
6.7.2.3 Central sodium tube cells 430
6.7.2.4 Central sulphur tube cells 431
6.7.2.5 Cell materials 432
6.7.2.6 The positive electrode 436
6.7.2.7 The negative electrode 441
6.7.2.8 Cell safety 440
6.7.2.9 Present status of sodium sulphur cells with beta alumina electrolyte 441
6.7.3 The sodium sulphur cell with glass electrolyte 441
6.7.3.1 The solid electrolyte 441
6.7.3.2 The negative electrode 443
6.7.3.3 The positive electrode 443
6.7.3.4 Cell materials 444
6.7.3.5 Outstanding problems 445
6.8 Sodium/antimony trichloride cell 445
6.8.1 Introduction 445
6.8.2 The negative electrode 447

6.8.3 The positive electrode 448
6.8.4 Cell design 449
6.8.5 Cell performance 451
6.9 Comparison of battery systems 454
6.9.1 Molten salt electrolyte batteries 454
6.9.2 Solid electrolyte batteries 455
6.9.3 Future prospects 457
Acknowledgements 457
References 458

7 Room temperature cells with solid electrolytes 464
T. Dickinson
7.1 Introduction 464
7.2 Silver ion conductors 465
7.2.1 Electrolytes 465
7.2.2 Power sources 465
7.3 Copper ion conductors 466
7.3.1 Electrolytes 466
7.3.2 Power sources 466
7.4 Proton conductors 467
7.4.1 Electrolytes 467
7.4.2 Power sources 467
7.5 Sodium ion conductors 467
7.5.1 Electrolytes 467
7.5.2 Power sources 468
7.6 Fluoride ion conductors 468
7.6.1 Electrolytes 468
7.6.2 Power sources 468
7.7 Lithium ion conductors 469
7.7.1 Electrolytes 469
7.7.1.1 Lithium halides 469
7.7.1.2 Other lithium compounds 470
7.7.2 Power sources 471
Acknowledgements 479
References 479

Preface

During the past 25 years or so, a number of monographs on electrochemistry and different battery systems have appeared. These are admirable publications, but anyone wishing to get comprehensive information about the fundamental principles and the advantages and disadvantages of the different systems must study a variety of different books. Moreover, it is important to bring the state of the art continually up to date by reference to the vast numbers of scientific and technical papers, presented at conferences and meetings and published in various international journals.

The prime intention of the present volume therefore was to bring together in one edition the outstanding, up-to-date features and applications of the most important battery systems, for the benefit of scientists and engineers engaged in research and development or in the technology and applications of the different power sources. The authors of these contributions are among the most eminent and experienced workers in these fields.

The volume opens with a short chapter on the history of the development of voltaic cells from the time of Volta's epoch-making discovery nearly 200 years ago, indicating how the advances were associated with the birth and growing understanding of the science of electrochemistry. Chapter 2 outlines in some detail the basic thermodynamic and kinetic principles underlying the conversion of chemical into electrical energy, on which all of these systems depend.

Chapter 3 deals with the most popular types of primary batteries for civilian use, including neutral air-depolarised, Leclanché, alkaline manganese batteries and mercury/zinc and silver/zinc button cell assemblies. Chapters 4 and 5 describe the two basic groups of rechargeable storage batteries, lead/acid and the alkaline systems, nickel/cadmium, nickel/iron, silver/zinc, silver/cadmium and nickel/zinc, respectively.

All of these systems work at ambient temperatures with aqueous electrolytes. Other types of electrolyte can be used, but in different circumstances, e.g. molten salts or solid ion-permeable diaphragms in high temperature cells, described in Chapter 6. To indicate the extent of this diversity, a brief Chapter 7 is included, describing cells with inorganic electrolyte operating at ambient temperatures.

Unfortunately, limitations of space made it impossible to include three other related aspects of interest, namely voltaic systems with organic electrolytes and anodes of metals, such as lithium, and the development and state of the art of fuel cells and metal/air cells. These must wait for another volume.

It is with the greatest regret that we have to record the sudden death of our colleague and co-author, Dr. T. Dickinson, author of Chapter 7, 'Room temperature cells with solid electrolytes' and we offer our belated sympathy to his family.

The editor wishes to express his appreciation and thanks to his co-authors for their co-operation and support and to Miss M. Hutchins and the book production staff at Peter Peregrinus Ltd. for their considerable help in preparing illustrations and drawings.

M. Barak
December 1979

Chapter 1

Primary and secondary batteries: fuel cells and metal–air cells

M Barak

1.1 Introduction

Primary and secondary batteries, fuel cells and metal-air cells convert the energy of electrochemical reactions directly into low voltage, direct current electricity. Since this conversion does not involve a heat stage, which would impose the constraints of the Carnot cycle, the thermodynamic efficiency can be at least twice that of a thermal power plant. Fig. 1.1 shows schematically the different stages in the operation of a diesel-powered MG set and an electrochemical power source. Some energy is lost as heat by the latter, but even at high rates of discharge this is not large. Electrochemical power sources have a number of other advantages. In many cases the active materials can be stored in their cells for long periods and the electrical energy can be immediately tapped simply by turning a switch and most of these devices operate without any pollution by noise or fumes.

In this book, because of limitations of space, it is possible to deal in detail only with primary and secondary batteries. However, because of the generic similarities of fuel cells and metal-air cells and the fact that their evolution and historical development have been closely associated with those of batteries, it is of interest to include them in this introductory chapter.

1.1.1 General principles

All of these devices contain three major components; a positive electrode or 'cathode', a negative electrode or 'anode', and an electrolyte. * Each electrode when immersed in the electrolyte displays its own characteristic potential or voltage. The significance of these electrode

*For the conventions regarding use of the terms anode and cathode refer to Section 4.3.1., p. 157f

potentials is fully described in Chapter 2. Suffice to say here that the voltage of the complete cell is equal to the difference between the potentials of the cathode and the anode. This cell voltage rarely exceeds 2V, but batteries of cells, connected in series, can be assembled to provide almost any desirable voltage. The electrochemical reactions which take place during a discharge are combined reduction-oxidation or 'redox' reactions. The electric current flows in the external circuit from the cathode to the anode. The cathode is reduced by the absorption of electrons, released by the oxidation of the anode. In the case of

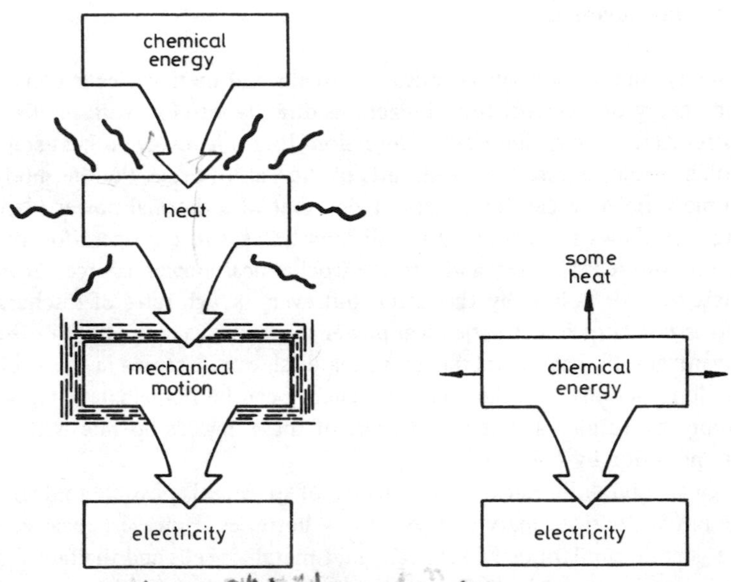

Fig. 1.1 *Diesel MG set and EC power source: conversion stages* [Courtesy B.J. Crowe]

rechargeable storage batteries, the flow of electrons is reversed during a charge, when the positive electrode is technically the anode, and the negative, the cathode, a condition which also exists in the processes of electrolysis and electroplating.

The electrolyte also plays a vital role, since it provides the ions on which the transfer of electrons inside the cell depends. To perform this function efficiently, the electrolyte must have a high ionic conductivity and this in turn requires a high concentration of ions, preferably of

Table 1.1 *Primary cells:- not rechargeable*

Cell	Electrodes positive	Electrodes negative	Electrolyte Aqueous Solution	Open-circuit EMF V
Leclaunché dry cell	MnO_2(C)	Zn	$NH_4Cl/ZnCl_2$	1·60
Alkaline Manganese cell	MnO_2(C)	Zn	KOH	1·55
Mercury cell	HgO	Zn	KOH	1·35
Lalande cell	CuO	Zn	KOH	1·10
Air-depolarised cell	O_2(C)	Zn	KOH or NaOH or NH_4Cl	1·45
Sea-water cell	AgCl	Mg	NaCl	1·30
Reserve electrolyte cells	$CuCl_2$	Mg	NaCl or KCl	1·20
Reserve electrolyte cells	PbO_2	Mg	NaCl or KCl	1·80
Reserve electrolyte cells	PbO_2	Zn	H_2SO_4	2·20
Torpedo cell	PbO_2	Pb	$HClO_4$	1·95

high mobility. Aqueous solutions of strong mineral acids, such as sulphuric, hydrochloric or perchloric acid or their salts and caustic alkalis, sodium or potassium hydroxide, fulfil most of the basic requirements, and, since they can be used at ambient temperatures, they form by far the most common group of electrolytes. Molten salts – chlorides and carbonates, – also have many of the required properties and are used in some electrochemical power sources, notably high temperature cells in which the reactions are based on lithium and sulphur or lithium and chlorine and in fuel cells.

Recently, also, some non-aqueous organic electrolytes, such as propylene carbonate, ethylene carbonate, acetonitrile, tetrahydrofuran and dimethoxyethane, have come into service for use particularly with the metal lithium, which reacts violently with water, but their high electrical resistivity, even when reduced by additions of salts such as lithium perchlorate, limits their use to applications calling for relatively low current drains. Some solid electrolytes also have the required ionic properties, e.g. silver rubidium iodide at ambient temperatures and sodium-doped beta aluminia, used in high temperature sodium/solphur cells.

As already mentioned, the open-circuit voltage of the cell E is equal to the difference between the potentials of the cathode and the *anode*, i.e. $E = E_c - E_a$. When the discharge current is flowing, both electrodes are affected in varying degrees by polarisation and their potentials fall from their open-circuit values by amounts referred to as 'over-voltages'. The cell voltage is further depressed by IR losses due to resistive factors in the electrodes and in the electrolyte. It is obviously an advantage to use a couple, having electrodes with potentials as widely spaced as possible on the potential scale; hence, the efforts to use lithium anodes and fluorine cathodes, for which the values of the standard electrode potentials are, respectively, − 3·0V and + 2·85V, giving for a Li/F cell a theoretical open-circuit voltage of 5·85V.

As shown in Table 1.1, the open-circuit voltage of the lead/acid cell is just over 2·0V, while that of a Ni/Cd alkaline cell is 1·3V. For any given battery voltage, therefore, where five alkaline cells would be needed, only three lead/acid cells of similar capacity (in ampere hours) would suffice.

Table 1.2 *Secondary or rechargeable cells*

				Open-circuit EMF	Energy density
				V	Wh/kg
Lead/acid Storage cell	PbO_2	Pb	H_2SO_4	2·10	25-35
Nickel/cadmium	NiOOH	Cd	KOH	1·30	25-35
Nickel/iron	NiOOH	Fe	KOH	1·35	28
Silver/zinc	Ag_2O	Zn	KOH	1·60	130
Silver/cadmium	Ag_2O	Cd	KOH	1·30	60

1.1.2 Common cells and batteries with aqueous electrolytes

Tables 1.1 and 1.2 list primary and secondary cells in common use. Primary cells are not rechargeable. When fully discharged, chemical changes have exhausted the positive or negative electrode or both and the active materials cannot be regenerated simply by passing an electric current through the cell in the reverse direction. In some cases, some of the capacity can be replaced for a limited number of cycles, but as this recharging process is not reliable for primary cells, it is not recommended.

In the case of secondary cells, however, the prime examples of which are the lead/acid and nickel/cadmium alkaline storage batteries, the capacity can be completely restored by recharging and this cycling operation can be repeated several hundred, or, in the case of some alkaline cells, several thousand times.

Most of these couples are described in detail in later chapters, but some common features should be noted at this stage. Cathodes are, for the most part, metallic oxides with some chlorides, and, of course, oxygen itself. The anodes are, without exception, metals, which are corrodable in varying degrees in the electrolyte. As mentioned previously, the cathode supplies the oxidant, which oxidises the metallic anode. In the secondary cells, the transfer of electrons is finally blocked by the growth of insulating layers of the highly resistant products of the chemical reactions. Mass transport of the electrolyte and the reaction products plays a vital role and persistent study of these factors has led to noteworthy gains in the performance of these cells.

1.1.3 Properties of materials for anodes and cathodes

Weight is generally a crucial factor in portable systems and much research and development effort is expended annually in attempts to reduce the weight of existing systems and in the search for new lightweight couples. The parameters under review are the 'energy density', generally expressed as watt-hours per kg or per dm^3 at any specified rate of discharge, and the 'power density' as watts/kg or /dm^3. The two factors involved are the capacity in ampere-hours (Ah) and the mean voltage during the discharge. Tables 1.3 and 1.4 show the mass density and the theoretical energy density of materials currently in use or of possible interest in new systems. In this case the energy density is expressed as Ah/kg, since the potential or voltage of the material depends on the electrolyte with which it is associated. Strictly speaking, the expression 'Ah/kg' represents the quantity of electricity involved in the chemical transformation of unit weight of the material concerned.

Outstanding features of these two tables are the relatively low energy densities of both the anodic and cathodic materials used in conventional lead/acid and nickel/cadmium alkaline storage batteries and the high values of the anodes, magnesium, sodium and lithium and the cathodes, oxygen, sulphur and chlorine.

Needless to say, these theoretical values for the energy densities are never realised in practice. For reasons described in later chapters, the coefficient of use of the active materials, i.e. the ratio of the capacity obtained to that which is theoretically available at any given rate, falls well below 100%, particularly at high rates of discharge.

Table 1.3 *Properties of anodes*

Material	Mass density	Equivalent weight	Energy density
	g/ml	g/faraday	Ah/kg
Pb	11·3	104	238
Cd	8·7	56	480
Zn	7·1	33	812
Fe	7·9	28	960
Mg	1·7	12	2230
Na	0·97	23	1170
Li	0·53	7	3830

Table 1.4 *Properties of cathodes*

Material	Mass density	Equivalent weight	Energy density
	g/ml	g/faraday	Ah/kg
MnO_2	5·0	43·5	616
HgO	11·0	108	248
CuO	6·0	72	312
AgCl	5·6	144	156
$CuCl_2$	3·0	67	400
PbO_2	9·4	120	223
NiOOH	7·0	91	295
Ag_2O	7·0	62	432
Sulphur	2·0	16	1675
Oxygen	gas	8	3350
Chlorine	gas	35·5	755
Bromine	3·1	80	355
Fluorine	gas	19	1410

1.1.4 Cells in the research and development stage or pilot plant production

Tables 1.5 and 1.6 list a number of systems under test or just emerging from the laboratory. Efforts are being made to use the most favourable materials and also to adopt techniques evolved in related technologies, e.g. porous gas electrodes, used in the development of fuel cells. The state of the art for most of these systems will be described in later chapters.

1.1.5 Historical Notes

1.1.5.1 Power sources and electrochemistry: The discoveries and inventions which led to the successful development of the many types of cells and batteries now in use were closely identified with the birth and growth of the science of electrochemistry. This science and the power sources concerned will always be associated with Luigi Galvani, lecturer in anatomy at the University of Bologna in 1790, and Alessandro Volta, professor of physics at Pavia University in 1800, who gave their names to the 'galvanic cell', 'galvanising', the 'voltaic pile' and the 'volt'. Using a copper probe, Galvani was trying to measure electric pulses which he believed originated in the muscles of the legs of a frog suspended from an iron hook. Volta correctly attributed the involuntary twitching, observed by Galvani, to the current between the unlike metals, copper and iron, with the animal's blood serving as the electrolyte. To prove his point, Volta built his celebrated 'pile', with alternate discs of silver and zinc, interleaved with absorbent paper or cloth, soaked with an electrolyte, such as caustic lye (sodium hydroxide) or brine, and in 1800 he described his invention in a letter to the Royal Society of Great Britain. Volta next constructed a multi-cell battery in the form of a crown, which he called his 'crown of cups', shown diagrammatically in Fig. 1.2. For the first time it was now possible to draw an electric current from a system at a controlled rate. In brief, this enabled Michael Faraday in 1834 to derive the quantitative laws of electrochemistry, which established the fundamental connection between chemical energy and electricity and opened the way to the development of the massive primary and secondary battery industries.

Following Volta's lead, J.F. Daniell (1836) produced his practical primary cell using the metals copper and zinc. He had observed that, if both metals were immersed in common sulphuric acid electrolyte, the

Table 1.5 *Primary cells*

Cell	Electrodes positive	negative	Electrolyte	Working temperature	open-circuit voltage V	Energy density Wh/kg
Lithium (Japan)	Polycarbon monofluoride, $(CF_4)n$	Li	organic, polypropylene carbonate, with $LiC1O_4$	room	2·80	230
Lithium (GTE)	$S0C1_2$(C)	Li	thionyl chloride	room	3·60	350
Lithium (Mallory)	$S0_2$(C)	Li	organic acetonitrile propylene carbonate	room	2·95	200/300
Cu0/Li (SAFT)	Cu0	Li	organic, with $LiCl0_4$	room	2·35 (1·50 on load)	600 Wh/dm^3
Lithium/C1 (GM)	C1	Li	molten salts	650°C	3·40	150/300
Zinc/air	oxygen	Zn	Ag. KOH	room	1·40	330

Table 1.6 *Secondary or reversible cells*

Cell	Electrode positive	Electrode negative	Electrolyte	Working temperature	Open-circuit voltage V	Energy density Wh/kg
Nickel/zinc	NiOOH	Zn	Aq. KOH	room	1·70	66
Nickel/hydrogen	NiOOH	H_2	aq. KOH	room	1·36	55/88
Silver/hydrogen	Ag_2O	H_2	aq. KOH	room	1·70	80/100
Zinc/air	oxygen	Zn	aq. KOH	room	1·40	80/100
Zinc/chlorine	chlorine	Zn	aq. $ZnCl_2$	room	2·12	66/144
Iron/air	oxygen	Fe	aq. KOH	room	1·35	77
Zinc/bromine	bromine in carbon on titanium	Zn	aq. $ZnBr_2$ with HBr	room	1·80	48
Sodium/sulphur	sulphur	Na	solid beta Al_2O_3	300/350	2·00	60/150
Lithium/sulpur	sulphur or FeS	Li or Li/Al alloy	Molten Halides	375/425	1·40 on load	60/100

copper cathode soon became polarised with a layer of gas. The metals were therefore held in different compartments, separated by a porous partition of pot, with the copper cathode in copper sulphate solution and the zinc anode, lightly amalgamated with mercury, in either a solution of zinc sulphate or in dilute sulphuric acid.

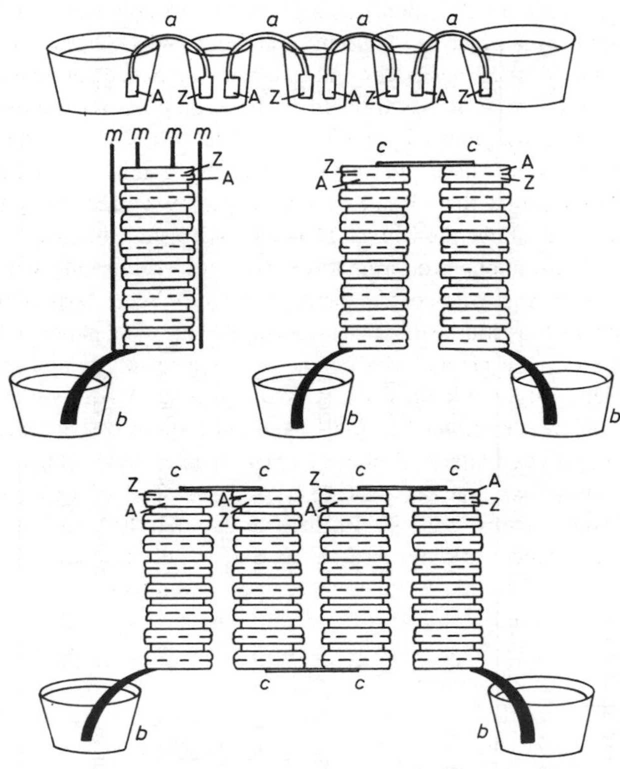

Fig. 1.2 *The Volta pile and 'couronne de tasses'. Copy of original plate 233*

William Grove, whose name is more closely identified with the discovery of practical gas electrodes and fuel cells, also used a two-compartment cell, with a platinum cathode immersed in dilute nitric acid and a zinc anode in dilute sulphuric acid electrolyte. The use of the costly metal, platinum, made this cell somewhat impractical, but Bunsen (1850) replaced the platinum with a carbon cathode, producing a practical cell which had an open-circuit voltage of about 2·0V and which was used effectively by Gaston Planté some years later in his experiments with lead electrodes.

1.1.5.2 Gas polarisation and the Leclanché cell: As previously mentioned, it had been observed that polarisation by gases, notably hydrogen on the cathode, limited the capacity available from some cells on discharge. Grove had used nitric acid as the de-polariser and it was this phenomenon which led Guiseppi Zamboni and Georges Leclanché to experiment with oxides for this purpose, in particular manganese dioxide. At the time it was thought that the de-polariser oxidised the hydrogen as it was formed. In the light of later knowledge, however, Heise and Cahoon[1] have pointed out that this is not so and that the de-polariser inhibits the formation of the gas by raising the potential of the electrode above the voltage at which hydrogen is formed.

In 1812, Zamboni produced a dry form of voltaic pile with thin discs of silver and gold-coated paper, the latter being coated on the metal side with a thin layer of manganese dioxide. The paper absorbed moisture from the air and this served as the electrolyte, and the pile gave a high voltage but only a very small current. A high voltage pile of this kind was assembled in the Clarendon Laboratory in Oxford in 1840 and 140 years later still shows its high open-circuit voltage.

It was, however, the cell of Leclanché, developed in 1866, which achieved the greatest success. In the wet model, Fig. 1.3, specially selected manganese dioxide served as the de-polariser and this was mixed with carbon black to improve the conductivity. The mixture was packed around a flat graphite plate, which served as the current collector, and held in an inner porous pot. The anode was a zinc rod, held in an outer jar, which was half filled with a saturated solution of ammonium chloride. The open-circuit voltage was 1·5V, the capacity at low and medium currents could be tailored to the need and the shelf-life of the cell was good. The demand was immediate and Leclanché, who had been exiled from France for political reasons[2], set up a factory in Belgium and during the next few years supplied many thousands of these cells for railroad signalling and telegraphic systems. The beneficial effect of keeping the upper portion of the cathode just moist was soon recognised and this led to the 'dry' assembly, attributed to Gassner in 1888. Other basic improvements followed; e.g. the manganese dioxide was mixed with extenders, such as starch or plaster of Paris, which helped to retain the electrolyte. The zinc electrode was made in the form of a cylindrical can, and, in addition to its function as an anode, served as a container for all of the other components of the cell. The unit could then be firmly closed at the top by means of a grommet, held in place with wax or some other sealant.

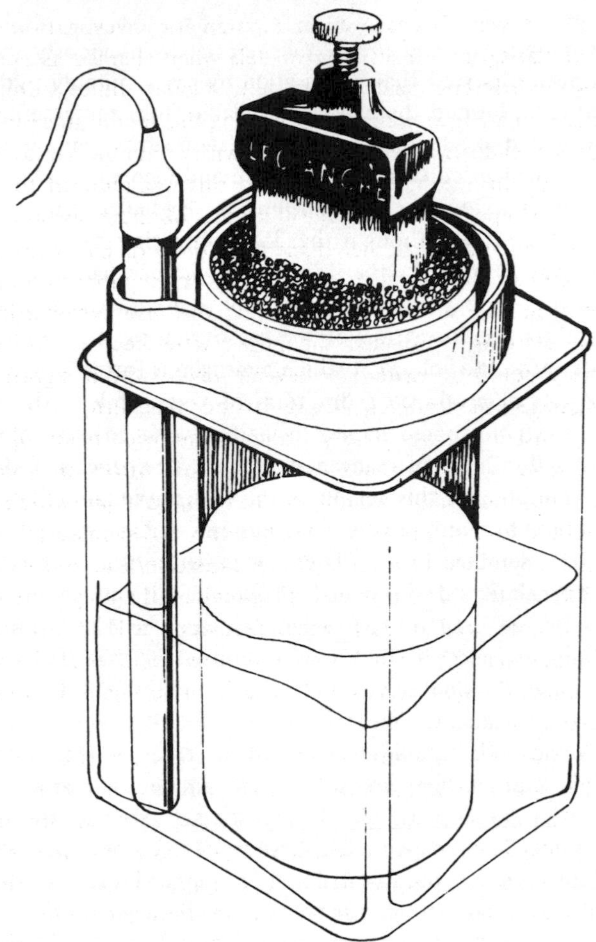

Fig. 1.3 *Leclanché cell of about 1870*

1.1.5.3 Origins of the lead/acid storage battery; the Planté process: Other workers had been preoccupied with the phenomenon of gas polarisation during charge. Gautherot (1802), for example, had observed that a feeble current was briefly obtained by the gases formed on platinum electrodes during the electrolysis of water and Ritter (1803) took this a stage further, using a number of different metals, from most of which similar effects were observed. Although nothing of a practical nature appears to have come from these observations, they sowed the seeds of reversibility, cultivated so effectively by Gaston Planté about 50 years later.

In 1859, Planté was also engaged in a systematic investigation into the effects of polarisation on different metals when charged as anodes and cathodes in different electrolytes, in particular dilute sulphuric acid. The behaviour of lead was unique. During the anodic stage of the charging cycle the lead sheet became covered with a thin brown layer of lead dioxide, and, during the cathodic stage, this was reduced to grey spongy lead. With repeated cycling the thickness of these layers could be steadily increased and Planté then made two other crucial discoveries. First, the capacity of the plates, as measured by the number of ampere hours available on the subsequent discharge, was also systematically increased, and secondly, provided the plates were kept in the acid electrolyte, they could be stored for several days without significant loss of charge, and so the lead/acid storage battery was born.

Cells were constructed by loosely coiling a sandwich made of two thin sheets of lead, separated by a thick layer of cotton or woollen cloth and inserting the coiled assembly into a cylindrical jar, which was then filled with dilute sulphuric acid. Once the active materials had begun to form, the charge voltage of the lead-dioxide/lead couple rose to about 2·70V. At that time primary cells provided the only source of direct current. Planté used at least two Bunsen-type cells in series and to obtain a useful capacity, the cells were submitted to reversal charging cycles for periods of several weeks and even months. Fig. 1.4 shows a typical coiled assembly of this type.

There were several objections to the process. The primary cells required for charging became exhausted fairly quickly and had to be thrown away. The process was therefore costly and very tedious. Also, the capacity available from the fully-charged cell was partly related to the total surface area of the active materials. It was an advantage, therefore, to use large areas of thin plates, but these often broke down through irregular anodic corrosion. Therefore, Planté turned to thicker plates with severely roughened surfaces, but these were heavy and incurred a weight penalty of excess metal. The tedium of the cycling process was also difficult to resolve. Some years later it was discovered that this could be greatly accelerated by the addition to the dilute sulphuric acid of small amounts of lead-corroding acids, such as nitric, chloric or perchloric acid. The duration of the formation process was ultimately reduced to less than 24 h, making the modern Planté process thoroughly viable. The final link in the chain was the development of a cheap and adaptable source of direct current. Michael Faraday laid the foundations for this by his discovery of electromagnetic induction in 1836, but 20 years were to elapse before the production of a practical electromechanical generator or dynamo, attributed to Werner von Siemens in 1857.

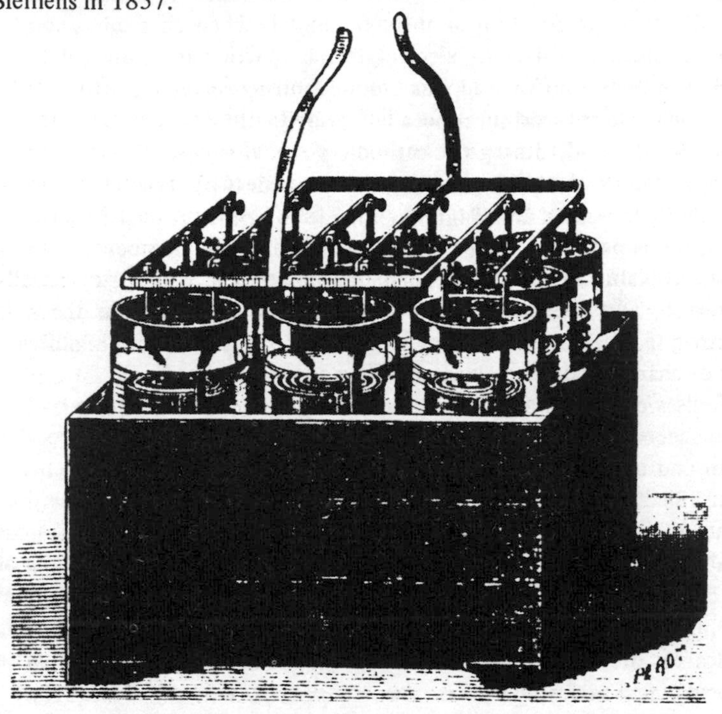

Fig. 1.4 *First lead/acid storage battery presented in 1860 to the French Academy of Sciences by Gaston Planté*

It is interesting to reflect that, in the long history of man, reaching back some thousands of years, during which the search for new and more powerful forms of energy has been a prime preoccupation, his ability to make and use the most adaptable form of all, electricity, has been crammed into a mere 150 years. In both electrochemistry and electricity the electron holds pride of place. Although its intimate habits were not to be fully revealed for another 40 or 50 years, this was indeed the era of the electron.

Planté type batteries were made for a variety of applications, primarily stationary because of the weight of the cells. There were inevitably many claims for a 'first', but undoubtedly one of the earliest examples of a house-lighting installation, with a dynamo for recharging was built at Rosport in Luxembourg by Henri Tudor in 1882. Tudor's patents and processes were later bought by the large German battery company,

Accumulatoren-Fabrik-Aktiengesellschaft (now Varta A.G.), who, during the next 30 years or so, built Tudor battery factories in several European countries, including Great Britain.

1.1.5.4 The Faure-Brush pasted plate: Among the many efforts to simplify the tedious Planté formation process the most noteworthy were those of Camille Faure in France in 1881 and Charles Brush in the USA. The starting material in both cases was finely-powdered lead oxide. In Faure's process, the oxide was first made into a stiff paste with sulphuric acid and applied to a flat lead sheet, which served as the current collector. Brush applied the oxide to plates, which had been scored, slotted or perforated by ramming the dry oxide into the cavities. The principle of using the current collector as a means of retaining the active materials led to the development of the open-mesh type of grid now in general use. In the course of time these processes were destined to revolutionise the storage battery industry, but, because of the complex patent position, other starting materials were tried. From about 1889, Clement Payen filed a number of patents in the USA for the use of lead chloride to which was added some cadmium chloride and zinc powder, and a year or so later yet another French chemical engineer, Francois Laurent-Cely, covered a similar process in the UK. These processes were abandoned a few years later, but they served as the corner-stones for the foundation of two battery manufacturing companies, destined to become two of the largest in the world; the Electric Storage Battery Company of the USA, now ESB-Ray-O-Vac Inc., and the Chloride Electrical Storage Syndicate of the UK, now the Chloride Group.

1.1.5.5 Standard cells: Other voltaic systems were also being developed at this time. Stable mercury/zinc and mercury/cadmium cells, with electrolytes composed of saturated mercury sulphate and zinc sulphate and mercury sulphate and cadmium sulphate, respectively, were produced by Clark in 1872 and Weston in 1892. These cells showed remarkable voltage stability, and, although they were of little use as sources of current, they have been adopted as regular standards for the determination of EMF, 1·434V at 15°C for the Clark Cell and 1·0183V for the Weston cell, the latter being regarded as the more stable.

1.1.5.6 Alkaline batteries: The next large commercial developments owed their origins to the period 1895 to 1905, when Waldemar Jungner in Sweden and Thomas Edison in the USA laid the foundations of the nickel/cadmium and nickel/iron alkaline storage battery industry. The advantages of these two systems and the way in which they were developed are given in Chapter 5. One special advantage ensured by an alkaline electrolyte lies in the wide choice of the materials of construction of electrodes, containers etc. Common metals, such as mild steel and nickel, that would be attacked by an acid electrolyte are widely used in alkaline batteries.

Other alkaline systems followed. In the 1930s André produced the first practical silver-oxide/zinc cell, which showed potentialities of giving the highest energy densities of systems then in existence, even at very high rates of discharge. In 1945, Samuel Ruben, in collaboration with the Mallory Company of the USA, developed the mercury-oxide/zinc cell, also using potassium hydroxide electrolyte, now generally known as the Ruben-Mallory or kalium cell, the latter designating the potassium or 'kalium' hydroxide.

1.1.5.7 Fuel cells: Fuel cells owe their origin to the classical experiments, carried out by William (later Sir William) Grove in 1839[3]. Using electrodes made of thin platinum foil, immersed in dilute sulphuric acid, he was studying the electrolysis of water. As expected, hydrogen and oxygen gases collected in the small tubes holding the electrodes. When the charge was stopped, Grove found that a current in the reverse direction was obtained owing to the re-combination of the gases on the platinum electrodes. To prove the point, Grove built a 50-cell 'gaseous voltaic battery' on lines shown in Fig. 1.5. In reporting his work, Grove showed remarkable prescience in observing three basic principles on which the whole structure of fuel cell technology has since been built. First, the platinum electrodes acted not only as current collectors, but also as a catalyst for the gas re-combination reaction; secondly, the reaction took place at the three-phase interface of gas-liquid-solid, and thirdly, to produce a current of any magnitude, it was necessary to have a 'notable surface of the electrode'.

Mond and Langer[4] extended the work of Grove on the oxidation of hydrogen, but the compelling attraction of these discoveries, with which the name 'fuel cell' had now become identified, lay in the direct conversion of the energy of fossil fuels into electricity, without the dirty and energy-wasteful thermal combustion stage. Prime movers in this aspect were Jacques[5] around the turn of the century and Baur[6] during 1910-1921. Jacques built a 1·5kW carbon/air battery of 100 cells, made of iron pots, containing the carbon anodes and the potassium hydroxide electrolyte, which was kept at about 500°C. Air was blown into the electrolyte through a tube passing to the bottom of the pots. Unfortunately, the alkaline electrolyte was rapidly contaminated by carbon dioxide with a serious decay in performance and nothing came of Jacques' ambitious proposals to build an electric power station and a ship, both powered by coal-burning fuel batteries.

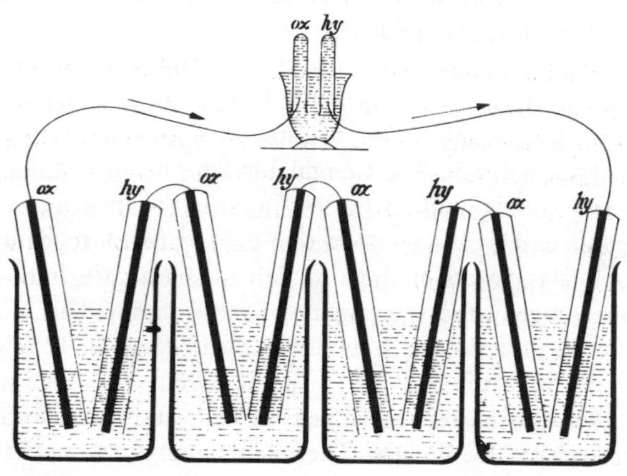

Fig. 1.5 *Four cells of Groves H_2/O_2 battery, used, in Grove's words, 'to effect the decomposition of water by means of its composition'*

Profiting from Jacques' difficulties, Baur[6] replaced the caustic potash by molten salts, such as carbonates and silicates. He also used silver as the air electrode, using its catalytic property of producing OH^- ions, but the working temperature had to be raised to 800-1000°C, which caused decomposition of the carbonate electrolyte. Work on the direct oxidation of carbonaceous fuels was, therefore, stopped, but as Liebhafsky and Cairns[7] have pointed out, the work of Jacques, Baur and co-workers was of particular interest to later chemists specialising in fuel cells with molten salt electrolytes.

In 1932, F.T. Bacon in the UK began to apply engineering principles more systematically than hitherto to the design and construction of multi-cell fuel batteries, using hydrogen and oxygen gas electrodes and caustic potash electrolyte. The Second World War interrupted this work, but by 1959, Bacon,[8] then at Cambridge University, had built and demonstrated a 5kW 'Hydrox' battery, working at 200°C and a pressure of 600 p.s.i. ($4{\cdot}14 \times 10^6$ N) and producing sufficient power to operate a pneumatic drill and an electric truck. By working at 200°C, Bacon avoided the use of expensive noble metal catalysts. At about the same time, Ihrig[9] of the Allis Chalmers Company of the USA, drove a farm tractor, powered by a 20 hp fuel battery, working at ambient temperatures with special catalysts.

The main principles of Bacon's battery were used by the Pratt and Whitney Company of the USA in building the hydrox fuel batteries which supplied the whole of the inboard power in the successful Apollo lunar spacecrafts. Other systems were also being developed at that time. Niedrach and Grubb[10] of the General Electric Company of the USA had demonstrated hydrogen/oxygen fuel cells, using cationic ion-exchange membranes (IEM) as electrolytes and batteries, developed on this principle by Douglas and Cairns[11], also gave a satisfactory performance in the Gemini spacecraft. The Gemini mission preceded Apollo, but the IEM batteries were found to be less durable than the hydrox models used in Apollo and it is understood that they were not repeated.

The complete success of these models in spacecraft, built regardless of cost, encouraged others to explore possible commercial applications, such as electric traction and standby power systems. Extensive tests have been made with variants in the materials of construction, the catalysts, the electrolytes, the design of cells and batteries and in the fuels and oxidants themselves. The main basic problem has been the cost, and so far no commercially viable system has emerged. Margins have, however, been significantly reduced and some organisations have kept their options open by supporting work on the fundamental problems, in particular the search for cheap catalysts. Recent problems in the supplies and the cost of petroleum fuels have re-awakened interest in the development of fuel cells.

1.1.5.8 Metal/air (or oxygen) cells: The work on fuel cells led to a much deeper understanding of both the behaviour of porous gas electrodes and electrochemical catalysis and this opened the way to the development of a variety of metal/air (or oxygen) systems, particularly with zinc anodes. The porous air cathode used in these cells was generally based on carbon. In his wet assembly Leçlanché recognised the advantage of leaving part of the carbon/manganese-dioxide cathode exposed to the air and so-called 'air-depolarised' cells, with porous carbon cathodes, zinc anodes and electrolyte, either ammonium chloride or caustic soda solution have been successfully used for many years on railway signalling duties. Experimental cells have also been tried with iron, aluminium or magnesium anodes. Fleischer[12] has made a comprehensive survey and analysis of the possible prospects of various metal/air systems. In this connection, the alkali metals, sodium and lithium, offer two special advantages; a high electrode potential, giving a favourable cell voltage, and a high energy density. Unfortunately, both metals react violently with water, and so they cannot be used in the usual way with aqueous electrolytes. In 1963, However, Yeager[13] described laboratory tests with cells having liquid sodium-mercury amalgam as the anode, porous silver or carbon, catalysed with silver, as the oxygen electrode and caustic soda electrolyte. These showed so much promise that plans were made to build a trial battery for a submarine, but the project was not pursued.

1.1.5.9 Metal/halogen cells: Other cathodes have been tested, notably chlorine and bromine, either in elemental form or as metallic salts, e.g. chlorides. In 1882, Warren de le Rue[14] built high voltage batteries of silver-chloride/zinc cells with ammonium chloride electrolyte, and for many years silver-chloride/magnesium batteries, activated by sea water, have been used for torpedo propulsion. Water-activated copper-chloride/magnesium batteries have been used for radarsonde and other meteorological apparatus.

In 1973, Symons[15] described promising experiments with chlorine/zine rechargeable batteries, having zinc chloride as electrolyte. A battery of this type was submitted to service trials in an electric automobile and plans have recently been announced in the USA for a test programme on a large battery as a standby electricity supply plant for load levelling and related services.

The favourable electrochemical properites of lithium are indicated in Table 1.3 and attempts have been made to couple this metal with chlorine using a mixture of molten sodium, potassium and lithium chlorides as electrolyte. The theoretical energy density of such a cell is estimated to be over 2000 Wh/kg and the open-circuit voltage 3·46V, but the tests were abortive because of the severe engineering problems caused by the high working temperature of 650°C.

Clerici *et al*[16] of the Magneti Marelli (FIAT) Company of Italy have reported the results of tests with a 30-cell, 3kW rechargeable zinc/bromine battery, and in the USA, Will[17] of the G.E. Co., and Bellows *et al*[18] of the Exxon Company have also published reports of their work on this couple.

1.1.5.10 High temperature batteries: Pride of place must go to the sodium/sulphur cell. Great interest in the use of sulphur as a cathode in an electrochemical cell was aroused by the inspired discovery by Kummer and Weber[19] of the Ford Motor Company of the USA that it could be used to oxidise sodium in a sealed rechargeable cell. The key to this discovery lay in the solid electrolyte, composed of fritted beta alumina doped with sodium oxide, which at about 350°C becomes permeable to sodium ions. Prototype batteries of this type are now being developed in the USA, the UK and Japan for electric traction and also for standby services. Metal sulphides have also been used as cathodes. Walsh and Shimotaki[20] of the Argonne National Laboratory of the USA have described promising performances of cells having iron sulphide cathodes, anodes composed of lithium-aluminium alloy and an electrolyte of mixed lithium and potassium chlorides, working at 400-450°C. The results have been so favourable that three large battery manufacturers in the USA have been invited to submit tenders for the production of practical models.

1.1.5.11 Cells with lithium anodes: Organic electrolytes, such as ethylene carbonate or polypropylene carbonate, provide a benign environment for lithium, and several authors have described cells operating at ambient temperatures with a variety of cathodes, e.g. sulphur dioxide,[21] copper chloride,[22] polycarbon-monofluoride[23] and oxides, copper oxide[24] and lead oxides,[25] most of which are listed in Table 1.3. Salts such as lithium perchlorate are added to the organic electrolyte to improve the ionic conductivity.

Encouraging results over a wide range of temperatures and rates of discharge have been described with cells having thionyl chloride as both the cathode and the electrolyte,[26,27] with additions of 1 to 1·8M $LiAlCl_4$. The above systems are all of the primary type. Gabano *et al.*[22] have claimed some success in cycling cells with copper chloride cathodes, but concluded that the cycling life was too short and the narrow limitations of the charge and discharge voltages would make their use in complex batteries virtually impossible.

1.1.5.12 Cells with solid electrolytes at room temperatures: The advantages of solid electrolytes have stimulated the search for materials which can perform this function at ambient temperatures. The two chief criteria are that the electrolyte should have a high ionic conductivity to specified ions and a high degree of stability. The complex inorganic salt, rubidium silver iodide, has been found to show good conduction of silver ions and practical sealed cells with this electrolyte, silver anodes and silver iodide cathodes have been proposed for special applications, calling for very small currents in the micro-amps range, such as heart-pacers. The cost was, however, prohibitive, and interest has now turned to electrolytes conductive to lithium ions. Information about these is given in Chapter 6.

References

1. HEISE, G.W., and CAHOON , N.C.: *The primary battery – Vol. 1*, 1971, (John Wiley & Sons Inc., New York), pp. 500
2. BARAK, M.: 'Georges Leclanché (183901852)', *Electron. & Power*, 1966, pp. 184-191
3. GROVE, W.R.: 'On gaseous voltaic battery', *Phil. Mag.* 1841, **S3 21**, p. 417
4. MOND, L., and LANGER, C.: *Proc. Roy. Soc.*, London, 1889, **46**, p. 296
5. JACQUES, W.W.: *Electr. Engr.*, 1896, **21**, p. 497
6. BAUR, E., and EHRENBERG, H.: *Zeitschr. Elektrochem.*, 1912, **18**, p. 1002
7. LIEBHAFSKY, H.A., and CAIRNS, E.J.: *Fuel cells and fuel batteries* 1968, (John Wiley and Sons Inc., New York), pp. 692
8. BACON, F.T.: *Fuel cells – Vol. 1*, YOUNG, G.J. (Ed.) (Reinhold Publishing Co., New York, 1960), Chap. 5
9. BARAK, M.: Fuel cells-present position and outstanding problems, *Advanced Energy Conversion*, 1966, **6**, pp. 29-55
10. NIEDRACH, L.W., and GRUBB, W.T.: *Fuel cells*, MITCHELL, W. Jun. (Ed.), (Academic Press, New York, 1963), p.253
11. DOUGLAS, D.L., and CAIRNS, E.J.: US Patent 3,134,696 (1964)
12. FLEISCHER, A.: Survey and analysis of metal-air cells, Technical Report AFAPL-Tr 68-6, March 1968, pp. 76, Res. and Tech. Div., US Air Force Systems Command, Wright Patterson Base, Ohio, USA
13. YEAGER, E.: *The sodium amalgam-oxygen continuous feed cell, fuel cells,* MITCHELL, W.Jun. (Ed.), (Academic Press, 1963), Chap. 7, pp. 299-328
14. DE LA RUE, W.: Phenomenon of the electric discharge with 14 400 chloride of silver cells, *Electrician*, 1882, **9**, p. 77
15. SYMONS, P.: 'Performance of zinc choloride batteries'. Paper presented at The Electric Vehicles Symposium, Washington DC, Electric Vehicle Council, USA, Feb. 1974, pp. 16
16. CLERICI G., DE ROSSI, M., and MARCHETTO, M.: *Zince-bromine storage battery for electric vehicles. Power sources 5,* COLLINS, D.H. (Ed.), (Academic Press, London, 1975), pp. 167-181
17. WILL, F.G.: *Recent advances in zinc-bromine batteries. Power sources 7,* THOMPSON, J. (Ed.), (Academic Press, London, 1979), pp. 313-328
18. BELLOWS, R.J., EUSTACE, D.J., GRIMES, P., SHROPSHIRE, J.A., TSIEN, H.C., and VENERO, A.F.: *Batteries 7,* THOMPSON, J. (Ed.), (Academic Press, London, 1979), pp. 301-302

19. KUMMER, J.T., and WEBER, N.: 'A sodium-sulphur secondary battery'. Paper presented at the Automotive Engineering Congress, Detroit, Michigan, USA, Society of Automotive Engineers, 1967, **670**, p. 179
20. WALSH, J., and SHIMOTAKI, H.: *Performance characteristics of lithium-aluminium/iron sulphide cells. Power sources 6*, COLLINS, D.H. (Ed.) (Academic Press, London, 1977), pp. 725-733
21. PER BRO, HOLMES, R., MARINCIC, N., and TAYLOR, H.: *The discharge characteristics of the Li-SO_2 battery system, Power sources 5*, COLLINS, D.H. (Ed.) (Academic Press, London, 1975), pp. 703-712
22. GABANO, J.P., LEHMANN, G., GERBIER, G., and LAURENT, J.F.: *Power sources 3*, COLLINS, D.H. (Ed.) (Oriel Press, England, 1971), pp. 297-308
23. FUKUDA, M., and IIJIMA, T.: *Lithium/poly-carbonmonofluoride cylindrical type batteries. Power sources 5*, COLLINS, D.H. (Ed.) (Academic Press, London, 1975), pp. 713-728
24. LEHMANN, G., BERBIER, G., BRYCH, A., and GABANO, J.P.: *The copper oxide-lithium cell. Power sources 5*, COLLINS, D.H. (Ed.) (Academic Press, London, 1975), pp. 695-701
25. BROUSSELY, M., JUMEL, Y., and GABANO, J.P.: *Lead oxides-lithium cells. Power Sources 7*, THOMPSON J. (Ed.) (Academic Press, London, 1979, pp. 637-646
26. AUBORN, J.J., and MARINCIC, N. (GTE, USA): *Inorganic electrolyte lithium cells. Power sources 5*, COLLINS, D.H. (Ed.) (Academic Press, London, 1975), pp. 683-694
27. DAY, A.N., and PER BRO (Mallory): *Primary Li/SOCl, cells 111. The effect of the electrolyte and electrode variables on the energy density. Power sources 6*, COLLINS, D.H. (Ed.) (Academic Press, London, 1977), pp. 493-510

Reversible cells with lithium negative electrodes and a copper chloride positive electrode. Power sources 3, COLLINS, D.H. (Ed.) (Oriel Press, England, 1971), pp. 297-308

Chapter 2

Definitions and basic principles

H R Thirsk

2.1 Introduction

Many electrochemical power sources have had so long a history that major developments have taken place with the most rudimentary application of basic electrochemical methods. Nevertheless, in recent years there has been an increased probing of classical systems and newer devices have been developed from the outset with much greater attention to fundamental problems. Because of the uneven application of basic studies to battery systems it would make the task of explaining the electrochemical methods by direct reference to the problems of real battery systems a lengthy and discursive operation. The alternative method of presenting an outline of the basic techniques, independent of illustrative reference to actual batteries, has, in general terms, been taken. It was also felt that this procedure would also incur a lesser chance of duplication in other chapters where various battery systems are treated.

The past decade has been a particularly active one in the development of electrochemical methods and it is important that battery technologists should have some knowledge of these newer skills, particularly in view of the fact that they have become increasingly proven by application to practical problems. In retrospect it is a matter of regret that a period of heavy financing of fuel cell research preceded significant developments in electrochemical methods. New battery systems can now be assisted in their evolution by much more substantial electrochemical investigation, but, and this is an important reservation, only if there is properly informed personnel to undertake this work.

Along with progress in electrochemical studies there have been substantial advances in tools for structural studies, combined with chemical information. Unfortunately, this chapter cannot attempt to deal with these matters excepting to emphasise that basic electrochemistry must be associated with structural investigations if a deep knowledge of battery problems is to be evolved.

2.2 Symbols

In this chapter we have to decide on the nature of the symbols and terminology that are essential to the subject. Recent recommendations have been approved[1] and it is these that are utilised.

An electrode is a condensed phase which has the property of electronic conduction: it can be a semi-conductor or a metallic conductor. It can take many forms; a liquid metal, an amalgam, a metal in any solid physical form, graphite or carbon conducting carbides, borides or nitrides, many oxides and sulphides. As battery terminology widens so does the diversity of the electrodes. Nevertheless, the following simple classification of a number of classical electrode systems is useful in identifying the operation of electrodes.

Electrodes are in contact with an electrolyte which is an ionic conductor and may be a solid, a melt or, most commonly, an electrolyte solution; water is the most common dielectric, but, of course, not the only one.

2.3 Electrodes and the galvanic cell

Electrodes in batteries with respect to their mode of operation take a number of differing forms but broadly fall into four groups.

(*a*) A metal electrode or alloy in contact with an electrolyte containing ions of the metal. The metal may freely dissolve as with a zinc/air cell when it may be considered to be a fuel. It is represented as

$$\mathrm{M} | \mathrm{M}^{n+}_{(\mathrm{sol})} \tag{2.1}$$

where by convention the vertical solidus represents the metal electrolyte phase boundary, a dashed vertical bar ($\vdots$) represents a junction between miscible liquids and a double dashed vertical line ($\vdots\vdots$) represents a liquid junction at which the potential is eliminated.

(*b*) A metal electrode at which an oxidation or reduction may take place with electron transfer:

(i) the reacting species may be a gas at a pressure p, e.g.

$$Cu_{(s)} \mid Pt_{(s)} \mid H_{2\,(g)} \mid HCl_{(aq)} \tag{2.2}$$

(ii) or a redox system in the solution, e.g.

$$Pt_{(s)} \mid Fe^{2+}_{(aq)} + Fe^{3+}_{(aq)} \tag{2.3}$$

$Pt_{(s)}$ represents the metal electrode which is assumed to be inert, i.e. it does not participate in the reaction other than acting as a source or sink of electrons.

(*c*) A metal in contact with a salt or oxide of the metal in contact with an electrolyte containing the anion of the salt or hydroxyl ions, e.g.

$$Cu_{(s)} \mid Ag_{(s)} \mid AgCl_{(s)} \mid Cl^{-}_{(aq)} \tag{2.4}$$

$$Cu_{(s)} \mid Pb_{(s)} \mid PbSO_{4\,(s)} \mid SO^{2-}_{4\,(aq)} \tag{2.5}$$

$$Pt_{(s)} \mid Hg_{(2)} \mid HgO_{(s)} \mid OH^{-}_{(aq)} \tag{2.6}$$

Electrodes of this type are often used as reference electrodes since the salt serves to stabilise the metal concentration in solution through an equilibrium of the type

$$AgCl \rightleftharpoons Ag^{+} + Cl^{-} \tag{2.7}$$

(*d*) An inert conductor in contact with a salt or oxide and totally immersed in this material which is in contact with an electrolyte containing an ion which can undergo an oxidation or reduction with electron transfer

$$Pt_{(s)} \mid MnO_{2\,(s)} \mid MnO_{4}^{-} \tag{2.8}$$

A galvanic cell is created by a suitable combination of two electrodes, e.g. from exprs. 2.2 and 2.4 a cell

$$Cu_{(s)} \mid Pt_{(s)} \mid \underset{p}{H_{2\,(g)}} \mid HCl_{(aq)} \mid AgCl_{(s)} \mid Ag_{(s)} \mid Pt_{(s)} \mid Cu_{(s)} \tag{2.9}$$

could be made for which the overall chemical reaction would be

$$\frac{1}{2} H_{2\ (g,\ \text{pressure}\ p)} + AgCl_{(s)} = H^+ + Cl^- + Ag_{(s)} \tag{2.10}$$

For the reaction to go to completion as written, a total charge F would have passed, where F is the Faraday constant (SI unit C/mol) being the product of the Avogadro constant and the charge of the proton.

In this simple cell (expr. 2.9) the charge number z, positive for cations and negative for anions, is equal to 1.

The charge number of the cell reaction $n(z)$ is the stoichiometric number equal to the number of electrons transferred in the cell reaction as formulated. It is a positive number, in this case 1.

The electric potential difference of the cell E (SI unit V), including the case when a current is flowing, is the difference of electric potential between a metallic terminal attached to the right hand electrode and an identical metallic terminal attached to the left hand electrode.

The value of E measured when the left hand electrode is at virtual equilibrium and acting as a reference electrode may be called the potential of the right hand electrode with respect to the left hand reference electrode.

The electromotive force E_{MF} is the limiting value of E when the current through the external circuit becomes zero and all local charge transfer equilibria across the phase boundaries and local chemical equilibria within phases is established.

In general there exists a difference in electrical potential ϕ (the inner potential) between the electrode and the electrolyte. This potential difference is called the Galvani potential and its absolute value cannot be measured. Thus one cannot study the electrical behaviour of an electrode in contact with an electrolyte in isolation. Two electrodes must be combined, as already described, for the system to be studied and its electric potential measured, which is the difference between the Galvani potentials of two identical metal terminals attached to the electrodes.

When an electrode potential is made sufficiently positive the electrode behaves as an anode. It may extract electrons from a species adjacent to the surface or pass into solution as an ion. In either case electrons move in the connecting circuit away from the interface. If the electrode is made sufficiently negative it will inject electrons into a species at the electrode or if there are metallic ions present the ion may deposit as a metal. The movement of the electrons in the external circuit is to the interface. Because of this an oxidation is said to take place at the anode and a reduction at the cathode.

2.4 Basic thermodynamics and the efficiency of cells

The application of thermodynamics to battery systems has a limited but important role. The limitations follow from two very important factors. First, it is the kinetics and problems of mass transfer which limit battery performance and these are not predictable by thermodynamics. Secondly, many important battery systems utilise chemical constituents which are not well identified and for which there is no appropriate thermodynamic data.

The limited role can be little more than, possibly with some intelligent guesses, a figure for the Gibbs free energy of systems that can be estimated and perhaps confirmed by measurement of an open-circuit potential. To go further and to deduce mechanisms of reactions without kinetic investigation is a hazard which fortunately is more and more appreciated. The exercise does, however, identify the basic feasibility of the chosen system as a potential source of electrical work and in some cases a thorough treatment of the reversible behaviour may be carried out as with the lead acid battery.[2]

To treat a battery system by thermodynamics it is assumed that the cell operates at constant temperature and pressure.

Represent the cell reaction by

$$\mathrm{A}/t + bB \rightleftharpoons cC + dD \tag{2.11}$$

For a complete thermodynamic treatment to be possible, enabling a calculation of the Gibbs free energy change to be made, appropriate data for the enthalpy and the entropy change for the process must be available.

For the reaction this is expressed as

$$\Delta G = \Delta H - T\Delta S \tag{2.12}$$

$T\Delta S$ is a measure of the unavailable energy for the process. ΔG is the energy change that can be extracted from the reversible reaction.

With a battery the criterion of reversibility is met by opposing the cell potential with an equivalent potential generated by a laboratory potentiometer. Under these conditions

$$\Delta G = nEF \tag{2.13}$$

where n is the charge number as defined above.

At constant pressure

$$-S = \left(\frac{\partial G}{\partial TP}\right) \tag{2.14}$$

Using the Gibbs-Helmholtz equation and

$$\Delta H = \Delta G - T\left[\frac{\partial(\Delta G)}{\partial T}\right]_P^{-} \tag{2.15}$$

ΔH, by substitution from eqn. 2.14, can be written as

$$\Delta H = -nF\left[E - T\left(\frac{\partial E}{\partial T}\right)_P\right] \tag{2.16}$$

The standard potential of the reaction in the cell E° (volts) is given by

$$E^\circ = -\Delta G^\circ/nF = (RT/nF)\ln K \tag{2.17}$$

when the activities of the products and reactants are unity, where ΔG° is the standard molar Gibbs free energy change for the reaction and K is the equilibrium constant of the reaction. R is the gas constant and T is the thermodynamic temperature.

For cell reactions in general, the potential of the cell can be expressed as

$$E_{cell} = E_0 - (RT/nF)\sum_i \nu_i \ln a_i \tag{2.18}$$

where a_i is the activities of the reacting species, which often for battery systems are equated to concentration of the species taking part in the reaction and ν_i is the stoichiometric numbers of the species taking part in the reaction, positive for species on the r.h.s. of the equation and negative for those on the l.h.s. Thus, for the reaction of eqn. 2.11

$$E = E^0 - (RT/nF)\ln\frac{(a_C)^c(a_D)^d}{(a_A)^a(a_B)^b} \tag{2.19}$$

If gaseous molecules participate in the reaction, the pressure, or, for non-ideal gases, the fugacity must be known. The reader should refer to standard texts in chemical thermodynamics for a detailed treatment concerning the definition and evaluation of activities in electrolyte solutions and of fugacity and pressure. The further development of the thermodynamic treatment is outside this chapter.

It will be easily realised that even in the absence of complete thermodynamic data a measurement of the cell potential using a potentiometer will enable ΔG for the cell reaction to be calculated. This will give a figure for the maximum electrical energy that the system might give and an experimental assessment of actual performance may be evaluated against this basic information.

There is a valuable classical text due to Latimer[3] on electrode potentials in aqueous systems which makes informative reading in connection with the energetics of electrode systems. Compilation of thermochemical data are also of value, e.g. Reference 4.

The search for effective combination of electrodes has been of interest since the earliest days and one should be aware of the massive compilation by Gibson and Sudworth.[5]

Furthermore, a series of micrographs on the electrochemistry of the elements edited by Bard[6] and in continuous production contains an enormous amount of material of great interest to the battery technologist.

2.5 Efficiency

Discussion concerning the free energy for the cell reaction leads directly to considerations of the efficiency of the system. If maximum efficiency could be realised the energetic performance for the cell is given by

$$\Delta H = \Delta G - TDS \tag{2.20}$$

For most battery systems ΔS is positive and the value, if gases are participating, substantial.

Two different approaches are made concerning cell efficiency. The first, based on the Gibbs free energy, is most reasonable to electrochemists; the second, based on the enthalpy of the reaction ΔH, which is used particularly in the dicussion of fuel cell performance, possibly because it is related to the performance of a heat engine which a fuel cell certainly is not.

Thus we can define the overall efficiency of a cell when a charge σ (coulombs) passes through an external circuit, for which q is the total charge given by nF for mole of cell reaction as

$$\text{overall efficiency} = \int_{q=0}^{q=nF} E\,d\sigma/\Delta G \tag{2.21}$$

The upper limit of nF is reached if there are no alternative chemical reactions in the cell which do not lead to a charge transfer. Such reactions allied to corrosion and chemical attack on cell materials are hopefully controllable through the skills of the cell designer.

Deviation of the cell voltage from E arises from resistive effects of the electrolyte through mass transfer, the materials of the cell and potentials which arise in general through electrochemical kinetic effects at the electrodes, to be considered later.

In the treatment of fuel cells, often an engineering concern rather than a purely electrochemical one, efficiency is dealt with in a manner that is an attempt to establish a comparison between a thermal convertor and an electrochemical device.

A thermal convertor has an upper limit of thermal efficiency given by

$$\text{thermal efficiency} = \frac{T_2 - T_1}{T_2} = \frac{\omega}{Q_2} \tag{2.22}$$

where Q_2 is the heat input at the higher temperature.

The equivalent thermal efficiency for a fuel cell is written as

$$\text{thermal efficiency} = \frac{\Delta G}{\Delta H} = (1 - T\Delta S)/\Delta H \tag{2.23}$$

The thermal efficiency may be regarded as a number for comparison of various battery systems. It should be used with some caution and due regard to the system. For example, in a fuel cell water may be a liquid or a vapour under differing operating conditions affecting the value of ΔH and the thermal efficiency, although in practice the two cells could be operating equally well.

2.6 The metal electrode solution interface

For a thorough study of this problem and its relation to electrochemical kinetics one reference only is made to the text by Delahay,[7] since this contains a substantial bibliography covering all earlier work and the subject is no longer in a vigorous state of development.

Species in solution are affected by the electrode through electrostatic forces, chemical or short range interactions and the existence of the phase boundary. Electrostatic interactions arise because the charge carried by the electrode surface [except at the unique potential of zero charge (p.z.c.)] promotes an alignment of polar molecules, attracts ions of a different charge sign and repels others. The solvent in electrochemical systems is invariably polar. Ions can be drawn towards the electrode until prevented from closer approach by their solvation shells. They are then said to be in the plane of closest approach or outer Helmholtz plane (o.h.p.) (Fig. 2.1). As anions have much larger crystal radii than cations with comparable atomic weights, they are less strongly solvated than cations, and, in consequence, anions can approach more closely to the electrode; the inner Helmholtz plane (i.h.p.) locates the centres of unsolvated or partially desolvated anions. Such anions are sometimes known to undergo specific adsorption because of their chemical interaction with the electrode metal. Some monovalent large cations, e.g. Tl^+, can also engage in specific interactions, as do certain types of neutral organic molecules. In some cases an organic molecule is dissociatively adsorbed (e.g. alcohols on platinum) whereas some aromatics lie flat on the surfaces of positively charged electrodes, their π-electron cloud interacting with the surface-metal ions. Furthermore, the solvent itself in general undergoes short-range interactions with the electrode so that some surface potential difference remains, even at the p.z.c. The inner Helmholtz plane defines the position of the adsorbed species.

The major problem with the theory of the interface is that classical electrostatics is used to describe forces between particles with atomic dimensions at small separations.

Again referring to Fig. 2.1, the diffuse layer lying outside the outer Helmholtz plane is treated by a classically based theory named after Gouy and Chapman. The Gouy-Chapman theory, which describes the charge distribution near to a charged planar electrode immersed in an electrolyte solution, is progressively more accurate in its predictions towards low concentrations. The excess ionic charge on the solution side of the interface exists as an ionic atmosphere: a space charge density which falls off with distance from the surface in concentrated electrolyte solutions the thickness κ^{-1} of the ionic atmosphere (or diffuse layer) at the electrode surface becomes as small as molecular dimensions; the Gouy-Chapman theory gives

$$\kappa^2 = 2z^2 F^2 c/\epsilon RT \tag{2.24}$$

where c is the concentration in mole/m^3 of a z-z electrolyte. If the bulk value of the dielectric constant ϵ is used, then κ^{-1} for a molar aqueous electrolyte solution is given as

$$\kappa \sim 3 \times 10^9 \text{ m}^{-1} \text{ if } \epsilon = \eta \times 10^{-10} \text{ Farad/m} \tag{2.25}$$

Thus, the diffuse layer, present in dilute ($< 10^{-2}$ mol/dm^3) electrolyte solutions, essentially contracts to the plane of closest approach in strong electrolytes (1 mole/dm^3) where the interface bears some resemblance to a parallel plate capacitor.

If the inner potentials within the metal and at the o.h.p. are ϕ_m and ϕ_2^2, respectively, relative to the bulk electrolyte solution, then we can divide the interfacial capacitance C into two parts

$$\frac{1}{C}=\frac{d\phi_m}{dq_m}=\frac{d(\phi_m-\phi_2)}{dq_m}+\frac{d\phi_2}{dq_m}=\frac{1}{C_1}+\frac{d\phi_2}{dq_m}+ \quad (2.26)$$

In the absence of specific adsorption, $q_m + q_d = 0$ where q_d is the net ionic charge in the diffusion layer. The Guoy-Chapman theory gives the result of a $z-z$ electrolyte

$$C_d=\frac{dq_d}{d\,\phi_2}=\frac{z}{F}\left[2\epsilon cRT\right]^{1/2}\cosh\frac{2F\phi_2}{2RT} \quad (2.27)$$

From the experimental value for C and the calculated value of C_d, the inner layer capacitance or Helmholtz capacitance C_i can be calculated by means of eqn. 2.26. In the absence of specific adsorption it is found that C_i is independent of c but dependent on q_m ; C_i increases with q_m owing to compression of the inner layer with increasing field strength. C_i has a value of the order of 0·2 F/m^2 for a smooth electrode. The experimental capacitance is determined almost entirely by C_i at concentrations liable to be met with in battery systems.

The mean value of the field strength can be calculated approximately within the inner layer by means of Gauss' theorem; for aqueous solutions at 25°C $d\phi/dx = -q_m/\epsilon = -1{\cdot}4 \times 10^9\ q_m$ (in V/m) where q_m is expressed in coulombs /m^2. As in many systems q_m can reach a value of 0·2 coulombs /m^2, $d\phi/dx$ can exceed 3 x 10^8 V/m. In this field an aligned water dipole has a potential energy of 3 x 10^3 J/mol below that of the randomly oriented molecule. This of course produces dielectric saturation, a decrease in ϵ, and, one would expect, a corresponding decrease in C_i. However, the increase in C_i due to dielectric compression tends to outweigh the effect of dielectric saturation.

2.7 Adsorption and kinetics

Adsorption has an important role to play in many electrode processes as would be appreciated intuitively. Apart from making the point, it is outside the scope of this chapter to attempt to treat the problem, particularly since a good deal of the literature relates to steady state experiments that are not particularly definitive and are not easily summarised. Reference 7 is of particular value in developing the problem without recourse to non-steady state methods of experimentation. This latter approach has, at the moment of writing, not been covered by an adequate review and there exists a substantial need for up-to-date reviews both of the kinetics of adsorption and desorption and indeed of the nature of adsorption itself.

In general terms in any proposed development of battery systems where adsorption is of considerable importance, fuel cells are an obvious example, investigation of the system by non-steady state techniques would still be a primary requirement. There is still a very considerable need for development of the related field of the catalytic activity of the electrode material.

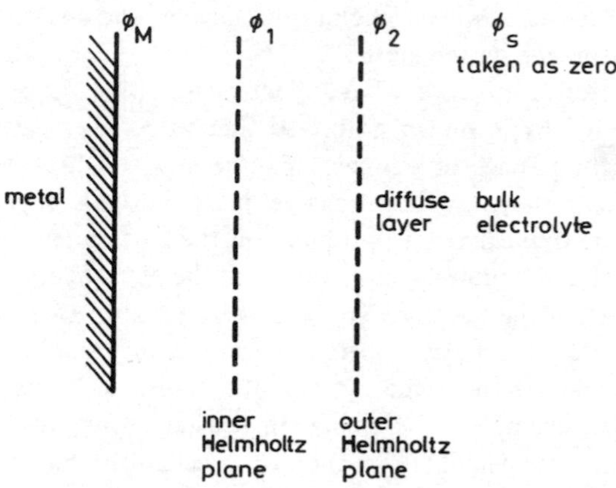

Fig. 2.1 *Properties of the interface inferred from measurements in the double layer region*

2.8 Electrochemical kinetics

It will be well understood that an extensive treatment of the kinetic problems of battery systems is impossible within the scope of the present short chapter. What can be attempted is some display of electrochemical method which can be given some coherence by associating the techniques with some of the electrode systems as outlined in section 2.3. If the reader has little or no knowledge of modern electrochemical kinetics, a text such as that by Albery[10] is helpful and provides a background for what follows.

Central to all electrochemical kinetics is the electron transfer controlled by the potential difference across the interface. If this is a slow step in the overall kinetics, the non-linear relationship between current density and potential, the Tafel law, may be examined experimentally and this would be particularly true of electrode systems of the type shown in exprs. 2.1, 2.2 and 2.3. Complicating factors would be the diffusion of reacting material up to and away from the surface, adsorption and, therefore, in some cases, double layer structure. The double layer, as previously discussed, is the region adjacent to the electrode surface which has a non-uniform ion distribution of a complex nature, on the solution side only with metal electrodes but within the electrode with semi-conductors. In thickness it is well within the normal distance from the electrode surface where concentration changes are controlled by effects of mass transfer. In many electrode systems the gross problems introduced by mass transfer and phase changes, e.g. exprs. 2.4, 2.5 and 2.6, overwhelm double layer and adsorption effects when evaluating the basic kinetics.

Many batteries, however, have electrodes with large surface areas and are often labyrinthic in nature so that the concentration changes induced by the passage of a current may be slow to adopt a new steady state. The comparatively large charge in the double layer, an open circuit, can be important on switching on the battery and to this end a knowledge of the double layer capacity of the electrode, depending on the use to which the battery may be put, is of value. Changes of phase during the use of a battery often mean the introduction of a non-conducting material into the system; the lead/acid battery is an obvious classical example. This change in resistance may not be of great importance in fundamental studies of mechanism but such factors must always be taken into account when considering the system as fabricated.

In systems such as exprs. 2.1, 2.4, 2.5 and 2.6, the problems, termed in electrochemistry 'electrocrystallisation', where phases are formed or disappear as current is passed, are paramount. In this case, steady state polarisation, whether with controlled current or controlled potential, are usually quite adequate by themselves to evaluate the system. Regrettably there are still quite modern textbooks on electrochemical kinetics which do not clarify this matter but it is being realised more and more that perturbation methods must be employed in battery research.

2.9 Steady state procedures

Almost invariably electrochemical kinetic investigations are carried out in cells with three electrodes. The electrode system of interest is a reference electrode in a separated compartment leading to a capillary with an end closely adjacent to the surface of the electrode; the Luggin capillary. This may take many forms physically and a rather complete description of the device as employed in battery electrode investigations is given in a text on primary batteries.[8]

The third electrode is often situated in a compartment separated from the compartment containing the electrode under investigation by means of a glass frit. It merely serves as a method of completing the cell through which the monitored current passes.

The working electrode and Luggin capillary reference electrode pair may either be used to monitor changes of potential across the interface when variable currents are passed through the cell, a galvanostatic experiment, or are used to impose a steady potential across the interface with an appropriate servo mechanism (potentiostat) altering the cell current to maintain the potential; a potentiostatic experiment.

The design of appropriate instrumentation has reached a very considerable degree of sophistication and information on this matter is widely disseminated through the electrochemical literature. Appropriate instrumentation is readily obtainable commercially. As an introduction there is a short modern review given by Greef.[9]

2.10 Simple electron transfer

The basic electrochemical process which occurs when an inert metal is in contact with a redox system, species O and R is

$$O + ne \underset{k_b}{\overset{k_f}{\rightleftharpoons}} R$$

Analogous to chemical kinetics the current that flows, the Faradaic current, at a particular potential, is

$$I = nFA\,(k_f C_0{}^s - k_b C_R{}^s) \tag{2.28}$$

The rate constants k_f and k_b are potential dependent and hence the great significance of potential control. A is the area of the electrode and the current is considered to be positive for the cathodic reaction going from left to right.

The main parameter determining the current is the potential E of the working electrode with respect to a reference electrode. k_f and k_b can be defined in terms of E as

$$k_f = k_1 \exp(-\alpha nfE)$$
$$k_b = k_{-1} \exp(1-\alpha)nfE \qquad (2.29)$$
$$nf = nF/RT$$

where k_1 and k_{-1} are potential independent constants.

To define k_f and k_b in a form which reflects the nature of the electrochemical process itself we can define them with respect to the standard potential E_0

$$k_f = k_{sh} \exp[-(E-E_0)\alpha nf] \qquad (2.30)$$
$$k_b = k_{sh} \exp[(E-E_0)(1-\alpha)nf]$$

where k_{sh} is the forward rate on the surface. Alternatively we can define k_f and k_b with respect to the potential E_{eq} at which $j = 0$.

Using $\eta = E - E_{eq}$

$$k_f = \frac{j_0}{nFC_0^s} \exp(-\alpha nf\eta)$$

$$k_b = \frac{j_0}{nFC_n^s} \exp[(1-\alpha)nf\eta \qquad (2.31)$$

If we include in this elementary discussion diffusion so that the concentration of reactant at the surface $C_0^s \neq C_0^b$

$$\text{current density} = \frac{I}{A} = j$$

$$= j_0 \left[\frac{C_0^s}{C_0^{sol}} \exp(-\alpha nf\eta) - \frac{C_R^s}{C_R^{sol}} \exp[(1-\alpha)nf\eta] \right] \qquad (2.32)$$

To use this equation to observe a stationary current for a given potential, the diffusion layer must be fixed and the most feasible device to do this is to fashion the working electrode as a rotating disc electrode. Because of the importance of this device its use will be described briefly in Section 2.9.

The non-linear nature of eqn. 2.32 gives considerable problems if a theoretical treatment of diffusion effects is required or in cases when it is desirable to examine overall reactions including diffusion, chemical steps and effects due to electrocrystallisation. This requires perturbation of the current or potential and the relevant theory develops with considerable mathematical complexity. In these cases a linearised form of the eqn. 2.32 for small potential excursions is used, e.g.

$$\left(\frac{\delta\eta}{\delta_j}\right)_{\eta\to 0} = \left\{\frac{1}{nf}\,\frac{1}{j_0} - \frac{1}{C_0^{sol}}\left(\frac{\delta C_0}{\delta j}\right)_{j\to 0} + \frac{1}{C_R^{sol}}\left(\frac{\delta C_R}{\delta j}\right)_{j=0}\right\} \tag{2.33}$$

The first term on the right hand side defines the charge transfer resistance and the remainder mass transfer resistance. The concept of resistance for these terms becomes of major importance in methods based on a.c. perturbations.

There is an important limiting case of eqn. 2.32 if the reaction is intrinsically irreversible or if high potentials are applied to the system

$$\ln j - \ln j_0 = -\alpha n f \eta \tag{2.34}$$

$$\frac{\delta \ln j}{\delta\eta} = -\alpha n f$$

which is the Tafel equation.

Familiarity with the many forms of relationships between current, voltage and concentrations is essential for electrochemical investigations and it is helpful to have the manipulation exposed deliberately. This has been attempted in a short text.[11]

2.11 Rotating disc electrode

There is a very substantial literature on disc electrodes and ring disc electrodes.[12]

Within the context of this chapter the use for extrapolating out the effect of diffusion is treated briefly as an important example of its application to kinetic problems.

It is assumed that the experiment being carried out is under constant potential conditions; constant current regimes are not usually advisable particularly in the preliminary investigations of unknown systems. The electrode is rotated at a strictly controlled and measurable angular velocity and the thickness of the diffusion layer depends on this velocity and is independent of time and the radial distance from the axis of revolution.

Consider again the electron transfer reaction

$$O + ne \underset{k_b}{\overset{k_f}{\rightleftharpoons}}$$

using eqn. 2.23 and the fact that for large exchange currents for the diffusion flux at the electrode surface

$$I = nFAD_O \frac{(C_O^b - C_O^s)}{\delta} = nFAD_R (C_R^s - C_R^b) \tag{2.35}$$

where D_O and D_R are the respective diffusion coefficients for species O and R and δ is the thickness of the diffusion zone. In the stationary state the concentration gradient is stationary as given by eqn. 2.35.

Eliminating surface concentration between eqns. 2.28 and 2.35

$$\frac{1}{I} = \frac{1}{nFA(k_f C_O^b - k_f C_R^b)} + \frac{\delta\left(\frac{k_f}{D_O} + \frac{k_O}{D_R}\right)}{nFA(k_f C_O^b - k_b C_R^b)} \tag{2.36}$$

Using the value of the diffusion layer thickness calculated by hydrodynamics[12a]

$$\delta = 1{\cdot}6 D^{\frac{1}{3}} \nu^{\frac{1}{6}} \omega_R^{-\frac{1}{2}} \tag{2.37}$$

where ω_R is the angular velocity of the disc in radians, then

$$\frac{1}{j} = \frac{1}{I} + \frac{K}{\omega_R^{1/2}} \tag{2.38}$$

where K is a constant depending only on potential and I is the current corrected for diffusion. The intercept of $1/j$ against $1/\omega_R^{1/2}$ gives I. I measured as a function of potential gives the Tafel slope.

If the electrode is perfectly reversible, i.e. $k_f \rightarrow \infty$ $k_b \rightarrow \infty$ it follows[11] from eqn. 2.38 that $1/j$ has the form

$$\frac{1}{j} = \frac{K^1}{\omega^{1/2}} \tag{2.39a}$$

and

$$\frac{1}{j} \text{ against } \frac{1}{\omega^{1/2}} \tag{2.39b}$$

are straight lines through the origin for all potentials.

The intelligent use of rotating disc electrodes extends the scope of simple galvanostatic and potentiostatic experiments, but it is, of course, equally useful for employment in non-steady state experiments.

A final point, many battery electrodes are porous and diffusion in these pores is a totally different problem. However, a.c. perturbation methods do permit the use of information from plane surfaces to be used with porous systems; this will be commented on briefly in Section 2.12.

2.12 Non-steady state methods

There are considerable mathematical problems involved in the analysis of transients produced by perturbing the steady state of an electrode system and a very extensive literature, much of which may not be particularly relevant to battery systems. Nevertheless, the variety of electrode systems employed as well as the diverse types of electrolytes used make great demands on theoretical and experimental skills.

There are a number of monographs which have attempted to treat *methods* rather than their application that do make something of a starting point in an understanding of the scope of the techniques available.

The first is a text by Damaskin,[13] which illustrates the use of various electrochemical methods as applied to the simple electron transfer redox process of eqn. 2.1. Thirsk and Harrison[11] cover a much wider range of reaction schemes and include treatments for electrocrystallisation; the emphasis is on technique. In the more recent book by McDonald[14] a similar framework is used but there is more illustrative material. Both experimental and mathematical techniques are treated in helpful detail.

In the latter two books the application of the perturbations shown in Fig. 2.2 is widely illustrated.

2.13 Nucleation and growth: potential step and a.c. perturbations

The foundations of studies of phase changes on electrodes by potential

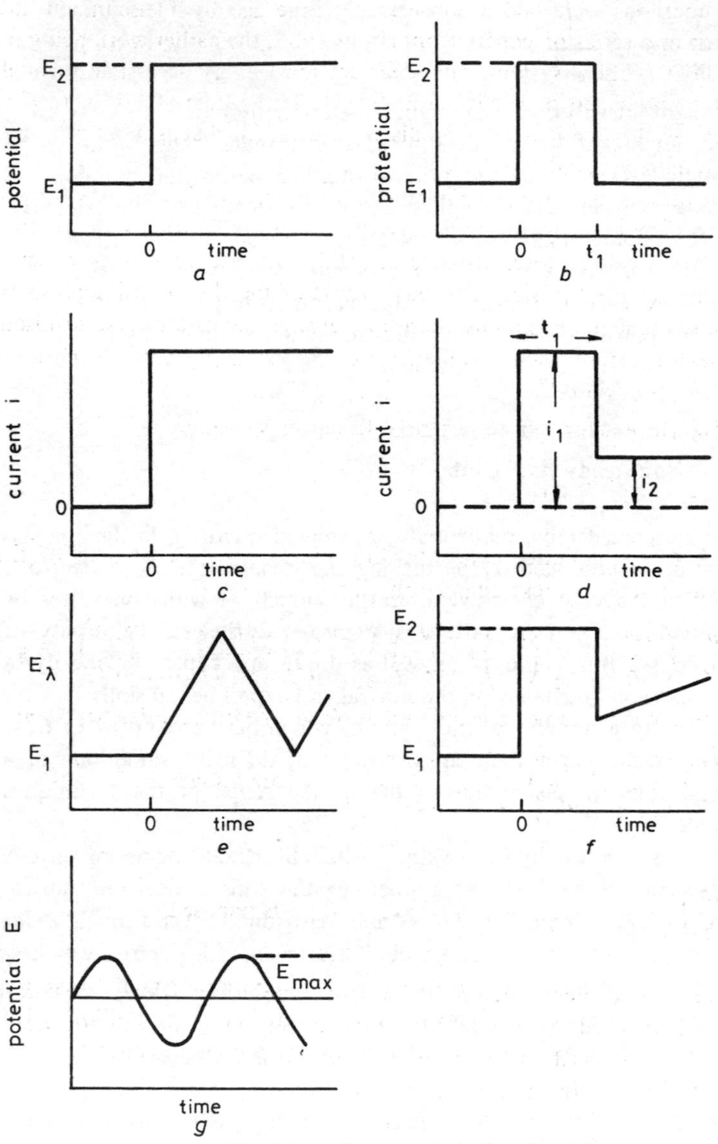

Fig. 2.2 *Some perturbations in common use*

- *a* Potentiostatic single pulse
- *b* Potentiostatic double pulse
- *c* Galvanostatic single pulse
- *d* Galvanostatic double pulse
- *e* Linear potential sweep
- *f* Potentiostatic pulse and sweep combined
- *g* Sinewave

step methods were laid a considerable time ago by Fleischmann and Thirsk in a series of papers from about 1953, the earlier work being on the $PbO_2/PbSO_4$ system. The ideas are most easily accessible through two review papers of 1963[15] and 1971[16]. At the time of writing (1978) much of this work is being rediscovered, and, what is more relevant, much more generally applied.

In these phase change problems there are inevitably complex surface changes. Mathematically it is possible to see similarities with reaction schemes involving intermediates and this can lead to some confusion of ideas in mechanisms. The basic roots of the theoretical analysis of experimental work should therefore always be carefully scrutinised. Some aspects of these problems are expressed in a recent Faraday Society discussion.[17]

The current for a given potential function is given by

$$j = f(C^b, \eta, A) \tag{2.40}$$

where C^b is a solution concentration and A the surface area of the growing phase. For a small perturbation

$$\frac{dj}{dt} = \left(\frac{\partial i}{\partial \eta}\right)_{C^b, A}\left(\frac{d\eta}{dt}\right) + \left(\frac{\partial i}{\partial C^b}\right)_{A, \eta}\left(\frac{dc}{dt}\right) + \left(\frac{\delta j}{\partial A}\right)_{C^b, \eta}\left(\frac{dA}{dt}\right) \tag{2.41}$$

The following comments can be made about the differentials involved:

$$\left(\frac{\partial i}{\partial \eta}\right)_{C^b, A}, \left(\frac{\partial j}{\partial C^b}\right)_{A, \eta}, \left(\frac{\partial j}{\partial A}\right)_{C^b, \eta} \quad \text{are known}$$

$$\left(\frac{d\eta}{dt}\right) \quad \text{is imposed}$$

(dc/dt) is in principle determinable from a solution of Fick's equation. For a.c. methods there are difficulties in setting up models for dA/dt since the rate constant and growth parameter is a function of t.

With a potential pulse the rate constant and growth parameter are independent of t and the calculation of models for surface area change with time are simplified. In view of the use of rotating disc electrodes, the case with a fixed diffusion layer is of interest.[18]

In outlining the approach used, consideration will only be given to three dimensional growth as being most relevant in the present context and as an example of the method of analysis. The basis of the overall problem has been discussed in review articles in considerable detail.[15,16]

The model for discrete centres assumes they are nucleated at a rate $A(\eta, t)$ and once nucleated grow outward in three dimensions. The rate of this outward growth can be described by a velocity normal to the electrode V_2 (η, t) and parallel to the electrode V_1. Hence the linear dimension of a centre, nucleated at a time u, at a time t is

$$r = \int_u^t v(\eta, t) dt \tag{2.42}$$

The problem is that of determining the total volume of such centres as a function of time under the specified experimental conditions. This is, however, a very difficult problem possible under potentiostatic conditions when it is a reasonable assumption that V is independent of time. Thus one can write

$$j = f(u)$$

for a single centre and

$$j' = \int_0^t f(u) A_{t=u} \, du$$

assuming no overlap. (2.43)

If centres overlap, use can be made of a general theorem due to Avrami. This distinguishes between an increment of volume ΔV_{ext} which would have occurred in the absence of overlap and the real increase ΔV which occurred when overlap was taken into account. It is convenient to consider these as fraction of a total volume V, i.e.

$$\Delta V = \Delta V_{ext} (1 - V)$$

available for the reaction. ΔV tends to zero, ΔV_{ext} remains finite and is being reduced by the fraction of the volume which remains to be filled.

By integration

$$V = 1 - \exp(- V_{ext})$$

and

$$\frac{dV}{dt} = \frac{dV_{ext}}{dt} \exp(- V_{ext}) \tag{2.44}$$

In some electrochemical experiments, i.e. $PbSO_4 \rightarrow PBO_2$, it is easy to identify V. In others it may require identification.

In discussing three dimensional growth, right circular cones have been considered as the growth centres and equations developed from two dimensional studies.

Applying the Avrami theorem to slices of the cones of height dx at a distance x from the electrode

$$dj = 2\pi V_1^2 N_0 (t - V_2/x) \exp [-\pi V_1^2 (t - V_2/x)^2] \frac{Q}{h} dx \qquad (2.45)$$

where q/h is the charge involved in the formation of unit thickness of deposit per unit area of electrode and N_0 the number of centres. Integration of eqn. 2.45 as a function of x gives[18]

$$j = \frac{Q}{h} V_2 \ [1 - \exp(-\pi N_0 V_1^2 t^2)] \qquad (2.45a)$$

for a fixed number of nuclei N_0, and

$$j = \frac{q}{h} V_2 \ [1 - \exp(-\pi A' V_1^2 t^3)] \qquad (2.45b)$$

for progressive nucleation given by the equation

$$N = N_0 \ [1 - \exp(-A't) \qquad (2.45c)$$

where A' is a constant.

At short times eqns. 2.45*a* and 2.45*b* become

$$j_{t\to 0} = \frac{q}{h} \pi N_0 V_1^2 V_2 t^2 \qquad (2.46a)$$

$$j_{t\to 0} = \frac{q}{h} \pi A' V_1^1 V_2 t^3 \qquad (2.46b)$$

and in the steady state from both

$$j_{t\to\infty} = \frac{q}{h} V_2 \qquad (2.46c)$$

Analysis of this type has been extended to examples of interaction with boundaries and with diffusion. For further reading Reference 11 should be helpful.

2.14 A.C. perturbations

There has been a very recent review[19a] of the application of a.c. perturbations to complex electrochemical reactions including reference to faradaic processes affecting the electrode itself.

The relevance to battery technology relates to many different problems: the dissolution and passivity of anodes, adsorption and the presence of intermediates with fuel cell electrodes; changes of phase, with the reservations expressed in the Section 2.13; the evaluation of porous electrodes; studies on solid electrolytes and over a very long period occasional studies of complete battery systems. A broad summary of these applications will be given with key references particularly to more recent studies.

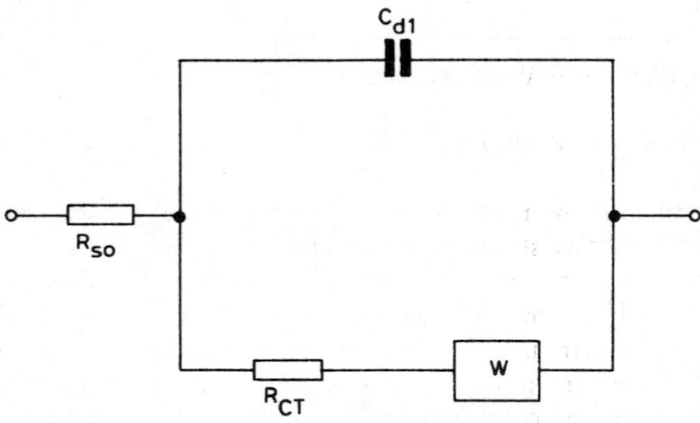

Fig. 2.3 *Randles' equivalent circuit*

In the classical work of Randles and Ershler considering rapid metal/metal-ion and redox reactions the a.c. perturbations were for small excursions from the equilibrium potential. The analysis relates to the Randles equivalent circuit (Fig. 2.3). R_{CT} is the charge transfer resistance and relates to the exchange current density and the electrochemical rate constant for the redox system. The double layer C_{dl} and the solution resistance R_{so} are removed by separate experiments carried out in the absence of the reacting system. The validity of this device has been questioned. W is the Warburg impedance associated with the diffusion of product and reactant.

Representing the residual impedance by a resistance capacity series network to evaluate k_s as a function of concentration and to eliminate the effect of diffusion

$$R_s = \frac{RT}{n^2 F^2 c}\left[\left(\frac{2}{\omega D}\right)^{1/2} + \frac{1}{k_s}\right];\ C_s = \frac{n^2 F^2 c}{RT}\left[\frac{D}{2\omega}\right]^{1/2} \qquad (2.47a,b)$$

Furthermore it can be shown that

$$R_{CT} = \frac{RT}{nF^2 k_s}\,(C_0{}^b)^a\,(C_R{}^b)^{i-a} \qquad (2.48a)$$

$$W = \sigma\omega^{1/2} - i\sigma\omega^{1/2} \text{ where } i = \sqrt{-1} \qquad (2.48b)$$

$$\sigma = \frac{RT}{n^2 F^2 \sqrt{2}}\left(\frac{1}{C_0{}^b D_0{}^{1/2}} + \frac{1}{C_R{}^b D_R{}^{1/2}}\right) \qquad (2.49a)$$

$$Z_f = R_{CT} + W = R_s + (1/i\,\omega C_s) \qquad (2.49b)$$

A second approach, which has the advantage that it may be automated to give an investigation over a very wide range of values of ω and hence the detection of relaxation times relevant to the electrode processes, follows from the analysis due to Sluyters and Sluyters-Rehbach.[20] Armstrong and co-workers working in fields of particular relevance to battery problems use equipment based on the Solartron 1170 frequency response analyser[19b] and Epelboin and collaborators used similar equipment described in a number of papers.

Using Fig. 2.3, the cell impedance including the double layer is given by

$$Z = R_{sol} + \left[1/i\omega C_{dl} + \left(\frac{1}{R_{CT}} + \sigma\omega^{-1/2} - i\sigma\omega^{-1/2}\right)\right] \qquad (2.50)$$

There are two useful limiting cases. The first when diffusion is unimportant

$$Z = R_{sol} + \frac{R_{CT}}{1 + \omega^2 C_{dl}{}^2 R_{CT}{}^2} - \frac{i\omega C_{dl} R_{CT}{}^2}{1 + \omega^2 C_{dl}{}^2 R_{CT}{}^2} \qquad (2.51)$$

If Z is plotted in the complex plane as a function of frequency a single semi-circle (Fig. 2.4) is obtained having a diameter equal to R_{CT}.

When the charge transfer resistance is small compared with diffusion

$$Z = R_{sol} + R_{CT} + \sigma\omega^{-1/2} - i(\sigma\omega^{-1/2} + 2\sigma^2 C_{dl}) \qquad (2.52a)$$

The impedance spectra (Fig. 2.5) is a straight line of 45° slope. Strictly this is true only if the a.c. diffusion layer is much smaller than the Nernstian diffusion layer. If the reaction is taking place under conditions where the two are comparable, for example this is a restriction which certainly applies as the process is taking place on a rotating disc electrode, the impedance is given by

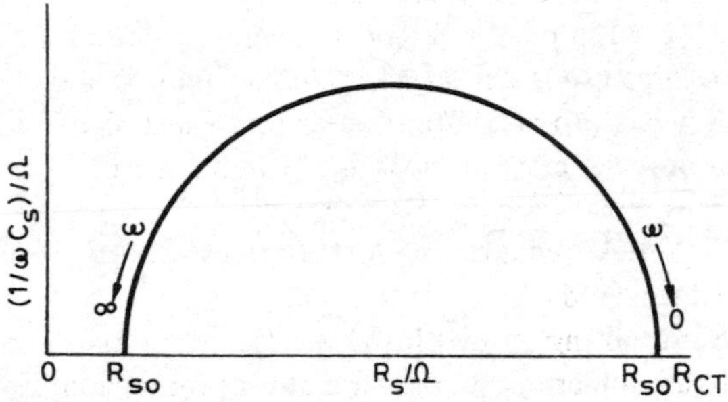

Fig. 2.4 *Complex-plane spectrum for the cell impedance described by Fig. 2.3 when diffusion is unimportant*

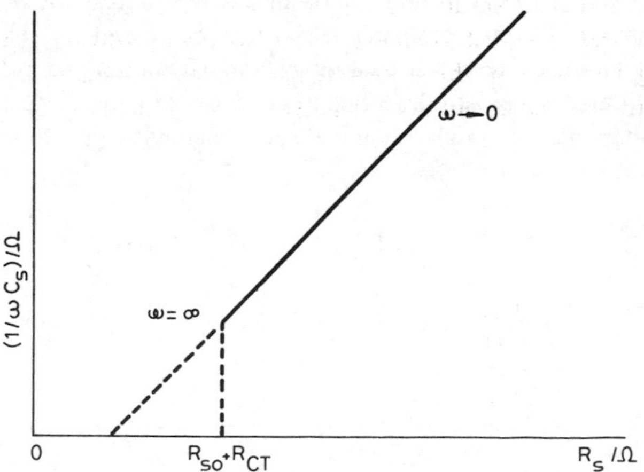

Fig. 2.5 *The Warburg impedance*

$$Z_{\delta N} = \left(\frac{\sigma}{\omega^{1/2}}\right) (1 - i) \tanh \; [\delta_N \, (i\omega/D)^{1/2} \qquad (2.53)$$

where $\delta_N = 1{\cdot}61 \; D^{\frac{1}{2}} \upsilon^{\frac{1}{6}} \; \omega_R^{-\frac{1}{2}}$, ω_R being the rotation speed.

Experimentally, these small a.c. excursions can be applied at any steady state condition of the operating of the electrode.

It is not within the context of this chapter to illustrate further the development of the application of these methods. They are, however, being increasingly employed in battery problems; one unexplored area in particular could be fruitful in investigation; that of the non-destructive testing of the condition of a battery.

A number of additional references could be found informative without encroaching too much on other chapters in this book.

It is twenty years ago at the time of writing that Euler and Dehmelt made a preliminary a.c. impedance study on a number of battery systems.[21]

Pilla has a review chapter on the application of perturbation methods to fuel cell studies.[22]

There is an interesting review chapter by Hamer[23] on cell resistances which refers to a number of established methods for examining resistive effects in batteries by a.c. methods and also several review references are given.

Some use has also been made in the applications of methods worked out by de Levie[24] for the study of porous electrodes.

One of the useful areas of application of impedance methods has been in the study of cells employing solid electrolytes. There is a growing literature on the subject to which a few references are given.[25] They are informative but not exhaustive.

References

1. International Union of Pure and Applied Chemistry. Manual of Symbols and Terminology for Physicochemical Quantities and Units. Appendix III, Electrochemical Nomenclature. Prepared for Publication by R. Parsons (1974), *Pure & Appl. Electrochem.*, 37, No. 4
2. BECK, W.H., and WYNNE-JONES, W.F.K. *Trans. Faraday Soc.*, 1954, **50**, p. 136
3. LATIMER, W.: *The oxidation states of the elements and their potentials in aqueous solutions* (Prentice-Hall, 1952, 2nd edn.)
4. BUCHOWSKY, F., and ROSSINI, F.: *The thermochemistry of chemical substances* (Rheinhold Publishing Co., New York, 1936)

5 GIBSON, J.G., and SUDWORTH, J.L.: *Specific energies of galvanic reactions and related thermodynamic data* (Chapman & Hall, 1973)

6 BARD, A.: *Encyclopedia of electrochemistry of the elements – Vol. 1* (Marcel Dekker, 1975) and subsequent volumes

7 DELAHAY, P.: *Double layer and electrode kinetics* (Interscience, 1965)

8 CAHOON, N.C., and HEISE, G.W. (Eds.): *The primary battery – Vol. 1* (American Electrochemical Society, John Wiley & Sons, (1971) p. 124 and LEVIE, R. de: *Advances in electrochemistry and electrochemical engineering – Vol. 6* (BARD, A. (Ed.), Marcel Dekker, 1967)

9 GREEF, R.: 'Instruments for use in electrode process research'. *J. Phys. E*, 1978, **VII**, p.1-12

10 ALBERY, W.J.: *Electrode kinetics* (Oxford Chemical Services, Clarendon Press, Oxford, 1975)

11 THIRSK, H.R., and HARRISON, J.A.: *A guide to the study of electrode processes* (Academic Press, 1972)

12*a* LEVICH, V.G.: *Physicochemical hydrodynamics* (Prentice-Hall, 1962)

12*b* RIDDIFORD, E.C.: *Advances in electrochemistry and electrochemical engineering - Vol. 4* (TOBIAS, C., and DELAHAY, P. (Eds.), Interscience, New York, 1966)

12*c* ALBERY, W.J., and HITCHMAN, M.L.: *Ring disc electrodes* (Oxford Science Research Papers, Clarendon Press, 1971)

13 DAMASKIN, B.B.: *The principles of current methods for the study of electrochemical reactions* (translated by G. Mamantov, McGraw-Hill, 1967)

14 MCDONALD, D.D.: *Transient techniques in electrochemistry* (Plenum Press, 1977)

15 FLEISCHMANN, M., and THIRSK, H.R.: *Advances in electrochemistry and electrochemical engineering* (DELAHAY, P. (Ed.), Interscience, New York, 1963)

16 HARRISON, J.A., and THIRSK, H.R.: *Electroanalytical chemistry – Vol. 5* (BARD, A.J. (Ed), Dekker, New York, 1971)

17 Faraday Society Discussions No. 56: 'Intermediates in electrochemical reactions'

18*a* HARRISON, J.A.: *J. Electroanal. Chem.*, 1972, **36**, p. 71

18*b* ARMSTRONG, R.D., FLEISCHMANN, M., and THIRSK, H.R.: *ibid*, 1966, **11**, p. 208

19*a* ARMSTRONG, R.D., BELL, M.F., and METCALFE, A.A.: 'The a.c. impedance of complex electrochemical reactions'. Specialist Periodical Reports, *Electrochemistry*, 1978, **6**. Senior Reporter THIRSK, H.R. Published by The Chemical Society, London (1978)

19*b* ARMSTRONG, R.D., BELL, M.F., and METCALFE, A.A.: 'A method for automatic impedance measurements and analysis', *J. Electro-anal. Chem.*, 1977, **27**, p. 287

20*a* SLUYTERS-REHBACH and SLUYTERS, J.H.: *Electroanalytical chemistry* (BARD, A.J., Ed., Dekker, New York Vol. 4, 1970)

20*b* SMITH, D.E.: *Electroanalytical chemistry* (BARD, A.J. (Ed.) Dekker, New York, Vol. 1, 1966)

21 EULER, J., and DEHMELT, Z.: *Electrokhim*, 1957, **61**, p. 1200

22 PILLA, A.A., and DIMASI, G.J.: *An equivalent electric circuit approach to the study of hydrocarbon oxidation techniques. Fuel cell systems II.*

Advances in chemistry (American Chemical Society Publications, 1969), p. 171
23 HAMER, W.T.: *The primary battery – Vol. II* (CAHOON, N.C., and HEISE, G.W. (Eds), John Wiley, 1976)
24 DE LEVIE, R.: *Advances in electrochemistry and electrochemical engineering – Vol. 6*, (BARD, A.J., Marcel Dekker, 1967), p. 329
25*a* GELLER, S. (Ed.): *Topics in applied physics – Vol. 21 Solid Electrolytes* (Springer-Verlag)
25*b* ARMSTRONG, R.D. (Ed.): *Solid ionic and ionic-electronic conductors* (Pergamon, 1977)

Chapter 3

Primary batteries for civilian use

F.L. Tye

3.1 Introduction

A battery is a device in which the free energy change of a chemical reaction is converted directly to electrical energy. The essential features are positive and negative active materials, electronic conduction between each active material and a terminal of the battery and ionic conduction between the active materials via the electrolyte and separator. If the active materials are used only once, and are not regenerated by electric current, the battery is a primary one. In this case the positive active material undergoes the electrochemical charge transfer goes that of anodic oxidation. For primary batteries, therefore, the positive and negative active materials can be referred to as cathodic and anodic reactants, respectively.
For primary batteries, therefore, the positive and negative active materials can be referred to as cathodic and anodic reactants, respectively.

It is usual not to classify fuel cells as primary batteries and consequently an additional requirement of the definition is that reactants and products are contained within the battery. A primary battery is thus a completely independent power source requiring neither an external power supply nor an external source of active materials. The continuing existence of primary batteries as commercial articles is due entirely to this independence. They are power sources of great convenience.

Recently, primary batteries have received comprehensive treatment in one and two volume works.[1,2,3] Clearly the same coverage cannot be achieved, neither is it now needed, in a single chapter. The purpose here is to emphasise aspects related to the civilian use of primary batteries. Furthermore, rather than describing each electrochemical system in isolation, an additional objective is to discuss the systems in parallel so that similarities and differences are more readily apparent. As these objectives have not previously been attempted it is hoped that this chapter complements, rather than repeats in abbreviated form, the major works already cited.

3.2 Standardisation and nomenclature

3.2.1 Standardisation

A vast industry exists predominantly to supply the populace with small primary batteries which provide the electrical power for a wide variety of portable appliances. It is taken for granted that the battery will fit into the appliance compartment and that the battery terminals will make appropriate contacts. Furthermore, such compatibility is expected irrespective of where batteries and appliances are manufactured, sold and used. In view of the portability of an appliance the owner would be rightly aggrieved if he were unable to obtain a suitable replacement battery in another country. Since batteries and appliances are usually manufactured by different companies, and there are many of both, the high degree of compatibility that exists must be counted as a remarkable achievement. It is, of course, no accident.

At the beginning of the 20th century, when the industry was emergent, companies manufactured both batteries and appliances and dimensions were maintained by internal documentation. As competition and the industry grew and diversified National Standards were issued. The first was the American National Standard[4] which was published in 1923 and which followed even earlier documents issued by the American National Bureau of Standards.[5] This National Standard, the most recent edition of which is 1972,[6] still commands wide respect, and much of the credit for similarity of dimensions is due to the early lead it provided. Eventually increases in international trade and travel brought the need for greater harmony among National Standards and for an international nomenclature. This task is the responsibility of the International Electrotechnical Commission (IEC). The first International Recommendation was published in 1957[7] and a new edition of this standard, which is now in two parts, has been published recently.[8]

3.2.2 Nomenclature

The IEC system of nomenclature is based on three series of unit cells. Each series is allocated a letter in accordance with a characteristic shape: cylindrically shaped cells have a letter R (round cells); cells of right-angled prismatic shape, with a square cross-section, have the letter S (square cells), and the letter F (flat cells) is used for right-angled prismatically shaped cells of square, oblong or circular cross-section in which the dimension normal to this cross-section is much the smallest. The shapes are illustrated in Fig. 3.1.

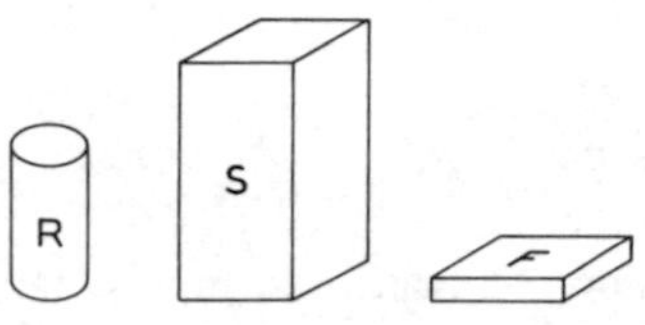

Fig. 3.1 *Unit cell shapes*

Each series consists of unit cells of different sizes. The size is defined by nominal dimensions to which the letter R, S or F and a following number specifically relates. The three series are given in Tables 3.1, 3.2 and 3.3. The American National Standard[6] uses identical designations for flat cells with the exception that the nominal values may differ slightly owing to conversion from imperial dimensions.

Each unit cell forms an integral part of the IEC designation system and for this purpose nominal dimensions are adequate. It is emphasised that these nominal values should not be used when battery dimensions are required. The word battery is used to describe a final product and *may* include, in addition to the cell, outer constructional features and terminals. Battery dimensions are discussed in the next section.

The IEC Standard[8] at present covers six electrochemical systems. The electrochemical system in a battery is denoted by a letter before the size designation. The only exception is the $MnO_2/NH_4Cl,ZnCl_2/Zn$ (Leclanché) system, which, being the original one standardised, has no additional letter. The letters used and the electrochemical systems covered are shown in Table 3.4, which gives, in order of decreasing volume, all the unit round cell batteries specified in the IEC Standard. It is interesting to note the importance of an electrochemical system in a particular size range only.

The letter A is used for the $O_2/NH_4Cl,ZnCl_2/Zn$ (neutral air depolarised) system. No unit round cell batteries in this electrochemical

Table 3.1 *Designations and nominal dimensions of round cells*

Nominal height	IEC round cell designations								
mm									
166									$\overline{\underline{\text{R40}}}$
150								R27	
105								R26	
91								$\overline{\underline{\text{R25}}}$	
83							R18[1]		
75								R22	
70							R15		
60						$\overline{\underline{\text{R12}}}$[2]		$\overline{\underline{\text{R20}}}$[3]	
50				$\overline{\underline{\text{R6}}}$	$\overline{\underline{\text{R51}}}$		$\overline{\underline{\text{R14}}}$[5]		
44		$\overline{\underline{\text{R03}}}$[6]							
37				R4[7]		$\overline{\underline{\text{R10}}}$			
30			$\overline{\underline{\text{R1}}}$[8]						
25				R3					
22		R06[6]							
19			R0[8]						
16					$\overline{\underline{\text{R50}}}$		R17[1,9]	R19[9]	
14			$\overline{\underline{\text{R01}}}$[8]						
11					$\overline{\underline{\text{R52}}}$				
6·0					$\overline{\underline{\text{R9}}}$		$\overline{\underline{\text{R53}}}$		
5·4	$\overline{\underline{\text{R48}}}$		$\overline{\underline{\text{R44}}}$						
4·2			$\overline{\underline{\text{R43}}}$						
3·6	$\overline{\underline{\text{R41}}}$	$\overline{\underline{\text{R45}}}$	$\overline{\underline{\text{R42}}}$						
3·0			R54[4]						
2·6	R59[4]	R57[4]	R56[4]						
2·1	R58[4]		R55[4]						
Nominal diameter (mm)	7·9	9·5	11·6	13·5	16	20	24	32	64

$\overline{\underline{\quad}}$ denotes sizes used for unit and/or multi-cell batteries currently specified in IEC Publication 86 (4th edn.)
(1) Nominal diameter 25·5 mm (2) Nominal Height 59 mm (3) Nominal height 61 mm (4) Standardisation under consideration (5) Nominal height 49 mm (6) Nominal diameter 10 mm (7) Nominal height 38 mm (8) Nominal diameter 11 mm (9) Nominal height 17 mm

Table 3.2 *Designations and nominal dimensions of square cells*

Nominal height	IEC square cell designations			
mm				
180			S8	S10
150		S6		
105	S4			
Nominal width and length, mm	50	57	75	95

═ denotes sizes used for unit and/or multi-cell batteries currently currently specified in IEC Publication 86, (4th edn.)

Table 3.3 *Designations and nominal dimensions of flat cells*

Nominal thickness	IEC flat cell designations							
mm								
10·4								F100
7·9						F90	F95[1]	
6·4						F80		
6·0		F22[3]	F25					
5·6						F70	F92[2]	
5·3				F40				
4·5	F16							
3·6					F50			
3·3				F30				
3·0	F15							
2·8		F20						
Nominal width, mm	14·5	13·5	23	21	32	43	37	45
Nominal length, mm	14·5	24	23	32	32	43	54	60

═ denotes sizes used for multi-cell batteries currently specified in IEC Publication 86 (4th edn.)
Also F24, nominal thickness 6·0mm and nominal diameter 23mm (1) Nominal width 37mm (2) Nominal thickness 5·5 mm (3) The F22 is given a thickness of 7·1mm in the American National Standard

Table 3.4 *Unit round cell batteries specified in the IEC standard*

IEC size designation	MnO_2 / NH_4Cl, $ZnCl_2$ / Zn	MnO_2 / alkali metal hydroxide / Zn	HgO, MnO_2 / alkali metal hydroxide / Zn	HgO / alkali metal hydroxide / Zn	Ag_2O / alkali metal hydroxide / Zn
R40	R40				
R20	R20	LR20			
R14	R14	LR14			
R12	R12[2]				
R10	R10[2]				
R51			NR51	MR51	
R6	R6	LR6	[1]	MR6	
R03	R03	LR03			
R50		[1]	NR50	MR50	
R1	R1	LR1	[1]	MR1	
R53		LR53	[1]		
R52			NR52	MR52	
R01				MR01[2]	
R9	R9	LR9	NR9	MR9	
R44		[1]	NR44[3]	MR44	SR44
R43			[1]	MR43	SR43
R42			NR42[3]	MR42	SR42
R54				MR54[3]	SR54[3]
R48			[1]	MR48	SR48
R45			[1]	MR45	[1]
R56					SR56[3]
R55					SR55[3]
R57					SR57[3]
R41				MR41	SR41
R59					SR59[3]
R58					SR58[3]

(1) Although not standarised by IEC an equivalent battery in this size and electrochemical system is available (2) Deletion of this battery from the IEC Standard is under consideration (3) Standardisation is under consideration

system are included in the IEC Standard although an AR40 battery is manufactured. The letter M is reserved for batteries in which only mercuric oxide is used as cathodic reactant. If any manganese dioxide is admixed with the mercuric oxide, the letter N must be used. This distinction is important as the voltage of M systems is always very close to 1·35V at ambient temperature and is fairly stable on discharge. Batteries in the N system have voltages which are higher and more

Table 3.5 *Descriptions of round and square multi-cell batteries*

IEC battery designation	Cell size	Electrochemical system	Number of cells in series	Number of series groups in parallel
5R40	R40	Leclanché	5	-
5AR40	R40	Neutral air-depolarised	5	-
6S6	S6	Leclanché	6	-
6AS6	S6	Neutral air-deplarised	6	-
4R40	R40	Leclanché	4	-
6S4	S4	Leclanché	6	-
6AS4	S4	Neutral air-depolarised	6	-
4R25-2	R25	Leclanché	4	2
R25-4	R25	Leclanché	-	4
4R25	R25	Leclanché	4	-
3R25	R25	Leclanché	3	-
3R20	R20	Leclanché	3	-
3R12	R12	Leclanché	3	-
2R10	R10	Leclanché	2	-
R6-2	R6	Leclanché	-	2

variable on open circuit and which fall initially on discharge. Some manufacturers incorrectly list batteries with open circuit voltages of 1·4V as being of the M electrochemical system.

The only unit square cell batteries specified in the IEC Standard[8] are Leclanché (S4, S8 and S10), although neutral air depolarised variants are also available. Flat cells are used only in multi-cell batteries.

In multi-cell batteries, identical cells may be connected in series, in parallel or in series-parallel. The number connected in series is given by the number preceding the letter denoting the electrochemical system, or, for Leclanché batteries, the letter denoting cell shape. The number of parallel connections is given by the number following the size number and connected to it by a hyphen. IEC nomenclature is thus very informative as regards multi-cell battery construction. The system is illustrated in Table 3.5, which describes, in order of decreasing volume, all the multi-cell batteries incorporating round or square cells that are included in the IEC Standard.[8]

Construction may similarly be deduced from the IEC designation for batteries with flat cells. In this case only, the American Standard[6] uses identical nomenclature. Table 3.13 shows the multi-flat cell batteries included in the IEC Standard.[8] For these, which may have many cells in series, it is useful to remember that their open circuit voltage is 1·6V

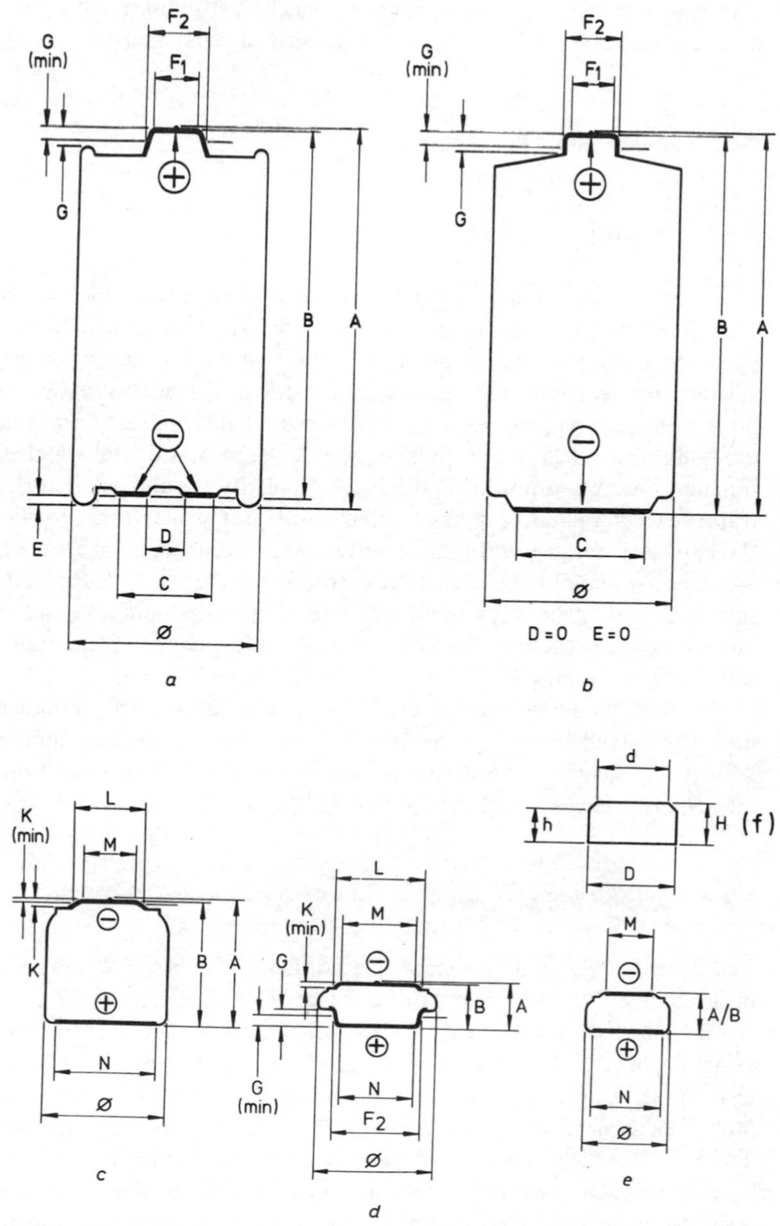

Fig. 3.2 *Profiles of unit round cell batteries*

(the approximate value for a single Leclanché cell) multiplied by the number preceding the letter F, i.e. the number of cells in series.

3.3 Terminals and dimensions

3.3.1 Polarity

Although the direction of electric current is unimportant in a torch, such freedom from the consequences of polarity is unusual. Motors operating cutting heads in razors, or cassette drives in tape recorders, are required to rotate in a particular direction. Transistors which are present in many appliances, e.g. radios, hearing aids and light emitting diode displays, such as are present in calculators and digital watches, function only when power of the correct polarity is supplied. For this reason each terminal of a battery has a different shape. This provides the appliance designer with an opportunity to construct compartments which accept only batteries that are correctly orientated. Unfortunately this facility is not always used, and one of the commonest causes of dissatisfaction is the incorrect placement in an appliance of one round cell battery of a set.

For unit round cell batteries four shapes of asymmetric terminals have been standardised by the IEC.[8] These shapes have been defined with the aid of the symbols and associated descriptions given in Table 3.6. The four shapes are shown in Fig. 3.2.

3.3.2 Cylindrical batteries

The dimensions of batteries made in accordance with Fig. 3.2*a* are given in Table 3.7. It is important to note that the dimensions apply to all electrochemical systems, e.g. R6 and LR6 batteries are made to the same dimensions. The shape and terminals represented by Fig. 3.2*a* are characteristic of Leclanché unit round cell batteries (see Table 3.4). MnO_2/alkali-metal-hydroxide/Zn (alkaline manganese) batteries (see Table 3.4), which were introduced into the IEC Standard at a later date, have been constructed to conform to the same dimensions and terminal polarity. In view of the importance of polarity it is axiomatic that batteries of the same shape but of opposite polarity should be discouraged. Such batteries are manufactured in 'R1' and 'R6' sizes but have not been included in the IEC Standard. An IEC battery designation thus confers a specific polarity to the terminals.

Table 3.6 *Symbols used to define unit round cell batteries*

Symbol	Description
A	Overal height of battery including pip if present
B	Distance between the flat contact surfaces of the two terminals
C or M	Outer diameter of the flat contact portion of the negative terminal
D	Inner diameter of the flat contact portion of the negative terminal
E	Recess of the flat contact portion of the negative terminal
F_1 or N	Outer diameter of the flat contact portion of the positive terminal
F_2	Diameter of the positive terminal measured at a distance G(minimum) from the flat contact portion of the terminal
G	Projection of the flat contact portion of the positive terminal
G(min)	Minimum value permitted for G
K	Projection of the flat contact portion of the negative terminal
K(min)	Minimum value permitted for K
L	Diameter of the negative terminal measured at a distance K(minimum) from the flat contact portion of the terminal
ϕ	Overall diameter of battery

Table 3.7 *Dimensions of round cell batteries (Fig. 3.2a)*

IEC	Battery dimensions, mm (see Table 3.6)									
size	A	B	C	D	E	F_2	F_1	G	ϕ	ϕ
designation	max	min	min	max	max	max	min	min	max	min
R40[1]	172[2]								67·0	63·0
R20	61·5	59·5	16·0	7·5	1·0	9·5	7·8	1·5	34·2	32·2
R14	50·0	48·5	12·0	5·0	0·9	7·5	5·5	1·5	26·2	24·7
R12	60·0	56·7	18·5	0	0·8	6·8	5·8	1·0	21·5	19·4
R10	37·3	35·75	9·0	5·0	0·8	6·8	5·8	1·0	21·8	20·0
2R10	74·6	71·5	9·0	5·0	0·8	6·8	5·8	1·0	21·8	20·0
R6	50·5	49·0	7·0	4·0	0·5	5·5	4·2	1·0	14·5	13·5
R03	44·5	42·5	4·0	0	0·5	3·8	2·0	0·8	10·5	9·5
R1	30·2	28·0	4·8	0	0·2	4·5	2·0	0·3	12·0	10·7

(1) R40 batteries are not made to Fig. 3.2*a*; screw terminals are specified[8] (2) Minimum overall height is 165mm

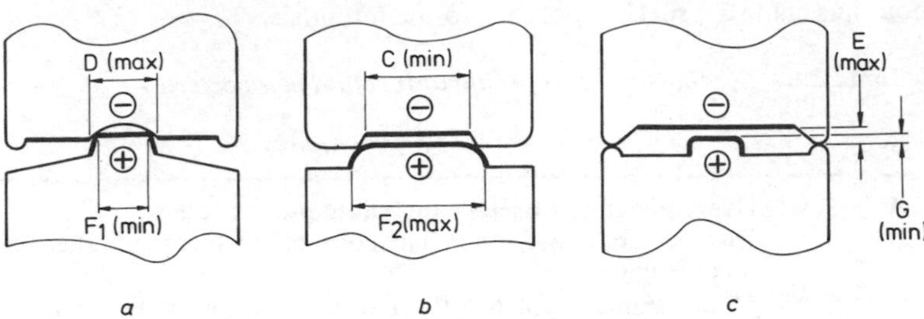

Fig. 3.3 *Some consequences of not adhering to the IEC terminal dimensions for cylindrical batteries*

Fig. 3.2*a* is illustrative only and a variety of top and bottom shapes is permissible within the mandatory dimensions of Table 3.7. Fig. 3.2*b* shows an example where the negative base terminal projects instead of being recessed and where the central concavity in this terminal is omitted. Relationships exist between certain dimensions: F_1 (min)> D(max), C(min)> F_2 (max) and G(min)>E(max). Fig. 3.3 illustrates the consequences of *not* keeping to these requirements. If D(max)> F_1 (min) (Fig. 3.3*a*), batteries nest when placed end to end in series so that the total contact height of the series is not an integral multiple of dimension B. In this circumstance the appliance contacts, if properly spaced, may not make contact with, or exert sufficient pressure on, the battery terminals of the end members. If F_2 (max)>C(min) (Fig. 3.3*b*), or E(max)>G(min) (Fig. 3.3*c*), the terminals of batteries placed end to end may be held apart and not make electrical contact. In the future it is likely that recesses will be permitted in the negative flat contact surface defined by dimensions C and D, provided that when batteries are placed end to end in series they make electrical contact with each other and that the total contact height of the series is an integral multiple of dimension B. Further points to note are that the external cylindrical sides of all the batteries listed in Table 3.7 are insulated from both battery terminals and that the pip (maximum height 0·4mm) on the positive terminal is optional.

The dimensions of batteries made in accordance with Fig. 3.2*c* are given in Table 3.8. This shape has been standardised only for batteries having mercuric oxide as cathodic reactant (M and N systems, see Table 3.4). The pip, this time on the negative terminal, is optional, and recession of the positive terminal, e.g. by a shrink sleeve overlapping the base, is not permitted. The projection of the negative terminal, dimension K, is measured from the next highest point, be that grommet, shrink sleeve or the lip or shoulder of the can. The tallest batteries of this group MR51 and NR 51 have external cylindrical sides which are insulated from both terminals. For the others, MR50, NR50, MR52, NR52 and MR01, the cans against which cathodic reactant is pressed internally are exposed, and so contact to the positive terminal may be made on the cylindrical sides instead of on the designated base area. This type of terminal is known as 'cap and case' to distinguish it from the type 'cap and base' discussed previously.

3.3.3 *Button batteries*

Fig. 3.2*d* shows a shape in which asymmetry of terminals is particularly emphasised. The dimensions of batteries made to this shape are given in Table 3.9. R9, LR9, MR9, NR9 and LR53 batteries have been standardised.[8] The pip is optional, and, as the terminal type is cap and case; side or base contact may be made to the positive terminal. Projection K is measured from the next highest point.

The shape depicted in Fig. 3.2*e* is for batteries which are becoming increasingly important in miniature electronic equipment. Dimensions are given in Table 3.10. The terminal type is cap and case. For this shape asymmetry is defined by means of profile gauges (Fig. 3.2*f*), the dimensions of which are included in Table 3.10.

Table 3.8 *Dimensions of round cell batteries (Fig. 3.2c)*

IEC size designation	Battery dimensions, mm (see Table 3.6) A max	B min	K min	L max	M min	N min	ϕ max	ϕ min
R51	50·0	49·0	0·2	8·5	6·6	13·0	16·5	15·5
R50	16·8	16·0	0·2	10·4	6·6	13·0	16·4	15·5
R52	11·4	10·6	0·5	8·5	6·1	13·0	16·4	15·5
R01	14·7[1]						12·0	11·0

(1) Minimum overall height is 14·2mm

Table 3.9 *Dimensions of round cell batteries (Fig. 3.2d)*

IEC size designation	Battery dimensions, mm (see Table 3.6) A max	B min	F_2 max	G min	K min	L max	M min	N min	ϕ max	ϕ min
R53	6·1	5·4	20·9	2·1	0·2	21·0	15·3	18·7	23·2	22·6
R9	6·2	5·6	13·5	2·0	0·2	12·5	10·0	10·0	16·0	15·0

Table 3.10 *Dimensions of round cell batteries (Fig. 3.2e and 3.2f)*

IEC size designation	Battery dimensions, mm (see Table 3.6) A/B max	A/B min	M min	N min	ϕ max	ϕ min	Gauge dimensions, mm H max	H min	h max	h min	D max	D min	d max	d min
R44	5·4	5·0	3·8	3·8	11·6	11·25	5·412	5·404	4·412	4·404	11·617	11·606	9·614	9·605
R43	4·2	3·8	3·8	3·8	11·6	11·25	4·212	4·204	3·212	3·204	11·617	11·606	9·614	9·605
R42	3·6	3·3	3·8	3·8	11·6	11·25	3·612	3·604	2·608[2]	2·602[2]	11·617	11·606	9·614	9·605
R54[1]	3·05	2·75	3·8	3·8	11·6	11·25				under consideration				
R48	5·4	5·0	[2]	[2]	7·9	7·55	5·412[2]	5·404[2]	4·612[2]	4·604[2]	7·914[2]	7·905[2]	6·314[2]	6·305[2]
R45	3·6	3·3	3·8	3·8	9·5	9·15	3·612	3·604	2·708	2·702	9·514	9·505	7·714	7·705
R56[1]	2·6	2·3	3·8	3·8	11·6	11·25				under consideration				
R55[1]	2·1	1·85	3·8	3·8	11·6	11·25				under consideration				
R57[1]	2·7	2·4	3·8	3·8	9·5	9·15				under consideration				
R41	3·6	3·3	3·8[2]	3·8[2]	7·9	7·55	3·612	3·604	2·808	2·802	7·914	7·905	6·314	6·305
R59[1]	2·6	2·3	[2]	[2]	7·9	7·55				under consideration				
R58[1]	2·1	1·85	[2]	[2]	7·9	7·55				under consideration				

(1) Standardisation is under consideration (2) Under consideration

3.3.4 Square cell batteries

Compared to unit round cell batteries the unit square cell batteries S4, S8 and S10 are large and are often used externally to equipment, so that it has not been necessary to define their shape so precisely. Basic dimensions are given in Table 3.11. Knurled nut and screw terminals are standardised,[8] one is centrally sited (positive) and the other is on the perimeter (negative). Other terminal arrangements are permitted and a common variant is the use of wire for the negative contact.

Table 3.11 *Dimensions of unit square cell batteries*

IEC size size designation	Height		Length and width	
	Max	min	max	min
	mm	mm	mm	mm
S10	210•0	200•0	110•0	103•0
S8[1]	200•0	190•0	85•0	80•0
S4	125•0	115•0	57•0	55•0

(1) Deletion of this battery from the IEC Standard is under consideration

3.3.5 Multi-cell batteries

Multi-round and multi-square cell batteries form a conglomerate group and are listed in order of decreasing volume in Table 3.12. Terminals are in a variety of forms and for the larger batteries their standardisation has not been found necessary. Sockets comprise two holes of different internal diameters, the larger being the positive terminal. One of the most important batteries of this group is the 3R12, which is shown in Fig. 3.4*a*.

The dimensions and terminals of flat cell batteries that are standardised by the IEC are given in Table 3.13. One of the most important is the 6F22 battery, which is shown in Fig. 3.4*b*. Snap fasteners, both in miniature and standard sizes, are an important class of terminal for these batteries. The stud, which is non-resilient, is the positive terminal and the resilient socket is the negative terminal (see Fig. 3.4*b*).

Table 3.12 *Multi-round and multi-square cell batteries*

IEC battery designation	Overall height, mm max	min	Length, mm max	min	Width, mm max	min	Terminals
5R40[12], 5AR40	190	-	184[1]	-	-	-	not standardised
6S6[12], 6AS6	162	-	192	-	128	-	not standardised
4R40X[2,12]	190	-	137	-	137	-	not standardised
4R40Y[2]	270	-	190	-	71	-	not standardised
6S4[12], 6AS4	114	-	168	-	113	-	not standardised
4R25-2	127[3]	-	136·5[4]	132·5	73[4]	69	screw posts and insulated nuts[5]
R25-4[6]	103	100	67	65	67	65	sockets[5]
4R25	102[7]	97	67[8]	65	67[8]	65	flat or spiral springs[9]
3R25[6]	106	101	102	99	35	33	sockets[5]
3R20X[10]	76	71	102	99	36	34	flat springs[5]
3R20Y[10,12]	88	84	102	99	36	34	screw posts and knurled nuts[5,11]
3R12	see Fig. 3.4*a*						flat springs, see Fig. 3.4*a*
2R10	see Fig. 3.2*a* and Table 3.7						
R6-2[12]	51	50	29	28	14·5	13·5	cap and base

(1) Diameter not length (2) the two variants are due to different physical arrangements of the four cells in the battery (3) Container height 109·5 - 114 mm (4) Corner radius 14·0 mm minimum (5) For further details see Reference 8 (6) No longer recommended for specific applications (7) Container height only, does not include terminals, an overall height of 108-115 mm including spiral spring terminals is under consideration (8) Rounded or bevelled corners and must pass freely through a circular gauge of internal diameter 87 mm (reduction to 82·6 is under consideration) (9) Usually spiral springs, see British Standard[9] (10) The two variants are due to different terminals (11) Other terminal arrangements are permitted (12) Deletion of this battery from the IEC Standard is under consideration

3.3.6 *Advantages of the IEC system*

A review of the advantages of the IEC system is worthwhile at this point. First, it is descriptive: the electrochemical system, the open circuit voltage, the cell shape and how the cells are combined in the

Table 3.13 *Flat cell batteries*

IEC battery designation	Overall height, mm max	Overall height, mm min	Length, mm max	Length, mm min	Width, mm max	Width, mm min	Terminals
60F40[1,9]	100	95	71	66	50	46	sockets[2]
45F40[1]	95	90	71	67	35	32	standard snap fasteners at one end[2]
30F40[1,9]	95	90	67	64	26	24	standard snap fasteners at one end[2]
20F20[1]	65	62·5	27[3a]	25	16[3a]	14	identical flat projecting contacts, one at each end[2]
15F20	51	48·5	27[3a]	25	16[3a]	14	identical flat projecting contacts, one at each end[2]
10F20[1]	37	35	27[3a]	25	16[3a]	14	identical flat projecting contacts, one at each end[2]
15F15[1]	51	48	16[3b]	14	15[3b]	13	identical flat projecting contacts, one at each end[4]
10F15[1]	35	33	16[3b]	14	15[3b]	13	identical flat projecting contacts, one at each end
6F100-3[9]	226	221	66	63	52	50	standard snap fasteners at one end[2]
6F100	81	78	66	63	52	50	standard snap fasteners at one end[5]
6F50-2[6]	70	68	36	34	34·5	32·5	miniature snap fasteners at orle end[7]
6F25[9]	51	48·5	25·5	24·5	25·5	24·5	standard snap fasteners, one at each end[2]
6F24	50	48	25·5[8]	24·5[8]			standard snap fasteners, one at each end[2]
6F22	see Fig. 3.4*b*						minature snap fasteners
4F16	20	19	16[3c]	14	16[3c]	14	identical flat projecting contacts, one at each end[4]

(1) No longer recommended for specific applications (2) For further details see Reference 8 (3) Must pass freely through a circular gauge of internal diameter (*a*) 29mm (*b*) 19·5mm (*c*) 18 mm (4) For further details see British Standard[9] (5) Plug in socket may be used, if so overall height 86-83mm (6) The modern construction of this battery uses only 6 cells: width and length are identical with the F50 cell but thickness is double (7) Plug in socket amy be used (8) Diameter (9) Deletion of this battery from the IEC Standard is under consideration

battery are immediately apparent from the battery designation. Secondly, should the need arise for a battery in an electrochemical system other than the one standardised, the nomenclature, dimensions and terminals which would be acceptable to the IEC are known. This helps to avoid proliferation of battery sizes and should prevent the

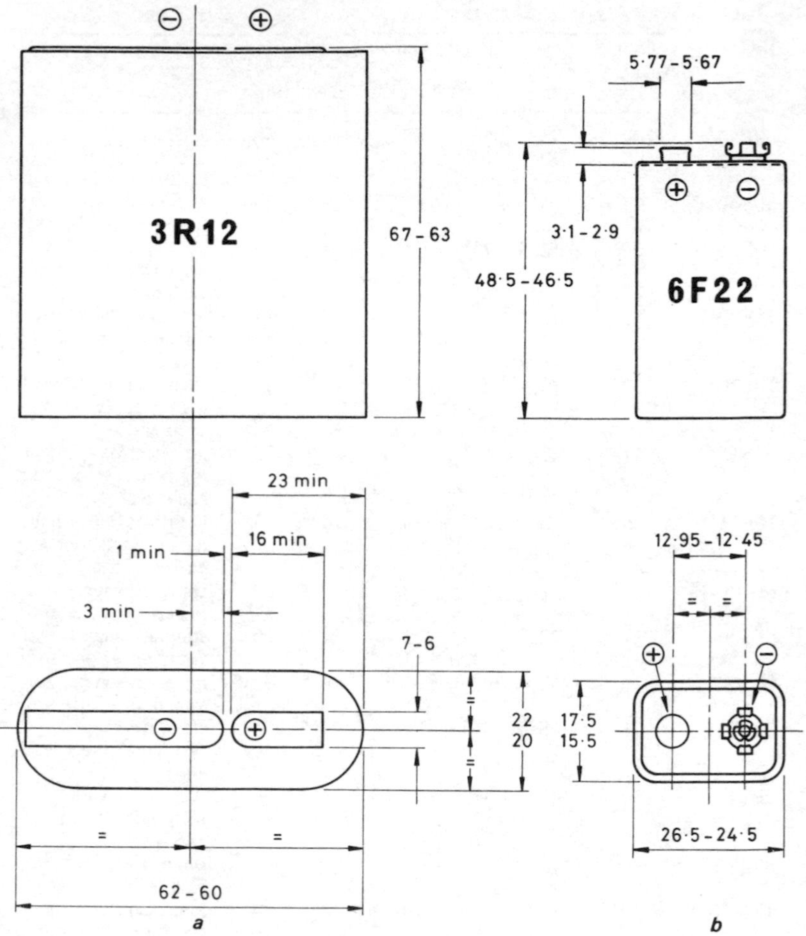

Fig. 3.4 *Two important multi-cell batteries (dimensions in mm)*

a **3R12**

b **6F22**

development of products which may be difficult to reconcile if the battery grows sufficiently in importance to warrant international standardisation. Inevitably, international standardisation takes time, and knowledge of the battery profile and dimensions that are likely to be specified can avoid costly tooling and machine modifications at a later stage. Thirdly, if an entirely new battery is required it may be possible to design this around one of the existing cell sizes (Tables 3.1 -

3.3), not all of which are implemented as batteries. Fourthly, batteries which have only national importance may be included in National Standards using the same system of nomenclature. Considerable use is made of this facility. The British Standard[9] has 16 batteries, e.g. NR1, NR43, 2R22 and 6F90, which are not specified in the International Standard. The IEC Standard is under constant review and reference should always be made to the latest edition and all its succeeding amendments.

3.4 Construction and electrode structure

The internal constructions of batteries are rightly not standardised, although general forms have evolved and will now be described.

3.4.1 Cylindrical Leclanché and alkaline manganese batteries

Figs. 3.5*a* and *b* illustrate and compare the internal constructions of cylindrical batteries made to the dimensions given in Table 3.7 and with the terminals shown in Fig. 3.2*a*. Fig. 3.5*a* shows a Leclanché battery and Fig. 3.5*b* an alkaline manganese battery. The first commercial production of alkaline manganese batteries (in a different form from Fig. 3.5*b*) was in 1949.[10]

The cathode of a Leclanché battery consists of manganese dioxide, ammonium chloride (solid) and a particular form of carbon known as acetylene black. The acetylene black, which has a chain-like structure,[11,12] provides the means by which electrons are transported from the carbon rod to the cathodic reactant, manganese dioxide. Graphite is occasionally used to improve the electronic conduction of Leclanché batteries intended for high current pulse applications, e.g. photoflash, but in general is not beneficial to performance. The chain-like structure of acetylene black also makes the moderately compressed mixture porous and the pores are occupied by aqueous $NH_4Cl/ZnCl_2$ electrolyte. Porosity is essential for good discharge performance. It facilitates diffusion of ammonium ions and ionic zinc species to the cathodic reaction sites and provides space to accommodate the reaction products which occupy a greater volume than the original reactants.

The cathode of an alkaline manganese battery is a highly compressed mixture of manganese dioxide and graphite, usually in the form of several annular tablets. For mixtures of manganese dioxide and carbon of given electronic conductivity, a greater manganese dioxide content is

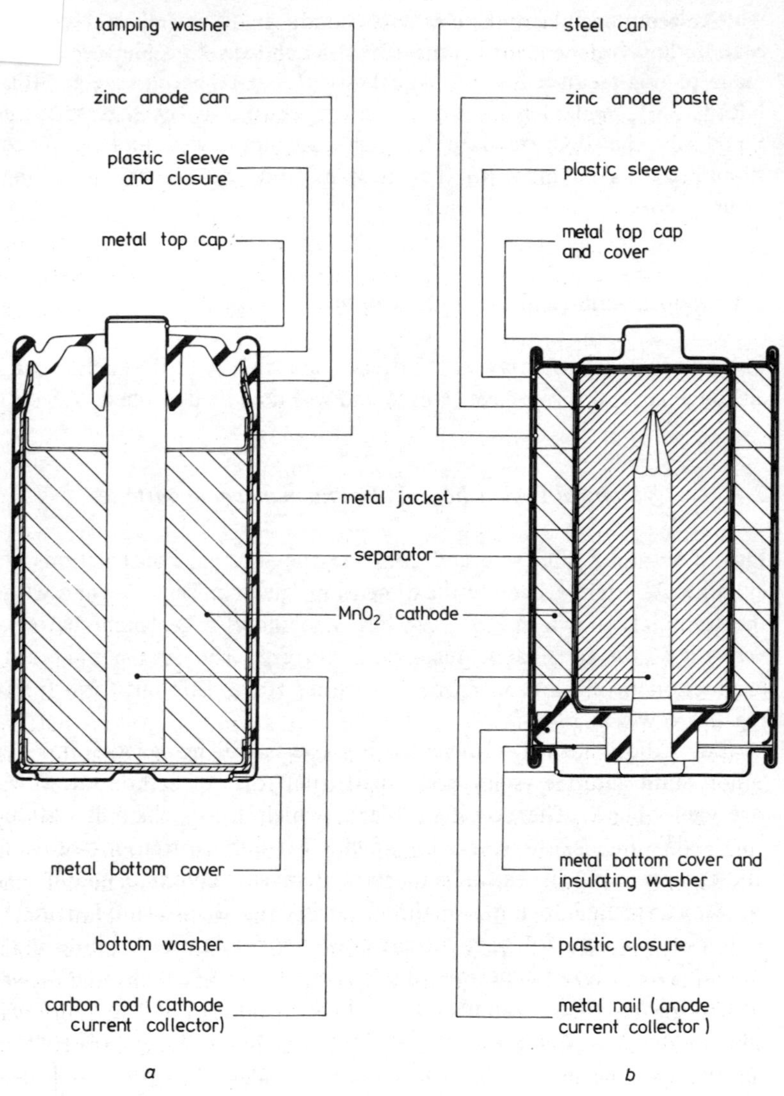

Fig. 3.5 *Constructions of Leclanché and alkaline manganese cylindrical batteries*

a Leclanché

b Alkaline manganese battery

attainable if the carbon is graphite rather than acetylene black.[13] Unlike acetylene black, the electronic conductivity of graphite, which is particulate, is dependent on the pressure applied to the mixture as well as on its volume fraction.[13] Fortunately, high compression is acceptable, as the cathodes of alkaline manganese batteries are able to discharge effectively at lower porosities than are necessary for the cathodes of Leclanché batteries. Diffusion of reactant ions is not required and the high transference number of the hydroxyl ion produced by the cathode reaction results in the product ZnO accumulating in the anode. The net effect of these factors and the absence of solid NH_4Cl is that 40% – 70% more manganese dioxide is present in alkaline manganese batteries than in Leclanché batteries despite a 20% – 40% lower cathode volume.

The anode of a round cell Leclanché battery is a zinc can of wall thickness 0·3 - 0·5mm. Zinc in this form has the minimum volume requirement and is satisfactory because solid discharge products do not usually accumulate close to the discharging zinc surface. In alkaline manganese batteries, because of the need to accommodate the discharge product ZnO and to minimise the extent to which ZnO blocks the discharging zinc surface, the anode is in a more dispersed form of higher surface area. Particulate zinc in the diameter range 10^{-2} to 10^{-1} cm is held in suspension in a solution of a gelling agent, e.g. sodium carboxymethyl cellulose and KOH.[14] In this form the anode occupies five times the space of the zinc used in a Leclanché battery and is responsible for the reduced volume that is available for the cathode in an alkaline manganese battery.

Zinc anodes are amalgamated[15,16] in order to reduce corrosion by increasing hydrogen overpotential. The amount of mercury used in Leclanché batteries ranges from $0{\cdot}5\times10^{-4}$ to 2×10^{-4} g/cm^2 of zinc surface. The mercury does not remain on the surface but diffuses[17] relatively rapidly (diffusion coefficient 10^{-5} cm^2/s)[18] along grain boundaries in the zinc and more slowly into the grains themselves (diffusion coefficient 10^{-12} cm^2/s).[19,20] In one study[18] mercury was shown to have pentrated at 30°C distances of 10^{-2} and $1{\cdot}5\times10^{-2}$ cm after 2 months and 2 years, respectively. Anodes of alkaline manganese batteries contain much greater quantities of mercury (3 - 12 w/w% of the zinc) than Leclanché batteries: the quantity per unit surface area of anode is an order of magnitude greater in alkaline manganese batteries than in Leclanché batteries. Substances present in the separator or electrolyte also have a beneficial effect in reducing corrosion: flour and starch,[21] or alkyl celluloses[22] and dissolved $ZnCl_2$ in Leclanché batteries, and zincate ions[23–26] in alkaline manganese batteries. Small proportions of lead (<1%) and camium (<0·35%) are added to the zinc[27,28,244] for Leclanché batteries to retard corrosion, to control grain size, and, in the case of cadmium, additionally to improve the resistance of the cans to deformation.[17] The latter is particularly important when batteries are assembled on modern high speed plant. Small additions of lead also reduce corrosion in alkaline electrolyte.[25,29]

The separator placed between the anode and cathode can take one of several forms in Leclanché batteries. The greatest availability for cathode mix volume is obtained with Kraft paper (of low metallic impurity content) thinly coated with flour and starch or an alkyl cellulose.[30] This is known as the paper-lined construction. Paper-lined batteries probably predominate, although cells with gelled flour and starch electrolyte layers 0·25 - 0·35cm thick, known as paste cells, are still made by many manufacturers. An important function of the separator is to keep the zinc can surface properly wetted. In alkaline manganese batteries the separator consists of two or three layers of an absorbent non-woven material.

Figs. 3.5*a* and *b* show both types of battery with centrally disposed current collectors. An impregnated carbon rod is always used as the cathode current collector in Leclanché round cell batteries and a corrosion resistant metallic 'nail' may be used as the anode current collector in alkaline manganese batteries. Kordesch has provided illustrations of other forms of anode current collector used in alkaline manganese batteries.[31] The zinc can itself and a steel can, which is usually nickel plated, are the current collectors for the other electrode.

Batteries must be well sealed for good storage life. Mechanical seals between metallic and plastic components sometimes aided by 'gap filling' materials are used, and in Leclanché batteries bitumen layer seals are frequently employed. There are many designs. The subject is inseparable from the outer packaging of a battery which is required to have consumer appeal and the functional purpose of containment of potential leakage products. Fuller discussions have been given by Huber[32] and Cahoon.[33]

3.4.2 Square cell Leclanché batteries

Unit square cell batteries of the Leclanché electrochemical system (Table 3.11) are constructed similarly to the unit round cell batteries. The cathode bobbin of square cross-section contains a centrally disposed carbon rod and is situated in a square zinc container. The cells are of paste construction and have bitumen seals. The cathode bobbins of these cells are large and the older procedure of wrapping them in muslin to maintain their integrity during assembly has been retained.

3.4.3 Neutral air-depolarised batteries

Neutral air-depolarised batteries are large (Table 3.12) and their construction is similar to that of Leclanché batteries, although there are interesting differences. The cathode structure usually comprises a moderately compressed mixture of activated carbon and graphite with minority proportions of manganese dioxide and ammonium chloride. The activated carbon provides the sites at which the cathode active material, oxygen, is reduced. Graphite conducts electrons from a centrally disposed carbon rod to the reduction sites. The porosity of the cathode structure is, however, not filled with electrolyte, as in Leclanché cells, but with air. Access of air to this porosity is via a tube in a bitumen seal. Cathodic reduction does not take place throughout the cathode structure but at the junction with the gelled electrolyte. The cathode structure is wrapped with muslin, which is treated to prevent excessive penetration of electrolyte into the structure prior to gelling. The anode is a zinc sheet which is shaped so that it can be placed around the cathode structure. The zinc is amalgamated and alloyed similarly to the zinc cans of Leclanché batteries. Plastic or hard rubber is used as the outer container. This enables the zinc to be completely submerged in the gelled electrolyte, thus minimising the oxygen corrosion of zinc that occurs when the battery is in its opened state. For the same reason the thickness of the gelled paste layer is greater than in Leclanché batteries: diffusion of oxygen from the porous cathode structure to the zinc is thus restricted. The composition of the electrolyte is similar to that used in Leclanché batteries.

3.4.4 Flat cell Leclanché batteries

Fig. 3.6 shows a Leclanché layer stack battery (6F22): the patent for this type of construction was filed in 1939.[34] It was introduced[35] and gained acceptance as a compact form for batteries generating 45 to 90V. Batteries giving 9V are more important at the present time. A characteristic feature is the duplex electrodes of zinc with either a non-porous conductive carbon paint coating or a graphite filled plastic sheet. The plastic envelopes are sealed to the paint coat or plastic sheet to prevent the loss of capacity that would result from an inter-cell electrolyte path.

The stack of cells is usually sealed by dipping it in wax. Alkaline manganese batteries in the 6F22 size are constructed from cylindrical cells.

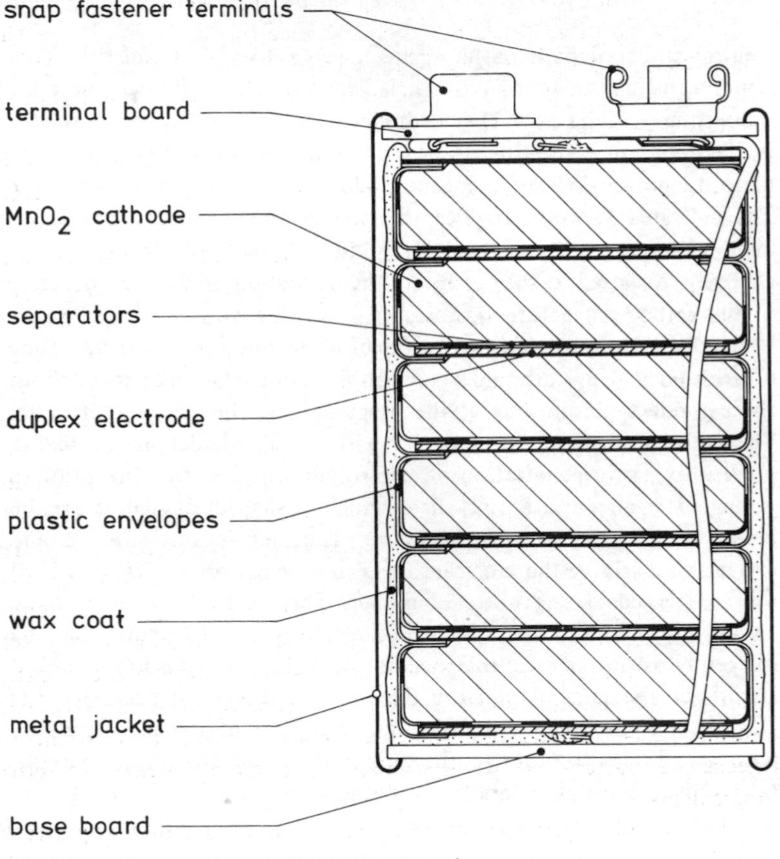

Fig. 3.6 *Construction of a flat cell Leclanché battery*

3.4.5 Comparison of electrode volumes

At this point it is useful to compare the theoretical/practical volume requirements of the various electrode structures under discussion. The volumes shown in Fig. 3.7 are theoretical in that only the reactions given in eqns. 3.2 to 3.6 (see Section 3.6) are taken into consideration and these reactions are assumed to occur at 100% coulombic efficiency.

No corrosion or alternative reactions are considered and the amounts of cathodic and anodic active materials are exactly balanced, i.e. each will deliver 1F of electricity. The volumes are practical in that associated volumes present in real electrode structures are included. Thus the total volumes include the volumes taken up by the electronic conductor, electrode porosity which is filled with electrolyte, and pores in the manganese dioxide.[36] Components such as current collectors, separators, seals and outer casings, which are more dependent on battery size and shape, are not included.

Fig. 3.7 shows clearly the large cathode porosity which is usual in Leclanché batteries and also the large volume taken up by solid NH_4Cl. The volume of solid NH_4Cl is not the total requirement for eqn. 2; NH_4Cl dissolved in the electrolyte is also consumed. Part of a manufacturer's art is the formulation of successful cathode mixtures with less solid NH_4Cl than is shown in Fig. 3.7.[37]

The advent of paper-lined batteries with their thinner separators has enabled electrolytes of lower conductivity to be used in Leclanché batteries. Zinc chloride Leclanché batteries contain an electrolyte which is 2 to 3 molar in $ZnCl_2$ and 0 to 1·5 molar in NH_4Cl, whereas the electrolyte in normal Leclanché batteries is saturated ammonium chloride (6·5 molar) and 1 to 2·5 molar in $ZnCl_2$ and has approximately four fold greater conductivity. Zinc chloride Leclanché batteries discharge in accordance with eqn. 3.3. Water is consumed in place of NH_4Cl. Fig. 3.7 shows that the extra porosity required to accommodate the water to be consumed (17·6ml) plus the additional space taken by the extra acetylene black that is necessary to generate this extra porosity add up to almost the volume occupied by solid NH_4Cl in normal Leclanché batteries. The total capacities of the two variants of Leclanché battery are thus similar for a given volume of active ingredients.

Fig. 3.7 illustrates the extent to which the greater total capacity of alkaline maganese batteries compared with Leclanché batteries is due to the smaller cathode porosity and to the absence of solid NH_4Cl. This volume saving is only partly offset by the anode porosity which is necessary with alkaline electrolytes. The volumes given in Fig. 3.7 are for a gelled anode which is the type usually used in LR20, LR14 and LR6 batteries and the reader may like to ponder on how the zinc anode functions in such a disperse form.

Fig. 3.7 shows that the use of Ag_2O and HgO as cathodic reactants gives a considerable space advantage, which is enhanced by using very highly compressed cathode structures of low porosity. A proportion of MNO_2 is frequently incorporated in Ag_2O and HgO cathode structures and Fig. 3.7 demonstrates that only small space penalties are incurred by such additions: this conclusion has been confirmed in published work.[38] The silver in the Ag_2O cathode structure is not dispersed but is confined to the outer surface and provides the necessary initial electronic conductivity. As the product of discharge is silver, good electronic conductivity throughout discharge is assured. Graphite in the Ag_2O cathode structure is present more as a tabletting aid. Graphite is necessary for electronic conductivity in HgO cathode structures and a greater quantity is therefore used than is present in Ag_2O cathodes.

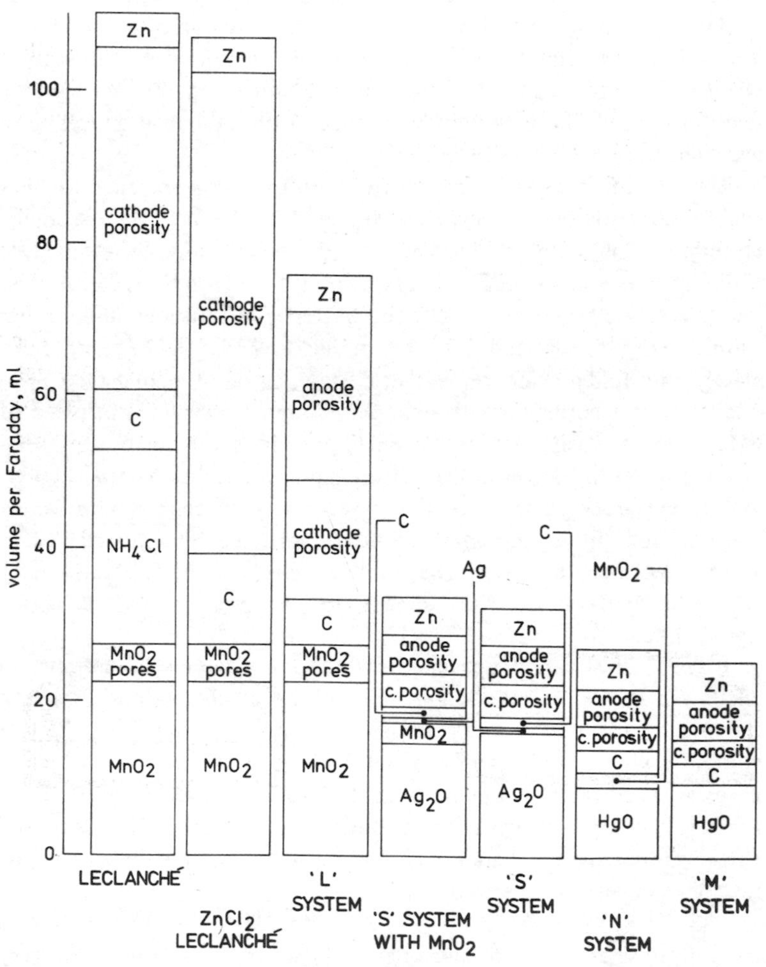

Fig. 3.7 *Comparison of electrode volumes*

It is obvious from Fi. 3.7 why Ag_2O and HgO cathodes are exclusively used in the smallest sizes of standardised batteries (see Tables 3.4 and 3.10). For these batteries (Table 3.10) powder anodes without a gelling agent may be employed. As illustrated in Fig. 3.7, these are more compact than the gelled anodes of alkaline manganese batteries and are an additional factor in maximising the total capacity attainable from small Ag_2O/Zn and HgO/Zn batteries.

3.4.6 *Button batteries: mercury (II) oxide/zinc and silver (I) oxide/zinc*

Button batteries, in which cathode, anode and separator are in parallel planes perpendicular to the axis of symmetry of the battery, are the most important constructional form of Ag_2O/Zn and HgO/Zn batteries. Two shapes are standardised for button batteries (Figs. 3.2*d* and *e*) and examples of constructions are given in Fig. 3.8.

The top cap is mild or stainless steel 0·2 - 0·3mm thick and is often plated with nickel and sometimes gold. The composition of the internal surface which is in contact with amalgamated zinc is chosen to minimise corrosion arising from the hydrogen evolution/zinc dissolution couple. Inner surfaces of gold, nickel, copper and tin are all used. The case is usually nickel plated mild steel 0·2 - 0·4mm thick, although stainless steel and nickel are employed. The cathode tablet of HgO or Ag_2O and graphite is consolidated against the inner surface of the case

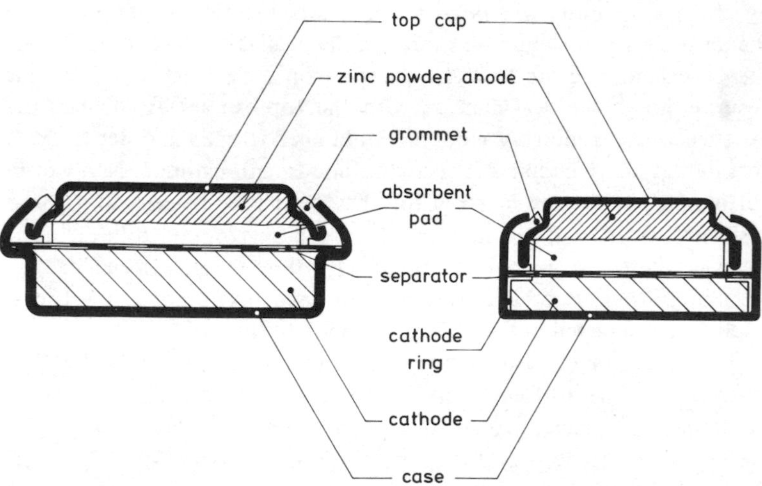

Fig. 3.8 *Constructions of Ag_2O/Zn and HgO/Zn button batteries*

to ensure good electronic contact and to provide a firm base for the mechanical seals of batteries made according to the shape of Fig. 3.2*e*. Additional support for sealing may be provided by a nickel plated mild steel cathode ring.

In terms of particle size range and amalgamation level there is no difference between the zinc employed in button batteries and cylindrical alkaline manganese batteries. Although button battery anodes are usually of the powder type, pellet anodes are used occasionally. Pellets can have an anode porosity as low as 1 ml/Faraday (cf Fig. 3.7). Although this appears desirable as regards space saving, the porosity may be insufficient to accommodate the discharge product ZnO which occupies 2•7 ml more space per Faraday than zinc. The penalty for insufficient anode porosity is likely to be poorer zinc utilisation and lower rate and pulse capability than is obtained with powder anodes. Anode capacity is usually designed to be less than cathode capacity; otherwise, on exhaustion of the cathode the Zn/H_2O couple can deliver current at a positive voltage with production of hydrogen and consequent development of pressure in the battery.[39]

As HgO and Ag_2O discharge to mercury and silver, cathode voidage increased by 2•4 and 6•0 ml/Faraday, respectively. The absorbent pad provides a reservoir of electrolyte to maintain adequate liquid as a vehicle for mass transfer in the structure and to keep the reacting cathode surface properly wetted. In addition, the cathodic reaction consumes 0•5 mol of water per Faraday, and although this is replaced by an equal quantity generated by the anode reaction, water present in the absorbent pad should be more readily available. Other functions of the absorbent pad are to exert pressure on a zinc pellet anode thus ensuring good electrical contact with the top cap and to increase the distance which a mercury globule would need to span in order to cause an internal short circuit between the anode and cathode. Non-woven cotton is the preferred material for absorbent pads because of its good resilience, high porosity and rapid absorption of electrolyte.[40] The latter property is an important consideration in high speed battery assembly. When absorbent pads are not used, extra porosity (at least double that depicted in Fig. 3.7) is provided in the cathode.

The separator consists of one or two membrane layers for HgO/Zn batteries and two to four layers for Ag_2O/Zn batteries. Particularly for the latter, cellophane, first introduced by André in 1941,[41] is still the preferred membrane. Ag_2O is slightly soluble in alkali (2×10^{-4} molar) and cellophane inhibits diffusion of the dissolved species to the zinc anode on which silver would deposit and promote corrosion.[42] Proprietary ion-exchange membranes, such as Acropor WA (Gelman

Hawksley), which has a crosslinked carboxylic ion-exchange resin held in a polymer film, and Permion (RAI Research Corporation), which has carboxylic groups grafted onto polyethylene, are also finding application. For HgO/Zn batteries the prime requirement of the separator is to resist penetration by mercury globules formed on discharge while allowing adequate mass transport.

The seals of button batteries are all mechanically formed by compressing a grommet between the top cap and the case. There are variations particularly in the shape of the top cap. Polyethylene, polypropylene, nylon and synthetic rubber are all used for the grommet, which may be a separate component or be overmoulded around the top cap. 'Gap fillers' are used. Alkali creeps readily along the negative surface of the top cap[240, 241] and care is needed in the design and mechanical formation of the seal for it to be an adequate obstruction. One patented method[43] of delaying the appearance of alkali on the external top cap surface is a double top cap which has an extended creepage path. The double top cap design is shown in Fig. 3.9.

The electrolyte usually used in alkaline primary batteries (systems L, M, N and S) is potassium hydroxide at a concentration of 7 to 10 molar, which is just above the concentration of maximum conductivity. Sodium hydroxide, which has a conductivity approximately two-thirds

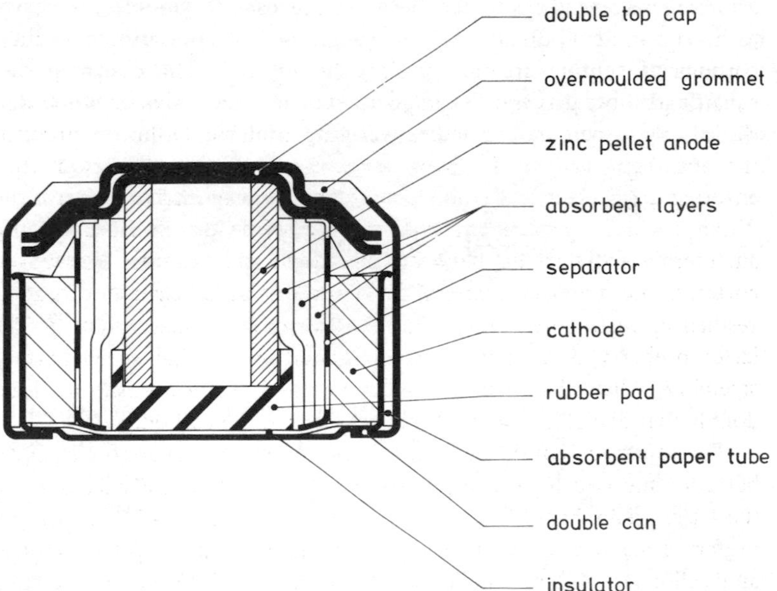

Fig. 3.9 *Construction of a cylindrical HgO/Zn battery*

of that of potassium hydroxide, can be used in batteries intended for low current drain applications, e.g. button batteries for watches having liquid crystal displays. Sodium hydroxide electrolyte has less tendency to creep on negative surfaces[240,241] and hence its use eases the task of sealing batteries.

Alkaline manganese batteries have also been standardised in the two largest button sizes (R53 and R9, Table 3.9) and their construction is similar to that already described (Fig. 3.8). A Leclanché button battery (R9) is also standardised, but this is of rather different construction.

3.4.7 *Cylindrical mercury batteries*

The largest standardised HgO/Zn batteries (Table 3.8) have a cylindrical construction as illustrated in Fig. 3.9. The illustration shows a pelleted zinc anode for which a separate current collector is not necessary. Good contact with the double top cap is maintained by pressure generated from the insulating rubber pad. Gelled anodes with current collectors as found in cylindrical alkaline manganese batteries (Fig. 3.5*b*) are also used. Another feature is the double can, the purpose of which is to allow the battery to vent, with expelled liquid being trapped by an absorbent paper tube situated between the cans.[43] Double cans have been used for cylindrical alkaline manganese batteries although other methods of venting are now usually employed.[31] The design of the insulating rubber pad and the disposition of the electrodes minimise the chance of mercury globules causing internal short circuits.

3.5 Capacity and electrical performance

3.5.1 *Total capacity*

Table 3.14 lists the nominal total capacity, volume, weight and open circuit voltage of unit batteries or cells for the sizes and electrochemical systems that have been standardised by the IEC. The products obtainable from manufacturers differ and the values given are intended only as a guide.

For a multi-cell battery in which cells are electrically connected in series, the capacity is still that given for a unit cell or battery, although its weight obviously approximates to that of the unit cell or battery multiplied by the number of cells in the battery. If the cells in a multi-cell battery are connected in parallel, then both its capacity and weight are obtained by multiplying the values listed in Table 3.14 by the number of cells in the battery.

The volume of a multi-cell battery is normally greater than the product of the volume of the unit cell or battery and the number of cells in the battery, owing to voidage included in the battery when the cells are packed together.

As flat cells are used only in multi-cell batteries, the weights and volumes given in Table 3.14 for single flat cells were obtained by dividing the volumes and weights of batteries by the number of cells incorporated.

3.5.2 Practical discharge capacities

The total coulombic capacity present in a battery is not available under all discharge conditions. The capacity given is dependent on the discharge regime. This is particularly true for all batteries containing a manganese dioxide cathode.

Fig. 3.10 shows typical on-load voltage curves for three types of R20 battery and an LR20 battery when each is discharged through a 3·9 Ω resistor for 1 h per day until the on-load voltage reaches 1·0V. At the commencement of each discharge period there is an immediate voltage drop from the open circuit voltage, shown as a dot, to the initial on-load voltage which is due to the internal resistance of the battery. During discharge the on-load voltage falls and the extent of this fall depends on battery composition. In the examples shown the average decrease in each discharge period is reduced from 0·30V to 0·21V by using electrodeposited manganese dioxide in place of active natural ore in the Leclanché battery (cf. Figs. 3.10*a* and *c*). The electrical performance of a Leclanché battery is primarily determined by the grade of manganese dioxide used in the cathode. Electrodeposited manganese dioxide is always used in zinc chloride Leclanché batteries and alkaline manganese batteries. For these, in the example illustrated, the average decreases in each discharge period are 0·18V and 0·14V, respectively.

The shape of the on-load voltage curve near the end of a discharge period is noteworthy. For alkaline manganese batteries, and to some extent for zinc chloride Leclanché batteries, the shape flattens, which means that polarisation would not increase substantially if the discharge

Table 3.14 *Nominal total capacities, volumes, weights and open circuit voltages of unit batteries and cells*

IEC size designation	Battery or cell volume	O_2 — NH_4Cl, $ZnCl_2$ — Zn		MnO_2 — NH_4Cl, $ZnCl_2$ — Zn		MnO_2 — alkali metal hydroxide — Zn		HgO and HgO(MnO_2) — alkali metal hydroxide — Zn		Ag_2O — alkali metal hydroxide — Zn	
	ml	Ah	g	Ah	g	Ah	g	Ah	g	Ah	g
S10	2 600	260	2 700	220	4 650						
S8	1 200	150	1 600	100	2 100						
S6	610	100	750								
R40	480	85	600	60	910						
S4	340	50	380	19	470						
R25	75			9·0	150						
R20	54			6·5	95	10	130				
F100	44			6·0	71						
R14	26			3·0	45	5·0	66				
R12	20			2·3	35						
R10	13			1·3	22						
R51	9·8							3·5	41		
R6	7·7			1·1	18	1·7	24	2·4	30		
F40	5·2			0·50	7·6						
F24	3·9			0·37	8·4						
F25	3·5			0·30	7·0						
R03	3·4			0·40	8·5	0·78	12				

R50	3·3							1·0	13		
F22	3·3			0·33	6·7						
R1	3·0			0·30	6·5	0·70	9·7	1·0	13		
R53	2·2					0·30	7·5				
R52	2·1							0·50	7·8		
R01	1·5							0·35	5·6		
F20	1·4			0·11	2·3						
R9	1·2			0·15	3·0	0·15	3·7	0·35	4·3		
F16	1·1			0·080	1·8						
F15	0·82			0·070	1·7						
R44	0·51							0·22	2·6	0·18	2·3
R43	0·38							0·15	2·0	0·13	1·7
R42	0·33							0·11	1·6	0·095	1·4
R54[1]	0·28							0·092	1·5	0·080	1·3
R48	0·24							0·086	1·2	0·070	1·1
R45	0·24							0·078	1·1		
R56[1]	0·23									0·060	1·1
R55[1]	0·18									0·045	0·90
R57[1]	0·17									0·044	0·75
R41	0·16							0·049	0·80	0·042	0·70
R59[1]	0·11									0·028	0·50
R58[1]	0·09									0·020	0·40
System		A		-		L		M, N		S	
open-circuit voltages		1·45		1·60		1·55		1·35, 1·40		1·60	

(1) Standardisation is under consideration

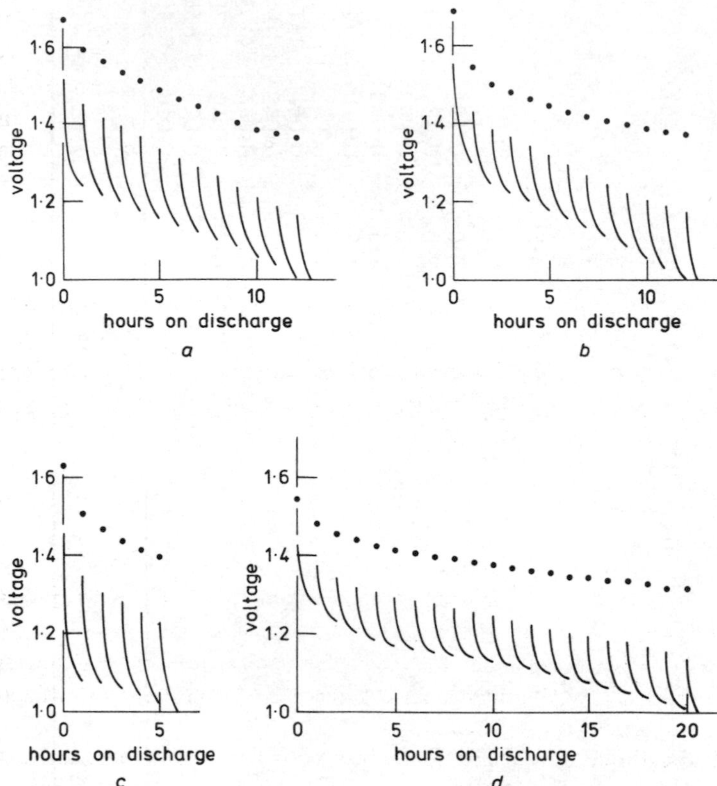

Fig. 3.10 *On-load voltage curves for R20 sized batteries on intermittent discharge (3·9Ω for 1h/day to 1·0V)*

a R20, electrodeposited MnO_2, normal Leclanché electrolyte

b R20, electrodepositied MnO_2, 'zinc chloride' electrolyte

c R20, active natural ore, normal Leclanché electrolyte

d LR20, electrodepositied MnO_2, alkaline electrolyte

period were lengthened. This is not the case for normal Leclanché batteries. Fortunately, in the period between discharges a battery recuperates so that the voltage at the commencement of the next discharge period is only a little lower than that of the preceding period.

Clearly, the percentage of the total capacity that can be withdrawn from a battery is dependent on the load resistance, the discharge time, the recuperation time and the on-load endpoint voltage. Fig. 3.11 shows how the capacity yield to an on-load endpoint of 0·9V varies

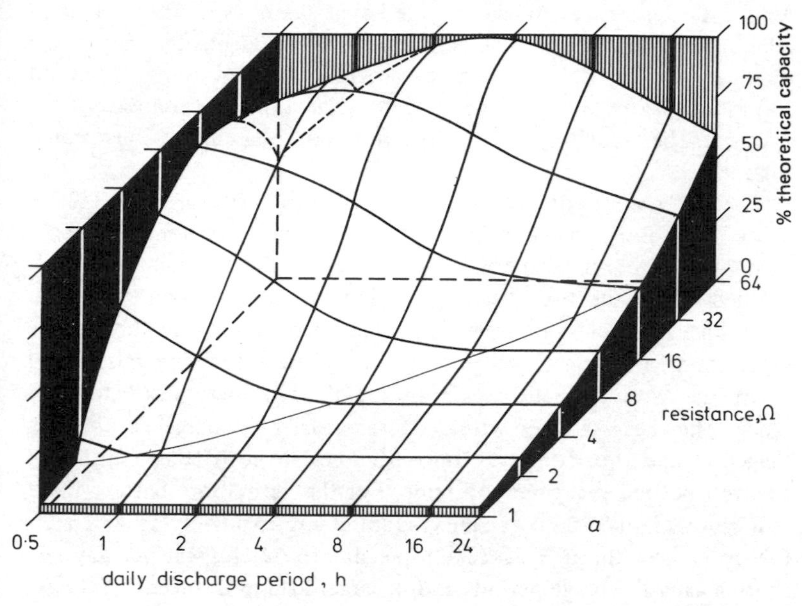

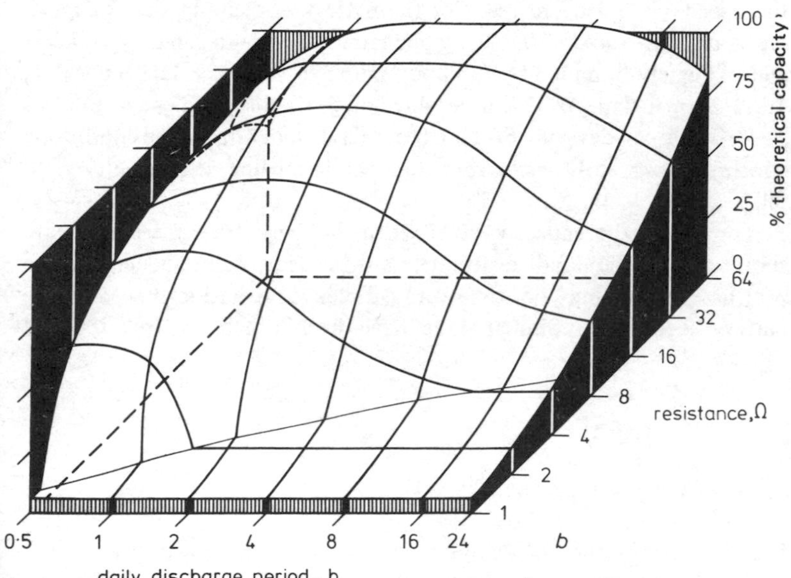

Fig. 3.11 *Capacity surfaces for R20 batteries*

a Active natural ore

b Electrodeposited MnO_2

with load resistance and daily discharge period for R20 batteries containing (*a*) active natural ore, and (*b*) electrodeposited manganese dioxide. The theoretical capacities were calculated for a reduction from $MnO_{1\cdot 95}$ to $MnO_{1\cdot 5}$ from the manganese dioxide contents of the batteries. The capacity yield is shown in the form of a three-dimensional surface.

The lower slopes of the surface at load resistances of 1 and 2 Ω are in the form of a parallel wedge with capacity yield apparently independent of daily discharge period. In this region the endpoint voltage is reached within the first discharge period so that there is no recuperation; capacity yields are very low. The discontinuity in the surface separates this region from that in which there is recuperation and where the capacity yield rises more rapidly with increasing load resistance. The rise is greater with decreasing daily discharge periods, i.e. longer recuperation times and for the Leclanché battery containing electrodeposited manganese dioxide. For the natural ore battery there is a ridge of maximum capacity running approximately from 16 Ω/0·5 h per day through 32 Ω/1 h per day to 64 Ω/2-4 h per day. At shorter daily discharge periods and/or larger load resistances than these, the capacity yield decreases. For the battery containing electrodeposited manganese dioxide there is a plateau of maximum capacity which is approximately bounded by the line 4Ω/0·5 h per day, 8 Ω/1 h per day, 16 Ω/2 h per day, 32 Ω/4 h per day and 64 Ω/8 h per day and the line 32 Ω/0·5 h per day and 64 Ω/1 h per day. Under discharge conditions which are apparently less severe than the latter line, the capacity yield falls.

The fall in the capacity yield when the daily discharge periods are short and/or the load resistance is high is a consequence of the protracted total time, i.e. time on load plus recuperation time, that the battery is on test. Similar falls have been noted by Huber[32] and Cahoon.[33]

3.5.3 *IEC application tests*

The amount of discharge testing required to establish a capacity surface is considerable and cannot normally be undertaken. Nevertheless, in view of the dependencies just noted it is clearly essential to test batteries on discharge regimes which approximate to the duty likely to be experienced in consumer use. It cannot be emphasised too strongly that continuous discharges, which may be preferred for testing convenience because they are of short duration and have no 'on/off' regime, are normally worthless as an assessment of Leclanché battery capability and for comparison with other electrochemical systems, such as alkaline manganese, for which capacity yield is less dependent on discharge regime. With the exception of some very low current drain applications, such as watches, commercial batteries are not used continuously, if only to avoid the expense incurred in replacing batteries every few days.

Recognition of the problem has led to the standardisation of a number of discharge tests which are intended to simulate the use of batteries in the more important applications. The application tests specified by the IEC[8] are given in Tables 3.15 to 3.18. The models available in any application do, of course, vary considerably in their electrical load characteristics and in the on-load voltage at which they cease to function adequately. Appliances may be used irregularly and certainly for lengths of time which are different from those specified in the IEC tests, and perhaps also in different ways, e.g. volume on a transistor radio. An application test is thus inevitably a compromise and the service duration given by a battery on such a test is unlikely to be the same as the service obtained by a particular consumer in his appliance. Nevertheless, these tests are the best means available of comparing battery performance.

For low current drain applications, such as watches and pocket calculators with liquid crystal displays, where batteries may last longer than one year in service, the application test also of course lasts longer than one year. In such instances application tests are only used for occasional reference and accelerated tests are formulated.

When establishing the performance of a given brand and grade of battery on an application test it is not, of course, sufficient to test one or even a few, as is the wont of consumer associations, since there are variations which arise inevitably from the production process. Fig. 3.12 shows the discharge durations obtained on the IEC radio test of over 300 Leclanché R20 batteries made to a constant specification with natural ore over a period of 1·5 years. The discharge durations are distributed normally with a coefficient of variation of 7·1%. The ranges in which an average discharge duration would fall in most cases (i.e. nine times out of ten) can be calculated for various sample sizes from the mean and standard deviation of the normal distribution which matches the results. The results of these calculations are given in Table 3.19. For the distribution illustrated a sample size of 10 will give, nine times out of ten, an average within 5% of the true average of the population.

Discharge durations on the IEC radio test depend essentially on the

Table 3.15 *Application discharge tests for unit cylindrical batteries*

IEC size designation	Discharge tests for applications											
	Calculator	Cine camera	Electric fence	Hearing aid	Lighting	Photoflash	Radio	Razor toothbrush	Tape recorder	Toy	Transistor clock	Quartz clock
R40			51 Ω continuous 0·9 V									
R20					5 Ω[1,3] 30 m/d 0·9 V	1 Ω[2] 0·75 V	40 Ω[3] 4 h/d 0·9 V	2·2 Ω 5 m/d 0·9 V	3·9 Ω 1 h/d 1·0 V	2·2 Ω 1 h/d 0·8 V		
R14	5·6 Ω 30 m/d 0·9 V				5 Ω[3] 10 m/d 0·9 V	1 Ω[2] 0·75 V	75 Ω[3] 4 h/d 0·9 V	2·2 Ω 5 m/d 0·9 V	6·8 Ω 1 h/d 1·0 V	3·9 Ω 1 h/d 0·8 V	33 kΩ[4] continuous 1·0 V	6·8 kΩ[4] continuous 1·3 V
R12					5 Ω 10 m/d 0·9 V							
R10					5 Ω 5 m/d 0·9 V							
R6	15 Ω 30 m/d 0·9 V	3 9 Ω 5 m/d 1·0 V		300 Ω 12 h/d 0·9 V	5 Ω[3] 5 m/d 0·9 V		75 Ω 4 h/d 0·9 V	3·9 Ω 5 m/d 0·8 V	10 Ω 1 h/d 0·9 V		33 kΩ[4] continuous 1·0 V	6·8 kΩ[4] continuous 1·3 V
R03				300 Ω 12 h/d 0·9 V	5 Ω[3] 5 m/d 0·9 V							
R1				300 Ω 12 h/d 0·9 V								
RO1				300 Ω 12 h/d 0·9 V								

(1) An alternative lighting test is 4 Ω for 4 min beginning at hourly intervals for 8 h/d to 0·9 V
(2) 15s/min for 1 h/d (3) Test under review (4) Test under consideration

Table 3.16 *Application discharge tests for unit button batteries*

Application	Application test	IEC battery designation
Hearing aid	300 Ω, 12 h/d, 0·9 V	R9, MR9
Hearing aid	625 Ω, 12 h/d, 0·9 V	MR44, NR44[1], MR42[2], NR42[1]
Hearing aid	1·5 Ω, 12 h/d, 0·9 V	MR48, SR48
Pocket calculator with liquid crystal display (l.c.d.)	18 Ω, 1 h/d, 1·2 V	test independent of size and system
Quartz analogue watch	470 Ω, continuous, 1·2 V	test independent of size and system
Quartz digital watch with l.c.d. display	680 Ω, continuous, 1·2 V	test independent of size and system
Quartz digital watch with l.c.d. display and battery powered backlight	680 Ω, continuous plus 100 Ω, 2 s/h for 8 h/d, 1·2 V	test independent of size and system
Quartz digital watch with light emitting diode display	470 Ω, continuous plus 47 Ω, 2 s/h for 8 h/d, 1·1 V	test independent of size and system

(1) Standardisation under consideration (2) Test under consideration for this battery

Table 3.17 *Application discharge tests for multi round and square cell batteries*

IEC battery designation	Discharge tests for applications		
	Electric fence	Lighting	Radio
5R40, 5AR40 240	240 Ω, continuous, 4·5V		
6S6, 6AS6	300 Ω, continuous, 5·4V		
4R40X, 4R40y	200 Ω, continuous, 3·6V		
6S4, 6AS4	300 Ω, continuous, 5·4V		
4R25-2		8·2 Ω, 30 m/d, 3·6V[3]	
4R25		20 Ω, 30 m/d, 3·6 V[2]	
3R20X		15 Ω, 30 m/d, 2·7 V[1]	
3R12		15 Ω, 10 m/d, 2·7V[2]	225 Ω, 4h/d, 2·7V
2R10		10 Ω, 5 m/d, 1·8V[2]	

(1) An alternative lighting test is 12Ω for 4 min, beginning at hourly intervals for 8h/d to 2·7V
(2) Test under review (3) Under consideration

Table 3.18 *Application discharge tests for flat batteries*

IEC battery designation	Discharge tests for applications		
	Calculator	Radio	Tape recorder
6F100-3		240 Ω, 4h/d, 5·4V	
6F100		450 Ω, 4h/d, 5·4V[2]	
6F50-2		450 Ω, 4h/d, 5·4V[1]	
6F25		900 Ω, 4h/d, 5·4V	
6F24		900 Ω, 4h/d, 5·4V	
6F22	180 Ω, 30m/d, 4·8V	900 Ω, 4h/d, 5·4V[2]	180 Ω, 1h/d, 5·4V

(1) An alternative radio test is 900 Ω, 4h/d, 5·4V
(2) Test under review

Table 3.19 *Range in which average discharge duration will fall, 9 times out of 10, for the normal distribution illustrated in Fig. 3.12*

Sample size	5	10	20	300	∞
Average discharge life, h	164·3 - 187·1	167·6 - 183·8	170·6 - 180·8	174·4 - 177·0	175·7

amounts of active material incorporated into the battery. On more severe discharge regimes other factors, such as manganese dioxide activity and cathode mix structure, also have an effect, and variations in these additional factors will increase the coefficient of variation so that the precision obtained from a sample of 10 is decreased.

3.5.4 Discharge durations of Leclanché and alkaline manganese batteries on IEC application tests

Fig. 3.13 shows distributions of the average discharge duration for over 800 brands, grades and sizes of battery manufactured throughout the world on some of the more important IEC application tests. A

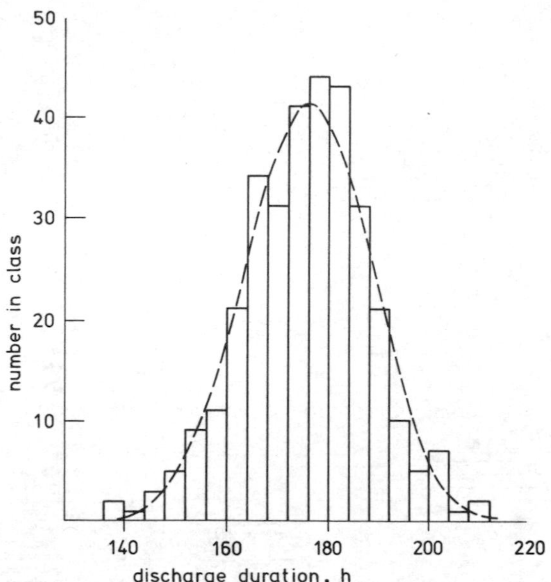

Fig. 3.12 *Distributon of discharge durations of one grade of R20 battery on the IEC radio test*
– – – – normal distribution curve matched to the data

manufacturer, wholesaler or consumer association can assess the standing of any given battery by reference to these histograms. On the less severe discharge regimes of radio and lighting tests the average discharge lives of Leclanché batteries show unimodal distributions with a ratio of approximately 1:2 between the worst and best performing products. On the more severe regimes of tape recorder tests the distributions in the R14 and R20 sizes are bimodal, which is a consequence of whether the manganese dioxide used is predominantly synthetic or natural ore. Batteries containing active natural ore are unsuitable for use in tape recorders, as is demonstrated by the very low discharge durations of some products.

Although zinc chloride Leclanché batteries, which are premium grade, appear in the upper part of the distributions there is little evidence that they give longer discharge durations on application tests than normal premium Leclanché batteries. This conclusion is not unexpected as the space requirements of the cathodes in the two types of battery are similar, as is illustrated in Fig. 3.7. In a controlled experiment with nine different manganese dioxides drawn from the bank of International Common Samples,[44] Ohta, Watanabe and Furumi[45]

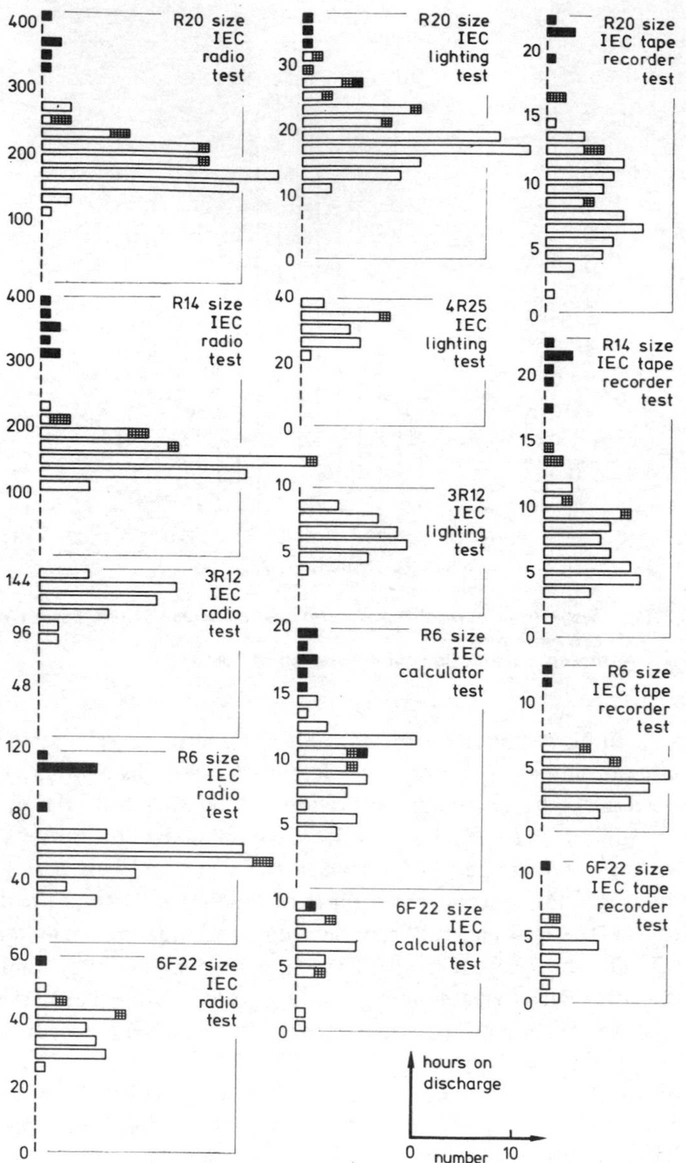

Fig. 3.13 *World survey of discharge durations on some important IEC application tests*

Leclanché batteries

zinc chloride Leclanché batteries

alkaline manganese batteries

concluded that there was no difference in discharge duration between zinc chloride and normal R14 Leclanché batteries on a simulated radio discharge but that the former were superior on a simulated lighting discharge. Also, as would be expected from Fig. 3.7, alkaline manganese batteries, for which there are fewer manufacturers, give longer discharge durations than Leclanché batteries. However, in some cases the differences between alkaline manganese and the best Leclanché batteries are quite small, e.g. R20 size on the IEC lighting test, R6 and 6F22 sizes on IEC calculator tests and the 6F22 size on the IEC radio test, and certainly do not justify the extravagant claims sometimes made for the former electrochemical system. These extravagant claims are usually based on continuous discharges.[31, 46, 47]

3.5.5 *Capacity yield of Leclanché batteries*

Discharge durations obtained on application tests do not reveal the extent to which the theoretical capacity (reduction from $MnO_{1.95}$ to $MnO_{1.5}$) of a Leclanché battery has been used. Fig. 3.14 shows the percentage of theoretical capacity that is used on a number of IEC, ANS (American National Standards) and Ever Ready application tests for R20 and R14 batteries. At least 30, and in some cases hundreds, of discharges have been used for each point. Interestingly, the plots show that capacity yields are approximately linear functions of the square root of the total time that a battery is on test, i.e. on-load time plus recuperation time. In view of the square root time dependence it is concluded that capacity yield is essentially controlled by diffusion.

For batteries containing only active natural ore the straight lines in Fig. 3.14 pass through the origin. This is interpreted as indicating that the diffusion limitation lies within or between the manganese dioxide particles. For batteries containing only electrodeposited manganese dioxide the straight lines are at a higher position, demonstrating the superiority of electrodeposited manganese dioxide relative to active natural ore, and do not pass through the origin. This is interpreted as indicating that the diffusion limitation is in the electrolyte and that under quite severe discharge regimes there is sufficient $ZnCl_2$ and/or NH_4Cl dissolved in the electrolyte within the cathode structure to sustain a reasonable capacity yield. A corollary to this is, of course, that the diffusion process in the manganese dioxide, which is limiting with active natural ore, must be faster with electrodeposited manganese dioxide. The nature of these diffusion processes is discussed later.

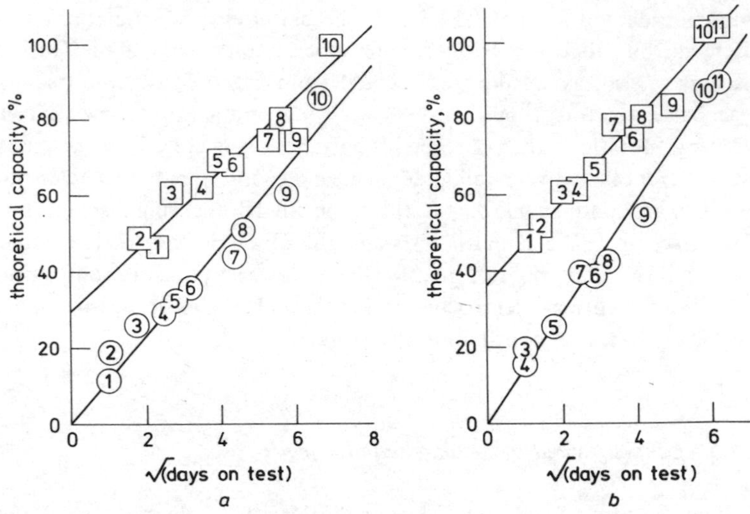

Fig. 3.14 *Capacity yields of Leclanché batteries on simulated application tests*

○ active natural ore □ electrodeposited MNO_2

a R20 battery

(1) 5 Ω, 2 h/d to 1·1V: Ever Ready tape recorder

(2) 0·4 A, 4 m/15 m for 8 h/d to 1·0V: ANS toy (test 46)

(3) 2·2 Ω, 1 h/d to 0·8V: IEC toy

(4) 3·9 Ω, 1 h/d to 1·0V: IEC tape recorder

(5) 2·25 Ω, 4 m/h for 8 h/d to 0·9V: ANS lighting (test 14)

(6) 2·5 Ω, 30 m/d to 0·9V: Ever Ready toy

(7) 2·5 Ω, 10 m/12 h to 0·9V: Ever Ready lighting

(8) 4 Ω, 4 m/h for 8 h/d to 0·9V: ANS lighting (test 15)

(9) 5 Ω, 30 m/d to 0·9V: IEC lighting

(10) 40 Ω, 4 h/d to 0·9V: IEC radio

b R14 battery

(1) 4 Ω, 4 m/15 m for 8 h/d to 0•75V: ANS lighting (test 16)1*a*

(2) 0•25 A, 4 m/15 m for 8 h/d to 1•0V: ANS toy (test 47)

(3) 3•9 Ω, 1 h/d to 0•8V: IEC toy

(4) 10 Ω, 2 h/d to 1•1V: Ever Ready tape recorder

(5) 6•8 Ω, 1 h/d to 1•0V: IEC tape recorder

(6) 5Ω, 30 m/d to 0•9V: Ever Ready toy

(7) 4 Ω, 4 m/h for 8 h/d to 0•9V: ANS lighting (test 15)1*b*

(8) 5•6 Ω, 30 m/d to 0•9V: IEC calculator;

(9) 5 Ω, 10 m/12 h to 0•9V: Ever Ready lighting

(10) 75 Ω, 4 h/d to 0•9V: IEC radio

(11) 83•3 Ω, 4 h/d to 0•9V: ANS electronic equipment (test 51)

The ANS Standard[6] specifies an endpoint of (*a*) 0•9V and (*b*) 0•75V for R14 batteries

It is worth emphasising that there appears to be only a secondary dependence (shown by the displacement of points from the straight lines in Fig. 3.14) on a particular condition of the discharge regime such as endpoint voltage, load resistance or daily discharge period. What is significant is the total time on test, which is, of course, a consequence of the combination of all three conditions. Fig. 3.14 shows that the total theoretical capacity is given by discharge regimes which last a total of 70 days (R20, active ore), 55 days (R20, electrodeposited MnO_2), 45 days (R14, active ore) and 30 days (R14, electrodeposited MnO_2). Obviously these values are only a guide, as they must be dependent on active material specifications and battery formulation.

For Leclanché batteries containing only active natural ore, the relationship between battery size and the total test time necessary to realise all the theoretical capacity has been examined by extrapolation of straight lines from the origin through the yields obtained on IEC radio tests in plots of the type shown in Fig. 3.14. The results given in Fig. 3.15 show direct proportionality between bobbin thickness, i.e. distance between carbon rod and separator and the total test time to obtain 100% capacity yield. By combining the results of Figs. 3.14 and 3.15 the following useful guide is obtained for natural ore batteries:

$$\text{percentage yield of theoretical capacity} = 40 \left(\frac{\text{total days on discharge test}}{\text{bobbin thickness in mm}}\right)^{\frac{1}{2}}$$

In order to gain a perspective on the significance of the times discussed in the preceding paragraphs it is necessary to know the periods over

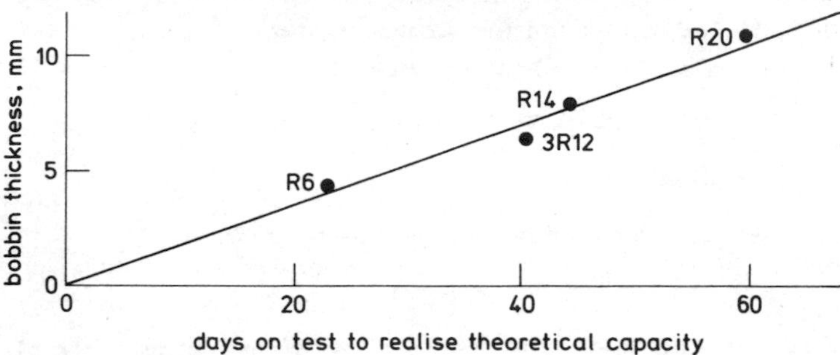

Fig. 3.15 *Total test time to realise theoretical capacity with active natural ore*

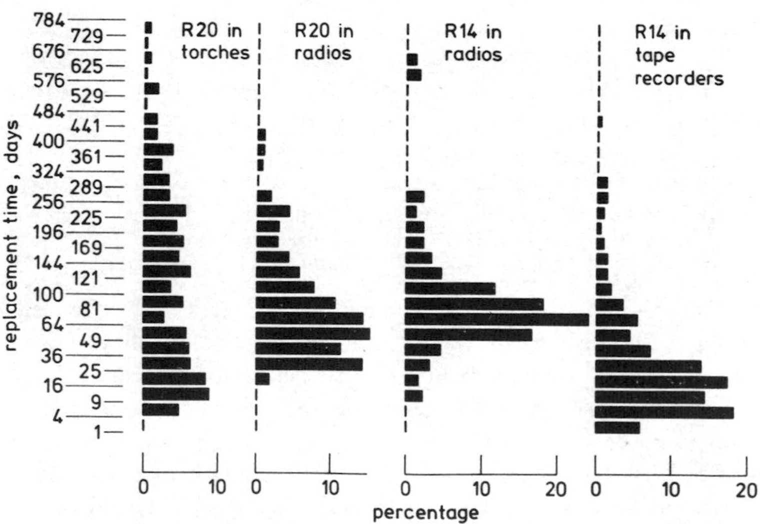

Fig. 3.16 *Battery replacement times*

which batteries are used in appliances. Information on this aspect has been gained from a study conducted at the writer's laboratory over

5 years on the battery usage pattern of 100 staff who were selected from various grades and to match the United Kingdom age distribution. Some results are shown in Fig. 3.16. The distribution of R20 battery usage in torches is widespread with replacement times varying from one week to two years. R20 and R14 batteries in radios show more classical distributions with peaks at replacement times that should guarantee high capacity yields from batteries used in this application. On the other hand, for R14 batteries in tape recorders the majority of replacement times are too short for the total capacity of batteries to be realised.

3.5.6 Capacity yield of alkaline manganese batteries

An examination of the capacity yield obtained from alkaline manganese batteries on application tests in terms of the square root of the total time on tests does not reveal significant correlations of the type applicable to Leclanché batteries. The capacity of alkaline manganese batteries is not dependent on medium term diffusion effects. This is a further indication of the rapidity of diffusion processes in electrodeposited manganese dioxide.

Fig. 3.17 shows the open-circuit voltage of the alkaline manganese system in various states of discharge. The falling shape of the curve is a usual feature of manganese dioxide cathodes and is discussed later. The present purpose of Fig. 3.17 is to show the open-circuit potential (after recuperation) at the end of various IEC application tests.

The open circuit voltage curve is in four parts: an initial rapid fall from $MnO_{1\cdot95}$ to $MnO_{1\cdot90}$, a fairly steady fall from $MnO_{1\cdot9}$ to $MnO_{1\cdot7}$, a second region of rapid fall from $MnO_{1\cdot7}$ to $MnO_{1\cdot5}$ and finally a horizontal portion. Significantly, the application tests finish in or beyond the second region of rapid fall. It is concluded that the major factor controlling the capacity yield of alkaline manganese batteries is the open-circuit potential of the manganese dioxide cathode. Two other factors influence the capacity yield

(*a*) the current drain at the endpoint voltage, which determines voltage losses in the battery due to internal resistance etc.
(*b*) the endpoint voltage itself.

When the load resistance is high as in radio application tests, current is low, voltage losses in the battery are low and discharge can be taken to an open circuit value of about 1·0V before the on-load voltage reaches the endpoint of 0·9V. When the load resistance is smaller, voltage losses in the battery are higher, the test finishes at a higher open circuit voltage and capacity yield is lower. This situation is aggravated if the endpoint voltage is high, as in IEC tape recorder tests for R14 and R20 size batteries, and alleviated if the endpoint is low, as in IEC toy tests. The performance of the zinc anode is also affected by the current drain, but the fact that the application tests finish on the rapidly falling portion of the open circuit voltage curve suggests that the potential of the manganese dioxide cathode has the greater influence on capacity yield.

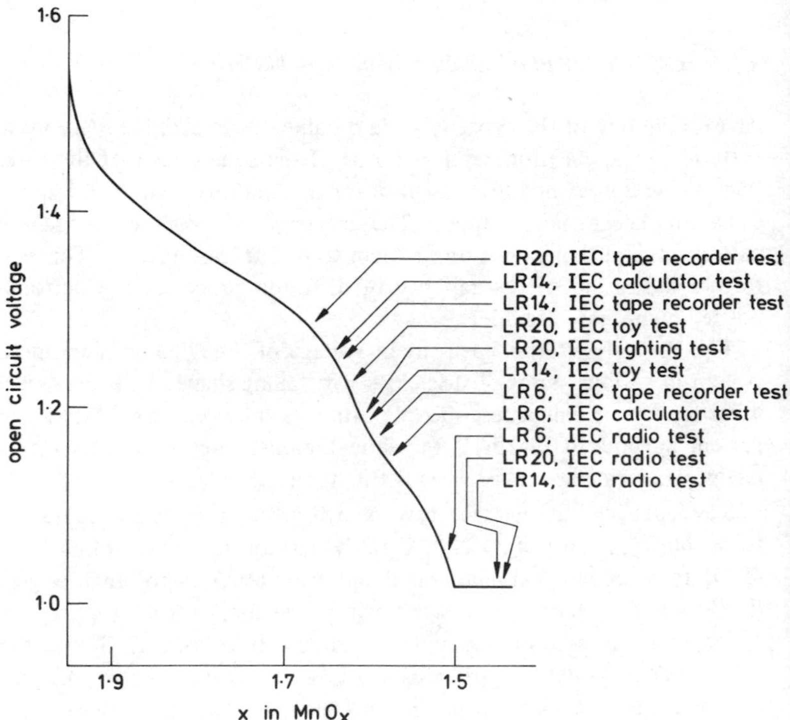

Fig. 3.17 *Capacity yields of alkaline manganese batteries on IEC application tests*

3.5.7 Neutral air-depolarised batteries

Fig. 3.18 shows the on-load voltage curve of a 5AR40 battery on the IEC electric fence test (Table 3.17). The initial portion of the curve at the higher and decreasing voltage is due to the participation of manganese dioxide in the cathode reduction process. The later flat portion of the discharge curve shows the true characteristics of oxygen reduction. Zinc is the most costly ingredient, as it is for Leclanché and alkaline manganese batteries, and the capacity of 'A' system batteries is usually limited by the amount of zinc incorporated. The full capacity is realised at low current densities, on both continuous and intermittent discharges, where the low voltage of the oxygen reduction process is only slightly depressed by voltage losses within the battery. Under these conditions the batteries operate equipment such as electric fences, railway circuits, telephone and telegraphy circuits and marine navigational aids for 3 months to 2 years. Despite the open condition, and hence the possibility of zinc loss by oxygen corrosion, the construction is such that capacity loss is minimal over long periods of operation. Because of their low voltage these batteries have only limited ability to generate high current pulses, and, clearly, judgement on their ability to sustain pulses in any particular application should not be made in the early part of the discharge when manganese dioxide is acting as cathodic reactant.

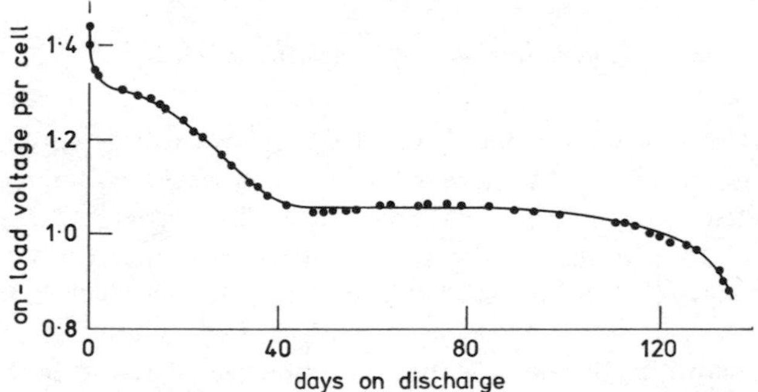

Fig. 3.18 *On-load voltage curve of a 5AR40 battery on the IEC electric fence test*

3.5.8 Button batteries

Fig. 3.19 shows the discharge curves of NR44 and SR44 batteries on the IEC hearing aid test (Table 3.16). The constant open-circuit voltages and the horizontal on-load voltage curves are in sharp contrast to the falling voltage curves which characterise batteries with manganese dioxide cathodes (Fig. 3.10). The initial open-circuit voltage and the on-load voltage in the first discharge period of the NR44 battery show the effect of small proportions of manganese dioxide in the cathode; subsequent behaviour is that of an MR44 battery. There is a sharp voltage drop when either the cathodic or, more usually, the anodic reactant is exhausted. This results in the capacity yield being independent of endpoint voltage, which is different from alkaline manganese

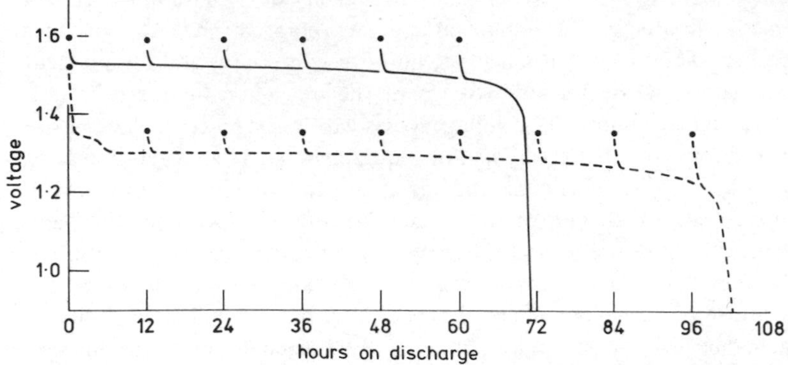

Fig. 3.19 *On-load voltage curves of R44 size batteries on IEC hearing aid test*

—— SR44

– – – NR44

● open circuit voltage prior to each discharge period

batteries. It is also clear from the shape of the on-load voltage curves that capacity is not affected by the length of the daily discharge period, which is different from Leclanché batteries. The capacity yield approaches 100% in terms of the limiting active material. On the discharge test illustrated, SR44 and NR44 batteries are exhausted in 6 and 9 days, respectively. Appliances which require more severe discharges exhaust the batteries in less time and are therefore both costly and inconvenient to the user. Such appliances are inappropriate applications for primary button batteries. In applications such as watches and calculators, particularly those with liquid crystal displays, battery replacement time is usually 1 year or more. For appropriate applications the yield from button batteries is usually close to their total capacity (see Table 3.14).

The small difference between open-circuit and on-load voltage, which is apparent in Fig. 3.19, indicates that for short periods SR44, NR44 and MR44 batteries are able to give currents considerably in excess of 2 – 2·5 mA taken in the hearing aid test. In watches, properly formulated button batteries can supply for 1 – 2 s (the time taken to read the digits) the 10 – 20 mA required to illuminate a liquid crystal display in the dark or the 10 – 50 mA necessary to operate a light emitting diode display. Potassium hydroxide solution is used as the electrolyte in preference to sodium hydroxide solution in batteries required to give high current pulses.

3.6 Overall cell reactions

The best simple representations of the overall cell reactions of the batteries under discussion are probably those given below:

Neutral air depolarised

$$O_2 + 4NH_4Cl + 2Zn \rightarrow 2H_2O + 2Zn(NH_3)_2Cl_2 \quad (3.1)$$

Leclanché

$$2MnO_2 + 2NH_4Cl + Zn \rightarrow 2MnOOH + Zn(NH_3)_2Cl_2 \quad (3.2)$$

Zinc chloride Leclanché

$$8MnO_2 + 8H_2O + ZnCl_2 + 4Zn \rightarrow 8MnOOH + ZnCl_2.4Zn(OH)_2 \quad (3.3)$$

Alkaline manganese

$$2MnO_2 + H_2O + Zn \rightarrow 2MnOOH + ZnO \quad (3.4)$$

Ag_2O/Zn

$$Ag_2O + Zn \rightarrow 2Ag + ZnO \quad (3.5)$$

HgO/Zn

$$HgO + Zn \rightarrow Hg + ZnO \quad (3.6)$$

In ammonium-chloride/zinc-chloride electrolytes the concentration of ammonium chloride is the major factor which controls whether $Zn(NH_3)_2Cl_2$ or $ZnCl_2.4Zn(OH)_2$ is produced. Cahoon[48] showed that $Zn(NH_3)_2Cl_2$ is obtained at ammonium chloride concentrations greater than 7 – 10 w/w%, the higher figure applying to 40 w/w% $ZnCl_2$ and the lower to 0 – 25 w/w% $ZnCl_2$. Friess[49] put the critical ammonium chloride concentration higher but his data are compatible with those of Cahoon. Unpublished work in the writer's laboratory, using X-ray diffraction to identify the compounds precipitated, has confirmed the conclusions reached by Cahoon with microscopy. In a complementary investigation, McMurdie, Craig and Vinal[50] showed that the addition of ZnO to ammonium-chloride/zinc-chloride electrolyte (which simulates the effect of discharge) caused first the precipitation of $Zn(NH_3)_2Cl_2$, and then, when the ammonium chloride concentration had been decreased sufficiently by the reaction

$$ZnO + 2NH_4Cl \rightarrow Zn(NH_3)_2Cl_2 + H_2O \quad (3.7)$$

the precipitation of $ZnCl_2.4Zn(OH)_2$. Bredland and Hull[51] have confirmed the results of these early investigations. Recent studies[242, 243] have indicated that the formula of the basic zinc chloride is $ZnCl_2.4Zn(OH)_2.H_2O$.

MnOOH is now generally accepted as the usual major cathodic reduction product of MnO_2. This conviction stems from the analytically demonstrable presence of MnOOH after chemical reduction, from X-ray diffraction studies and from the ease with which such a product fits into the explanations of recuperation and homogeneous reduction which are discussed later. The analytical evidence is that for manganese dioxide partially reduced with hydrazine or cinnamyl alcohol the manganese oxidation state and the water evolved between 120°C and 250°C[52] correspond to compositions of the general formula $(1-n)$ $MnO_2.n$ MnOOH[52–55] (loss of water from MnOOH can create micropores[56] in the manganese oxide structure[57], which, incidently, invalidate[58] BET estimates of transitional pore area). X-ray diffraction studies showed that the products of cathodic reduction of electrodeposited manganese dioxide (γ–MnO_2) progressively approached the crystal structure of the mineral Groutite (a–MnOOH).[59–61]

Many studies[62–67] in excess electrolyte at a near neutral pH have found that dissolved Mn^{2+} ions are a major product of discharge, but, with the limited amount of electrolyte that is present in Leclanché batteries, dissolved Mn^{2+} does not occur in significant quantities.[59] Mn^{2+} ions are in equilibrium with the manganese oxyhydroxide in accordance with the following equation:[50, 64, 68]

$$MnO_2 + Mn^{2+} + 2H_2O \rightleftharpoons 2MnOOH + 2H^+ \tag{3.8}$$

Similarly, although cathodic reduction can be taken to $Mn(OH)_2$ in excess alkaline electrolyte,[60, 69–73] $Mn(OH)_2$ is not a significant product of discharge in alkaline manganese batteries.[74]

As the solubility of ZnO in both potassium and sodium hydroxide solutions is about one half of that of $Zn(OH)_2$,[75, 76] ZnO is the expected ultimate product of the anodic oxidation of zinc in alkaline electrolyte. In the type of alkaline batteries under discussion, i.e. those with limited electrolyte, ZnO is indeed the product.[76–78] As regards button batteries, this is fortunate as the provision of more water to produce $Zn(OH)_2$ would seriously reduce the total capacities of this type of battery. $Zn(OH)_2$ is a product in large CuO/Zn batteries which have excess 6 molar NaOH as electrolyte.[79]

To conclude this short discussion of overall cell reactions, it is remarkable to note that, of the many alternatives proposed for the Leclanché system, Divers[80] suggested the reaction represented by eqn. 3.2 in 1882.

3.7 Open-circuit cell potentials

3.7.1 Heterogeneous and homogeneous reactions

The thermodynamically reversible potential E of an electrochemical cell is given by the equation

$$E = \frac{\sum_r N_r(\mu_r^0 + RT \ln a_r) - \sum_p N_p(\mu_p^0 + RT \ln a_p)}{nF} \qquad (3.9)$$

where μ^0 is the Gibbs free energy of formation per mole in the chosen standard state, a is the activity relative to the standard state, N is the number of moles of each participant in the cell reaction, subscripts r and p refer to reactants and products, respectively, R is the gas constant, T is absolute temperature, F is the Faraday and n is the number of equivalents of charge transferred from one electrode to the other in the cell reaction. If the reactants and products are in their standard states, i.e. activities are unity, the cell potential is called the standard potential (E^0).

It is conventional with solids to choose the pure solid as the standard state. If, in the course of discharge, as reactants change to products, all participants remain as distinct phases (even though mixed in a macro sense) the Gibbs free energy per mole of each participant cannot change. It follows from eqn. 3.9 that in this case the reversible cell potential must remain unaltered throughout discharge. The cell reaction is termed heterogeneous. Fig. 3.19 shows clearly that the cell reactions of Ag_2O/Zn and HgO/Zn batteries represented by eqns. 3.5 and 3.6 are heterogeneous.

For batteries with manganese dioxide cathodes there is little doubt that the products $Zn(NH_3)_2Cl_2$, $ZnCl_2.4Zn(OH)_2$ and ZnO of cell reactions 3.1, 3.3 and 3.4 occur as distinct phases. The manganese dioxide is responsible for the difference between the behaviour of these batteries and that of Ag_2O/Zn and HgO/Zn batteries. Manganese dioxide and its reduction product MnOOH form a solid solution such that the Gibbs free energy per mole of one participant is not independent of the presence and concentration of the other. The activity term $RT \ln a_{MnO_2}/a_{MnOOH}$ in eqn. 3.9 decreases continuously throughout discharge and this is the cause of the characteristic sloping open circuit voltage curve of batteries containing manganese dioxide cathodes. Manganese dioxide is reduced in homogeneous phase.

Chaney[81] in 1916 was perhaps the first to suggest the possibility of solid solutions between manganese oxides. In 1931 the concept was more clearly expressed by Keller,[82] who attributed the idea to Schreiber.[83] These authors, however, considered the reduction product to be Mn_2O_3 and not until MnOOH became accepted did the formation of a solid solution become conceptually simple, as homogeneity then requires only the diffusion and equilibration of protons and electrons in the host MnO_2 lattice. Brenet, Malessan and Grund[84] made an important advance when they showed, using X-ray diffraction, that on initial cathodic reduction of γ-MnO_2 the basic crystallographic structure did not change but only expanded as OH^- (r = 1·53Å) and Mn^{3+} (r = 0·62Å) ions took the place of O^{2-} (r = 1·40Å) and Mn^{4+} (r = 0·52Å) ions. Progressive dilation of the structure of γ-MnO_2 with reduction has been confirmed for cathodic reduction in ammonium chloride,[61, 65] ammonium-chloride/zinc-chloride[59] and alkaline electrolytes,[60, 73, 74, 85] and for chemical reduction.[53, 86, 87]

Fig. 3.20 shows the open-circuit voltage of R20 batteries containing electrodeposited manganese dioxide on the IEC radio test immediately prior to each daily discharge period. The open-circuit voltage falls continuously and indicates homogenous reduction between MnO_2 and $MnO_{1.5}$, i.e. MnOOH. Active natural ores of ρ crystallinity behave similarly although the open circuit voltages are lower by 0·05 – 0·10V. The conclusion that the phase width of homogenous reduction extends from MnO_2 to $MnO_{1.5}$ for γ-MnO_2 has been reached previously, for chemical reduction on the basis of a continuous change in the volume of the unit cell[53, 86] and from potential measurements,[87, 88] and also from the open-circuit voltages of batteries.[89]

Homogenous reduction does not occur with all varieties of manganese dioxide. With a natural ore of β crystallinity the phase width of homogeneous reduction is only from MnO_2 to $MnO_{1.96}$[60, 87] and possibly between $MnO_{1.8}$ and $MnO_{1.6}$, although the latter region was not observed in the earlier work.[87] With electrodeposited manganese dioxide which had been transformed from γ to β crystallinity by a heat treatment Bode, Schmier and Berndt[87] found that homogeneous reduction was also restricted to the range MnO_2 to $MnO_{1.96}$ while Kozawa and Powers[71] concluded that it occurred over the full range MnO_2 to $MnO_{1.5}$. Dam'e and Mendzheritskii[90, 91] have established that some manganese dioxides give horizontal discharge curves due to the formation of hetaerolite by the following heterogeneous cell reaction

$$2MnO_2 + Zn \rightarrow ZnOMn_2O_3 \tag{3.10}$$

Copeland and Griffith[92] concluded that reaction 3.10 was the chief cell raction of Leclanché batteries, and Gabano, Laurent and

Marignat[59] considered that hetaerolite type compounds are formed in the later stages of discharge. Both groups of workers used X-ray diffraction without complementary voltage measurements.

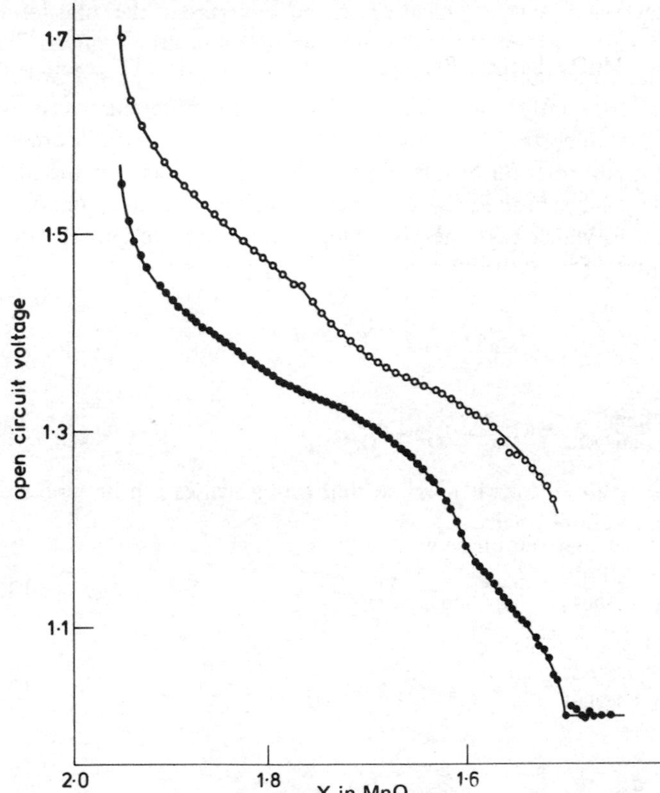

Fig. 3.20 *Open-circuit potential of R20 sized batteries containing electrodeposited MnO_2 on the IEC radio test*

○ R20

● LR20

3.7.2 *Effect of composition on the potential of manganese oxyhydroxy electrodes*

Attention has been focused on the problem of predicting the electrode potential of a $MnO_2/MnOOH$ solid solution from its composition. The usual proposition[71, 88, 93–95] is equivalent to replacing the activity terms in the expression $RT \ln a_{MnO_2}/a_{MnOOH}$ by mol fractions X_{MnO_2}, X_{MnOOH} calculated from the stoichiometry of the solid solution. This approach gives curves of the correct qualitative form but the rate of fall of potential with increasing reduction is insufficient.[71, 96] This deficiency caused the activity coefficients deduced by Neumann and von Roda[88] to depart excessively from unity; a sure sign of the inadequacy of the original proposition.

Kozawa and Powers[95] have presented a picture of the solid solution which hints at a mobile equilibrium with the positions of Mn^{4+} and Mn^{3+} ions constantly changing by electron exchange and with the protons constantly attaching themselves to different oxygens as favourable opportunities arise for O–H bonds to break during the vibration and rotation of OH^- ions. This idea suggests that the MnO_2/MnOOH solid solution might be better regarded as an ordinary ionic solution, in which case the following expressions are appropriate for activities:[68]

$$a_{MnO_2} = a_{Mn^{4+}} a^2_{O^{2-}} \tag{3.11}$$

$$a_{MnOOH} = a_{Mn^{3+}} a_{O^{2-}} a_{OH^-} \tag{3.12}$$

If the mixture of ions is ideal so that ion activities can be replaced by ion mol fractions, then

$$a_{MnO_2} = \frac{27}{4} X_{Mn^{4+}} X^2_{O^{2-}} \tag{3.13}$$

$$a_{MnOOH} = 27 X_{Mn^{3+}} X_{O^{2-}} X_{OH^-} \tag{3.14}$$

and

$$RT \ln \frac{a_{MnO_2}}{a_{MnOOH}} = RT \ln \frac{X_{Mn^{4+}} X_{O^{2-}}}{4X_{Mn^{3+}} X_{OH^-}} \tag{3.15}$$

The numerical terms are necessary for a_{MnO_2} and a_{MnOOH} to be unity for the compositions MnO_2 and MnOOH, respectively. Eqn. 3.15 gives rise to a greater rate of potential fall with reduction than the older proposition

$$RT \ln \frac{a_{MnO_2}}{a_{MnOOH}} = RT \ln \frac{X_{MnO_2}}{X_{MnOOH}} \tag{3.16}$$

and, although not yet tested over the full range of solid solutions, eqn. 3.15 does lead to a better representation than eqn. 3.16 of the potential of electrodeposited manganese dioxide (γ crystallinity) for compositions close to MnO_2.[68] Interestingly, the converse is true for a β- MnO_2 ore.[68]

Table 3.20 *Gibbs free energies of formation at 25° C*

Ion or compound	State	μ°	Reference
		k cal/m ol	
Electrodeposited MnO_2	solid	- 109·2	68
MnOOH	solid	- 132·3	68
$Mn(OH)_2$	solid	- 146·9	97
Ag_2O	solid	- 2·586	97
HgO	solid	- 13·990	97
ZnO	solid	- 76·40	98
$Zn(NH_3)_2Cl_2$	solid	- 120·65	this work
$ZnCl_2.4Zn(OH)_2$	solid	- 633·1	this work
NH_4Cl	solid	- 48·73	99
H_2O	liquid	- 56·690	97
Zn^{2+}	aqueous	- 35·184	97
NH_4^+	aqueous	- 19·00	97
Cl^-	aqueous	- 31·350	97
H_2O_2	aqueous	- 31·47	97

3.7.3 *Calculated standard cell potentials*

The Gibbs free energies of formation per mole of compounds and ions at 25°C in their usual standard states which are relevant to this discussion are given in Table 3.20. By definition the Gibbs free energies of elements in their normal state, and of the H^+ ion in its standard state in aqueous solution, are zero. No values were found in the literature for $Zn(NH_3)_2Cl_2$ and $ZnCl_2.4Zn(OH)_2$.

$\mu^0_{Zn(NH_3)_2Cl_2}$ was calculated from the equilibrium

$$Zn^{2+} + 2NH_4Cl\ (solid) \rightleftharpoons Zn(NH_3)_2Cl_2\ (solid) + 2H^+ \qquad (3.17)$$

whereby

$$\mu^0_{Zn(NH_3)_2Cl_2} = \mu^0_{Zn^{2+}} + RT \ln a_{Zn^{2+}} + 2\mu^0_{NH_4Cl} + 4{\cdot}606\ RT\,pH \qquad (3.18)$$

Then, expressing $\mu^0_{Zn^{2+}} + RT \ln a_{Zn^{2+}}$ in terms of the potential of a zinc electrode E_{Zn} the following expression is obtained

$$\mu^0_{Zn(NH_3)_2Cl_2} = 2FE_{Zn} + 2\mu^0_{NH_4Cl} + 4{\cdot}606\ RT\,pH \qquad (3.19)$$

Thus the Gibbs free energy of $Zn(NH_3)_2Cl_2$ was calculated from the potential of a zinc electrode immersed in a zinc chloride solution saturated with ammonium chloride and the pH at which $Zn(NH_3)_2Cl_2$ is precipitated. These data are available from the works of Sasaki and Takahashi,[100] McMurdie, Craig and Vinal[50] and Cahoon.[33]

In a similar manner the Gibbs free energy of $ZnCl_2.4Zn(OH)_2$ was calculated from the equilibrium

$$5Zn^{2+} + 2Cl^- + 8H_2O \rightleftharpoons ZnCl_2.4Zn(OH)_2 \text{ (solid)} + 8H^+ \qquad (3.20)$$

using the expression

$$\mu^0_{ZnCl_2.4Zn(OH)_2} = 10FE_{Zn} + 2(\mu^0_a- + RT \ln a_a-) +$$
$$8(\mu^0_{H_2O} + RT \ln a_{H_2O}) + 18{\cdot}42\, RT\, pH \qquad (3.21)$$

The chloride ion and water activity date for zinc-chloride/ammonium-chloride solutions measured by Sasaki and Takahashi[100] were used for the calculation.

Table 3.21 *Calculated standard cell voltages*

Electrochemical system	Overall cell reaction (equation)	Calculated standard cell voltage	Observed open-circuit voltage
Neutral air-depolarised	3.1	1·73	1·45
Leclanché	3.2	1·50	1·60
Zinc chloride Leclanche	3.3	1·44	1·57
Alkaline manganese	3.4	1·43	1·55
Ag_2O/Zn	3.5	1·60	1·60
HgO/Zn	3.6	1·35	1·35

Table 3.21 lists standard cell voltages calculated from the overall cell reactions (eqns. 3.1 to 3.6) using the Gibbs Free energies given in Table 3.20 and eqn. 3.9. Where the cell reaction is heterogeneous and the reactants and products are in their standard states, as is the case for Ag_2O/Zn and HgO/Zn batteries, there is excellent agreement between the calculated standard cell voltage and the observed open-circuit voltage of batteries. The standard cell voltages of batteries with manganese dioxide cathodes which reduce in homogeneous phase are hypothetical, since it is not possible for product and reactant to be present together at unit activity. In these circumstances the observed

open-circuit voltages of batteries are higher than the calculated standard cell voltage because the activity of MnOOH is far removed from its standard state. Eqns. 3.13 and 3.14 can be used to estimate the effect of MnOOH and MnO_2 activity terms. If the initial composition of the manganese dioxide is $MnO_{1 \cdot 95}$ or $0 \cdot 9MnO_2 . 0 \cdot 1MnOOH$ then insertion of appropriate activity terms for MnOOH and MnO_2 in eqn. 3.9 would change the calculated cell voltages by +0·102V and −0·005V, respectively, and would bring the calculated values much closer to the observed open-circuit voltages. For the Leclanché system the Gibbs free energy of solid NH_4Cl was used in the calculation as the electrolyte is initially saturated with this compound.

For neutral air-depolarised batteries the calculation is unsuccessful for two reasons. First, the overall cathode reaction

$$O_2 + 4H^+ + 4e \rightarrow 2H_2O \, , E^0 = 1 \cdot 229V \tag{3.22}$$

does not take place reversibly on activated carbon. If the standard potential of the reversible reaction

$$O_2 + 2H^+ + 2e \rightarrow H_2O_2 \, , E^0 = 0 \cdot 682V \tag{3.23}$$

is taken in place of that for reaction 3.22 then the calculated cell voltage becomes 1·19V, which is more in line with the observed on-load voltage of 'A' system batteries in the plateau region (see Fig. 3.18).

Secondly, the initial open-circuit voltage of an 'A' system battery does not depend on the potential of the oxygen electrode but on the potential of the manganese dioxide that is incorporated into the cathode structure. Initially, neutral air-depolarised and Leclanché batteries have the same overall cell reaction. It is, therefore, of interest to examine why the initial open-circuit voltage of an 'A' system battery is only 1·45V. Part of the reason is the grade of manganese dioxide used and part is the establishment of a redox balance between the $MnO_2/MnOOH$ electrode and surface groups on the activated carbon. The existence of groups containing oxygen on the surface of carbons is well known.[101–103] Garten and Weiss[104] postulated quinone/hydroquinone groups in order to interrelate and explain various aspects of the behaviour of activated carbon. The electrochemical redox behaviour of carbon surfaces has been investigated[105–108] and has usually been discussed in terms of quinone/hydroquinone groups. Caudle, Summer and Tye[109] showed that the potential of MnO_2/carbon mixtures decreased with increasing carbon content, in confirmation of earlier work,[110, 111] and with increasing ability of the carbon to absorb acid. The redox couple shown in eqn. 3.24 was postulated.[109]

$$\text{COO}^- + 2H^+ + 2e \rightleftharpoons \text{COOH} \qquad (3.24)$$

Postulation of a zwitterion enabled both acid absorption and the slow evolution of carbon dioxide from mixtures wetted with electrolyte to be explained in terms of eqns. 3.25 and 3.26

$$\text{COO}^- + HCl \rightleftharpoons Cl^- \ \text{COOH} \qquad (3.25)$$

$$\text{COO}^- + HCl \longrightarrow Cl^- + CO_2 \qquad (3.26)$$

Surface groups are almost completely absent from acetylene black[109, 112] and graphite,[109] with the result that the open-circuit potentials of Leclanché and alkaline manganese batteries are only very slightly depressed by the admixed carbon. This is not true of blacks which have been tried as alternatives to acetylene black.[109]

3.7.4 *Equilibrium pH effects*

It is clear from eqns. 3.5 and 3.6 that there is no change in the average composition of electrolyte during the discharge of Ag_2O/Zn and

HgO/Zn batteries. Although water is consumed during the discharge of alkaline manganese batteries (eqn. 3.4) the pH of the electrolyte is essentially unchanged. The pH of the electrolyte is, however, changed in the discharge of Leclanché batteries.

In accordance with a potential-determining reaction

$$MnO_2 + H_2O + e \rightarrow MnOOH + OH^- \qquad (3.27)$$

the potential of a manganese dioxide electrode should decrease by 0·0592V at 25°C for a unit increase in pH. Many investigators who have attempted to establish this point found slopes numerically greater than −0·0592V/pH.[62, 113–116] Johnson and Vosburgh[117] suggested that this discrepancy is connected with the ion-exchange property of manganese dioxides.[117–119] Benson, Price and Tye[120] showed that there is indeed an approximate correlation between the extent of the discrepancy and the extent to which ions such as K^+, Li^+, Cu^{2+} etc. are ion-exchanged onto the manganese dioxide surface. If, as suggested by Kozawa,[118] water molecules physisorbed onto the manganese dioxide surface are the sites for ion exchange, then the process as represented by eqn. 3.28 causes an increase in the concentration of hydroxyl ions at the surface

$$MnO_2.H_2O + K^+ \rightleftharpoons MnO_2.OH^-, K^+ + H^+ \qquad (3.28)$$

As ion-exchange increases steadily with pH,[119, 120] the concentration of hydroxyl ions at the manganese dioxide surface also increases steadily with pH, even though the ratio of Mn^{4+} and Mn^{3+} ions in the dioxide is unchanged. Vosburgh and co-workers[121, 122] have repeatedly stated that the potential of a manganese oxyhydroxide electrode is determined by the composition of the surface, and the rectitude of this has been clearly demonstrated by Kozawa who showed that the potential of a manganese dioxide that had been heat treated to produce a surface layer of Mn_2O_3 approached that of an Mn_2O_3 electrode.[123] If, following Benson, Price and Tye,[120] the potential is expressed in surface Gibbs free energy terms as in eqn. 3.29, then it becomes clear that an increase in surface hydroxyl ions with pH as a result of ion-exchange must increase μ_{OH^-} (surface) and cause the potential/pH slope to be numerically greater than −0·0592 V/pH at 25°C

$$E = -0{\cdot}0592\text{pH} + \Big[\mu_{Mn^{4+}}(\text{surface}) - \mu_{Mn^{3+}}(\text{surface}) + \mu_{O^{2-}}(\text{surface}) - \mu_{OH^-}(\text{surface})\Big]/F \qquad (3.39)$$

In the absence of all cations except H^+, when cation exchange is not possible, a slope of −0·0595V/pH at 25°C has recently been obtained.[124] Reaction 3.27 is now considered to be the potential-determining reaction even in the presence of Mn^{2+} ions.[125]

3.7.5 Influence of compositional changes during discharge on open-circuit voltage

As already discussed, the cell reactions of Ag_2O/Zn and HgO/Zn batteries take place heterogeneously so there is no change in open-circuit voltage during the course of discharge. The situation with Leclanché batteries is more complicated and is now considered in detail as it places in perspective some of the preceding discussion.

Fig. 3.21 has been synthesised on the following basis:

(*a*) the solid ammonium chloride present initially is 50% of that depicted in Fig. 3.7
(*b*) the electrolyte is of sufficient volume to fill all the porosity shown in Fig. 3.7
(*c*) the electrolyte is initially saturated with NH_4Cl and is 2·67 molal in $ZnCl_2$
(*d*) no hetaerolite is formed on discharge
(*e*) no $ZnCl_2.4Zn(OH)_2$ appears until the NH_4Cl concentration falls to 7 w/w%
(*f*) pH behaviour is described by eqn. 3.17) with NH_4Cl in either solid or dissolved form and eqn. 3.20
(*g*) the activities of MnO_2 and MnOOH are given by eqns. 3.13 and 3.14, respectively
(*h*) the pH dependence of the manganese oxyhydroxide electrode is -0·0592V/pH
(*i*) zinc potentials and chloride activity data were taken from Cahoon[35] and Saski and Takahashi[100]
(*j*) a discharge of 1 Faraday which reduces the manganese oxyhydroxide from $MnO_{1.95}$ to $MnO_{1.5}$.

The behaviour divides into three stages. In the first stage, from A to B, NH_4Cl is consumed and $Zn(NH_3)_2Cl_2$ produced in accordance with eqn. 3.2. NH_4Cl removed from solution is replaced by dissolution of solid NH_4Cl. The net effect is consumption of solid NH_4Cl; the electrolyte is invariant and pH constant. Zinc potential is constant but the potential of the manganese oxyhydroxide electrode falls as the composition of the solid solution changes towards MnOOH. At B, all the solid NH_4Cl has been consumed and from B to C $Zn(NH_3)_2Cl_2$ is produced at the expense of a decreasing NH_4Cl concentration in the electrolyte. The fall in NH_4Cl concentration causes a slight rise in pH (in accordance with the equilibrium represented by eqn. 3.17) which results in a slight downward inflection of the potential of the manganese oxyhydroxide. The major effect of the decrease in NH_4Cl concentration is on zinc potential. This can be understood on the basis of Sasaki and Takahashi's excellent investigation[100] as the consequence of the equilibrium

$$Zn^{2+} + 4Cl^- \rightleftharpoons ZnCl_4{}^{2-} \tag{3.30}$$

shifting to the left as the chloride ion concentration decreases. In the final stage, C to D, the consumption of dissolved NH_4Cl and the production of $Zn(NH_3)_2Cl_2$ continues but at a reduced level. The major cell reaction is now the consumption of $ZnCl_2$ and H_2O and the production of $ZnCl_2.4Zn(OH)_2$ in accordance with eqn. 3.3. The net effect of these changes on electrolyte composition is to reverse the zinc potential trend and arrest the more rapid fall in overall cell potential that occurred in stage B to C.

Fig. 3.20 shows the open-circuit voltage of an R20 battery (electrodeposited MnO_2) on the IEC radio test and the step in the central region of discharge is clearly discernable. No step in this region is apparent with an LR20 battery for which the electrolyte is essentially invariant during discharge (Fig. 3.20).

Fig. 3.20 also shows another interesting difference between Leclanché and alkaline manganese batteries. Whereas the open-circuit voltage curves suggest that a compound of oxidation value $MnO_{1.5}$, e.g. MnOOH, is the most reduced composition for the solid solution in Leclanché electrolyte, a compound of oxidation value $MnO_{1.6}$ is indicated in alkaline electrolyte. Similar data presented by Bell and Huber [60] and by Gabano, Morignat and Laurent[74] could also be taken as indicating an end-member of oxidation value $MnO_{1.6}$ in alkaline electrolyte. On the other hand, Kozawa and Powers[71,95] concluded that the end-member was $MnO_{1.5}$ and it is not obvious why the nature of the external electrolyte should affect the course of a solid state reaction.

A possible explanation of these differences is that the solid state reduction is essentially identical in the two types of electrolyte but that the solid solution becomes increasingly metastable as the oxidation value decreases and the structure dilates. It must then be pre-

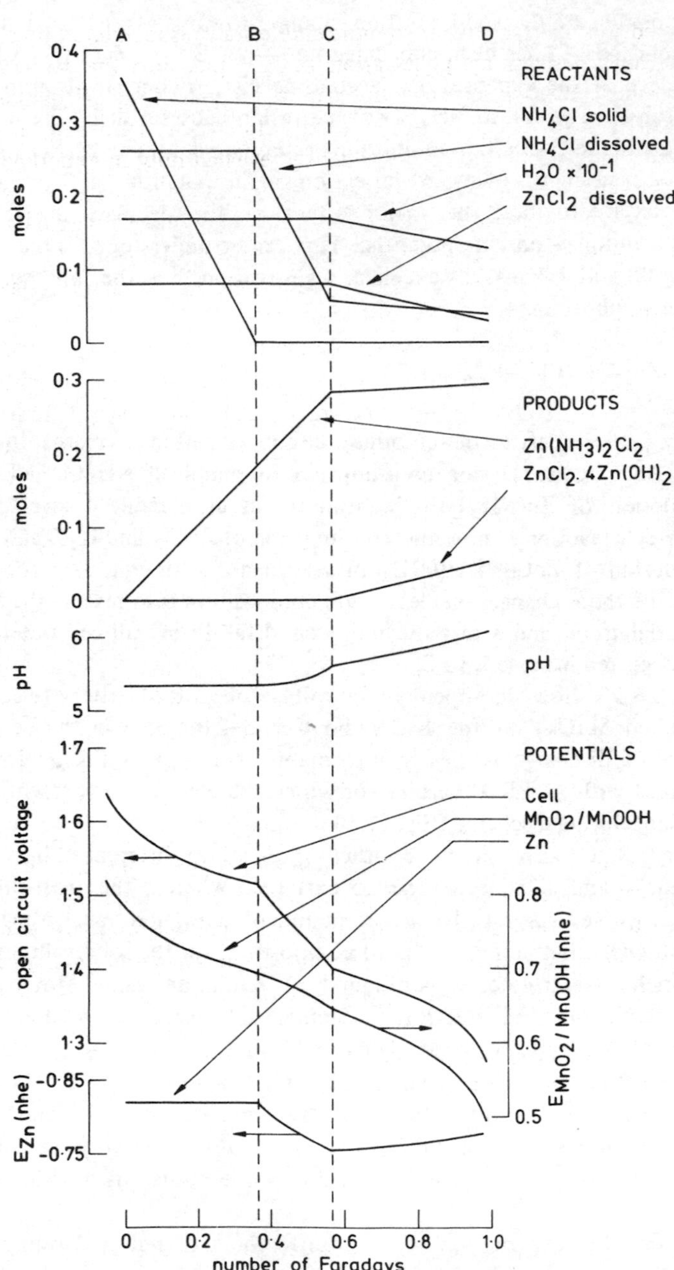

Fig. 3.21 *Calculated composition and potential changes during slow discharge of Leclanché cells*

sumed that rearrangement to more stable states is facilitated by alkaline solution (perhaps by penetration of K^+ ions into the structure, as Bell and Huber[60] suggested, to account for new X-ray diffraction lines which appeared in electrodeposited MnO_2 on mere immersion in alkali) and is a slow process which is revealed only in experiments of long duration. The IEC radio discharge tests take 90 days to complete with LR20 batteries (Fig. 3.20) and Bell and Huber[60] waited at least 6 weeks before making open-circuit voltage measurements. Perhaps it is pertinent to mention that a compound Mn_5O_8, i.e. $MnO_{1 \cdot 6}$ has been synthesised.[126, 127]

If solid solutions of lower oxidation value are metastable, it is clear that recourse to heating during their recovery for X-ray diffraction may lead to erroneous conclusions. Kang and Liang[128] found that quite mild vacuum heating changed the electrochemical behaviour of a manganese oxyhydroxide of oxidation value about $MnO_{1 \cdot 5}$. Boden, Venuto, Wisler and Wylie[73, 129] commendably avoided washing and drying in their investigations.

The open-circuit voltage plateau that occurs at just above 1·0V[60, 74, 95] with the manganese-oxhyhydroxide/zinc couple in alkaline solution (see Fig. 3.20) is due, as first proposed by Kozawa and Yeager,[70] to the heterogeneous overall cell reaction

$$2MnOOH + H_2O + Zn \rightarrow 2Mn(OH)_2 + ZnO \qquad (3.31)$$

The standard potential of reaction 3.31, calculated from Gibbs free energies given in Table 3.20, is 1·06V. As shown by Fig. 3.17, reaction 3.31 plays little part in the electrical energy delivered on various application tests. It is, however, brought into use when a battery is inadvertently left on load, as the MnOOH present is normally sufficient to exhaust the zinc and thus prevents discharge of the Zn/H_2O couple which would produce undesirable hydrogen.

3.8 Cathode discharge mechanisms

3.8.1 Leclanché cells

Electrons are conveyed from the positive terminal via acetylene black or graphite to the manganese dioxide surface where the potential determining charge transfer reaction given in eqn. 3.27 takes place. In Leclanché and zinc chloride Leclanché batteries the hydroxyl ion generated reacts further by either eqn. 3.32 or 3.33, depending on the ammonium chloride concentration

$$2OH^- + Zn^{2+} + 2NH_4^+ + 2Cl^- \rightarrow Zn(NH_3)_2Cl_2 + 2H_2O \qquad (3.32)$$

$$8OH^- + 5Zn^{2+} + 2Cl^- \rightarrow ZnCl_2.4Zn(OH)_2 \qquad (3.33)$$

(although the dissolved zinc species is represented as Zn^{2+} for simplicity in eqns. 3.32 and 3.33, $ZnCl_4^{2-}$ is the predominant species in solutions which are concentrated in ammonium chloride[100] and various cationic and anionic species are present in zinc chloride solutions). These changes perturb the concentration at the manganese oxyhydroxide-electrolyte interface in a way that is represented schematically in Fig. 3.22.

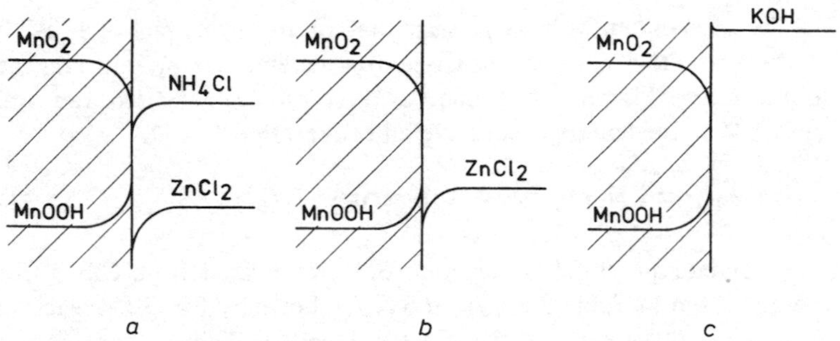

Fig. 3.22 *Schematic representation of concentration polarisation*

a Leclanché cell

b $ZnCl_2$ Leclanché cell

c Alkaline manganese cell

Although concentration polarisation sets in on both sides of the interface, most attention has been directed at the manganese oxyhydroxide phase, probably because differences in the battery performance of various sources of manganese dioxide[45, 130, 131] indicate that diffusion in this phase is the process which limits the electrical energy obtainable from Leclanché batteries. Although this is undoubtedly correct for natural ores, the data presented earlier in Fig. 3.14

and the performance of alkaline manganese batteries suggest that the diffusion in electrodeposited manganese dioxide is fast enough to change the limiting process to diffusion in the electrolyte phase.

Concentrations of MnO_2 and MnOOH are given in Fig. 3.22, although the situation might be better regarded as a variation of r in a general composition $MnOOH_r$. Discharge increases the value r at the surface above that for the bulk and thereby decreases electrode potential. As first appreciated by Coleman,[132] potential recovery or recuperation is accomplished by diffusion of electrons and protons from the more populated surface to the interior.

Pertinent evidence in support of this mechanism is experimental observation pointing to a slower diffusion process in deuterated solutions.[95, 133] A point not often remarked on is that a solid solution capable of existing in a range of compositions is a necessary adjunct to the establishment of a concentration gradient for diffusion.[134] The low battery activity of manganese dioxides of β crystallinity could be ascribed to the restricted composition range over which they can exist as a single phase. Potential and equilibrium data for small concentrations of MnOOH in host β-MnO_2 and γ-MnO_2 lattices are distinguishable and have been interpreted in terms of MnOOH being associated and dissociated in the former and latter lattices, respectively.[68] Association could be a precursor to separation of an MnOOH rich phase and also suggests a low concentration of the diffusing species, i.e. protons, even in the single phase region.

Diffusion coefficients should be dependent on the bulk properties of the diffusing medium and so it is not surprising that battery activity correlates with bulk properties such as crystallinity and thermal stability. Manganese dioxides which are battery active have lower crystallinity and thermal stability[130] than other manganese dioxides. Increasing activity also correlates approximately with increasing surface area.[36] This is seen as a consequence of diffusion taking place across a greater area; the surface regions of higher potential, which are remote from carbon contacts, being reduced by a coupled discharge-charge

Such a mechanism is represented in eqn. 3.34

$$e \quad \left[\begin{array}{l} \left\{\begin{array}{l} MnO_2 \\ MnO_2' \end{array}\right. \\ \quad\uparrow \\ \left\{\begin{array}{l} MnOOH \\ MnOOH \end{array}\right. \end{array}\right. \quad + \quad \begin{array}{c} H_2O \\ \\ OH^- \end{array} \quad \rightarrow \quad \left[\begin{array}{l} \left\{\begin{array}{l} MnO_2 \\ MnOOH \end{array}\right. \\ \\ \left\{\begin{array}{l} MnO_2 \\ MnOOH \end{array}\right. \end{array}\right. \quad + \quad \begin{array}{c} OH^- \\ \\ H_2O \end{array} \qquad (3.34)$$

mechanism with partially reduced regions of lower potential, which are adjacent to carbon contacts.

Informative mathematical treatments of the mechanism of the manganese dioxide cathode have been published. These are resonably successful in accounting for the capacity of paste Leclanché batteries containing active ore at light and moderate continuous constant current drains,[132] the shape of potential-time recuperation curves with electrodeposited MnO_2 cathodes,[135, 136] and initial current-time and potential-time relationships under potentiostatic and galvanostatic discharge conditions, respectively, with active ore and electrodeposited MnO_2.[133]

Scott's calculations[135] drew attention to the possibility of the manganese oxyhydroxide surface approaching saturation in MnOOH in the vicinity of the carbon contacts. This is more likely with manganese dioxides of lower battery activity, at high currents particularly if continuously withdrawn, and of course as the average value r in $MnOOH_r$ approaches unity. If the surface approaches saturation in MnOOH then cathodic reduction takes place by reaction 3.35 instead of reaction 3.27.

$$MnOOH + H_2O + e \rightarrow Mn^{2+} + 3OH^- \qquad (3.35)$$

Compared with reaction 3.27, reaction 3.35 produces three times the number of hydroxyl ions per Faraday and its potential per unit rise in pH decreases at three times the rate. The onset of reaction 3.35 thus causes rapid depletion of NH_4Cl and/or $ZnCl_2$ at the manganese oxyhydroxide surface due to reaction 3.32 or 3.33 and a rapid drop in potential resulting from pH increases caused by the change in electrolyte composition. In short, saturation of the manganese oxyhydroxide surface with MnOOH in the vicinity of carbon contacts usually presages the end of useful discharge in Leclanché batteries. Such saturation occurs with natural ores and limits the capacity given by them on IEC application tests.

McMurdie, Craig and Vinal[50] showed that if the pH of a solution containing manganous ions and dissolved zinc species is increased in the presence of MnO_2, hetaerolite is formed by the reaction

$$MnO_2 + Mn^{2+} + Zn^{2+} + 2H_2O \rightarrow ZnO.Mn_2O_3 + 4H^+ \qquad (3.36)$$

It is probable that this reaction takes place after discharge by reaction 3.35 has generated soluble Mn^{2+} ions and during the recuperation period, when the oxidation state of the manganese oxyhydroxide surface has increased above MnOOH. As reaction 3.36 buffers the electrolyte between pH4 and pH5, which is lower than the pH values at which $Zn(NH_3)_2Cl_2$ and $ZnCl_2.4Zn(OH)_2$ are precipitated[48, 50] (see Fig. 3.21), the resultant effects of the formation of hetaerolite are the gradual dissolution first of $ZnCl_2.4Zn(OH)_2$

$$ZnCl_2.4Zn(OH)_2 + 8H^+ + 8Cl^- \rightarrow 5ZnCl_2 + 8H_2O \tag{3.37}$$

and then of $Zn(NH_3)_2Cl_2$

$$Zn(NH_3)_2Cl_2 + 2H^+ + 2Cl^- \rightarrow ZnCl_2 + 2NH_4Cl \tag{3.38}$$

with regeneration of zinc chloride and ammonium chloride.

The diffusion of protons and electrons in electrodeposited MnO_2 is considered to be sufficiently fast for the manganese oxyhydroxide surface not to approach saturation in MnOOH on IEC application tests until the bulk composition approaches MnOOH, i.e. until the capacity yield is close to 100%. The capacity limitations shown in Fig. 3.14 are due to polarisation on the electrolyte side of the interface. An insight into the factors which are then important can be obtained by considering a model in which the interface is planar, the electrolyte medium is semi-infinite with account taken of tortuosity and porosity,[40] and with ion migration occurring by electrical transference and diffusion but not by convection.[137–139] For this model the concentration (c_m mol/cm^{-3}) of species m at the interface after time t (s) is given by eqn. 3.39

$$c_m = \bar{c}_m - \frac{2i\theta t^{1/2}(B - t_m)}{z_m F\pi^{1/2} V_e D_m^{1/2}} \tag{3.39}$$

z_m is the valency and is always positive, $\bar{c}_m$ (mol cm^{-3}) is the bulk concentration, t_m is the transference number towards the interface and D_m (cm^2/s) the diffusion coefficient in free electrolyte solution of species m.i (A/cm^2) is the current density of charge transference at the interface. B is the gram equivalent of species m removed (negative if added) from the electrolyte by, or immediately subsequent to, a charge transfer of 1 Faraday. Eqn. 3.39 is a modified form of Sand's equation,[138] which strictly requires t_m to be zero or constant in the diffusion zone. Combination of charge transfer reaction 3.27 with the subsequent chemical reaction 3.32 leads to B values of unity for Zn^{2+} and NH_4^+. If the electrolyte composition were such that the subsequent chemical reaction was reaction 3.33, then B would be $^{5}/_{4}$ for Zn^{2+} and 0 for NH_4^+. If, however, the charge transfer reaction were reaction 3.35, as might be the case with an active ore, then B for Zn^{2+} would be 3 and $^{15}/_{4}$ for chemical reactions 3.32 and 3.33, respectively. The tortuosity in the electrolyte medium θ is created by all undissolved substances and is defined as the mean path length of diffusion channels per unit length normal to the interface. V_e is the volume fraction of electrolyte.

The transition time τ_m (s) until species m reaches zero concentration is obtained by rearrangement of eqn. 3.39 with c_m equal to zero

$$\tau_m = \pi D_m \left[\frac{z_m \bar{c}_m V_e F}{2i\theta\,(B-t_m)} \right]^2 \qquad (3.40)$$

A useful approximation is to equate θ with $1/V_e$ [40, 140] so that

$$\tau_m = \pi D_m \left[\frac{z_m \bar{c}_m F}{2i\,(B-t_m)} \right]^2 V_e^4 \qquad (3.41)$$

The appearance of V_e in eqn. 3.41 at the fourth power shows immediately the importance of porosity in delaying the time at which a species, e.g. $ZnCl_4^{2-}$, in Leclanché electrolyte reaches zero concentration at the manganese oxyhydroxide surface. This is the underlying reason why acetylene black is preferred to graphite as the electronic conductor in the cathodes of Leclanché batteries. Obviously, the amount of porosity is a compromise since too much reduces the manganese dioxide content and therefore, capacity.

In normal Leclanché batteries during discharge V_e is increased by dissolution of solid NH_4Cl and reduced by precipitation, first of $Zn(NH_3)_2Cl_2$ and later of $ZnCl_2.4Zn(OH)_2$, as shown in Fig. 3.21. The latter effects are overriding and the endpoint on an application test with a Leclanché battery containing electrodeposited MnO_2 is often caused by V_e decreasing to such an extent that the daily discharge period exceeds the transition time for a component of the Leclanché electrolyte. In zinc chloride Leclanché batteries there is no solid NH_4Cl to dissolve and reduce the rate at which V_e decreases with discharge. Thus, as already noted in Section 3.4.5, the cathodes for this variant have greater initial porosity than the cathodes of normal Leclanché batteries.

In order to establish which of NH_4^+ and $ZnCl_4^{2-}$ is more likely to be depleted to zero concentration at the manganese oxyhydroxide surface, it is convenient to express D_m of eqn. 3.41 in terms of some known bulk properties of a Leclanché electrolyte.[140] This can be accomplished using the relationship

$$|t_m| = F\rho z_m \bar{c}_m U_m \tag{3.42}$$

and the Nernst-Einstein approximation

$$D_m \simeq \frac{RT\,U_m}{z_m\,F} \tag{3.43}$$

U_m (Cm/s) is the mobility of species m in a field of 1V/cm and ρ(2/cm) is the specific resistance of the electrolyte. Substituting in eqn. 3.41

$$\tau_m \simeq \frac{\pi RT\,V_e^{\,4}\,\bar{c}_m|t_m|}{4\,i^2\,\rho\,(B-t_m)^2} \tag{3.44}$$

The group of parameters in eqn. 3.44 which differs for NH_4^+ and $ZnCl_4^{2-}$ is $c_m|t_m|\,(B-t_m)^{-2}$. Using B=1, $c_{NH_4^+}$ = 0·0066 and $c_{ZnCl_4^{2-}}$= 0·0015 and published data[140] of $t_{NH_4^+}$ = 0·59 and $t_{ZnCl_4^{2-}}$= -0·14, this quantity is 2·3x10^{-2} for NH_4 and two orders of magnitude lower at 1·6x10^{-4} for $ZnCl_4^{2-}$. Even allowing for the inaccuracy of the model, other assumptions and the decreasing concentration of ammonium ions (Fig. 3.21) it is probable that the species most significantly depleted in the charge transfer region under all circumstances is zinc. The underlying reasons are the presence of zinc in Leclanché electrolyte as an anion which is transported away from the charge transfer region and the low rate of diffusion of that anion in comparison with NH_4^+[140] into the depleted region.

If soluble zinc species are absent from the cathodic charge transfer region, hydroxyl ions cannot be consumed by either reaction 3.32 or 3.33. It is well known in the industry that under certain discharge conditions the cathodes of Leclanché cells smell of ammonia. When the transition time for soluble zinc species is exceeded, hydroxyl ions accumulate at the charge transfer surface, the potential of the manganese oxyhydroxide electrode falls by about 0·06V per unit increase in pH and diffusion of hydroxyl ion from the charge transfer region commences. The diffusion path is short, as hydroxyl ions react with solid $Zn(NH_3)_2Cl_2$ precipitated earlier in the discharge to form soluble $Zn(NH_3)_4^{2+}$. The reaction

$$2OH^- + Zn(NH_3)_2Cl_2 + 2NH_4^+ \rightarrow Zn(NH_3)_4^{2+} + 2Cl^- + 2H_2O \tag{3.45}$$

buffers the electrolyte at pH 7·5 – 8·0.[48] As this is about two pH units higher than the buffering level provided by reaction 3.32, the cathode potential changes by at least 0·12V when the transition time for soluble zinc species is exceeded.

In application tests, during the recuperation period soluble zinc species of course continue to diffuse into the charge transfer region. However, the linear dependence of capacity yield on the square root of the total time that a battery is on test, i.e. on-load time plus recuperation time, can be understood on the basis that the $ZnCl_2$ distribution prior to a daily discharge period may not be completely restored to the distribution prior to the preceding daily discharge period. A $ZnCl_2$ deficiency thus builds up gradually in the charge transfer region until the transition time for zinc species is exceeded in a daily discharge period and then the events described in the preceding paragraph cause battery voltage to fall, probably to the endpoint voltage of the application test.

Although zinc chloride Leclanché batteries do not give longer discharge durations on application tests than normal premium Leclanché batteries, they can yield up to 100% more capacity on continuous discharges lasting less than a few days. This is partly due to the higher initial porosity (Fig. 3.7) of the cathodes of zinc chloride Leclanché batteries and partly due to a changing zinc transference number. In normal Leclanche electrolyte, $t_{ZnCl_4^{2-}}$ changes from −0·14 towards zero as the concentration of zinc chloride decreases and an increasing proportion of the current is carried by the ions of ammonium chloride. The effect of this is to change $(B-t_m)^{-2}$ in eqn. 3.44 from 0·8 towards 1·0, i.e. a relatively small increase in transition time. In comparison, $t_{Zn^{2+}}$ for zinc chloride solutions changes from zero towards 0·4 as the concentration decreases from 2·0 molal to zero,[141] due to the equilibrium balance of zinc-containing species shifting towards cationic varieties. This changes $(B-t_m)^{-2}$ from 0·6 towards 1·4, which must result in a substantial increase in transition time. A major reason for the superiority of zinc chloride Leclanché batteries on continuous discharge is thus an increasing electrical transference of dissolved zinc species into the charge transfer region as the zinc chloride concentration in this region decreases.

3.8.2 Alkaline manganese cells

In alkaline electrolyte, the hydroxyl ions generated by the charge transfer reaction 3.27 are not removed by a subsequent chemical reaction. Hydroxyl ions therefore accumulate at the charge transfer interface. However, the high mobility of hydroxyl ions, both in electrical transference and concentration diffusion, means that the concentration perturbation at the interface is only small, as depicted schematically in Fig. 3.22. Water, which is consumed in the charge transfer reaction, is abundantly present: in 10 molar KOH solution the water molarity is 47. It is concluded that the performance of the cathodes in alkaline manganese batteries cannot be limited by concentration polarisation in the electrolyte. Transition times do not arise and porosity (V_e) is not so critical to cathode performance as is the case for the cathodes of Leclanché batteries. This is the underlying reason why graphite, which at a given conductivity permits a greater MnO_2 content at the expense of porosity than does acetylene black,[13] can be used in alkaline manganese batteries.

In the later stages of discharge, cathodic charge transfer may occur at least in part by the reaction

$$MnOOH + H_2O + e \rightarrow Mn(OH)_2 + OH^- \quad (3.46)$$

Kozawa and Yeager[70,72] have demonstrated that this reaction, which is heterogeneous, occurs by a dissolution - precipitation mechanism. Mn^{3+} is soluble in potassium hydroxide solutions;[142] in 9·0 molar KOH the saturation concentration of Mn^{3+} is 0·0044 molar[142] but in 1·0 molar KOH the solubility is probably an order of magnitude lower. It is, therefore, significant that, under the conditions employed by Kozawa and Yeager,[70] reaction 3.46 proceeded in 9·0 molar KOH but not in 1·0 molar KOH. It could, however, be made to take place in 1·0 molar KOH by the addition of triethanolamine which complexes with Mn^{3+}, thereby increasing its solubility. The mechanism of reaction 3.46 is dissolution

$$MnOOH + H_2O \rightarrow Mn^{3+} + 3OH^- \quad (3.47)$$

diffusion of Mn^{3+} ions to the electronic conductor, graphite, charge transfer at the graphite

$$Mn^{3+} + e \rightarrow Mn^{2+} \quad (3.48)$$

and precipitation of manganous hydroxide

$$Mn^{2+} + 2OH^- \rightarrow Mn(OH)_2 \quad (3.49)$$

On the basis of an increasing limiting current density with decreasing particle size of the manganese oxyhydroxide, Kozawa and Yeager concluded that the limiting process was the dissolution reaction 3.47. A dissolution - precipitation mechanism explains the cementation of the electrode structure observed by Bell and Huber[60] when discharge reached the stage at which reaction 3.46 set in and also makes comprehensible the inability of reduced manganese oxyhydroxides to be anodically oxidised back to MnO_2 from oxidation states less than $MnO_{1{\cdot}5}$.[128]

3.8.3 Mercury cells

Mendzheritskii and Bagotski,[143] and Ruetschi[144] have concluded that the cathodic reduction of mercuric oxide

$$HgO + H_2O + 2e \rightarrow Hg + 2OH^- \qquad (3.50)$$

which is heterogeneous, occurs by dissolution of the mercuric oxide

$$HgO + H_2O \rightleftharpoons Hg(OH)_2 \qquad (3.51)$$

migration of the dissolved species to an electronic conductor which is joined electrically with the positive terminal and charge transfer at the surface of the electronic conductor

$$Hg(OH)_2 + 2e \rightarrow Hg + 2(OH)^- \qquad (3.52)$$

The most compelling evidence in support of this mechanism is the appearance of the small mercury globules, i.e. those which have just nucleated on the surface of the electronic conductor at positions remote from HgO particles.[143,144] Markov, Boynov and Toschev have studied the cathodic growth of mercury globules from mercurous nitrate solution.[145]

The solubility of HgO in concentrated alkali is low: Garrett and Hirschler[146] report a molarity of only 0·00032 in 5m NaOH. A Leclanché cathode requires a concentration of dissolved zinc species at least three orders of magnitude greater than this in order to sustain discharge without excessive solution concentration polarisation at the charge transfer surface. The explanation of this difference is probably to be found in the area of the charge transfer surface. In a Leclanché cathode this area is confined to regions close to actual carbon-MnO_2

contact points and is therefore small, particularly since the mixed entities in the structure are probably agglomerates of MnO_2 or carbon black[147] so that the contacts are limited to the surfaces of the agglomerates. In the mechanism proposed for HgO cathodes all the surface area of the electronic conductor is available for charge transfer. Obviously this could reduce the charge transfer current density relative to the Leclanché cathode by orders of magnitude. It is clear from eqn. 3.40 that current density has equal and inverse status to concentration in determining transition times.

The identity of the dissolved mercury species is not well established. The solubility of HgO is fairly insensitive to alkali concentration and therefore Ruetschi[144] suggested $Hg(OH)_2$ in preference to an anionic species.

Under severe conditions of discharge, not normally experienced in civilian applications, it has been suggested[143] that the phenomenon limiting the capacity of HgO cathodes is perturbation of hydroxyl ion concentration to such an extent that crystallisation sets in. This seems unlikely, as hydroxyl ions have high mobility and a high transference number out of the charge transfer zone. It is more probable that a restriction in discharge capacity is caused by the concentration of dissolved mercury species falling to zero at the charge transfer surface.

3.8.4 *Silver (Ag_2O) cells*

Ag_2O has some solubility in alkaline solutions.[148–153] Although other formulas have been proposed,[148–150] the conclusion of Antikainen, Hietanen and Sillen[154] from a critical potentiometric study that the dissolved species is $Ag(OH)_2^-$ has received general assent. These workers also concluded that $Ag(OH)_2^-$ is the only dissolved silver species on the evidence that solubility is proportional to hydroxyl ion activity. This may need modification in view of the broad solubility maximum found subsequently by Amlie and Ruetschi[151] at about 6 molar KOH. Miller[155] has provided support for a maximum in solubility from a ring-disc electrode study, although earlier investigations did not find a maximum.[150,153] There is disagreement on whether the solubility of Ag_2O is affected by zincate.[151,153]

The concentration of $Ag(OH)_2^-$ in alkaline solutions of battery interest is about 0·0004 molar at ambient temperatures.[148–151] In terms of gram equivalents per litre this solubility is two-thirds that reported for HgO.[146] Thus, if a dissolution-precipitation mechanism is operative for HgO cathodes it is also likely for Ag_2O cathodes. The probable reaction sequence is dissolution of Ag_2O

$$Ag_2O + 2OH^- + H_2O \rightleftharpoons 2\,Ag(OH)_2^- \tag{3.53}$$

migration of the dissolved species to an electronic conductor which is joined electrically with the positive terminal and charge transfer at the surface of the electronic conductor

$$Ag(OH)_2^- + e \rightarrow Ag + 2OH^- \tag{3.54}$$

As discharge proceeds the surface area of the electronic conductor increases as a result of deposition of silver.

Dissolution-precipitation and solid state mechanism should be distinguishable by visual comparison of the shape and size of discharge products in relation to the shape and size of the original reactant. No similarity is likely if discharge occurs by a dissolution-precipitation mechanism while a close relationship is expected for discharge by a solid state mechanism. It is thus significant that Wales and Simon[156] found the size of deposited silver particles to be controlled, not by the size of oxide particles, but by the magnitude of the discharge current; the larger particles being obtained at the lower currents. Wales has confirmed this important finding several times.[157–160] Wales and Simon[156] also noted a cavity around a growing silver particle and attributed this to the lower volume occupied by Ag relative to Ag_2O and to the dissolution of oxide prior to charge transfer. Wales has also confirmed this observation.[160] Effects found in an X-ray diffraction study of a planar working electrode are also more in accord with a dissolution-precipitation mechanism.[161]

The larger silver particles obtained at the lower current densities[156] may be a consequence of nucleation overpotential with relatively few growing centres being nucleated at low current densities. Although on galvanostatic discharge at ambient temperature initial voltage depressions, which might be due to nucleation overpotential, are not readily apparent,[162] Gagnon and Austin[162,163] have found such depressions at $-40°C$. Recent fundamental studies have considered the cathodic deposition of silver from soluble salts onto single crystal silver,[164] graphite[165] and platinum.[166] Miller[155] applied voltammetric sweeps to a rotating ring-disc electrode in an informative investigation. Using cathodic sweeps with electroformed Ag_2O on the disc he showed that prior to Ag_2O reduction at the disc there was constant response at the ring over a potential range of 0·2V; this response corresponded to reduction of a saturated solution of $Ag(OH)_2^-$. Precisely at the point when disc current begins to rise for the sharp reduction peak of Ag_2O, response at the ring appears to fall to zero (see Reference 155, Fig. 1). This would be strong evidence that Ag_2O reduction proceeds via a dissolved species but for the sharpness of the disc reduction peak which gives rise to some uncertainty. A slight shift in the position at which the ring current commences to decrease, so that it corresponded to the completion of the disc reduction peak would merely indicate a fall of ring current, due to removal of Ag_2O from the disc (see Reference 155, Fig. 2). Clarification is obviously desirable.

More attention has been directed to the reverse process: the formation of Ag_2O by anodic oxidation of silver. There is substantial evidence that this process, particularly in the early stages, proceeds by the dissolution-precipitation route.[155, 167–171] The oxidation shows similarities to the anodic dissolution of zinc, which will be discussed shortly, in that solutions supersaturated in the dissolved species are produced,[155] transition times to passivation conform to Sand's equation,[172] and, depending on the convective conditions, films with two types of appearance are formed.[171,173]

In conclusion, it seems certain that cathodic reduction of Ag_2O to Ag can proceed via the soluble species $Ag(OH)_2^-$. What is less certain is whether the reduction proceeds only by this route or whether there is also a parallel solid state mechanism, as suggested by some authors.[167,174] Migration of O^{2-} ions through the Ag_2O phase has been suggested in the solid state mechanism.[174,175]

3.9 Anode discharge mechanisms

3.9.1 Leclanché cells

The simplest case to consider is the discharge of planar zinc in zinc chloride solutions. As a result of the charge transfer reaction at the zinc surface, dissolved zinc species are produced in the adjacent solution at a rate which is faster than their removal by electrical transference and diffusion. In consequence, the concentration of dissolved zinc species at the zinc surface increases. The model on which eqn. 3.39 is based is particularly appropriate for this situation. Agopsowicz *et al.*[140] have confirmed that eqn. 3.39 with B equal to -1 described experimentally observed behaviour reasonably well. With the approximation $\theta = 1/V_e$, the concentration of dissolved zinc species at the charge transfer surface c_{Zn} (mol/cm^3) is given by

$$c_{Zn} = \bar{c}_{Zn} + \frac{i(1 + t_{Zn})}{FV_e^2} \left(\frac{t}{\pi D_{Zn}} \right)^{1/2} \tag{3.55}$$

where $\bar{c}_{Zn}$ (mol/cm^3) is the bulk concentration of dissolved zinc species, t_{Zn} is an 'average' zinc transference number towards the zinc in the diffusion zone, D_{Zn} (cm^2/s) is the diffusion coefficient in free solution, V_e is the volume fraction of electrolyte, i (A/cm^2) is the current density of charge transference and t(s) is the time the current has been flowing. Eqn. 3.55 shows the importance of V_e, in other words the porosity of the separator, in controlling the increase in zinc chloride concentration.

The development of the concentration profile of dissolved zinc has been established experimentally by analysis of five separator layers placed against a dissolving zinc surface,[140] as is shown in Fig. 3.23.

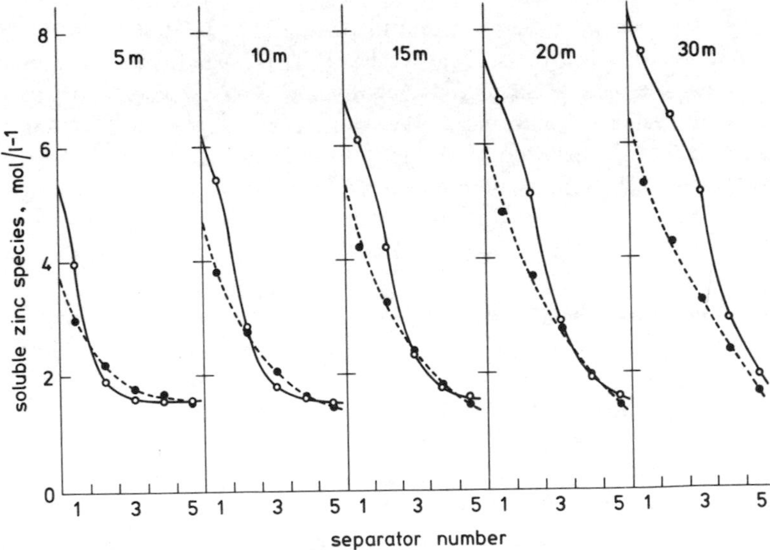

Fig. 3.23 *Concentration profiles of dissolved zinc species in Leclanché cells discharging at 0·022 A/cm^2 (from reference 140)*

—●— zinc chloride electrolyte

—○— normal Leclanché electrolyte

[Courtesy Academic Press Inc., London]

Fig. 3.23 also shows that the increase in concentration of dissolved zinc species adjacent to the zinc surface is greater in normal Leclanché electrolyte ($NH_4Cl/ZnCl_2$) than it is in zinc chloride electrolyte. This is due to differing values of t_{Zn}. In Leclanché electrolyte the dissolved zinc is present as $ZnCl_4^{2-}$, which results in zinc being electrically transported into the diffusion zone.[140] In zinc chloride electrolyte, zinc is present as both anionic and cationic species and the net effect is electrical transport of zinc out of the diffusion zone.[140]

A further difference between the two electrolytes is the shape of the concentration profile of the dissolved zinc species. In zinc chloride electrolyte the shape is that resulting from a classical diffusion controlled situation. In normal Leclanché electrolyte the profile is more complex with two inflections. This complexity is caused by the behaviour of NH_4^+ ions which move away from the zinc surface in the manner of an autogenic moving boundary experiment in which NH_4^+ and cationic zinc species are the leading and following ions, respectively. This is illustrated in Fig. 3.24, which indicates that the inflections correspond with the start and finish of the diffuse moving boundary.

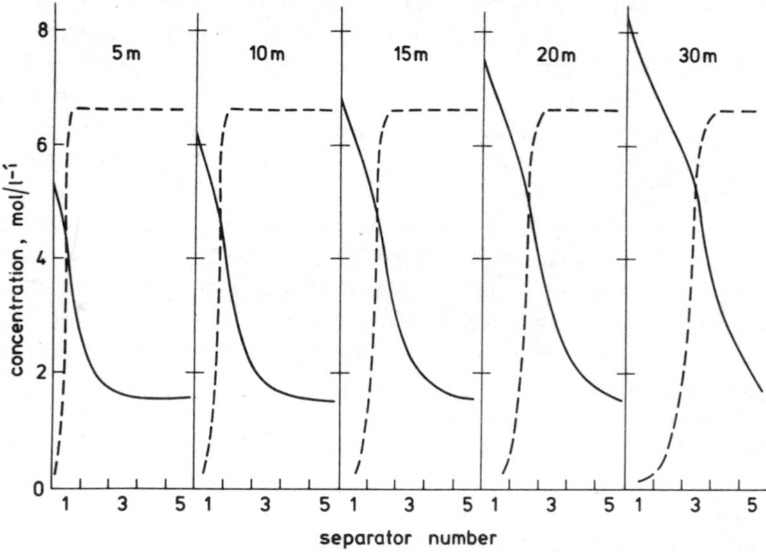

Fig. 3.24 *Concentration profiles in Leclanche cells discharging at 0·022A cm^{-2} (Reference 40)*

— dissolved zinc species

-- dissolved NH_4^+ ions

[Courtesy Academic Press Inc., London]

3.9.2 Alkaline cells

In alkaline solutions the predominant dissolved zinc species is $Zn(OH)_4^{2-}$. This has been established by equilibrium potential measurements,[176–178] solubility studies,[178,179] nuclear magnetic resonance,[180] Raman and infrared reflection spectra.[181,182] As shown by Dirkse[177] and confirmed by Boden, Wylie and Spera[178] the potential determining reaction is

$$Zn + 4OH^- \rightarrow Zn(OH)_4^{2-} + 2e \tag{3.56}$$

High concentrations of $Zn(OH)_4^{2-}$ can exist in concentrated alkali solutions. The solubilities of ZnO and $Zn(OH)_2$ increase with increasing concentrations of alkali;[75,179,183] the concentrations of $Zn(OH)_4^{2-}$ derived from solutions saturated with $Zn(OH)_2$ being approximately double those derived from solutions saturated with ZnO.[75] The saturation concentration of $Zn(OH)_4^{2-}$ with respect to ZnO is remarkably independent of temperature.[76] Supersaturation of alkali solutions with respect to ZnO by electrochemical dissolution of zinc is well established.[182–187] The concentrations of $Zn(OH)_4^{2-}$ that can be generated electrochemically are similar to those obtained by saturation with $Zn(OH)_2$. This may be understood if the dissolution reaction 3.56 proceeds via a transient $Zn(OH)_2$ intermediate, as has been concluded by a number of workers on the basis of detailed mechanistic studies.[188–190]

Solutions of $Zn(OH)_4^{2-}$ which are supersaturated with respect to ZnO are surprisingly stable. Zinc oxide precipitates only slowly and Dirkse[184] has reported that in some instances standing for more than a year was necessary before the concentration of $Zn(OH)_4^{2-}$ decreased to the saturation composition. This stability has prompted suggestions that the extra zinc may not be present as $Zn(OH)_4^{2-}$ and in support of this Jackovitz and Langer found that supersaturation increased the intensity of Raman spectra less than expected.[182] However, they failed to find evidence of other dissolved species. Colloidal species[185] are ruled out on the basis that conductivity changes progressively with increasing supersaturation[184] and that nuclear magnetic resonance behaviour shows a smooth transition from unsaturated to supersaturated solutions.[191] As potential response is in conformity with eqn.3.56, it seems reasonable to conclude that $Zn(OH)_4^{2-}$ is the species present even in supersaturated solutions.

Neither seeding nor shock hastens the precipitation of ZnO from supersaturated solutions.[192] Clearly there is an activation energy barrier to the decomposition of zincate ions by the process

$$Zn(OH)_4^{2-} \rightarrow ZnO + 2OH^- + H_2O \tag{3.57}$$

which is not lowered by the presence of ZnO. The activation energy barrier thus probably relates to rearrangement of the tetrahedrally symmetrical zincate ions to form an activated complex rather than to the nucleation of ZnO. The observed slow rate of exchange of labelled zinc between solid ZnO and zincate ions[193] is in accord with an activation energy barrier.

On continued electrochemical dissolution of zinc, a stage is reached at which precipitation commences in the electrolyte.[184,187] For Russian workers this stage marks the dividing line between the primary and secondary phases of discharge.[185,187] The precipitate which is loose and flocculent and may settle onto the zinc surface is the type I film of Powers and Breiter.[194] It is likely that precipitation commences when the $Zn(OH)_4^{2-}$ concentration exceeds saturation with respect to $Zn(OH)_2$ which implies that the decomposition

$$Zn(OH)_4^{2-} \rightarrow Zn(OH)_2 + 2OH^- \tag{3.58}$$

is less inhibited than decomposition direct to ZnO by reaction 3.57. The critical $Zn(OH)_4^{2-}$ concentration at which reaction 3.58 commences must be reached first of all at the zinc surface. However, the flocculent non-blocking nature of the precipitate indicates that reaction 3.58 does not occur immediately and so reaction 3.58 may also take place via an activated complex which allows time for some diffusion away from the zinc surface before there is deposition.

The precipitate is normally ZnO,[184,195,196] indicating a reasonably facile dehydration of $Zn(OH)_2$ in concentrated alkaline solutions. In some circumstances, notably at low temperatures and in less concentrated alkaline solutions, the precipitate may be $Zn(OH)_2$.[185,195,197] Huetig and Moeldner[198] found that the instability of $Zn(OH)_2$ with respect to ZnO increased with increasing concentration of alkali. There are amorphous and several crystalline forms of $Zn(OH)_2$.[199] Dietrich and Johnston[200] reported that the most stable form, ϵ - $Zn(OH)_2$, does not transform readily into ZnO. Both ϵ and γ crystalline forms have been identified after secondary phase discharge[185,197] and a plausible mechanism is that γ -$Zn(OH)_2$ is precipitated, and, according to conditions, this remains or is transformed into ZnO or ϵ -$Zn(OH)_2$

If the primary phase of discharge ends when the concentration of $Zn(OH)_4^{2-}$ reaches a critical concentration c'_{crit} mol/cm^3 at the zinc surface, the factors affecting the primary phase capacity may be gauged by application of eqn. 3.39. Thus the primary phase capacity per cm^2 of charge transfer interface $i\tau'$ is

$$i\tau' = \frac{\pi D_{Zn(OH)_4^{2-}}}{i} \left(\frac{FV_e}{\theta}\right)^2 \left(\frac{c'_{crit} - \bar{c}_{Zn(OH)_4^{2-}}}{1 + t_{Zn(OH)_4^{2-}}}\right)^2 \qquad (3.59)$$

where τ'(s) is the time for the concentration of zincate at the zinc surface to increase from the initial bulk value $\bar{c}_{Zn(OH)_4^{2-}}$ mol/cm^3 to the critical value and B in eqn. 3.39 is -1 in accordance with reaction 3.56. The capacity in the primary discharge phase is inversely related to current density, as is clearly apparent from the results of Arkhangel'skaya, Andreeva and Mashevich.[187] Their results, however, do not show inverse proportionality, and the inadequacy of eqn. 3.59 is attributed to the failure of the semi-infinite diffusion model on which eqn. 39 is based. The maximum capacity in the primary phase of discharge is obtained at low current densities and may be estimated from the amount of charge transfer required to increase the concentration of $Zn(OH)_4^{2-}$ in the electrolyte throughout the anode structure to the critical value. A reasonable maximum estimate of the critical concentration is 3 molar,[186,187] which, in conjunction with the data in Fig. 3.7, leads to the conclusion that primary phase discharge constitutes less than 3% and 14% of the total capacity of powder and gelled zinc anodes, respectively.

At the beginning of the secondary phase of discharge, active dissolution continues and the situation at the charge transfer surface is similar to that existent during the primary phase. Zincate ions are produced in the electrolyte adjacent to the zinc surface and their concentration increases toward a steady state in which the amount of $Zn(OH)_4^{2-}$ removed by diffusion, convection, and, in the secondary phase, by deposition, balances the amount produced by charge transfer and by ionic transport through the electrolyte to the surface layer. In battery electrodes, convection is usually assumed absent, although careful observations[194,201] have revealed the presence of small convection currents even on horizontal planar electrodes which are used for model studies of quiescent conditions. It is worthy of note that at this stage no limitation of charge transfer current can arise as a consequence of the $Zn(OH)_4^{2-}$ diffusion. The greater the charge transfer current density the higher must be the $Zn(OH)_4^{2-}$ concentration at the zinc surface and this enables a greater quantity of $Zn(OH)_4^{2-}$ to migrate by diffusion. However, the total charge transfer current, at a given potential, is affected by the nature of the zinc surface. On pure zinc, polycrystalline or single crystal, the charge transfer current is localised with zinc dissolution taking place at grain boundaries,[201] kink sites or emergent dislocations. More uniform dissolution occurs with alloys[201] or by amalgamation, and higher total active dissolution currents are achieved in consequence.[202-204] The accumulation of ZnO during the secondary phase of discharge may play an important role in providing a route along which electrons are removed from dispersed zinc particles after charge transfer. Its other effect is undesirable: tortuosity (θ) and volume fraction of electrolyte (V_e) in eqn. 3.39 are increased and decreased, respectively, and in consequence, under galvanostatic conditions, the concentration of $Zn(OH)_4^{2-}$ at the zinc surface increases, causing an increase in the Nernst potential of reaction 3.56. The amount of type I precipitate formed in a pelleted zinc anode discharged to 25% of its total capacity has been shown to increase towards the cathode and be more non-uniform the higher the discharge current.[205] The next stage is reached at a critical potential which permits a solid state reaction to take place so that there is direct electrochemical formation of an oxide or hydroxide on the zinc surface. This is the type II film of Powers and Breiter.[194] A change in the nature of the anodic process can be clearly demonstrated by linear sweep voltammetry. Under quiescent conditions a current peak is produced,[194,196,201,203,205-211] while in a convective situation there is a distinct change in the slope of the voltammogram.[194,203,209,212,213] Powers has described conditions for observing the skinlike nature of the type II film.[201]

Despite the presence of type II film, voltammetric sweep studies show that pure zinc electrodes are able to sustain substantial anodic currents, both under convective[203,206,209,212] and quiescent conditions,[194,196,201,203,205,207,209-211] up to potentials which are more positive by about 0·15V than the potential of initial passivation. Armstrong and Bulman[206] have referred to this as the region of passive dissolution. Interestingly, Hull and Toni[203] found that the passive dissolution region was absent if the zinc was amalgmated and that active dissolution occurred in the region normally associated with passive dissolution until final passivation set in at a potential of −1·05V (Hg/HgO, 1N KOH). The passive dissolution region may be explained on the basis of a spectrum of Gibbs free energy changes for the reaction

$$Zn + 2OH^- \rightarrow ZnO + H_2O + 2e \qquad (3.60)$$

The spectrum arises from the differing Gibbs free energies of various surface zinc atoms and the differing degrees of strain energy incorporated into ZnO formed directly on zinc[209] in different locations. A number of authors have discussed the strain energy that occurs on deposition of a coherent crystalline phase on the surface of another crystal when the two have different lattice parameters.[214–217] Vermilyea has commented[214] that, if a film nucleus is coherent and strained 1%, then the potential where it forms may differ from the normal equilibrium potential by tenths of a volt. Evans[218] considers that wrinkling of a film after separation from the substrate crystal is evidence of strain in the film as deposited. So the observation by Powers[196] of 'cobweb' formation during dissolution of type II film may be significant. The near perfect reversibility found by Hull, Ellison and Toni[212] for voltammograms under some conditions suggests an explanation in energetic terms. A small peak indicative of a nucleation process is present in their voltammograms[212] at the start of passivation. Thus, in the passive dissolution region, part of the surface is passivated by a type II film while the remainder undergoes active dissolution. Indeed, Powers[201] has observed situations in which type II film was present only at grain boundaries. The correspondence between anodic and cathodic currents in ring-disc experiments[212] shows that under convective conditions little anodic current is expended in thickening the type II film in the passive dissolution region. Final passivation occurs when a type II film covers the whole zinc surface. This explanation is supported by the observation, on open-circuit recovery after anodic polarisation, that transient potential arrests indicative of zinc/film potentials occurred only if the polarisation was taken beyond the point of complete passivation.[207,219] Current oscillations noted near to the point of final passivation[212,219,220] are also explained in terms of completion of a film and its partial dissolution when the drop in current allows a change in the composition of the electrolyte at the anodic surface. At potentials more positive than the final passivation potential, anodic current can only be sustained by the diffusion of ions through the type II film. Hull, Ellison and Toni[212] noted that at final passivation the film became black, which suggests that the diffusing species may be zinc.

Huber,[197] Powers,[196] and Sato, Niki and Takamura[221] found the passivating film (type II) to be ZnO as represented in eqn. 3.55 while Iofa, Mirlina and Moiseeva[185] concluded that passivation was caused by ϵ-$Zn(OH)_2$. ZnO with 1:1 ratio of Zn:O should be able to form as a contiguous layer more readily than $Zn(OH)_2$. On the basis that the variation in the Gibbs free energy of reaction 3.55 resides entirely in the strain energy of the contiguous ZnO layer, the Gibbs free energy per mole of ZnO in the layer is readily calculated from the potentials of intial and final passivation. Published data show variations[194,196,201,203,205–207,212] but representative values are 1•23V and 1•06V versus HgO/Hg in the same electrolyte for initial and final passivation, respectively, which lead to Gibbs free energies of −70•7 and −62•9 k cal/mol, compared to −76•4 k cal/mol for normal ZnO.

A number of investigations into the passivation times of zinc in alkaline electrolyte have been carried out galvanostatically.[186,208,221–233] Studies of this type can relate closely to conditions during battery discharge. At passivation the potential of planar electrodes increases quite sharply and substantially[221,223,226,234,235] so that a distinction between initial and final passivation is of little significance in regard to passivation time. No detail is discernible in potential-time plots prior to the increase at passivation,[221,223,226,234,235] and so it is probably correct to assume that the advent of initial passivation increases current density in those regions of zinc still undergoing active dissolution, with the result that type II film is soon generated on the remaining active regions and passivation is complete. Initial passivation thus signifies the end of useful discharge in batteries.

Passivation commences when the concentration of $Zn(OH)_4^-$ at the active charge transfer surface reaches a critical concentration such that the Nernst potential of reaction 3.51 is at the potential at which direct electrochemical formation of a strained contiguous surface layer of oxide (or hydroxide) occurs. This second critical $Zn(OH)_4^{2-}$ concentration c''_{crit} mol/cm^3 (the first being at the onset of the secondary

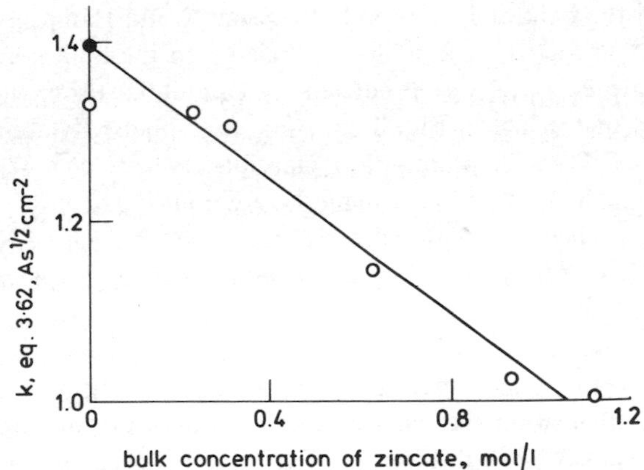

Fig. 3.25 *k (equation 3.61) as a function of zincate concentration*

● reference 223

○ reference 228

phase of discharge) is in effect the saturation solubility of the strained surface oxide. On the above basis the semi-inifinite diffusion model eqn. 3.39, leads to the following relationship between passivation time τ_p and charge transfer current density $i(\mathrm{A/cm^2})$

$$i\tau_p^{1/2} = k \tag{3.61}$$

where

$$k = \frac{F\pi^{1/2} V_e \, D^{1/2}_{Zn(OH)_4^{2-}} \left(c''_{crit} - \bar{c}_{Zn(OH)_4^{2-}}\right)}{\theta \left(1 + t_{Zn(OH)_4^{2-}}\right)} \tag{3.62}$$

The transference number $t_{Zn(OH)_4^{2-}}$ is probably small compared to unity so that its variation with concentration of $Zn(OH)_4^{2-}$ does not invalidate the derivation of eqn. 3.61. A number of authors have cited eqn. 3.62 (without V_e and θ terms) with the omission of $t_{ZN(OH)_4^{2-}}$,[186,219,221,223,228,229,231,232] which is one way of dealing with its rather ill-defined value in a diffusion zone. The inverse proportionality of $\tau_p^{1/2}$ and i required by eqn. 3.61 has been confirmed when experimental conditions are chosen so that convection effects are minimal and the total amount of discharge is insufficient for the precipitation of type I film to cause significant departures of tortuosity (θ) and volume fraction of electrolyte (V_e) from unity.[221,223,227,228,232] Fig. 3.15, plotted from the data of Dirkse and Hampson[223] and Hampson, Shaw and Taylor,[228] shows that k is linearly related to the bulk concentration of zincate $\bar{c}_{Zn(OH)_4^{2-}}$ as predicted by eqn. 3.62. From the slope and intercept of the line in Fig. 3.25, c''_{crit} for 7 molar KOH has been calculated as 3·7 molar which is a plausible value.[219,228] With this value of c''_{crit} and assuming V_e and θ are unity, $D_{Zn(OH)_4^{2-}}/(1 + t_{Zn(OH)_4^{2-}})^2$ has been calculated from a $\tau_p^{1/2}$, i plot in 7 molar KOH[223] as $4{\cdot}8 \times 10^{-6}$ cm²/s. This is in accord with the directly determined value of $D_{Zn(OH)_4^{2-}}$ in 7 molar KOH[236] of 6×10^{-6} cm²/s and a value of $t_{Zn(OH)_4^{2-}}$ of about 0·1. In other words, short time passivation behaviour on planar electrodes under quiescent conditions can be completely explained on the basis of semi-infinite diffusion theory and the onset of passivation at a critical $Zn(OH)_4^{2-}$ concentration at the charge transfer surface.

By analogy with the behaviour of ZnO and $Zn(OH)_2$[75,179,183] the saturation solubility of the strained surface oxide c''_{crit} should increase more than proportionately with increase in alkali concentration and up to 7 molar KOH changes in k[223] do appear to reflect such presumed solubility changes as required by eqn. 3.62. At higher concentrations k decreases,[223,227,228] probably due to $D_{Zn(OH)_4^{2-}}$ decreasing at the higher concentrations of alkali.[236,237] The broad maximum in k representing maximum passivation times in the concentration region 7-12 molar KOH has been noted several times,[222,223,226,227,234] including a study with porous zinc[224] and has obvious significance for battery performance.

Sato, Niki and Takamura[221] have shown that passivation times in 8 molar NaOH are approximately half of those in 8 molar KOH and that this is largely due to the effect of viscosity on $D_{Zn(OH)_4^{2-}}$ in eqns. 3.61 and 3.62. Passivation times are substantially reduced by the addition of K_2CO_3 and this is similarly explained.[221]

Hampson, Shaw and Taylor[228] pointed out that at passivation the molar zincate concentration at the zinc surface c''_{crit} was approximately half that of the molarity of KOH in the bulk solution. Their conclusion has been confirmed recently.[238] The inference drawn from this and the reverse of reaction 3.57 was that the hydroxyl ion concentration at the zinc surface is zero at passivation and the zero concentration is the real cause of passivation. This argument neglects hydroxyl ion diffusion which commences immediately there is depletion of hydroxyl ion at the zinc surface due to formation of zincate. Application of eqn. 3.39 to both hydroxyl and zincate ions leads to the relationship

$$\frac{2\Delta c_{Zn(OH)_4^{2-}}}{\Delta c_{OH^-}} = -\left(\frac{D_{OH^-}}{D_{Zn(OH)_4^{2-}}}\right)^{1/2}\left(\frac{1 + t_{Zn(OH)_4^{2-}}}{2 - t_{OH^-}}\right) \tag{3.63}$$

where Δc is the difference between surface and bulk molarity. By inserting diffusion coefficients for 7 molar KOH[236] and assuming reasonable values for $t_{Zn(OH)_4^{2-}}$ of 0·1 and for t_{OH^-} of 0·7, the following useful approximation is obtained:

$$\Delta c_{OH^-} \simeq - \Delta c_{Zn(OH)_4^{2-}} \tag{3.64}$$

From eqn. 3.64 it is apparent that in 7 molar KOH hydroxyl ion concentration at the zinc surface is not zero when the zincate concentration has risen to the critical level of 3·5 - 3·7 molar.

Under convective conditions it is possible for a pseudo-steady state to be established whereby the $Zn(OH)_4^{2-}$ concentration at the charge transfer surface never increases to the critical concentration c''_{crit} and hence passivation does not occur. Empirical modification[239] of eqn. 3.61 to

$$(i - i_L)\tau_p^{1/2} = k \qquad (3.65)$$

has enabled such data to be represented,[186,225,226,230,231,235] where i_L (A/cm^2) is the limiting current density below which there is no passivation. As expected the magnitude of the limiting current density is dependent on the convective conditions pertaining.[186,235] k appeared independent of convective conditions in a study with flowing electrolyte,[235] but this cannot be strictly true, as the values of k obtained are approximately half of those obtained under quiescent conditions with horizontal electrodes.[228]

Acknowledgments

The author's debt to those workers who have published is readily acknowledged in the long list of references. His debt to colleagues and other scientists with whom he has had discussions during 15 years in the battery industry is less easily discharged. It is possible to name only a few.

The author's understanding of international standardisation derives almost entirely from association with G.S. Bell (Varta, Batterie AG), A. Gilbert (formerly of S.A.F.T.) and R.B. Jay on the editing committee of IEC TC 35. A. Agopsowicz and F.D.S. Baker have added perspective and perception to internal discussions over many years and have often brought the author back to paths of logic and reality from his greater flights of fancy. Together with R. Barnard they kindly read and commented on the manuscript.

The author thanks the Directors of Berec Group Limited for permission to publish the article. Much unpublished data has been used and the author appreciates the help of the following in obtaining the data and in processing it to the form required: D. Ashton, P.A. Gardiner, D.S. Freeman, J. Neate, S.W. Overall and D.B. Ring. The author is also in debt to D.J. Parfitt and H. Thornton, both of S.A.F.T. (UK) Ltd., for augmenting his meagre knowledge of neutral air-depolarised batteries.

Penultimately, the author thanks R.C. King and his staff for the excellent drawings, V.J. Overall for struggling to improve the author's written style, D.H. Spencer for checking all references and J.C. Griffiths for typing, checking and general assistance on the project. Finally, the author apologises to his family for his considerable underestimate of the extent to which writing this chapter would impinge on his home life.

References

1 KORDESCH, K.V. (Ed.): *Batteries – Vol. 1. Manganese dioxide* (Marcel Dekker Inc., New York, 1974)

2 HEISE, G.W., and CAHOON, N.C. (Eds.): *The primary battery – Vol. 1* (John Wiley & Sons Inc., London, 1971)

3 CAHOON, N.C., and HEISE, G.W. (Eds.): *The primary battery – Vol. 2,* (John Wiley & Sons Inc., London, 1976)

4 US Government Standard Specification, No. 58, 1923

5 Bureau of Standards Circular, No. 79, 1919, p.39; No. 79, 2nd edn., (1923), p.51

6 American National Standard C18.1, 1972

7 International Electrotechnical Commission, Publication 86, 1957

8 International Electrotechnical Commission, Publications 86-1, 4th edn., 1976 plus Amendment No. 1, 1978, and 86-2, 4th edn., 1977, plus Amendments No. 1, 1978, and No. 2, 1979

9 British Standards Institution, BS 397, 1976

10 HERBERT, W.S.: 'The alkaline manganese dioxide dry cell', *J. Electrochem. Soc.,* 1952, **99**, pp.190-191c

11 MRGUDICH, J.N., and CLOCK, R.C.: 'X-ray and electron microscope evaluation of carbon black', *Trans. Electrochem. Soc.,* 1944, **86**, pp.351-364

12 WATSON, J.H.L.: 'Observations of crystal structure and particle shape in electron micrographs of several carbon blacks', *ibid.,* 1947, **92**, pp.77-90

13 CAUDLE, J., RING, D.B., and TYE, F.L.: 'Physical properties of mixtures of carbon and manganese dioxide, *in* COLLINS, D.H.(Ed.) *Power sources 3* (Oriel Press, 1971), pp.593-606

14 KING, B.H.: US Patent 2 593 893, 1952

15 KEMP, K.T.: 'Description of a new kind of galvanic pile and also of another galvanic apparatus in the form of a trough', *Edinburgh New. Phil. J.,* 1823, **6**, pp.70-77

16 STURGEON, W.: *Recent experimental researches on electromagnetism and galvanism* (Sherwood, Gilbert and Piper, 1830)

17 AUFENAST, F.: 'The quality of zinc for dry batteries'. Proceedings of the International symposium on batteries, Christchurch, 1958, Paper (b)

18 SWIFT, J., TYE, F.L., WARWICK, A.M., and WILLIAMS, J.T.: 'Amalgamation of zinc anodes in Leclanché dry cells' *in* COLLINS, D.H.(Ed.) *Power sources 4* (Oriel Press, 1973), pp. 415-435

19 PLETENAVA, N.A., and FEDOSEEVA, N.P.: 'The effect of temperature on The diffusion of mercury in zinc', *Doklady Akad. Nauk, SSR,* 1963, **151**, pp. 384-386

20 BATRA, A.P., and HUNTINGDON, H.B.: 'Anisotropic diffusion of mercury in zinc', *Phys. Rev.*, 1967, **154**, pp.569-571

21 MOREHOUSE, C.K., HAMER, W.J., and VINAL, G.W.: 'Effect of inhibitors on the corrosion of zinc in dry cell electrolytes', *J.Res. Nat. Bur. Standards,* 1948 **40**, pp.151-161

22 CAHOON, N.C.: US Patent 2 534 336, 1950

23 RUETSCHI, P.: 'Solubility and diffusion of hydrogen in strong electrolytes and the generation and consumption of hydrogen in primary batteries', *J. Electrochem. Soc.*, 1967, **114**, pp.301-305

24 DIRKSE, T.P., and TINIMER, R.: 'The corrosion of zinc in potassium hydroxide solutions', *ibid.*, 1969, **116**, pp.162-165

25 MANSFIELD, F., and GILMAN, S.: 'The effect of several electrode and electrolyte additives on the corrosion and polarisation behaviour of the alkaline zinc electrode', *ibid.,* 1970, **117**, pp.1328-1333

26 BOCKRIS, J.O'M., NAGY, Z., and DAMJANOVIC, A.: 'On the deposition and dissolution of zinc and alkaline solutions', *ibid.*, 1972, **119**, pp.285-295

27 KRUG, H., and BORCHERS, H.: 'Ergebuisse aus Korrisionsantersuchungen an Zinklegierungen', *Electrochim. Acta,* 1968, **13**, pp. 2203-2205

28 BELL, G.S.: 'The stability of zinc in zinc chloride/ammonium chloride electrolytes', *ibid.,* 1968, **13**, pp.2197-2202

29 MOSHTEV, R.V., and STOICHEVA, R.: 'Corrosion in alkali hydroxide solutions of electrolytic zinc powder concerning co-deposited lead', *J. Appl. Electrochem.,* 1976, **6**, pp.163-169

30 CAHOON, N.C., and KORVER, M.P.: 'A film lining for high-capacity dry cells', *J. Electrochem. Soc.,* 1958, **105**, pp.293-295

31 KORDESCH, K.V.: 'Alkaline manganese dioxide zinc batteries' *in* KORDESCH, K.V. (Ed.) *Batteries - Vol. 1. Manganese dioxide* (Dekker, 1974), pp.241-384

32 HUBER, R.: 'Leclanché batteries', *in* KORDESCH, K.V. (Ed.) *Batteries – Vol. 1. Manganese dioxide* (Dekker, 1974), pp.1-240

33 CAHOON, N.C.: 'Leclanché and zinc chloride batteries' *in* CAHOON, N.C., and HEISE, G.W. (Eds.) *The primary battery – Vol. 2* (Wiley, 1976), pp.1-147

34 FRENCH, H.F.: US Patent 2 272 969, 1939

35 FRENCH, H.F.: 'Improvements in B-battery portability', *Proc. IRE,* 1941, **9**, pp.299-303

36 ST. CLAIRE-SMITH, C.R., LEE, J.A., and TYE, F.L.: 'Pore characteristics of a variety of manganese dioxides', *in* KOZAWA, A., and BRODD, R.J. (Eds.) *Manganese dioxide symposium – Vol. 1* (U.C.C.I.C. Sample Office, 1975), pp.132-158

37 BELL, G.S.: 'The relation between service life of Leclanchë type dry cells and the composition of geometry of the cathode', *Electrochem. Technol.,* 1967, **5**, pp.513-517

38 TAKAHASHI, K., KANEDA, Y., and NARUISHI, T.: 'Effects of addition of I.C. MnO_2 samples on performance of silver oxide (Ag_2O) and mercury oxide (HgO) zinc anode button cells' *in* KOZAWA, A., and BRODD, R.J. (Eds.) *Manganese dioxide symposium – Vol. 1* (U.C.C.I.C. Sample Office, 1975), pp.202-229

39 RUBEN, S.: 'Balanced alkaline dry cells', *Trans. Electrochem. Soc.*, 1947, **92**, pp.183-193

40 LEE, J.A., MASKELL, W.C., and TYE, F.L.: 'Separators and membranes in electrochemical power sources' *in* MEARES, P (Ed.) *Membrane separation processes* (Elsevier, 1976), pp. 400-467

41 ANDRE, H.: 'L'accumulator argent-zinc', *Bull. Soc. Fr. Electr.*, 1941, **1**, pp.132-146

42 DIRKSE, T.P., and DE HAAN, F.: 'Corrosion of the zinc electrode in the silver-zinc-alkali cell', *J.Electrochem. Soc.*, 1958, **105**, pp.311-315

43 MALLORY BATTERIES LTD.: British Patent 753 090, 1956

44 KOZAWA, A.: 'Electrochemistry of manganese dioxide' *in* KORDESCH, K.V. (Ed.) *Batteries – Vol. 1. Manganese dioxide* (Dekker, 1974), p.504

45 OHTA, A., WATANABE, J., and FURUMI, I.: 'I.C. $Mn0_2$ samples tested for both Leclanché and zinc-chloride type practical cells', *in* KOZAWA, A., and BRODD, R.J. (Eds.) *Manganese dioxide symposium – Vol. 1*, (U.C.C. I.C. Sample Office, 1975), pp.159-183

46 DALEY, J.L.S.: 'Alkaline zinc-$Mn0_2$ batteries', *Proc. Annual Power Sources Conf.*, 1961, **15**, pp.96-98

47 DANIEL, A.F., MURPHY, J.J., and HOVENDON, J.M.: 'Zinc-alkaline $Mn0_2$ dry cells' *in* COLLINS, D.H. (Ed.) *Batteries* (Pergamon, 1963), pp.175-169

48 CAHOON, N.C.: 'Electrolyte equilibria in relation to dry cell performance', *Trans. Electrochem. Soc.*, 1947, **92**, pp.159-172

49 FRIESS, R.: 'The reactions of ammonia on the system $ZnCl_2$ - NH_4Cl-H_2O', *J. Am. Chem. Soc.*, 1930, **52**, pp.3083-3087

50 MCMURDIE, H.F., CRAIG, D.N., and VINAL, G.W.: 'A study of equilibrium reactions in the Leclanché dry cell', *Trans. Electrochem. Soc.*, 1940, **90**, p.509-528

51 BREDLAND, A.M., and HULL, M.N.: 'Leclanché electrolyte compositional studies for thin film batteries', *J. Electrochem. Soc.*, 1976, **123**, pp.311-315

52 HITCHCOCK, J.L., and PELTER, P.F.: 'A thermo-analytical study of electrodeposited manganese dioxide' *in* BUZAS, I. (Ed.) *Thermal analysis – Vol. 1. Proceedings of the fourth international conference on thermal analysis* (Akademiai Kiado, 1975), pp.979-989

53 GABANO, J.P., MORIGNAT, B., FIALDES, E., EMERY, B., and LAURENT, J.F.: 'Etude de la réduction chimique du bioxyde de manganese γ', *Z. Phys. Chem. Abt. A*, 1965, **46**, pp.359-372

54 BROUILLET, P.H., GRUND, A., and JOLAS, F.: 'Sur la constitution et la decomposition thermique sous vide des bioxydes de manganese varieté γ' *C.R.Acad. Sci.*, 1964, **257**, pp.3166-3169

55 COEFFIER, G., and BRENET, J.: 'Contribution a L'étudedes mechanismes de réduction chimique d'un bioxyde de Manganese type gamma', *Bull. Soc. Chim. Fr.*, 1964, pp.2835-2839

56 LIPPEUS, B.S., and DE BEER, J.H.: 'Studies on pore systems in catalysts', *J. Catal.*, 1965, **4**, pp.319-323

57 LEE, J.A., NEWNHAM, C.E., STONE, F.S., and TYE, F.L.: 'Temperature programmed desorption studies on γ phase manganese dioxide in static water vapor environments', *J. Colloid Interface Sci.*, 1973, **45**, pp.289-294

58 BROWN, A.J., ST.CLAIRE-SMITH, C.R., TYE, F.L., and WHITEMAN, J.L.: 'The dependence of the absorptive properties of a γ phase electro-deposited manganese dioxide on the out gassing temperature', *ibid.*, 1975, **51**, pp.516-521

59 GABANO, J.P., LAURENT, J.F., and MORIGNAT, B.: 'Etude des composes formes dirant la décharge de cellules electrochimiques du type Leclanché', *Electrochim. Acta,* 1964, **9**, pp.1093-1117

60 BELL, G.S., and HUBER, R.: 'On the cathodic reduction of manganese dioxide in alkaline electrolyte', *J. Electrochem. Soc.,* 1964, **111**, pp.1-6

61 BELEY, M, and BRENET, J.: 'Etude des mecanismes de formation de varietes de bioxyde de manganese a haute reactivité electrochimique electrocatalytiques and catalytiques et comparison des proprietes electrochimiques avec les varietes β et γ $Mn0_2$', *Electrochim. Acta,* 1973, **18**, pp.1003-1011

62 CAHOON, N.C.: 'An electrochemical evaluation of manganese dioxide for dry battery use', *J. Electrochem. Soc.,* 1952, **99**, pp.343-348

63 VOSBURGH, W.C., JOHNSON, R.C., REISER, J.S., and ALLENSON, D.R.: 'A further study of electrodeposited manganese dioxide electrodes', *ibid.,* 1955, **102**, pp.151-155

64 CHREITZBERG, A.M., ALLENSON, D.R., and VOSBURGH, W.C.: 'Formation of manganese (II) ion in the discharge of the manganese dioxide electrode', *ibid.,* 1955, **102**, pp.557-561

65 NEUMANN, K., and FINK, W.: 'Die chemische reaktion un Braunsteinelement', *Z.Elektrochem.,* 1958, **62**, pp.114-122

66 VOSBURGH, W.C., PRIBBLE, M.J., KOZAWA, A., and SAM, A.: 'Formation of manganese (II) ion in the discharge of the manganese dioxide electrode II.Effect of volume and pH of the electrolyte', *J. Electrochem. Soc.,* 1958 pp.1-4

67 VOSBURGH, W.C., and PAO-SOONG, L.: 'Experiments on the discharge mechanism of the manganese dioxide electrode', *ibid.,* 1961, **108**, pp.485-490

68 TYE, F..L.: 'Manganese dioxide electrode – III, Relationship between activities of stoichiometry for compositions near to MnO_2', *Electrochim. Acta,* 1976, **21**, pp.415-420

69 CAHOON, N.C., and KORVER, M.P.: 'The cathodic reduction of manganese dioxide in alkaline electrolyte', *J. Electrochem. Soc.,* 1959, **106**, *pp.* 745-750

70 KOZAWA, A., and YEAGER, J.F.: 'The cathodic reduction mechanism of electrolytic manganese dioxide in alkaline electrolyte', *ibid.,* 1965, **112**, pp.959-963

71 KOZAWA, A., and POWERS, R.A.: 'Cathodic reduction of β $-Mn0_2$ and γ $-Mn0_2$ in NH_4C1 and KOH electrolytes., *Electrochem. Technol.,* 1967, **5**, pp.535-541

72 KOZAWA, A., and YEAGER, J.F.: 'Cathodic reduction mechanism of Mn00H to $Mn(0H)_2$ in alkaline electrolyte', *J. Electrochem. Soc.,* 1968, **118**, pp.1003-1007

73 BODEN, D., VENUTO, C.U., WISLER, D., and WYLIE, R.B.: 'The alkaline manganese dioxide electrode', *ibid.,* 1967, **114**, pp.415-417

74 GABANO, J.P., MORIGNAT, B., and LAURENT, J.F.: 'Variation of phys-

icochemical parameters in an alkaline MnO_2–Zn cell during discharge. *in* COLLINS, D.H. (Ed.) *Power sources 1966* (Pergamon, 1967), pp.49-63

75 SCHUMACHER, E.A.: 'Primary cells with caustic alkali electrolyte' *in* HEISE, G.W., and CAHOON, N.C. (Eds.) *The primary battery – Vol. 1,* (Wiley, 1971), pp.169-189

76 DIRKSE, T.P.: 'Chemistry of zinc/zinc oxide electrode' *in* FLEISCHER, A., and LANDER, J.J. (Eds.) *Zinc-silver oxide batteries* (Wiley, 1971), pp.19-29

77 RUBEN, S.: 'The mercuric oxide:zinc cell' *in* HEISE, G.W., and CAHOON, N.C. (Eds.) *The primary battery – Vol. 1* (Wiley, 1971), pp.207-223

78 CAHOON, N.C., and HOLLAND, H.W.: 'The alkaline manganese dioxide: zinc system' *in* HEISE, G.W., and CAHOON, N.C. (Eds.): *The primary battery - Vol. 1* (Wiley, 1971), pp.239-263

79 SCHUMACHER, E.A.: 'The alkaline copper oxide:zinc cell' *in* HEISE, G.W., and CAHOON, N.C. (Eds.) *The primary battery – Vol. 1* (Wiley, 1971), pp.191-206

80 DIVERS, E.: 'On the Leclanché cell and the reactions of manganese oxides with ammonium chloride', *The Chemical News,* 1882, **46,** pp.259-260

81 CHANEY, N.K.: 'Discussion comment' *Trans. Am. Electrochem. Soc.,* 1916, **29,** pp.318-321

82 KELLER, A.: 'Uber das Leclanché-Element', *Z. Electrochem.,* 1931, **37,** pp.342-348

83 ACHREIBER,: Dissertation, Dresden, 1923

84 BRENET, J., MALESSAN, P., and GRUND, A.: 'Sur la dépolarisation dans les cellules electrochimiques au bioxyde de manganese', *C.R. Acad. Sci.,* 1965, **242,** pp.111-112

85 MCBREEN, J.: 'The electrochemistry of manganese oxides in alkaline electrolytes' *in* COLLINS, D.H. (Ed.) *Power sources 5* (Academic Press, 1975), pp.523-534

86 FEITKNECHT, W., OSWALD, H.R., and FEITKNECHT-STEINMANN, U.: 'Uber die topochemische emphasiqe reduction r–MnO_2', *Helv. Chim. Acta,* 1960, **43,** pp.1947-1950

87 BODE, H., SCHMIER, A., and BERNDT, D.: 'Zur Phasenanalyse von Mangandioxyd', *Z.Electrochem.,* 1962, **66,** pp.586-593

88 NEUMANN, K., and VON RODA, E.: 'Untersuchungen uber das Braunsteinpotential un Ammonium chloridlosung', *Ber. Bunsenges, Phys. Chem.,* 1965, **69,** pp.347-358

89 HUBER, R., and BAUER, J.: 'The analysis of discharge curves of Leclanché type dry cells', *Electrochem. Technol.,* 1967, **5,** pp.542-548

90 D'AME, V.N., and MENDZHERITSKII, E.A.: 'Manganese-zinc cells with a horizontal discharge curve', *Elektrokhim, Margantsa,* 1967, **3,** pp.173-178

91 D'AME, V.N., and MENDZHERITSKII, E.A.: 'Manganese dioxide with stable potential. III, Ion-exchange characteristics of manganese dioxides', *Elektrokhimiya,* 1968, **4,** pp.280-285

92 COPELAND, L.C., and GRIFFITH, F.S.: 'The reaction of the Leclanché dry cell', *Trans. Electrochem. Soc.,* 1946, **89,** pp.495-506

93 LUKOVTSEV, P.D., and TEMERIN, S.A.: 'The nature of the potential and the electrochemical behaviour of real oxide electrodes', *Trudy*

Soveschaniya Elektrokhim. Acad. Nauk SSSR, Otdel Khim Nank 150, 1953, pp.494-503

94 JOHNSON, R.S., and VOSBURGH, W.C.: 'Electrodes of mixed manganese dioxide and oxyhydroxide', *J. Electrochem. Soc.,* 1953, **100,** pp. 471-472

95 KOZAWA, A., and POWERS, R.A.: 'The manganese dioxide electrode in alklaline electrolyte; the electron-proton mechanics for the discharge process from $Mn0_2$ to $Mn0_{1\cdot5}$', *ibid.,* 1966, **113,** pp.870-878

96 ATLUNG, S.: 'A theory for the dependence of gamma-manganese dioxide on the degree of reduction' *in* KOZAWA, A., and BRODD, R.G. (Eds.) *Manganese dioxide symposium – Vol. 1* (U.C.C. I.C. Sample Office, 1975), pp.47-65

97 LATIMER, W.M.: *Oxidation potentials* (Prentice-Hall, 1952, 2nd edn.)

98 SCHINDLER, P., ALTHAUS, H., and FEITKNECHT, W.: 'Solubility products of metal oxides and hydroxides. IX. Solubility products and free enthalpies of formation of zinc oxide, amorphous zinc hydroxide, β_1–, β_2–, γ–, δ– and ϵ– zinc hydroxide', *Helv. Chim. Acta,* 1964, **47,** pp.982-991

99 ROSSINI, F.D., WAGMAN, D.D., EVANS, W.H., LEVINE, S., and JAFFE, I.: *Selected values of chemical thermodynamic properties* (National Bureau of Standards, 1952)

100 SASAKI, K., and TAKAHASHI, T.: 'Studies on the electrolyte for dry cells', *Electrochim. Acta,* 1959, **1,** pp.261-271

101 DONNET, J.B.: 'The chemical reactivity of carbons', *Carbon,* 1968, **6,** pp. 161-176

102 PURI, B.R.: 'Surface complexes on carbons' *in* WALKER, P.R. (Ed.) *Chemistry and physics of carbon – Vol. 6* (Dekker, 1970), pp.191-282

103 RIVIN, D.: 'Surface properties of carbon', *Rubber Chem. Technol.,* 1971, **44,** pp.307-343

104 GARTEN, V.A., and WEISS, D.E.: 'The quinone-hydroquinone character of activated carbon and carbon black', *Austr. J. Chem.,* 1955, **8,** pp.68-95

105 HALLUM, J.V., and DRUSHEL, H.V.: 'The organic nature of carbon black surfaces', *J. Phys. Chem.,* 1958, **62,** pp.110-117

106 JONES, I.F., and KAYE, R.C.: 'Polography of carbon suspensions', *J. Electroanal. Chem.,* 1969, **20,** pp.213-221

107 KINOSHITA, K., and BETT, J.A.S.: 'Potentiodynamic analysis of surface oxides on carbon blacks', *Carbon,* 1973, **11,** pp.403-411

108 EPSTEIN, B.D., DALLE-MOLLE, E., and MATTSON, J.S.: 'Electrochemical investigations of surface functional groups on isotropic pyrolytic carbon', *ibid.,* 1971, **9,** pp.609-615

109 CAUDLE, J., SUMMER, K.G., and TYE, F.L.: 'Evidence of interaction between carbon and manganese dioxide' *in* COLLINS, D.H. (Ed.) *Power sources 6* (Academic Press, 1977), pp.447-467

110 KRIVOLUTSKAYA, M.S., TEMERIN, S.A., and LUKOVTSEV, P.D.: 'Effect of absorbed oxygen on the potential and the kinetics of discharge of the carbon-manganese dioxide electrod', *Zh. Fiz. Khim.,* 1947, **21,** pp.313-333

111 JENNINGS, C.W., and VOSBURGH, W.C.: 'A proposed mechanism for self discharge of the Leclanché cell', *J. Electrochem. Soc.,* 1952, **99,** pp. 309-316

112 CAUDLE, J., SUMMER, K.G., and TYE, F.L.: Ion exchange of carbon blacks. *Third conference on industrial carbons and graphites* (Society of Chemical Industry, 1971), pp.168-171

113 SASAKI, K.: 'Studies on the manganese dioxide for dry cells', *Mem. Fac. Eng. Nagoya Univ.*, 1951, **3**, pp.81-101

114 HOLLER, H.D., and RITCHIE, L.M.: 'Hydrogen ion concentration in dry cells', *Trans. Am. Electrochem. Soc.*, 1920, **37**, pp.607-616

115 DANIELS, F.: 'Physical-chemical aspects of the Leclaunché dry cell', *ibid.*, 1928, **53**, pp.45-69

116 THOMPSON, B.M.: 'Effect of hydrogen-ion concentration on the voltage of the Leclanché dry cells', *Ind. Eng. Chem.*, 1928, **20**, p.1176

117 JOHNSON, R.S., and VOSBURGH, W.C.: 'The reproducibility of the manganese dioxide electrode and change of electrode potential with pH', *J. Electrochem. Soc.*, 1952, **99**, pp.317-322

118 KOZAWA, A.: 'On an ion-exchange property of manganese dioxide', *ibid.*, 1959, **106**, pp.552-556

119 MULLER, J., TYE, F.L., and WOOD, L.L.: 'Ion-exchange of manganese dioxides' *in* COLLINS, D.H. (Ed.) *Batteries 2* (Pergamon, 1965), pp.201-217

120 BENSON, P., PRICE, W.B., and TYE, F.L.: 'Potential-pH relationships of gamma manganese dioxide', *Electrochem. Technol.*, 1967, **5**, pp.517-523

121 CHREITZBERG, A.M., and VOSBURGH, W.C.: 'The overpotential of the manganese dioxide electrode', *J. Electrochem. Soc.*, 1957, **104**, pp.1-5

122 YOSHIZAWA, S., and VOSBURGH, W.C.: 'The overpotential of the manganese dioxide electrode – II acid electrolytes', *ibid., 1975,* **104**, *pp.* 399-406

123 KOZAWA, A.: 'The potential of the manganese dioxide electrode and the surface composition of the oxide', *ibid.*, 1959, **106**, pp.79-82

124 CAUDLE, J., SUMMER, K.G., and TYE, F.L.: Manganese dioxide electrode. Pt. I – Hydrogen ion response in the absence of other cations', *J. Chem. Soc. Faraday Trans. 1*, 1973, **69**, pp.876-884

125 CAUDLE, J., SUMMER, K.G., and TYE, F.L.: 'Manganese dioxide electrode. Pt. II – Hydrogen ion response in the presence of manganese (II) ions', *ibid.*, 1973, **69**, pp.885-893

126 FEITKNECHT, W.: 'Einfluss der Teilchengrosse auf den Mechanismus von Festkorperreaktionen', *Pure Appl. Chem.*, 1964, **9**, pp.423-440

127 YAMAMOTO, N., KIYAMA, M., and TAKADA, T.: 'A new preparation method of Mn_5O_8', *Jpn. J. Appl. Phys.*, 1973, **12**, pp.1827-1828

128 KANG, H.Y., and LIANG, C.C.: 'The anodic oxidation of manganese oxides in alkaline electrolytes', *J. Electrochem. Soc.*, 1968, **115**, pp.6-10

129 BODEN, D., VENUTO, C.J., WISLER, D., and WYLIE, R.B.: 'The alkaline manganese dioxide electrode', *ibid.*, 1968, **115**, pp.333-338

130 FREEMAN, D.F., PELTER, P.F., TYE, F.L., and WOOD, L.L.: 'Screening of manganese dioxides for battery activity by thermogravimetric analysis', *J. Appl. Electrochem.*, 1971, **1**, pp. 127-136

131 UETANI, Y., TOGO, T., IWAMURA, T., and TOCHICUBO, I.: 'I.C. MnO_2 samples tested for practical cells-zinc chloride' *in* KOZAWA, A., and BRODD, R.J. (Eds.) *Manganese dioxide symposium – Vol. 1* (U.C.C. I.C. Sample Office, 1975), pp.183-201

132 COLEMAN, J.J.: 'Dry cell dynamics: the bobbin', *Trans. Electrochem. Soc.*, 1946, **90**, pp.545-583

133 GABANO, J.P., SEGURET, J., and LAURENT, J.F.: 'A kinetic study of the electrochemical reduction of manganese dioxide in a homogenous phase', *J. Electrochem. Soc.*, 1970, **117**, pp.147-151

134 ANDERSON, J.S.: 'Non-stoichiometric compounds', *Chem. Soc. Annual Rep.*, 1946, **43**, pp.104-120

135 SCOTT, A.B.: 'Diffusion theory of polarisation and recuperation applied to the manganese dioxide electrode', *J. Electrochem. Soc.*, 1960, **107**, pp.941-944

136 KORNFEIL, F.: 'On the polarisation of the manganese dioxide electrode', *ibid.*, 1962, **109**, pp.349-351

137 DEWHURST, D.J.: 'Concentration polarisation in plane membrane-solution systems', *Trans. Faraday Soc.*, 1960, **56**, pp.599-605

138 SAND, H.J.S.: 'On the concentration at the electrodes in a solution with special reference to the liberation of hydrogen by electrolysis of a mixture of copper sulphate and sulphuric acid', *Philos. Mag.*, 1901, **1**, pp.45-79

139 TOBIAS, C.W., EISENBERG, M., and WILKE, C.R.: 'Diffusion and con-*vection in electrolysis – a theoretical view',* J. Electrochem. Soc., 1952, **99**, pp.359C-365C

140 AGOPSOWICZ, A., BRETT, R., SHAW, J.E.A., and TYE, F.L.: 'Mass transport in the separator region of Leclanché cells' *in* COLLINS, D.H. (Ed.) *Power sources 5* (Academic Press, 1975), pp.503-524

141 HARRIS, A.C., and PARTON, H.N.: 'The transport numbers of zinc chloride from EMF measurements', *Trans. Faraday Soc.*, 1940, **36**, pp. 1139-1141

142 KOZAWA, A., KALNOKI-KIS, T., and YEAGER, J.F.: 'Solubilities of Mn(II) and Mn(III) ions in concentrated alkaline solutions', *J. Electrochem. Soc.*, 1966, **113**, pp.405-409

143 MENDZHERITSKII, E.A., and BAGOTSKI, V.S.: 'Cathodic reduction of the mercuric oxide electrode', *Elektrokhimiya*, 1966, **2**, pp.1312-1316

144 RUETSCHI, P.: 'The electrochemical reactions in the mercuric oxide-zinc cell' *in* COLLINS, D.H. (Ed.) *Power sources 4* (Oriel Press, 1972), pp.381-400

145 MARKOV, I., BOYNOV, A., and TOSCHEV, S.: 'Screening action and growth kinetics of electrodeposited mercury droplets', *Electrochim. Acta* 1973, **18**, pp.377-384

146 GARRETT, A.B., and HIRSCHLER, A.E.: 'The solubilities of red and yellow mercuric oxides in water, in alkali and in alkaline salt solutions. The acid and basic dissociation constants of mercuric hydroxide', *J. Am. Chem. Soc.*, 1938, **60**, pp299-306

147 CAUDLE, J., BETTS, C.A., and TYE, F.L.: 'Physical properties of mixtures of carbon and manganese dioxide' *in* COLLINS, D.H. (Ed.) *Power sources 2* (Pergamon, 1969), pp.319-334

148 LANE, E.: 'The amphoteric nature of silver oxide', *Z.Anorg. Allg. Chem.*, 1927, **165**, pp.325-363

149 JOHNSTON, H.L., CUTA, F., and GARRETT, A.B.: 'The solubility of silver oxide in water and in alkaline salt solutions. The amphoteric character of silver oxide', *J. Am. Chem. Soc.*, 1933, **55**, pp.2311-2325

150 PLESKOV, Y.V., and KABANOV, B.N.: 'The composition of the complex ions of silvers in strong alkaline solutions', *Zh. Neorg. Khim.*, 1957, **2**,

pp. 1808-1811

151 AMLIE, R.F., and RUETSCHI, P.: 'Solubility and stability of silver oxides in alkaline electrolytes', *J. Electrochem. Soc.*, 1961, **108**, pp.813-819

152 DIRKSE, T.P., VANDER LUGHT, L.A., and SCHNYDERS, H.: 'The reaction of silver oxides with potassium hydroxide', *J. Inorg. Nuc. Chem.*, 1963, **25**, pp. 859-865

153 KOVBA, L.D., and BALASHEVA, N.A.: 'Measurement of the solubilities of oxides of silver in alkali solutions by radioactive tracer methods', *Russ. J. Inorg. Chem.*, 1959, **4**, pp.94-95

154 ANTIKAINEN, P.J., HIETANEN, S., and SILLEN, L.G.: 'Studies on the hydrolysis of metal ions. 27. Potentiometric study of the argentate (I) complex in alkaline solution', *Acta Chem. Scand. Ser. A.*, 1960, **14**, pp.95-101

155 MILLER, B.: 'Rotating ring-disk study of the silver electrode in alkaline solution', *J. Electrochem. Soc.*, 1970, **117**, pp.491-499

156 WALES, C.P., and SIMON, A.C.: 'Changes in microstructure of a sintered silver electrode after repeated cycling at a low current', *Ibid.*, 1968, **115**, pp.1228-1236

157 WALES, C.P.: 'Effects of KOH concentration on morphology and electrical characteristics of sintered silver electrodes' *in* COLLINS, D.H. (Ed.) *Power sources 4* (Oriel Press, 1973), pp.163-183

158 WALES, C.P.: 'The microstructure of sintered silver electrodes', *J.Electrochem. Soc.*, 1969, **116**, pp.729-734

159 WALES, C.P.: 'The microstructure of sintered silver electrodes. II. At the end of 1-hour rate discharges' *ibid.*, 1969, **116**, pp.1633-1639

160 WALES, C.P.: 'Reduction of the silver oxide electrode in CsOH solutions', *ibid.*, 1974, **121**, pp.727-734

161 WALES, C.P., and BURBANK, J.: 'Oxides on the silver electrode. II. X-ray diffraction studies of the working silver electrode', *ibid.*, 1965, **112**, pp.13-16

162 GAGNON, E.G., and AUSTIN, L.G.: 'Low-temperature cathode performance of porous Ag/Ag_2O electrodes', *ibid.*, 1971, **118**, pp.497-506

163 GAGNON, E.G., and AUSTIN, L.G.: 'A description of the discharge mode of porous Ag/Ag_2O electrodes at low temperature in KOH', *ibid.*, 1972, **119**, pp.807-811

164 VITANOV, T., POPOV, A., and BUDEVSKI, E.: 'Mechanism of electrocrystallisation', *ibid.*, 1974, **121**, pp.207-212

165 MORCOS, I.: 'On the electrochemical nucleation of silver on different crystal orientations of graphite', *ibid.*, 1975, **122**, pp.50-53

166 TOSCHEV, S., MILCHER, A., and VASSILEVA, E.: 'Electronmicroscope investigations of the initial stages of silver electrodeposition', *Electrochim. Acta.*, 1976, **21**, pp.1055-1059

167 TILAK, B.V., PERKINS, R.S., KOZLOWSKA, H.A., and CONWAY, B.E.: 'Impedance and formation characteristics of electrolytically generated silver oxides – I, Formation and reduction of surface oxides and the role of dissolution processes', *Electrochim. Acta,* 1972, **17**, pp.1447-1469

168 AMBROSE, J., and BARRADAS, R.G.: 'The electrochemical formation of Ag_2O in KOH electrolyte', *ibid.*, 1974, **19**, pp.781-786

169 GILES, R.D., HARRISON, J.A., and THIRSK, H.R.: 'Anodic dissolution of silver and formation of Ag_2O in hydroxide solutions using single crystal electrodes. A Faradaic impedance study', *J. Electroanal. Chem. Interfacial Electrochem.*, 1969, **22**, pp.375-388

170 GILES, R.D., and HARRISON, J.A.: 'Potentiodynamic sweep measurements of the anodic oxidation of silver in alkaline solutions', *ibid.*, 1970, **27**, pp.161-163

171 GIBBS, D.B., RAO, B., GRIFFIN, R.A., and DIGNAM, M.J.: 'Anodic behaviour of silver in alkaline solutions', *J. Electrochem. Soc.*, 1975, **122**, pp.1167-1174

172 DIRKSE, T.P., and DE ROOS, J.B.: 'The formation of thin anodic films of silver oxide', *Z. Phys. Chem. (Frankfurt am Main)*, 1964, **41**, pp.1-7

173 DIGNAM, M.J., BARRETT, H.M., and NAGY, G.D.: 'Anodic behaviour of silver in alkaline solutions', *Can. J. Chem.* 1969, **47**, pp. 4253-4266

174 NAGY, G.D., and CASEY, E.J.: 'Electrochemical kinetics of silver oxide electrodes' *in* FLEISCHER, A., and LANDER, J.J. (Eds.) *Zinc-silver oxide batteries* (Wiley, 1971), pp.133-151

175 YOSHIZAWA, S., and TAKEHARA, Z.: 'Electrode phenomena of silver-silver oxide system in alkaline battery', *J. Electrochem. Soc. Jpn.*, 1963, **31**, pp.91-104

176 KUNSCHERT, F.: 'Untersuchung Komplexer Zinksalze', *Z. Anorg. Allg. Chem.*, 1904, **41**, pp.337-358

177 DIRKSE, T.P.: 'The nature of the zinc containing ion in strongly alkaline solutions', *J. Electrochem. Soc.*, 1954, **101**, pp.328-331

178 BODEN, D.P., WYLIE, R.B., and SPERA, V.J.: 'The electrode potential of zinc amalgam in alkaline zincate solutions', *ibid.*, 1971, **118**, pp.1298-1301

179 DIRKSE, T.P., POSTMUS, C., and VANDENBOSCH, R.: 'A study of alkaline solutions of zinc oxide', *J. Am. Chem.Soc.*, 1954, **76**, pp.6022-6924

180 NEWMAN, G.H., and BLOMGREN, G.E.: 'NMR studies of complex ions in the aqueous ZnO-KOH systems', *J. Chem. Phys.*, 1965, **43**, pp.2744-2747

181 FORDYCE, J.S., and BAUM, R.L.: 'Vibrational spectra of solutions of zinc oxide in potassium hydroxide', *ibid.*, 1965, **43**, pp.843-846

182 JACKOVITZ, J.F., and LANGER, A.: 'A spectroscopic investigation of the zinc-hydroxy system' *in* FLEISCHER, A., and LANDER, J.J. (Eds.) *Zinc-silver oxide batteries* (Wiley, 1971), pp.29-36

183 LANGER, A., and PANTIER, E.A.: 'A coulogravimetric investigation of the zinc electrode in potassium hydroxide', *J. Electrochem. Soc.*, 1968, **115**, pp.990-993

184 DIRKSE, T.P.: 'Electrolytic oxidation of zinc in alkaline solutions', *ibid.*, 1955, **102**, pp.497-501

185 IOFA, Z.A., MIRLINA, S.Y., and MOISEEVA, N.B.: 'The investigation of processes occurring at the zinc electrode of a cell with an alkaline electrolyte', *Zh. Prikl. Khim.*, 1949, **2**, pp.983-994

186 EISENBERG, M., BAUMAN, H.F., and BRETTNER, D.M.: 'Gravity field effects on zinc anode discharge in alkaline media', *J. Electrochem. Soc.*, 1961, **108**, pp. 909-915

187 ARKHANGEL'SKAYA, Z.P., ANDREEVA, G.P., and MASHEVICH, M.N.: 'Anodic dissolution of zinc electrodes with porous active materials in

voltaic cells with alkaline electrolyte', *J. Appl. Chem. USSR,* 1968, **41**, pp.1640-1645

188 GERISCHER, H.: 'Kinetik der entladung einfacher und komplexes zink-ionen', *Z. Phys. Chem.,* 1953, **202**, pp.302-317

189 HAMPSON, N.P.: 'Kinetics of the zinc electrode' *in* FLEISCHER, A., and LANDER, J.J. (Eds.) *Zinc-silver oxide batteries* (Wiley, 1971), pp.37-61

190 PAYNE, D.A., and BARD, A.J.: 'The mechanism of the zinc (II) – zinc amalgam electrode reaction in alkaline media as studied by chronocoulometric and voltammetric techniques', *J. Electrochem. Soc.,* 1972, **119**, pp.1665-1674

191 VAN DOORNE, W., and DIRKSE, T.P.: 'Supersaturated zincate solutions' *ibid.,* 1975, **122**, pp.1-4

192 FLEROV, V.N.: 'The ageing process of supersaturated zincate', *Zh. Fiz. Khim.* 1957, **31**, pp.49-54

193 DIRKSE, T.P., VANDER LUGHT, L.A., and HAMPSON, N.A.: 'Exchange in the Zn, zincate, Zn0 system', *J. Electrochem. Soc.,* **1971, 118,** pp, 1606-1609

194 POWERS, R.W., and BREITER, M.W.: 'The anodic dissolution and passivation of zinc in concentrated potassium hydroxide solutions', *ibid.,* 1969, **116**, pp.719-729

195 NIKITINA, Z.I.: 'Passivation of zinc electrodes in galvanic cells with alkaline electrolytes', *Zh. Prikl. Khim.,* 1958, **31**, pp.209-216

196 POWERS, R.W.: 'Anodic films on zinc and the formation of cobwebs', *J. Electrochem. Soc.,* 1969, **116**, pp.1652-1659

197 HUBER, K.: 'Anodic formation of coatings on magnesium, zinc and cadmium', *ibid.,* 1953, **100**, pp.376-382

198 HUETIG, G.F., and MOELDNER, H.: 'Die spezifischen Warmen des Kristallisierten Zinkhydroxyds und die Berechnung der affinitaten Zwishen Zinkoxyd und Wasser', *Z.Anorg. Allg. Chem.,* 1933, **211**, pp.368-378

199 FEITKNECHT, W.: 'Investigations into the transformations of solids into liquids. 2nd communication. On the various modifications of zinc hydroxide', *Helv. Chim. Acta,* 1930, **13**, pp.314-345

200 DIETRICH, H.G., and JOHNSTON, J.: 'Equilibrium between crystalline zinc hydroxide and aqueous solutions of ammonium hydroxide and of sodium hydroxide', *J. Am. Chem. Soc.,* 1927, **49**, pp.1419-1431

201 POWERS, R.W.: 'Film formation and hydrogen evolution on the alkaline zinc electrode', *J. Electrochem. Soc.,* 1971, **118**, pp.685-699

202 DIRKSE, T.P., DEWIT, D., and SHOEMAKER, R.: 'The anodic behaviour of zinc in KOH solutions:, *ibid,* 1968, **115**, pp. 442-444

203 HULL, M.N., and TONI, J.E.: 'Formation and reduction of films on amalgamated and non-amalgamated zinc electrodes in alkaline solutions', *Trans. Faraday Soc.,* 1971, **67**, pp.1128-1136

204 DIRKSE, T.P.: 'A comparison of amalgamated and non-amalgamated zinc electrodes: *in* COLLINS, D.H. (Ed.) *Power sources 2*(Pergamon, 1970), pp.411-422

205 BREITER, M.W.: 'Anodic films and passivation of zinc in concentrated potassium hydroxide solutions', *Electrochim. Acta,* 1971, **16**, pp.1169-1178

206 ARMSTRONG, R.D., and BULMAN, G.M.: 'The anodic dissolution of zinc

in alkaline solutions', *J. Electroanal. Chem. Interfacial Electrochem.,* 1970, **25**, pp.121-130

207 VOZDVIZHENSKII, F.S., and KOCHMAN, E.D.: 'Voltamperographic investigation of the anodic dissolution and passivation of zinc in alkaline solutions', *Russ. J. Phys. Chem.,* 1965, **39**, pp.374-350

208 ELDER, J.P.: 'The electrochemical behaviour of zinc in alkaline media', *J. Electrochem. Soc.,* 1969, **116**, pp.757-762

209 LEWIS, R.W., and TURNER, J.: 'The effect of silicate ion on the anodic behaviour of zinc in alkaline electrolyte', *J. Appl. Electrochem.,* 1975, **5**, pp.343-349

210 HUTCHISON, P.F., and TURNER, J.: 'Some aspects of the electrochemical behaviour of zinc in the presence of acrylate and methacrylate ions', *J. Electrochem. Soc.,* 1976, **123**, pp.183-186

211 DIRKSE, T.P., and HAMPSON, N.A.: 'The anodic behaviour of zinc in aqueous KOH solutions – II. Passivation experiments using linear sweep voltammetry', *Electrochim. Acta,* 1972, **17**, pp.387-394

212 HULL, M.N., ELLISON, J.E., and TONI, J.E.: 'The anodic behaviour of zinc electrodes in potassium hydroxide electrolytes', *J. Electrochem. Soc.,* 1970, **117**, pp. 192-198

213 POPOVA, T.I., SIMONOVA, N.A., and KABANOV, B.N.: 'Mechanism of passivation of zinc in strong zincate solutions of alkali', *Sov. Electrochem.,* 1966, **2**, pp.1347-1350

214 VERMILYEA, D.A.: 'Anodic films' *in* DELAHAY, P. (Ed.) *Advances in electrochemistry and electrochemical engineering* (Interscience, 1963), pp.211-286

215 HOLLOMON, J.H., and TURNBULL, D.: 'Nucleation' *in* CHALMERS, B. (Ed.) *Progress in metal physics* (Pergamon, 1961), pp.333-388

216 FRANK, F.C., and VAN DER MERWE, J.H.: 'One dimensional dislocations. II. Misfitting monolayers and oriented overgrowth', *Proc. R. Soc., London, Ser. A.* 1949, **198**, pp.216-225

217 TURNBULL, D.: 'Phase changes', *Solid State Phys.,* 1956, **3**, pp. 225-306

218 EVANS, U.R.: 'The mechanism of oxidation and tarnishing', *Trans. Electrochem. Soc.,* 1947, **91**, **pp.547-572**

219 DIRKSE, T.P.: 'Voltage decay at passivated zinc anodes', *J. Appl. Electrochem.* 1971, **1**, pp.27-33

220 BREITER, M.W.: 'Dissolution and passivation of vertical porous zinc electrodes in alkaline solution', *Electrochim. Acta,* 1970, **15**, pp.1297-1304

221 SATO, Y., NIKI, H., and TAKAMURA, T.: 'Effects of carbonate on the anodic dissolution and the passivation of zinc electrode in concentrated solution of potassium hydroxide', *J. Electrochem. Soc.,* 1971, **118**, pp.1269-1272

222 DIRKSE, T.P., and KROON, D.J.: 'Effect of ionic strength on the passivation of zinc electrodes in KOH solutions', *J. Appl. Electrochem.,* 1971, **1**, pp.293-296

223 DIRKSE, T.P., and HAMPSON, A.: 'The anodic behaviour of zinc in aqueous KOH solution – I. Passivation experiments at very high current densities', *Electrochim. Acta,* 1971, **16**, pp.2049-2056

224 ELSDALE, R.N., HAMPSON, A., JONES, P.C., and STRACHAN, A.N.: 'The anodic behaviour of porous zinc electrodes', *J. Appl. Electrochem.,*

1971, **1**, pp.213-217

225 IVANOV, E.A., POPOVA, T.I., and KABANOV, B.N.: 'Passivation of zinc in KOH solutions supersaturated with zincate. Impedance determination', *Elektrokhimya,* 1969, **5**, pp.353-356

226 HAMPSON, N.A., TARBOX, M.J., LILLEY, J.T., and FARR, J.P.G.: 'The passivation of vertical zinc anodes in potassium hydroxide solution', *Electrochem. Technol..* 1964, **2**, pp.309-313

227 HAMPSON, N.A., and TARBOX, M.J.: 'The anodic behaviour of zinc in potassium hydroxide solution', *J. Electrochem. Soc.,* 1963, **110,** pp.95-98

228 HAMPSON, N.A., SHAW, P.E., and TAYLOR, R.: 'Anodic behaviour of zinc in potassium hydroxide solution. II. Horizontal anodes in electrolytes containing Zn(II)', *Br. Corros. J.,* 1969, **4**, pp. 207-211

229 POPOVA, T.I., BAGOTSKII, V.S., and KABANOV, B.N.: 'The anodic passivation of zinc in alkali. I. Measurements at constant current density', *Russ. J. Phys. Chem.,* 1962, **36**, pp.766-770

230 LANDSBERG, R., and BARTLETT, H.: 'Bedeckungs vorgange an Zinkanoden in Natroniauge', *Z. Elekrochem.,* 1957, **61**, pp.1162-1168

231 FARMER, E.D., and WEBB, A.H.: 'Zinc passivation and the effect of mass transfer in flowing electrolyte', *J. Appl. Electrochem,* 1972, **2**, pp. 123-126

232 BUSHROD, C.J., and HAMPSON, N.A.: 'The anodic behaviour of zinc in KOH solution. V. Galvanostatic polarisation with an interruption', *ibid.,* 1971, **1**, pp.99-101

233 COATES, G., HAMPSON, N.A., MARSHALL, A., and PORTER, D.F.: 'The anodic behaviour of porous zinc electrodes. II. The effects of specific surface area of the zinc compact material', *ibid.,* 1974, **4**, pp.75-80

234 DIRKSE, T.P., and HAMPSON, N.A.: 'The anodic behaviour of zinc in aqueous solution – III. Passivation in mixed KF-KOH solutions', *Electrochim. Acta,* 1972, **17**, pp.813-818

235 BROOK, M.J., and HAMPSON, N.A.: 'The anodic behaviour of zinc in KOH solution – IV. Anodic experiments in flowing electrolyte', *ibid.,* 1970, **15**, pp.1749-1758

236 DIRKSE, T.P.: 'Passivation studies on the zinc electrode', *in* COLLINS, D.H. (Ed.) *Power sources 3,* (Oriel Press, 1971), pp.485-495

237 MCBREEN, J.: 'Study to improve the zinc electrode for spacecraft electrochemical cells' 2nd Quarterly Report on Contract NAS 5-10231, 1967

238 MARSHALL, A., and HAMPSON, N.A.: 'The point of passivation of a zinc electrode in alkali', *J. Appl. Electrochem,* 1977, **7**, pp.271-273

239 LANSBERG, R.: 'Zum anodischen Verhallen des Zinks in Natronlauge', *Z. Phys. Chem. (Leipzig),* 1956, **206**, pp.291-301

240 HULL, M.N., and JAMES, H.I.: 'Why alkaline cells leak', *J. Electrochem. Soc.,* 1977, **124**, pp.332-339

241 BAUGH, L.M., COOK, J.A., and LEE, J.A.: 'A mechanism for alkaline cell leakage', *J. Appl. Electrochem.,* 1978, **8**, pp.253-263

242 BELL, G.S.: 'The performance characteristics of zinc chloride dry cells – a survey', *Extended Abstracts, The Electrochem. Soc.,* 1975, **75-2,** pp.15-16

243 POUSSARD, B., DECHENAUX, V., CROISSANT, P., and HARDY, A.:

'Study of electrolyte in zinc chloride Leclanché cells' *in* THOMPSON, J. (Ed.) *Power sources* 7 (Academic Press, 1979), pp. 445-462

244 MEEUS, M.L., CROCQ, G.G., and LAMAITRE, C.A.: 'Metallurgical structural and electrochemical factors influencing the production of optimal zinc cans for Leclanché dry cells by impact extrusion' *in* THOMPSON, J. (Ed.) *Power sources* 7 (Academic Press, 1979), pp. 463-484

Chapter 4

Lead-acid storage batteries

M. Barak

4.1 Introduction

During the past two decades, several promising portable power sources have appeared, e.g. fuel cells, metal/air cells, high temperature cells using materials of relatively low density, such as sodium, lithium, sulphur and so on. So far, however, none of these has posed a real threat to existing practical systems. On the other hand, the lead/acid storage battery has not only extended its uses in established fields, but, because of its great versatility, has opened the way to new applications and is now by far the most widely used portable power source. One statistician has claimed that there are at least 95 different types of service in which storage batteries are used.

Storage batteries represent the largest single consumer of lead. By 1975/76, the total annual consumption of lead in the Western World and in Japan was running at just under 3 million tonnes, of which 1·2m tonnes, or about 40%, was used in the production of batteries.[1] In most of the countries in Western Europe and in Japan, the amount of lead used in batteries now exceeds that used in cables, for many years the main usage of lead. While the advent of new plastics has probably been the largest single factor in the declining use of lead for cables, innovatory uses of plastics in the storage battery industry, for thin-walled containers with heat-sealed covers, microporous separators, fibrous tubes for traction batteries and so on, have undoubtedly enhanced the growth of the battery industry.

In terms of numbers of units or of watt-hours of capacity, the usage of the lead/acid battery is probably over twenty times as large as that of its nearest rival, the nickel/cadmium (or iron) alkaline storage battery.

There are several contributory factors to this success:

(*a*) great versatility; the battery can supply on instant demand high or low currents over a wide range of temperatures

(*b*) good storage characteristics or shelf life, particularly in the dry-charged condition

(*c*) a very high degree of reversibility: it is capable of giving hundreds of discharge-charge cycles with great reliability. As indicated later, each gram of positive active material in a traction battery gives an aggregate output over its lifetime of over 100 Ah

(*d*) lead, the basic material of construction, has a low melting point and the various metallic components, grids, busbars, terminal posts, inter-cell connectors, can be easily cast and grouped together by simple low temperature welding techniques

(*e*) high cell voltage, due to the high potential of the lead dioxide electrode in sulphuric acid, namely E_0 = 1·685V, giving a cell voltage of 2·04V

(*f*) the metal is relatively cheap, when compared with nickel, cadmium and silver used in other storage batteries.

The great disadvantage of the lead/acid storage battery lies in its weight. The energy densities of PbO_2 and Pb are the lowest of all the cathodic and anodic materials listed in Tables 1.3 and 1.4 in Chapter 1, and the performance of the lead/acid battery is further reduced by the low coefficient of use of the active materials. At the lowest rates of discharge, this rarely exceeds 60%, at the 5h rate it may reach 30% and at the highest rates, 10 to 15 mins, only about 10%. The reasons for the inefficiency and the steps which have been taken to improve it are described later in this chapter. Suffice to say here that, during the past 25 years or so, gains of about 50% in energy and power densities have been achieved in portable units, with no significant loss and sometimes a gain in cycling lifetimes, and, in some stationary types, the gain in output per unit of volume has exceeded 200%.

4.2 Classes of storage batteries

Batteries are classified according to the service which they provide.

4.2.1 Automotive batteries

These supply power for engine starting, lighting and ignition (SLI) of vehicles propelled by internal combustion engines, such as automobiles, buses, lorries and other heavy road vehicles, motor cycles and so on. For obvious reasons these batteries are referred to as 'portable' and make up by far the largest field of application. Detailed statistics are not available for many of the industrialised countries, but those for the USA and Japan are of particular interest. Fig. 4.1[2] shows the total annual shipments for these countries during the period 1965 to 1975. In the USA, from 1965 to 1973 the average annual rate of increase

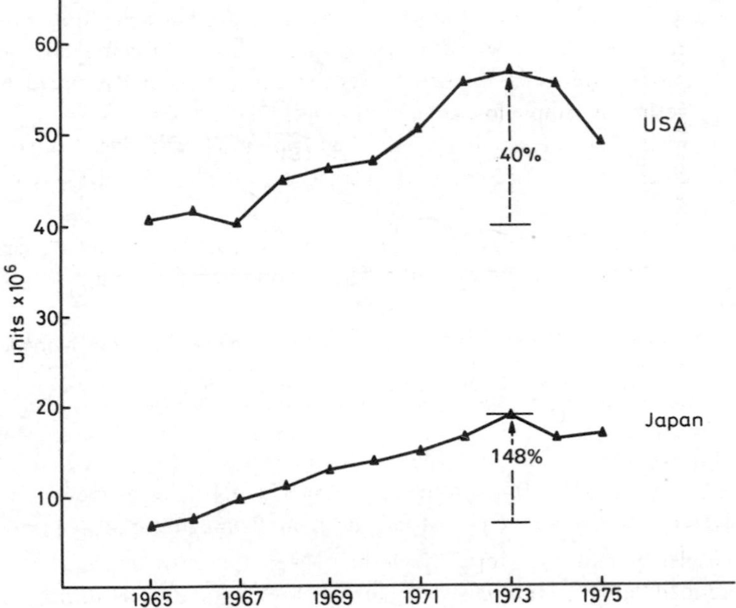

Fig. 4.1 *Annual shipment of SLI batteries in USA and Japan (including motorcycles)*

was about 5%, while in Japan it reached the exceptionally high figure of about 18%. This was followed by a drop in production in the USA and in Japan and the figure more or less levelled out. The rapid growth in Japan during the early years is shown graphically in Fig. 4.2,[3] which indicates that by 1974 the lead used for batteries represented over 80% of the total consumption of the metal.

The figures for the production of SLI batteries naturally followed the growth in the population of internal combustion engined vehicles, although this pattern has been somewhat altered by improvements in the average battery lifetime. During the period 1962 to 1970, claims

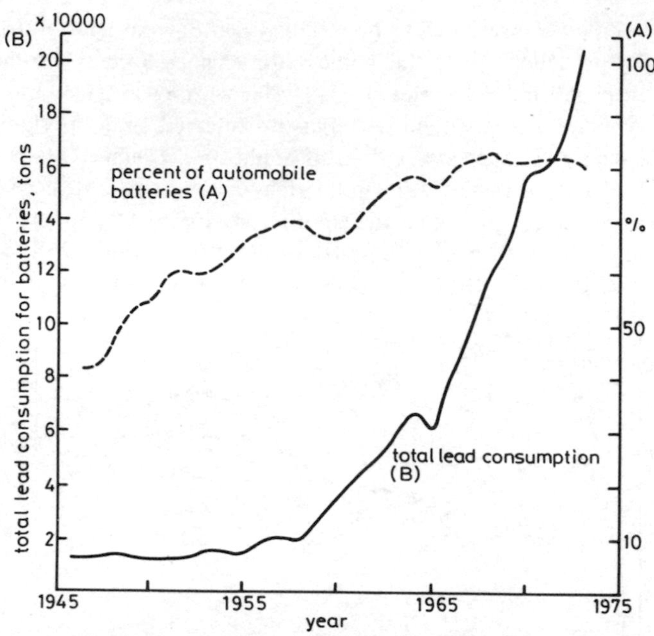

Fig. 4.2 *Total lead consumption for batteries and percentage of automobile batteries in Japan*
[Courtesy A. Kozawa and T. Takagaki, *Japanese lead-acid battery industry*, p. 37, Electrochemical Society of Japan, 1977]

were made that the average life of SLI batteries in the USA had increased by about 20%, from 34 to 41·7 months.[4] Surveys also indicated that the main cause of failure lay in disintegration of the positive grids.

4.2.2 *Traction batteries*

Traction batteries also come in the 'portable' category. They provide the motive power for a wide variety of vehicles, notably industrial trucks for the mechanical handling of raw materials and finished goods in factories, warehouses and railway stations and commercial road vehicles, such as delivery vans, milk floats, mobile shops and so on. Batteries are used to drive mining locomotives and submarines when submerged and during the last War they were used in electricially-propelled torpedoes.

Great Britain has led the world in the development and use of battery-powered electric vehicles. Fig. 4.3[5] shows the annual production of industrial trucks of all types in the UK for the decade from 1965 to 1976. Up to 1971, the proportion of electrics exceeded 60% of the total. During the past 5 years it has been running at a lower level of about 50%, but now appears to be climbing again, stimulated, no doubt, by the growing general interest in electric vehicles. There are

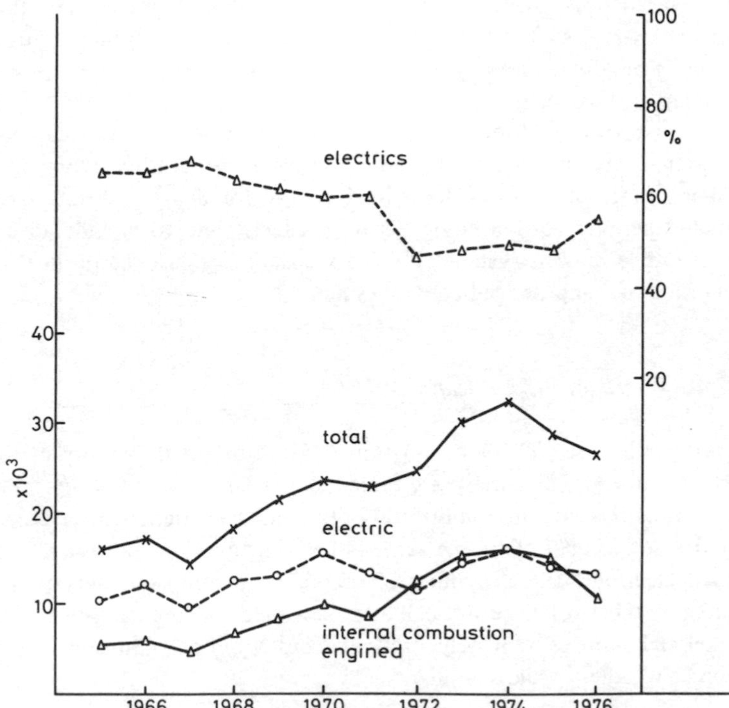

Fig. 4.3 *Annual production of industrial trucks in UK: internal combustion engine and electric*

now about 75 000 battery-powered trucks in the UK. Statistics for electric commercial road vehicles are not so complete, but the number now in service in the UK is considered to exceed 40 000.

D'Arcy[6] has estimated the population of electric trucks in West Germany to be about 42% of the total, with a growth rate of about 2·5% per year. In some other European countries, and in the USA, the proportion is reckoned to be around 30% and slightly less in France and Japan. In West Germany, more than 200 battery-powered rail cars have been operating successfully on branch lines for many years. In the UK there are over 250 000 battery-powered lawn-mowers; in the USA, golf-carts provide an annual market of over 100 000 batteries.

4.2.3 Stationary batteries

One long-dated service has been the supply of standby power in power stations. This meant supplying fairly heavy loads for short periods, in the event of a power failure. During recent years the pattern of service has changed from a deep cycling regime to switch operation and other ancilliary services. Stationary batteries also supply emergency lighting in public buildings of all types, theatres, cinemas, hospitals and power for telephone exchanges.

Two massive developments currently in progress, also referred to in Chapter 1, are the uses of storage batteries in competition with other systems to store off-peak base load energy for delivery during peak demand periods; load-levelling, as it is called,[7] and to provide an uninterruptible power system (UPS) to ensure absolute continuity of operation of computerised control systems.[8]

4.2.4 Miscellaneous applications

Several other uses, all in the portable class, do not fall in any of the aforementioned categories, e.g. miners' cap-lamp batteries, batteries for the lighting and air-conditioning of trains, batteries for aircraft and for fire and burglar alarms. A recent promising development is a sealed, coiled electrode cell of cylindrical shape,[9] as a complementary power source to the LeClanché dry cell and the sealed alkaline manganese and nickel/cadmium cells, widely used in hand torches, radio sets, instruments and other cordless appliances.[10]

4.3 Active materials and electrochemical reactions

4.3.1 Active materials

4.3.1.1 Positive electrode or 'cathode', lead dioxide (PbO_2): The properties of this material are described in some detail in Section 4.5.3. It should be noted here that the stoichiometric ratio of oxygen atoms to lead seldom reaches 2 and is generally nearer 1·95. For convenience, however, the value of 2 is used in equations showing the electrochemical reactions.

4.3.1.2 Negative electrode or 'anode', spongy lead (Pb): The convention regarding the terms 'cathode' and 'anode' arises from the fact that the discharge reaction is the basic function of a battery. This is the reverse of that generally used in the processes of electrolysis and electroplating, where the flow of electrons is away from the positive electrode, which is called the 'anode'.

4.3.2 Electrolyte

Dilute sulphuric acid (H_2SO_4) is used as the electrolyte, and, since this takes part in the electrochemical reactions, it is technically also an active material. The dilute acid is made by adding the concentrated acid, which may contain up to 98% of H_2SO_4, to water. Considerable heat is generated in this operation, which should never be done the other way round for this reason and there is a volume change in the final product. Full details of the physical and electrochemical properties of sulphuric acid have been listed by Vinal[11] and Bode.[12] Only those of direct relevance are noted here.

In storage battery technology, concentrations are generally expressed in terms of the specific gravity (SG). This is defined as the ratio of the density in g/cm^3 or kg/l at any given temperature to that of water at the same temperature. Strictly speaking, the datum line for water should be 4°C, but, for convenience, values are usually expressed at 15°C or 25°C. Table 4.1 gives the concentrations of electrolyte used in batteries for different types of service and Fig. 4.4 shows the relation of the SG at 25°C to the concentration in g/l. Corrections are, of course, necessary when the temperature varies from the norm, and Fig. 4.4[11] also shows the temperature coefficients, based on the SG at 15°C.

Electrical resistivity is another important characteristic of the electrolyte and Fig. 4.5 shows the relationships between resistivity, concentration and temperature, derived from values given by Vinal[11] (Table 15, p. 110) and Bode[12] (Tables 2.27 and 2.28, pp. 75-76).

4.3.2.1 Purity standards for sulphuric acid: Table 4.2 lists the most important impurities included in US Federal Specification 0-S-801-b, 4.14.65 and British Standard Specification BS 3031, 1972. The units in these specifications have been retained. Vinal[11] has given details of the effects of these impurities and brief notes only must suffice here.

Chloride: the chloride ion may be oxidised at the positive electrode during overcharge to chlorate (ClO_3^-) or perchlorate (ClO_4^-). Both of

Table 4.1 *Electrolyte for various applications*

Service	SG/25°C kg/l	H_2SO_4 g/l	Electrochemical equivalent Ah/l
Automotive SLI: temperate climate	1·270 to 1·285	460 to 488	125 to 133
tropical climate	1·240	404	110
Traction: flat plate or tubular	1·250 to 1·280	423 to 480	115 to 131
Stationary: Planté types	1·210	351	95
Discharged: depending on rate	1·140 to 1·080	233 to 120	63 to 33

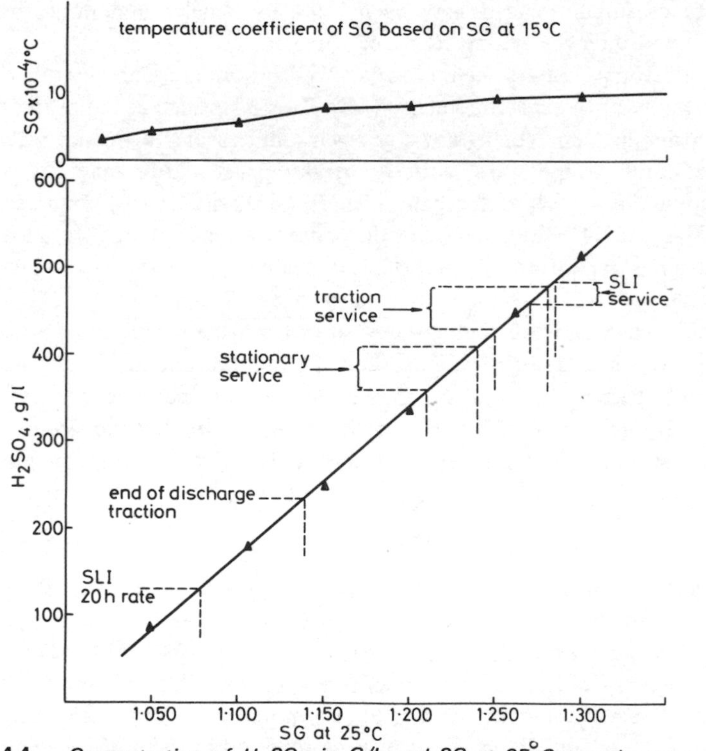

Fig. 4.4 *Concentration of H_2SO_4 in G/L and SG at 25°C over the range in general use*

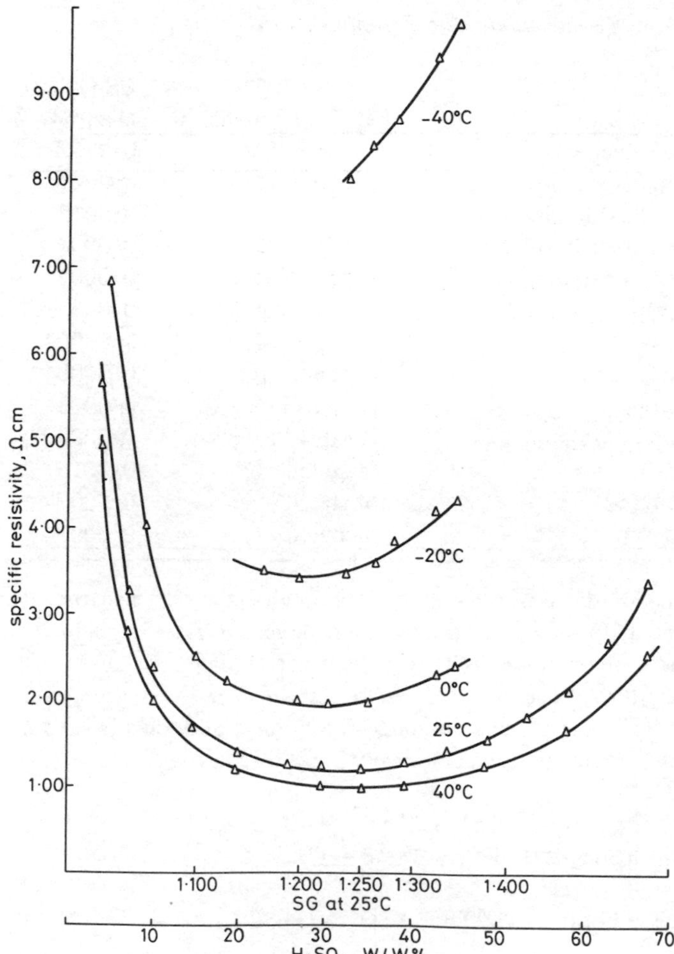

Fig. 4.5 *H_2SO_4: Specific resistivity at different concentrations and temperatures*

these oxy-acids form soluble lead salts, but these reactions are inhibited by a high concentration of SO_4^{2-} ions. If the concentration of sulphuric acid is low at the end of discharge, particularly in the inner areas of the active materials, these oxy-acids cause corrosion of the grids when the recharge is started. If they accumulate to any degree in the main bulk of the electrolyte they may oxidise and thereby discharge the negative material.

Nitrogen oxides: these may be oxidised at the positive plates to form nitrate ions (NO_3^-), which, like the oxy-acids of chlorine, may cause

Table 4.2 *Purity standards for sulphuric acid*

	BS 3031 Ref. 1·215 SG/20°C	US O-S801-b Max. %W/W
Fixed residue	0·015%	0·075%
Chloride (Cl)	7 PPM	0·004%
Sulphur dioxide (SO_2)	5 PPM	0·0015
Nitrogen as ammonium (NH_4)	50 PPM	0·0004
Nitrogen as oxides	5 PPM	0·0002
Iron (Fe)	12 PPM	0·003
Arsenic (As)	2 PPM	0·00004
Antimony (Sb)	not stated	0·00004
Copper (Cu)	7 PPM	0·0025
Manganese (Mn)	0·4 PPM	0·000007
Zinc (Zn)	not stated	0·0015
Selenium (Se)	not stated	0·0007
Nickel (Ni)	not stated	0·00004

corrosion of the positive grid or loss of charge of the negative active material. With repeated gassing overcharges, nitrate is gradually reduced to ammonium ion (NH_4^+) at the negative plates.

Iron: during cycling, iron will pass through the redox reaction, $Fe^{2+} \rightleftharpoons Fe^{3+} + e$, causing discharge reactions at both positive and negative plates, represented schematically as follows:

$$\text{At the positive} \quad Pb^{4+} + 2Fe^{2+} = Pb^{2+} + 2Fe^{3+} \tag{4.1}$$

$$\text{At the negative} \quad Pb + 2Fe^{3+} = Pb^{2+} + 2Fe^{2+} \tag{4.2}$$

followed in both cases by $Pb^{2+} + SO_4^{2-} = PbSO_4$

Arsenic: Vinal[11] has bracketed arsenic with antimony in its effects on negative plates. This could arise from the transition As(3) $\rightleftharpoons$ As(5), which, as in the case of other multi-valent cations, might be expected to discharge both the positive and negative active material. No evidence of such reactions has, however, been recorded, and the chief objection to arsenic lies in the possible conversion to poisonous arsine, during gassing overcharges. The standard potential of the reaction

$$As + 3H^+ + 3e = AsH_3(g) \tag{4.3}$$

is −0·61V, which is well below the potential at which hydrogen is evolved at the negative electrode and arsine is, therefore, a possible product. The gas is, however, rapidly decomposed by the oxygen and water vapour present in the overcharge gases and is difficult to detect under these conditions.

Antimony: the deleterious effects of antimony on the negative plate are much more significant. These are described in some detail in Section 4.6.23 dealing with antimony-bearing lead alloys used for grids. Suffice to say here that the metal is electrodeposited on the negative plates, and, since the hydrogen over-voltage of antimony is appreciably lower than that of lead, hydrogen is liberated and the active material discharged by the process

$$Sb/Pb + H_2SO_4 = Sb/H_2(g) + PbSO_4 \tag{4.4}$$

Copper and Nickel: these metals are also electrodeposited on the negative plate during charging, and, since their over-voltages are lower than that of lead, their action is similar to that of antimony. Nickel is more active than copper in this respect and may cause a permanent reduction of the cell voltage on charge. In fact, proposals have been made to add to the active material small amounts of powdered nickel-bearing glass, calculated to keep the nickel concentration fairly steady and thereby improve the watt-hour efficiency of the discharge/charge process. This is, however, a somewhat doubtful procedure, since the lowering of the hydrogen over-voltage makes the lead electrode more susceptible to poisoning by other metals, notably antimony.

Tellurium: although not included in either of the specifications listed already, tellurium is regarded by some manufacturers as a dangerous impurity, calling for a strict limit, similar to that for antimony. It is also plated out on the negative plate during charging and, having a lower hydrogen over-voltage, like antimony it produces hydrogen and discharges the spongy lead.

Manganese: this metal also forms multi-valent ions, and, during charge, may be converted at the positive plate to manganate or permanganate. In this form it will attack any organic materials, separators, tubular retainers etc.

4.3.2.2 Purity standards for water: Standards are equally necessary for the water used to dilute concentrated acid and for topping up batteries in service.* Distilled water has for many years been the first choice, but water of similar purity may now be prepared by de-mineralising processes, using organic ion-exchange resins. British Standard BS 4974:1975 specifies limits for two grades of water. The main requirement for Grade A is that the electrical conductivity should not exceed 10 Ω^{-s}/cm at 20°C. Broadly speaking, the total amount of inorganic salts should not exceed 100 PPM and metals, chloride and nitrate ions and organic materials, such as acetic or related acid, should not separately exceed 5-10 PPM.

4.3.3 *Electrode potentials*

The behaviour of electrodes in contact with an electrolyte is described in Chapter 2. With the immediate formation of an electrical double layer on the surface, the electrode exerts its own characteristic redox potential. In the absence of any purely chemical side reactions, this steady state condition may persist for long periods and the electrode is said to have a good 'shelf life'. As mentioned in Chapter 2, single electrode potentials cannot be measured directly, but they can be determined indirectly by forming a galvanic cell with another electrode, the potential of which is known. Reference electrodes, such as the hydrogen electrode or the mercury sulphate electrode, are used for this purpose. With these, liquid junctions are established with a luggin capilliary tube, placed in close proximity to the electrode in question. In a simpler, although less accurate procedure, often used in the practical factory processes, solid metal electrodes, such as zinc or cadmium, lightly amalgamated with mercury may be used. Some notes on the use of cadmium reference electrodes are included in Section 4.4.5.

The standard hydrogen electrode (SHE) with hydrogen gas at unit fugacity or pressure in a solution of a hydrogen acid of unit ionic activity is regarded as the reference electrode with $E^0 = 0$.

The redox potential of an electrode is defined by the Nernst expression

$$E = E^0 + \frac{RT}{nF} \ln \frac{a_1}{a_2} \tag{4.5}$$

where E^0 is the potential of the electrode under standard conditions of temperature and unit concentration of the reacting species, with reference to the SHE, R is the gas constant, 8·314 J/mole and T, absolute temperature, n is the number of electrons involved in the redox re-

*Once a battery has been primed, it should not be necessary to add any further acid. Water is lost through evaporation and electrolysis and will continually need to be replenished. Over the cycling life of the battery, therefore, the total volume of water added will in general exceed considerably the original volume of the electrolyte. Levels of impurities in water should, therefore, be lower than for acid.

action, F is the faraday, 96 500 coulombs and a_1 and a_2 are the activities of the ions on the right-hand and left-hand sides of the redox equation.

In a galvanic cell, the overall EMF in the steady state E^0_C is equal to the difference between the potentials of the cathode E^0_c and anode E^0_a

$$E^0_C = E^0_c - E^0_a \tag{4.6}$$

When the external circuit is closed, electrons flow freely from the anode to the cathode outside the cell. The steady state is disturbed and the double layers dicharged. The electrochemical redox reactions then become irreversible and the electrode potentials change from their steady state values by amounts referred to as the over-potential or over-voltage (η_c, η_a). The voltage of the cell then becomes

$$E_C = E^0_c - \eta_c) - (E^0_a + \eta_a) - IR \tag{4.7}$$

The third component IR takes account of the internal resistance.

4.3.4 *Electrochemical reactions*

The discharge/charge reactions are shown in the following equations:

The electrolyte

$$H_2O \rightleftharpoons H^+ + OH^- \tag{4.8}$$

$$H_2SO_4 \rightleftharpoons 2H^+ + SO_4^{2-} \tag{4.9}$$

or

$$H_2SO_4 \rightleftharpoons H^+ + HSO_4^- \tag{4.10}$$

At the cathode

$$PbO_2 + SO_4^{2-} + 4H^+ + 2e \underset{c}{\overset{d}{\rightleftharpoons}} PbSO_4 + 2H_2O; E^0_c = 1{\cdot}685V \tag{4.11}$$

At the anode

$$Pb + SO_4^{2-} \underset{c}{\overset{d}{\rightleftharpoons}} PbSO_4 + 2e; \; E^0 = -0{\cdot}356V \tag{4.12}$$

Overall

$$PbO_2 + 2H_2SO_4 + Pb \underset{c}{\overset{d}{\rightleftharpoons}} 2PbSO_4 + 2H_2O; E_C^0 = 2{\cdot}041\text{V} \quad (4.13)$$

where d and c represent the discharge and charge conditions.

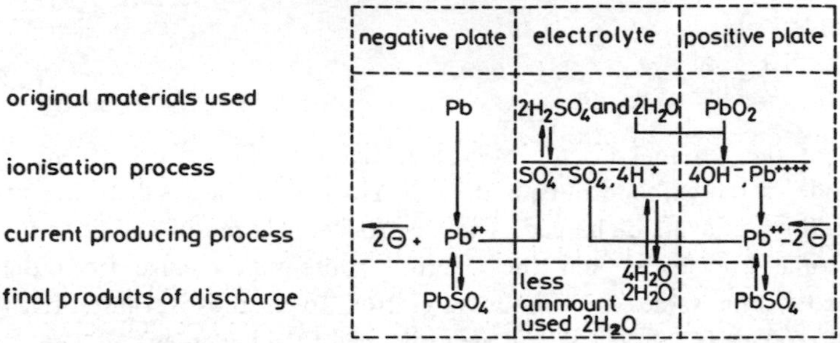

	negative plate	electrolyte	positive plate
original materials used	Pb	$2H_2SO_4$ and $2H_2O$	PbO_2
ionisation process		SO_4^{--} SO_4^{--}, $4H^+$	$4OH^-$, Pb^{++++}
current producing process	$2\ominus$ + Pb^{++}		Pb^{++} - $2\ominus$
final products of discharge	$PbSO_4$	less ammount used $2H_2O$; $4H_2O$ $2H_2O$	$PbSO_4$

Fig. 4.6 ***Discharge reactions***
[Courtesy G. Vinal, ***Storage batteries***, 4 edn., copyright Wiley & Sons, 1955]

	negative plate	electrolyte	positive plate
final products of discharge	$PbSO_4$	$4H_2O$	$PbSO_4$
ionisation process	Pb^{++}, SO_4^{--}	$2H^+$, $4OH^-$, $2H^+$	SO_4^{--}, Pb^{++}
process produced by current	$+2\ominus$		$-2\ominus$ Pb
original materials restored	Pb	H_2SO_4 $2H_2O$ H_2SO_4	PbO_2

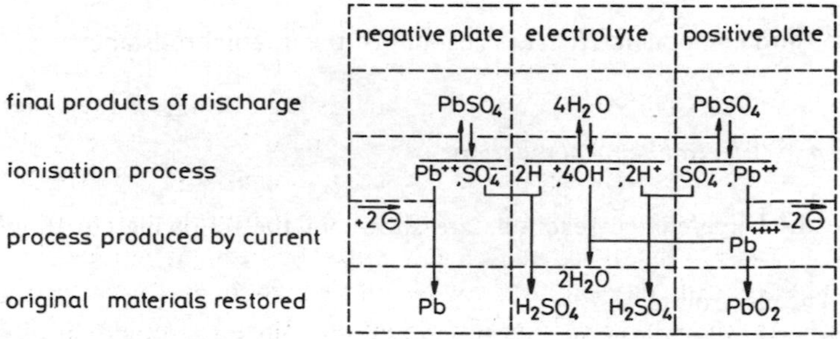

Fig. 4.7 ***Charge reaction***
[Courtesy G Vinal, ***Storage batteries,*** 4 end., copyright Wiley & Sons, 1955]

Figs. 4.6 and 4.7 from Vinal[11] show in more detail the different stages of these reactions. The possible involvement of HSO_4^- ions must also be taken into account. Bode[12] and Berndt[13] have proposed the following reactions:

$$PbO_2 + HSO_4^- + 3H^+ + 2e \rightleftharpoons PbSO_4 + 2H_2O \quad (4.14)$$

$$Pb + HSO_4{}^- \rightleftharpoons PbSO_4^- + H^+ + 2e \quad (4.15)$$

Overall

$$PbO_2 + 2HSO_4^- + 2H^+ + Pb \rightleftharpoons 2PbSO_4 + 2H_2O \quad (4.16)$$

Using values for the changes in free energy content of the substances taking part. Berndt has shown that at unit concentration of the electrolyte, 1 mol/l, the cell voltage should be 1·928V, and, from the change in entropy, the temperature coefficient of the voltage, about 0·2mV/°C, values in close agreement with those reported by Vinal. Using rotating disc, ring-disc and potentiostatic pulse measurements, however, Fleming *et al.*[14] reported difficulties in identifying the two sulphate ions HSO_4^- and $SO_4^{=}$. In any case, the end products of the discharge reaction are the same by both routes. The conclusion that $PbSO_4$ is produced in equal quantities at both cathode and anode, referred to as the 'double sulphate' theory was first recorded by Gladstone and Tribe over 90 years ago and since abundantly confirmed by various authors, notably Craig and Vinal[15] and Beck, Lind and Wynne Jones.[16]

From the above equations, it can readily be estimated that the amounts of the active materials consumed per ampere-hour are PbO_2, 4·45g; Pb, 3·86g; H_2SO_4, 3·68g. The stability of the lead-acid battery is largely determined by three critical factors:

(*a*) The high hydrogen over-voltage of the lead negative electrode. Kabanov *et al.*[17] reported a figure of about −0·95V at a CD of 0·1 mA/cm² in 2N H_2SO_4, which is slightly higher than similar values found by Gillibrand and Lomax.[18] This value is about 0·6V above the reversible $Pb/PbSO_4$ potential (E^0 = −0·356V), which means that the evolution of hydrogen by a reaction between the metal and the acid would be strongly inhibited.

(*b*) The high oxygen over-voltage of the PbO_2 positive electrode. As Ruetschi *et al.*[19] have pointed out, since the potential of the $PbO_2/PbSO_4$ electrode (E^0 = 1·685V) is about 0·45V above the voltage at which H_2O decomposes (1·229V), oxygen should normally be evolved. The oxygen over-voltage on porous PbO_2 electrodes is, however, appreciably higher than the $PbO_2/PbSO_4$ potential. Barak, Gillibrand and Peters[20] reported figures of about 1·95V at CD around 1 mA/cm² and Ruetschi and Cahan[21] 2·0V at 3 mA/cm².

(*c*) The low solubility of the main product of the discharge reaction, $PbSO_4$. Fig. 4.8 shows the solubility of $PbSO_4$ in H_2SO_4 over the range of concentrations likely to be met in storage batteries, reported by Craig and Vinal[22] and Afifi, Edwards and Hampson.[23] The value passes through a maximum of about 6·8 mg/l in 1·070 SG and at 1·250 SG it is in the range 2 to 2·5 mg/l. As a result of the low solubility, there is no significant migration of $PbSO_4$ during cycling and this ensures a high degree of reversibility. The solubility is, however, high enough to promote the dissolution-precipitation mechanisms, described in Section 4.5.4, on which the performance of the accumulator depends.

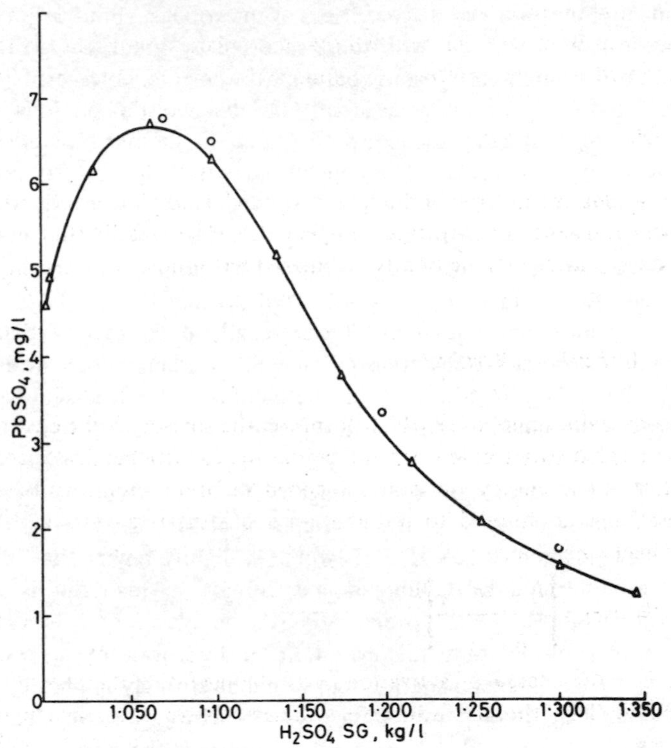

Fig. 4.8 *Solubility of $PbSO_4$ in sulphuric acid at 25°C*

△ Craig and Vinal[22]
○ Afifi, Edwards and Hampson[23]

4.4 Polarisation and kinetic aspects

The energy changes that take place during electrochemical reactions are described in Chapter 2. Eqn. 2.16 defines the partition of electrical energy and heat evolved and eqn. 2.17 [$E^0 = -\Delta G/(nF) J$] the standard potential of the cell reaction. Practical discharges at finite rates do not meet the conditions of thermodynamic reversibility.[24] Electrode and cell voltages are less than the reversible values and the resulting decrease in electrical energy appears as heat. These changes in the potentials, referred to in Section 4.3.3 as over-voltages, are also termed 'polarisation'. Since the main function of voltaic cells is to supply energy (watt-hours) or power (watts) to do useful work, the current-time-voltage relationships under different conditions of discharge and charge and the manner in which these are affected by polarisation of the electrodes are of prime importance.

Agar and Bowden[25] have shown that the polarisation of an electrode is composed of three different over-voltages

$$\Delta E = \eta_a + \eta_c + \eta_r \tag{4.17}$$

where η_a stands for activation over-voltage and η_c and η_r for concentration and resistance over-voltages, respectively. The contribution which each factor makes may be briefly summarised as follows

4.4.1 Activation polarisation

The electrical double layer which forms on the surface of the electrode acts somewhat as a condensor, and, before the electrochemical reaction can start, some energy must be expended in overcoming this barrier. Bockris[26] has attempted to put this on a quantitative basis with the following expression:

$$I = A \exp \left[\frac{E - \alpha F \eta_a}{RT} \right] \tag{4.18}$$

where E is the energy of activation, A is a constant, α the transfer coefficient and η_a the activation over-voltage, the value of which then becomes

$$\eta_a = A' + \frac{RT}{\alpha F} \ln I \tag{4.19}$$

where $A' = (E - \ln A)/\alpha F$

At constant temperature and current density, the value of the activation polarisation is small for most solid electrode systems, but accounts for the small drop in voltage immediately the current is switched on.

4.4.2 Concentration polarisation

As soon as the charge-transfer reaction starts, the concentration of ions reacting with the electrode falls rapidly in the immediate vicinity and progressively less as the distance increases at a rate dependent on the current. The electrode potential then falls by an amount defined by the Nernst equation, eqn. 4.5. Migration and diffusion processes operate to reduce the concentration gradients by mass transport of the reactants.

Figs 4.9A and B illustrate schematically how concentration polarisation operates during the discharge of a lead electrode.[26a] SO_4^{2-} ions are removed at the surface of the electrode at a rate equal to $I/2F$ gmoles/s and replenished from the bulk of the electrolyte at $It^-/2F$ gmoles/s, while protons in the form of H_3O^+ move into the bulk of the electrolyte at a rate of It^+/F gmoles/s. t^- and t^+ are transport numbers of SO_4^{2-} and H_3O^+ ions, respectively.

The concentration gradients set up in the diffusion layer accelerate the diffusion of SO_4^- ions but retard the diffusion of H_3O^+ ions by amounts equal to $2K \cdot dc/dx$, where K is the diffusion constant and dc/dx the rate of change of concentration c at a distance x from the surface. Fig. 4.9B shows how changes in the concentration, which occur in the vicinity of the electrode at times T_0 to T_5 from the outset of the

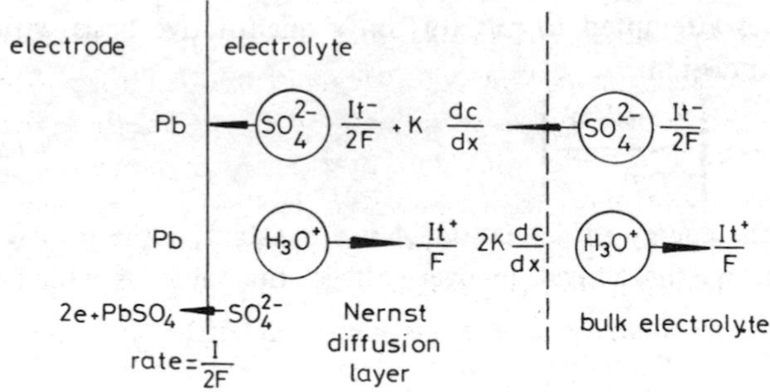

Fig. 4.9A *Concentration polarisation during discharge*

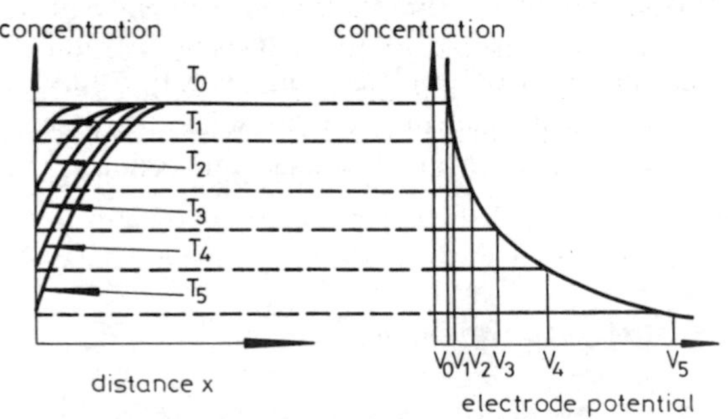

Fig. 4.9B *Electrode potentials V_0/V_5 at times T_0/T_5*

discharge, affect the electrode potential, as defined by the expression $V = V_0 - RT/nF \cdot \ln c$. A similar model can be drawn for the PbO_2 positive electrode, although this is more complex. In this case, the formation of H_2O, as indicated by eqn. 4.11, causes further dilution of the electrolyte in the diffusion layer.

4.4.3 Resistance polarisation

During discharge, the concentration of the electrolyte falls steadily with a corresponding increase in electrical resistance. As shown in Table 4.3, this amounts to over 60% for a fall in SG from 1·250 to 1·070, which might be expected in practice with a deep discharge at a low rate. The high resistance of $PbSO_4$ puts it almost in the category of a non-conductor and the combined effect of these two factors adds another vector to the over-voltage of the cell. It is of interest to note that the resistivity of massive PbO_2 is only 10 times higher than that of metallic lead, but less than that of storage battery electrolyte by a factor of 10^{-4}.

Problems of mass transport, caused by restricted diffusion, are mainly responsible for the low coefficients of use of both positive and negative active materials and to limit the polarisation, these materials are prepared in a highly porous state by methods described in Section 4.6.7. The average pore size is, however, small, and the channels for circulation of the electrolyte are tortuous, presenting additional obstacles to diffusion. For this reason, plate thicknesses are kept as low as possible, particularly where discharges at high rates are mainly needed.

As the discharge proceeds, the particles of active material become isolated from one another and also from the current-collecting grids by films of lead sulphate. The molecular volume of the $PbSO_4$ is appreciably higher than that of either the PbO_2 or the Pb, from which it is formed (Table 4.5) and the porosity of the active material falls steadily during the discharge. There is, however, a practical limit to the porosity of the active materials, which must retain electrical contact between particles and sufficient mechanical strength to prevent disruption by dimensional changes, leading to premature shedding and a loss of capacity.

Table 4.3 *Specific resistivities and densities at 20° C*

		Specific resistivity Ωcm	Density g/ml
PbO_2 [27]	massive	2×10^{-4}	9·4
	porous active material	74×10^{-4}	
Pb	dense metal	$0{\cdot}2 \times 10^{-4}$	11·3
H_2SO_4 solution	SG 1·250 or 4·3 Molar	1·230	1·250
	1·070 or 1·1 Molar	2·000	1·070
$PbSO_4$ [28]		$0{\cdot}3 \times 10^{10}$	6·3

4.4.4 Discharge-voltage characteristics

Following earlier work by Gillibrand and Lomax,[18] Baikie, Gillibrand and Peters[29] have reported the results of tests with standard SLI positive and negative plates, having dimensions and weights of active materials as follows:

	Dimensions	Active materials
Positive plates	11x11x0·19cm	153g of PbO_2
Negative plates	11x11x0·16cm	142g of Pb

Each plate was assembled between two plates of opposite polarity in an excess of electrolyte of 1·250 SG, without separators. These three-plate cells were discharged at constant current at three current densities 21, 42 and 165 mA/cm² and at three different temperatures 16°C, 0°C and -18°C. The potential of the central plate was measured against a standard mercury sulphate electrode and expressed with reference to a standard hydrogen electrode by the addition of 0·601V to the reading.

4.4.4.1 Effect of current density on capacity: Fig. 4.10*a* shows average discharge-voltage curves for both plates at the three current densities at 16°C and 10*b*, at 21 mA/cm² and the three temperatures. There are three main portions of interest

(*a*) An immediate small, but significant, drop in potential. As Ruetschi[30] has pointed out, this is due partly to resistive effects in the electrolyte and the plates and partly to activation polarisation, caused by the discharge of the double layer capacity on the surfaces of the electrodes. The latter effect initiates the charge-transfer processes, which supply the current in the external circuit.

(*b*) The middle portion of the curves, more or less horizontal at the longest rate, but becoming steeper as the current density increases. At the longest rates, diffusion processes are able to keep the concentration of SO_4^{2-} ions fairly steady, and, therefore the potential, as defined by the Nernst equation, eqn. 4.5.

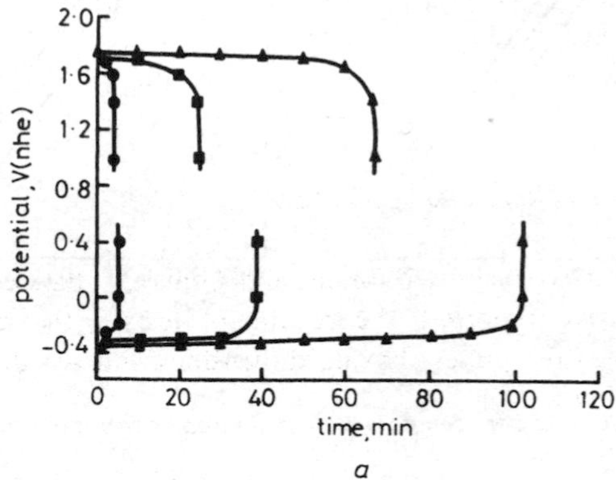

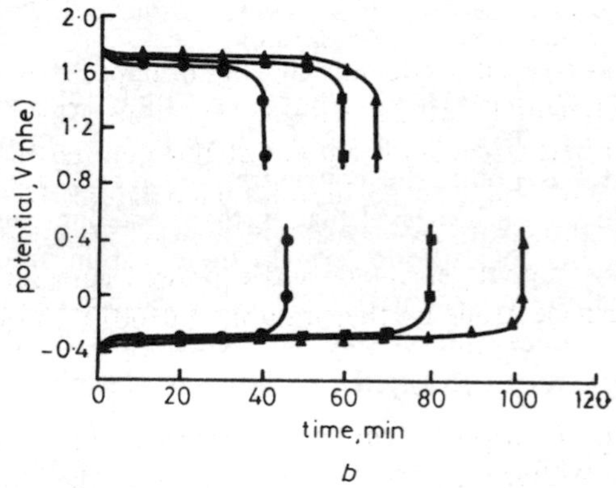

Fig. 4.10 *Discharge characteristics of battery plates*
(*a*) At 16°C[29]

▲ 21mA/cm² ■ 41 mA/cm² ● 165mA/cm²

(*b*) At 21 mA/cm²

▲ 16°C ■ 0°C ● −18°C

[Courtesy P.E. Baikie, K. Peters and M.I.G. Gillibrand, *Electrochemica Acta,* **17**, 1972, Pergamon Press Ltd.]

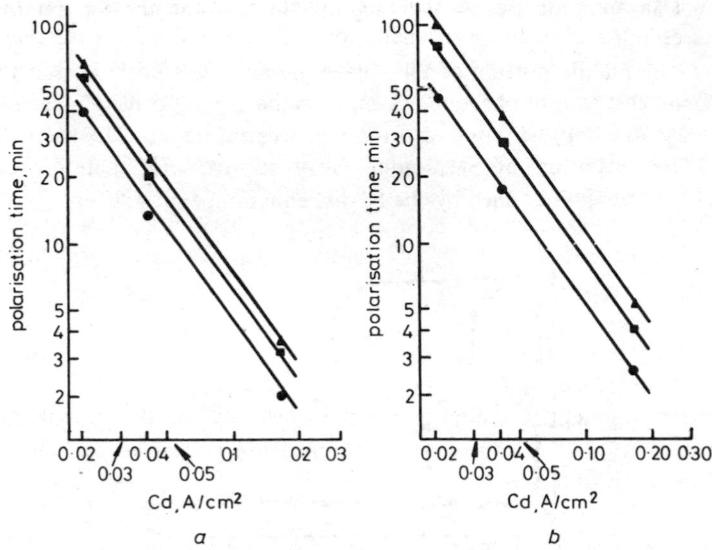

Fig. 4.11 *Relationship between polarisation time and current density*
▲ 16°C
■ 0°C
● 0 - 18°C
a Positive plates
b Negative plates
[Courtesy P.E. Baikie, K. Peters and M.I.G. Gillibrand, *Electrochemica Acta,* **17**, 1972, Pergamon Press Ltd.]

(*c*) The sharp fall in voltage at the end, generally referred to as the 'knee' of the curve, beyond which little further capacity can be drawn. At this point, the concentration of SO_4^{2-} ions has been reduced to such a low level that further depletion causes a sharp change in the logarithmic term in the equation and an almost vertical fall in the potential curve. All three polarisation factors, resistance, concentration and re-crystallisation of the reaction products are then exerting their maximum effects.

Discharge currents are generally identified by their duration to the knee of the curve, e.g. the 10 h, 5 h, 1 h, 10 min rate etc. Those used in these tests corresponded to the approximate 70 min, 25 min and 3 min rates for the positive plates and 100 min, 40 min and 5 min rates for the negatives. Under these conditions, the rate of polarisation of both plates was largely controlled by diffusion, in the manner indicated for negative plates in Fig. 4.9A and B. In practical multi-plate cells, diffusion processes are further hindered by the close packing of the plates and the porous separators. The slopes of the discharge voltage curves become steeper as the rate increases and the knee less pronounced. As already mentioned, towards the end of the discharge layers of highly resistant sulphate tend to encapsulate the particles of active material. Some observations on this mechanism of sulphate formation are made in Section 4.5.4.

Many attempts have been made to deduce from first principles an expression defining the relation between the discharge current and the capacity. The most widely accepted formula, and one of the simplest, was derived impirically by Peukert (1897) from the results of tests on complete batteries

$$t = \frac{K}{I^n} \text{ or } C = \frac{K}{I^{(n-1)}} \tag{4.20}$$

where t represents the duration of the discharge, I is the current, C is the capacity and K and n are constants. This expression may also be written logarithmically

$$\log t = \log K - n \log I \tag{4.21}$$

Since $\log K$ is also a constant, there is a linear relationship between $\log I$ and $\log t$, with n numerically equal to the slope of the line and K the intercept on the ordinate. Gillibrand and Lomax[31] have recorded a value of 1·39 for n for both positive and negative plates at 25°C and values for K of 0·35 for positive plates and 0·42 for negative plates.

Figs. 4.11A and B, recorded by Baikie *et al.*[29] confirm this linear relationship over the range of current densities and temperatures described. The slopes of the lines were practically the same for both positive and negative plates giving a value for n of 1·44, in reasonable agreement with the earlier figure quoted.

4.4.4.2 Effect of temperature on capacity: Figs. 4.12 *a* and *b* show that the capacity in ampere-hours and the temperature also obeyed a linear relationship for both plates, the slopes of the lines for negative plates being somewhat higher than for positive plates. This was borne out by temperature coefficients of capacity, determined by the expression

$$C_1 = C_2 \, [1 + \alpha(\theta_1 - \theta_2)] \tag{4.22}$$

where C_1 is the capacity in ampere-hours at temperature θ_1, C_2 is the

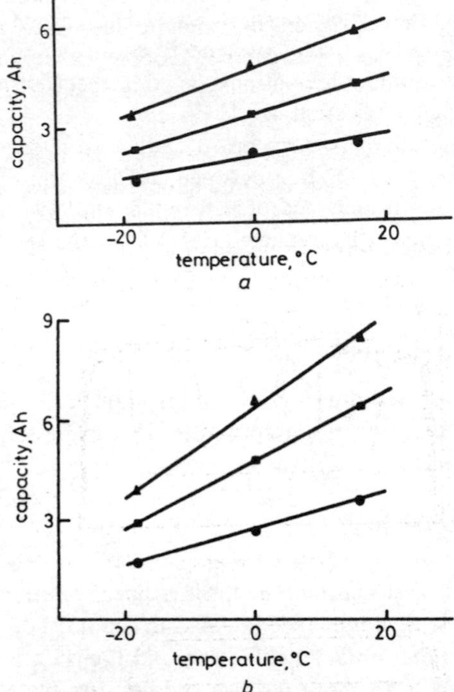

Fig. 4.12 *Variation in plate capacity with temperature*
▲ 21mA/cm²
■ 41 mA/cm²
● 165 mA/cm²
a Positive plate
b Negative plate

capacity of θ_2 and α_1 the temperature coefficient in Ah/1°C. Estimated values for the temperature coefficients over the whole range from 16°C to −18°C and the three current densities recorded were

Positive plates $1{\cdot}20$ to $1{\cdot}30 \times 10^{-2}$ Ah/1°C
Negative plates $1{\cdot}50$ to $1{\cdot}60 \times 10^{-2}$ Ah/1°C

It follows from exp. 4.20 that K should have a similar linear temperature relationship to C and eqn. 4.20 can therefore be extended to cover the full range of temperatures, as well as current densities, by the expression

$$C_1 = \frac{K_1\ [1 + \alpha(\theta_1 - \theta_2)]}{I^{n-1}} \tag{4.23}$$

4.4.4.3 Effect of plate thickness on the coefficient of use of the active materials: The diffusion of sulphate ions into the tortuous channels in the porous active materials becomes more and more restricted as the thickness of the plate increases, and this effect is increasingly significant as the rate of discharge increases.

From measurements with an X-ray micro-probe, Bode, Panesar and Voss[32] have plotted the contours of $PbSO_4$ in positive and negative plates when discharged at various rates from the 20 h to the 10 min rate. Their results for plates, respectively 2mm and 1·5mm in thickness, are shown in Figs. 4.13*a* and *b* and it is evident that similar processes

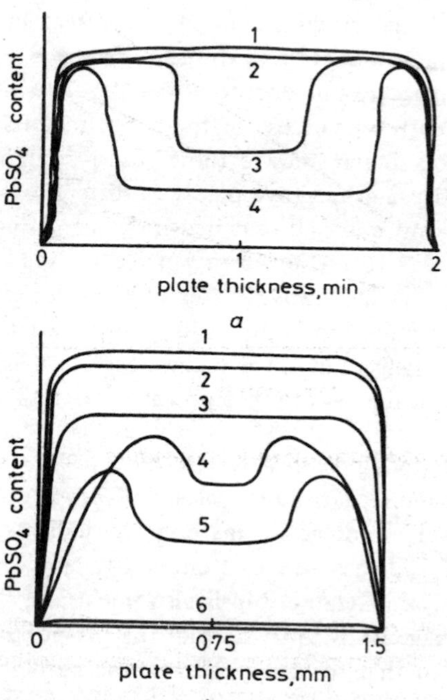

Fig. 4.13 ***$PbSO_4$ distribution across the thickness of discharged plates as a function of current density (Bode, Panesar and Voss, 1968 and 1969)***

a Positive plates, discharged at:
(1) 1·5mA/cm² (20h)
(2) 6 mA/cm²₂ (5h)
(3) 30mA/cm (1h)
(4) 180mA/cm² (10min)

b Negatives plates, discnarged at:
(1) 1·5mA/cm
(2) 6·0mA/cm
(3) 30mA/cm
(4) 90mA/cm
(5) 180mA/cm
(6) Fully charged

[Courtesy H. Bode, *Lead-acid batteries*, copyright Wiley & Sons Inc.]

operate in both plates. Bode *et al.* concluded that the acid in the inner zones made the greatest contribution to the capacity at the higher rates of discharge. This is certainly true during the early stages of the discharge, but the high peaks near the surfaces of the plates indicate that the surface layers of the active material, which are in direct contact with a small reservoir of electrolyte, become fully involved in the later stages of the discharge. Because the concentration of SO_4^{2-} ions is rapidly changing in the different zones, concentration cells are continually formed. These will create local potential differences, defined by the Nernst equation, and this will tend to equalise the concentration of $PbSO_4$. If chemical analysis is attempted, this effect will be aggravated by washing plates in water before analysis.

Peters *et al.*[33] have examined the effect of plate thickness on the capacity and the coefficient of use of the active materials in a series of tests, carried out on similar lines to those described in Sections 4.4.4.1 and 4.4.4.2. Positive and negative plates of different thicknesses were discharged at eight different current densities, increasing stepwise by a factor of 2 from 5·2 to 661 mA/cm^2 in electrolyte of 1·280 SG at 20°C. Both positive and negative plates obeyed Peukert's expression, showing a linear relationship between log I and log t over the bulk of the current density range, but at the highest current densities, there was some convergence of the lines for the positive plates, giving a final slope of about 2.

In Fig. 4.14 the relationship between the capacities and the rates of discharge are shown on logarithmic scales for two pairs of plates, having thicknesses and weights of active materials as indicated in Table 4.4. Plate areas were, in all cases, 11 x 11 cm or 242 cm^2 for both sides. As might be expected, this relationship is also linear and the slopes of the lines fall in the range 0·28 to 0·33, that for the negative plates being slightly higher than that for the positives. Included in Table 4.4 are estimated values for the coefficients of use of the active materials at the 5min, 1h and 5h rates of discharge. The coefficients of use represented the ratios, expressed as percentages of the measured capacity to the Faradaic capacity, calculated on a basis that 4·45g of PbO_2 and 3·86g of Pb are each equivalent to 1 Ah.

The relationship between the capacity and the plate thickness for the three different rates is shown in Fig. 4.15. Observations made by Bode,[12] confirmed this relationship was not linear, as reported by Vinal. At the highest rate, the capacity approached an asymptotic value as the thickness increased.

It should be noted that these tests with single plates were carried out with a greater excess of free electrolyte than would be present in a

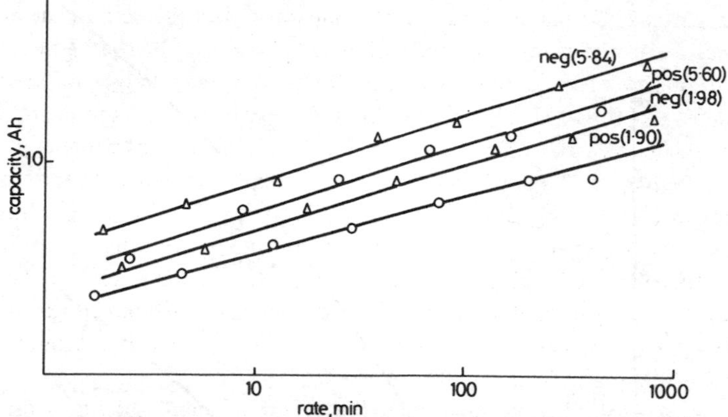

Fig. 4.14 *Capacity against rate of discharge for positive and negative plates of different thickness (plate thicknesses, in mm, are shown in brackets: see Table 4.4)*

Table 4.4 *Coefficients of use of active materials*

	PbO_2/Pb	Plate thickness	5 min rate capacity coefficient		1h rate capacity coefficient		5h rate capacity coefficient	
	g	mm	Ah	%	Ah	%	Ah	%
Positive	81·3	1·90	3·0	16·4	6·0	32·9	9·4	51·5
	245·7	5·60	4·5	8·2	10·20	18·5	17·0	30·8
Negative	80·4	1·98	3·9	18·7	8·3	39·9	13·7	65·8
	238·5	5·84	6·3	10·2	14·0	22·7	23·5	38·0

practical cell and so the coefficients of use are likely to be somewhat on the high side.

4.4.5 *Characteristics during charge*

Fig. 4.16 indicates graphically the voltage changes which take place during the recharging of positive and negative plates after a complete discharge to a final cell voltage of 1·75V. In practice, the output is limited by one or other of the plate groups; generally the positive, or, in some cases, the electrolyte. It is unusual for both plate groups to fail simultaneously, but this does not materially alter the picture. In the Figure, all single potential values are shown with respect to the standard hydrogen electrode as zero. As soon as the discharge current is cut off, the potentials of the two electrodes move along BC and B′C′, until the open-circuit voltage of the cell reaches about 2·0V, symptomatic of the small amounts of residual PbO_2 and Pb still in the

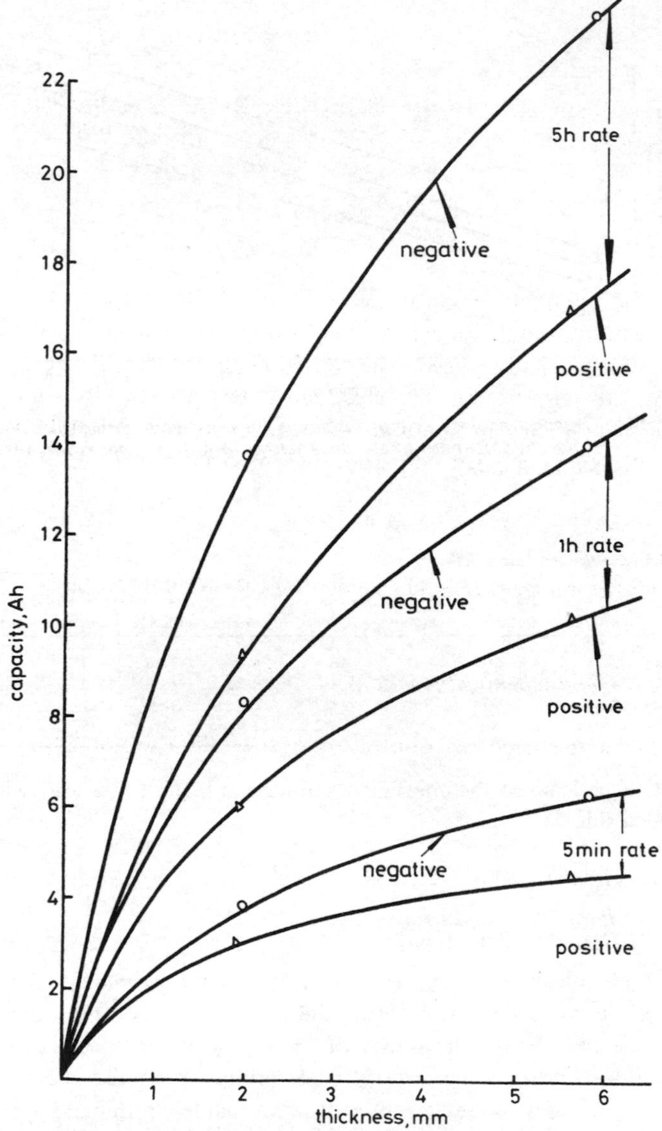

Fig. 4.15 *Capacity against plate thickness at different rates*

active materials. When the recharge is started, the plates assume overvoltages CD and C′D′, corresponding to the reactions $PbO_2/PbSO_4$ and $Pb/PbSO_4$. At points E and E′, practically all of the $PbSO_4$ has been converted to PbO_2 of Pb, respectively, and the potentials move up

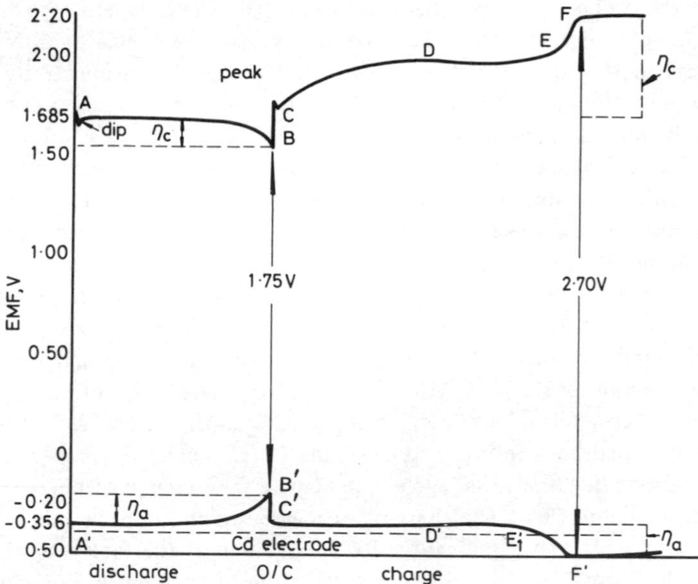

Fig. 4.16 *Voltages of positive and negative electrodes on discharge and charge*

to F and F′, where electrolytic decomposition of H_2O in the electrolyte takes place, according to the following reactions:

at the positive

$$SO_4^{2-} + H_2O = H_2SO_4 + \frac{1}{2}O_2 + 2e \tag{4.24}$$

at the negative

$$2H^+ + 2e = H_2 \tag{4.25}$$

At this point, the combined voltage of the two electrodes at currents commonly used in practice would be about 2·70V.

The *cadmium reference electrode* provides a convenient means of checking the state of charge of individual electrodes or plate groups in the factory or in the test room. For this purpose a short rod of cadmium, about 6mm in diameter, encased in an open-ended perforated hard rubber sheath is used, with a voltmeter having a high resistance, to eliminate any discharge current through the reference electrode. A pointed probe is used to make contact with the plate or plate group.

The standard potential of this electrode, $Cd^{2+} + 2e = Cd$, is −0·403V and this value (uncorrected for the pH of the electrolyte) is shown by the dotted line in Fig. 4.16. The voltages of the plates at different stages of the discharge/charge cycle can be roughly estimated by the sum or difference shown in the Fiigure. There is some advantage in lightly amalgamating the electrode with mercury.

The efficiency of the charging process is obviously important, particularly at low temperatures, which have an adverse effect on the viscosity of the electrolyte and on the rates of the electrochemical reactions. Peters, Harrison and Durant[33] have reported the results of measurements of charge acceptance at various rates and temperatures from 0°C to 40°C, on small SLI type cells. Charging currents were designated in terms of the capacity at the 10h rate of discharge C and varied from 0•1C to 0•8C. With negative plates, the efficiency of charge acceptance was high, being 100% until almost 99% of the capacity had been returned at 0•1C and 25°C. Under the most adverse conditions of these tests, current 0•8C and 0°C, charge acceptance did not fall below 70%. The charging efficiency of the positive plates was more variable and significantly lower and one of the factors involved was the formation and subsequent decomposition of persulphuric acid. At the lowest charging rate, 0•1C, full charge acceptance of 100% decreased from about 90% conversion to PbO_2 at 40°C to only 50% at 0°C and at a charging rate of 0•8C, these figures fell from about 70% at 40°C to only 34% at 0°C. The following equations indicate the probable mode of formation and decomposition of persulphuric acid:

$$H_2SO_4 \rightleftharpoons H^+ + HSO_4^-$$

$$HSO_4^- + HSO_4^- = H_2S_2O_8 + 2e \tag{4.26}$$

$$PbO_2 + H_2S_2O_8 = PbSO_4 + H_2SO_4 + O_2 \tag{4.27}$$

It was concluded from these tests that further improvements in the charge acceptance of this type of cell depended on the positive rather than the negative plates.

4.4.6 *Over-voltage and the Tafel expression*

The over-voltage or polarisation of an electrode and the current involved are related by the expression first recorded by Tafel in 1905

$$\eta = a - b \ln I \tag{4.28}$$

where η is the over-voltage, I the current density, usually expressed as A/cm^2, and a and b are constants with the following values:

$$a = \frac{2{\cdot}304\, RT}{zF}; \log I_0 \tag{4.29}$$

$$b = \frac{2{\cdot}304\, RT}{zF} \tag{4.30}$$

where z is the charge transfer coefficient, having a value roughly midway between zero and 1, R is the gas constant, T^0 the absolute temperature, F the Faraday constant and I_0 the exchange current of the reversible electrode reaction. There is obviously a linear relationship between η and $\log I$, the slope of which is represented by the expression b. Many authors have recorded this linear relationship under both anodic and cathodic conditions on both smooth and porous lead dioxide and lead electrodes, respectively. Relatively smooth lead dioxide electrodes have been made by electrodeposition on suitable metal substrates. It is impossible to eliminate a certain roughness factor, but the increase in surface area over an exactly planar surface would, with care, probably not exceed a factor of 2. Two other points must be borne in mind, first, the possible formation of layers of oxide at the interface of metal and PbO_2, which could cause secondary electrochemical side effects, and secondly, the different electrochemical properties of the two polymorphic modifications of PbO_2, known as alpha and beta, described briefly in Section 4.5.3.2.

Ruetschi *et al.*[34] have reported results of over-voltage tests on alpha and beta PbO_2, deposited on pure lead substrates from alkaline and acidic solutions of lead salts. The measurements were carried out in H_2SO_4 of concentration 4·4M or 1·255 SG at 30°C and the results are recorded in Fig. 4.17. The slopes of the Tafel lines were 0·15 for the alpha and 0·08 for the beta PbO_2. As indicated by Barak, Gillibrand and Peters,[20] however, the linearity achieved on these tests was not always so reproducible with the highly porous positive active material prepared for battery plates. The divergences were attributed to a number of factors, the prime one being uncertainty over the exact surface area and therefore the current density. The retention of pockets of adsorbed oxygen in the inner pore structure would aggravate this effect. The presence of adsorbed $SO_4^=$ ions in the oxide matrix and the formation of lead oxide at the interface between the PbO_2 and the lead current collector, as indicated by Burbank,[35] could also have deleterious effects.

Departures from linearity were less significant with porous lead negative electrodes, as Gillibrand and Lomax[36] have found, probably because hydrogen is less strongly adsorbed on lead and pore dimensions are generally larger in spongy lead matrices than in lead dioxide, allowing the gas to diffuse away more readily. Several authors have reported values for b in the range 0·10 to 0·15.

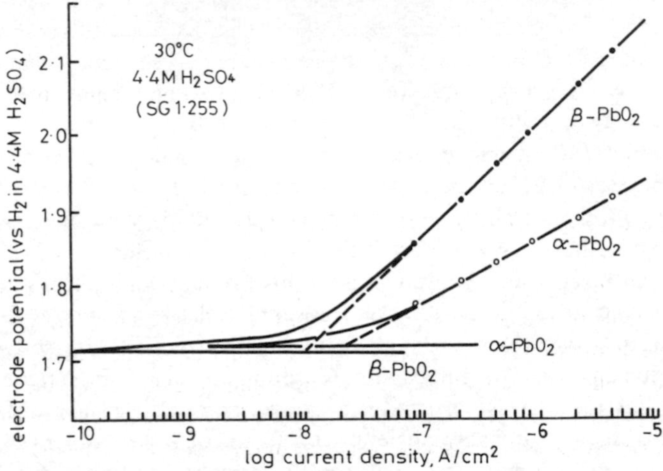

Fig. 4.17 *Tafel lines for oxygen overvoltages on alpha and beta PbO_2 at 30°C, with extrapolations to the open-circuit potentials of the two modifications. Current densities refer to 'true' surface areas, obtained by charging curves and BET area determinations*
[Courtesy P. Ruetschi, J. Sklarchuck and R.I. Angstedt, *Electrochimica Acta*, **8**, Pergamon Press Ltd., 1963]

Theoretically, Tafel lines offer a neat method for the estimation of the real surface area of an electrode. If the true surface of the porous electrode is related to the planar surface by a factor A, it can readily be deduced from the Tafel expression that:

$$\log A = \frac{\eta_1 - \eta_2}{b} \tag{4.31}$$

where η_1 and η_2 are the over-voltages of the planar and porous electrodes at unit current density, respectively.

4.5 Oxides of lead

As these oxides form one of the main materials of storage battery technology, relevant properties will be briefly considered. In its atomic structure, lead has four valency electrons in the *p* shell, two of which are in the 6*p* and two in the 6*s* orbitals. Because of the partially filled orbitals, the metal can exist in more than one state of oxidation and a number of different stoichiometries are therefore possibe. Curiously enough, however, two of the valency electrons are available for sharing, but not for ionisation. Lead, therefore, behaves predominantly as a divalent metal in its polar compounds,[37] e.g. PbO, $PbCl_2$, $PbSO_4$ and so on, all of which yield plumbous ions, Pb^{2+}, in electrolytes. In covalent organic compounds, on the other hand, all four valency electrons are active and the metal is quadrivalent in compounds such as lead tetra-methyl, $Pb(CH_3)_4$.

4.5.1 Lead monoxide PbO. Ratio Pb: O = 1:1

The lowest form exists in two polymorphic modifications, generally distinguishable by their colours. The red polymorph, known as 'litharge' or beta lead monoxide, has a rutile or tetragonal crystalline form and the bright yellow 'massicot' has an ortho-rhombic structure. The red form is regarded as stable at lower temperatures and the yellow form at high temperatures, but the transition from red to yellow is reported by Hoare[38] to take place over a fairly wide temperature band, 486 to 586°C, indicating the relative stability of both polymorphs. The transition temperature is, however, of importance in the preparation of 'grey oxide', the material used in the paste-making process and described in Section 6.7. Although the colours of the oxides are distinctive, conclusive evidence of the two polymorphs can be obtained only by X-ray diffraction measurements. Even then, there are difficulties in mapping the much lighter oxygen in the crystal lattice and neutron diffraction studies yield more precise results. The two forms also have slightly differing chemical properties. For example, the solubility of the yellow form in water or alkali solution is about twice that of the red form, the density of the former is slightly higher than that of the latter, but the electrode potential, $E_{Pb/PbO}$ in 1M NaOH at 25°C of the red form exceeds that of the yellow by about 7 mV.

4.5.2 Red lead or minium Pb_3O_4. Ratio Pb:O = 1:1·33

Red lead is formed when lead monoxide is heated in a stream of air at about 540°C. If the temperature is held much above this, decomposition sets in and the oxide reverts to the monoxide. Red lead has a unique tetragonal crystal structure, established by both X-ray and neutron diffraction examination. It can be prepared in a finely-divided form and when treated with sulphuric acid it reacts as though its stoichiometric composition is PbO_2.2PBO, the lead monoxide forming lead sulphate, leaving the PbO_2 unaffected. When used in a positive plate paste mix, therefore, about one third of the oxide is converted directly into PbO_2 before the charging process is started and this assists the formation process.

4.5.3 *Lead dioxide PbO_2. Ratio Pb:O = 1:2 (nominally)*

In their comprehensive review of the lead dioxide electrode, Carr and Hampson[39] have described the chemical and electrical modes of preparation, the structure and physical and electrochemical properties, with references to 157 papers on the subject. Hoare[38] has also provided information on many of these aspects. Only salient points will be mentioned here.

4.5.3.1 Stoichiometry: A cardinal point concerns the ratio of oxygen atoms to lead. The formula is generally written PbO_2 and eqns. 4.11 to 4.16 are universally accepted for its behaviour during discharge and charge. In practice, the ratio rarely reaches 2:1. Depending on the way it is prepared and treated, the ratio may vary from a maximum of 1·98 to about 1·90, when, according to Katz and LeFaivre,[40] a change in phase occurs. As shown in Table 4.3, PbO_2 has a high electronic conductivity. Like metals, it has a characteristic electrode potential, and, in contact with electrolytes, it can be polarised both anodically and cathodically. As Pohl and Rickert[41] have pointed out, the PbO_2 electrode behaves in a dual capacity, both the oxygen and the metal ions taking part in the electrochemical reactions. These two ions differ appreciably in size and in electronegativity, and, as a result, bonding electrons are not shared equally. The imbalance of electrons in the space lattice of the crystal gives lead dioxide the properties of a highly-doped semiconductor. It has been shown, for example, that a thin layer of the dioxide, electrodeposited on a metal substrate, such as platinum, nickel or steel, has the property of rectifying alternating current.[42] There are, however, conflicting views as to whether the electron imbalance is associated with the excess of lead or the deficiency in oxygen.

The high electrical conductivity was considered by Thomas[27] to be due to an excess of lead and he reported a Hall coefficient which indicated that the charge carriers were electrons at concentrations approaching 10^{21} electrons per millilitre. Frey and Weaver,[43] on the other hand, concluded that the semiconductor properties were due to a deficiency in oxygen. Ruetschi and Angstadt[44] have also pointed out that an oxygen deficiency produces free electrons and this would account for the high conductivity when the ratio of oxygen to lead atoms approaches a value of 2. However, they refer to an observation by Kittel that, as the ratio falls, the conductivity decreases due to the formation of compounds of higher resistivity and this was confirmed by a decrease in the Hall coefficient. Ruetschi and Cahan[21] also found that the secondary phase, observed by Katz and Le Faivre, with an oxygen to lead ratio of about 1·90 had a significantly lower conductivity than oxides with a higher ratio.

4.5.3.2 Polymorphism: PbO_2 crystallises in two distinct polymorphic species, known as the 'alpha' and 'beta' forms, both readily identified by X-ray diffraction spectroscopy. Kameyama and Fukumoto[45] were the first to describe the alpha polymorph, later explored, along with the beta form, by many authors, notably Zaslavski *et al.*[46], Bode and Voss,[47] Burbank[48,49] Ruetschi and Cahan.[21] The alpha form has an ortho-rhombic or columbite structure and the beta form, a tetragonal or rutile structure. As Carr and Hampson[39] have pointed out, in both cases, each metal ion is at the centre of a distorted octohedron. In the alpha form, neighbouring octohedra share non-opposing edges in such a way that zigzag chains are formed. In the beta form, neighbouring octohedra share opposite edges, which results in the formation of linear chains, which are connected by sharing corners. These structures, shown schematically in Fig. 4.18, hold the key to the slight, but significant, differences in the properties of the two polymers. For example, as might be expected from the packing of the crystals, the alpha form is harder and more compact than the beta form.

The alpha form is precipitated electrochemically in an alkaline medium and the beta form in an acidic environment. Ruetschi and Cahan[50] found that the electrode potential of the alpha form is 7 to 10 mV more positive than the beta form, but the latter, when discharged in sulphuric acid, has a slightly higher capacity per unit weight than the former, probably because it is more porous. The alpha form is more readily passivated by lead sulphate, but, oddly enough, the beta form appears to adsorb SO_4^{2-} ions to a greater degree. As indicated in Section 4.4.6 and Fig. 4.17, beta PbO_2 has a steeper Tafel slope than alpha, 0·15V/decade of CD compared with 0·08V/decade. When lead sulphate is recharged, even if originally formed from the alpha polymorph, the product is beta PbO_2, since the medium is then acidic. Both species have been observed in formed positive plates. Initially the amount of alpha may be as high as 30%, but within a few cycles of discharge/recharge, the bulk of this is converted to beta. The explanation for the initial formation of alpha lies in the fact that, after seasoning, the dried paste may contain up to 5% of metallic lead,

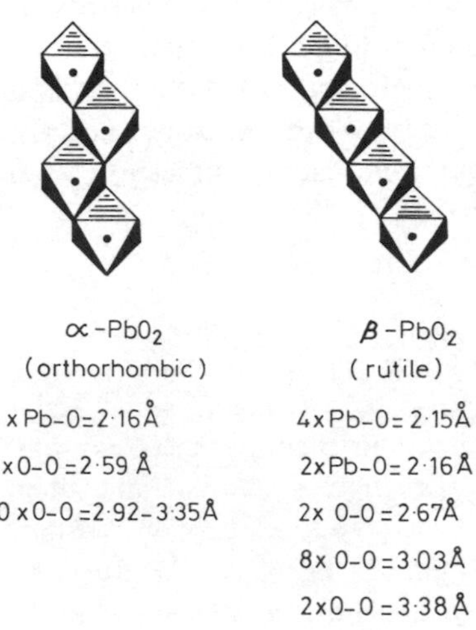

Fig. 4.18 *Octahedral structures of alpha and beta PbO_2, with inter-atomic spacing of Pb-O and O-O*[39]
[Courtesy N. Hampson and J.P. Carr, *Chemical Reviews,* **72**, The American Chemical Society, 1972]

as well as oxide, hydrate, basic sulphates and so on. The metal of the grid mesh is also distributed throughout the mass of the paste. As the electrolyte diffuses into the porous matrix, the acid is neutralised, the concentration of SO_4^{2-} ions may fall to a very low value and the water will form lead hydrate with the residual lead and the embedded grid members. The pH of the innermost regions may move into an alkaline condition and, when the current is switched on, conditions are favourable to the formation of alpha PbO_2. Voss and Freundlich[51] have confirmed that, during the first few cycles, alpha PbO_2 was present in the interior of positive plates, while the beta form predominated in the surface layers, in contact with the electrolyte, where the pH was very low and the concentration of SO_4^{2-} high, an observation also recorded by Bagshaw and Wilson.[52]

4.5.4 *Mechanism of the formation of lead sulphate during discharge*

The low solubility of $PbSO_4$ in the electrolyte (8×10^{-6} gM/L in 1·250 SG) has somewhat masked the steps in its formation during the discharge process and these difficulties are magnified when the reactions take place within the porous matrix of a standard battery plate. The recent development of the scanning electron microscope (SEM) has made it possible to observe at high resolutions the same particles of active material during a succession of discharge-charge cycles. Some excellent work on this aspect has been done by the Japanese group, HATTORI *et al.*[53] at the Yuasa Battery Co., Ltd., sponsored by the International Lead Zinc Research Organisation (ILZRO) of the USA. As Ruetschi[30] has indicated, visual evidence of this kind has helpd to verify conclusions reached by other techniques.

4.5.4.1 The lead anode: Vetter[54] has examined this problem in some detail and concluded that the discharge-charge reactions follow dissolution-precipitation mechanisms of the types

$$Pb - 2e \rightleftharpoons Pb^{2+} \qquad (4.32)$$

followed by

$$Pb^{2+} + SO_4^{2-} \rightleftharpoons PbSO_4 \qquad (4.33)$$

Vetter derived the following expression for the change in concentration of the lead ions (Δc) in the electrolyte during discharge or charge:

$$\Delta c = \frac{I\,2p}{A^2 \times \rho^2\,(1-p)^2\,d\,n\,F\,D} \qquad (4.34)$$

where I = superficial CD ($0{\cdot}1\,A/cm^2$)
p = bulk porosity of the material (0·5)
A = BET surface area of the porous mass. (10^2/m/g)
ρ = density of the solid material ($10 g/cm^3$)
d = thickness of the plate (0·1 cm)
n = number of electrons involved, the valency (2)
F = Faraday constant (say 10^5)
D = diffusion constant ($5 \times 10^{-4}\ cm^2/s$)

Using the figures shown in brackets, Vetter obtained a value of Δc of 4×10^{-12} mol/cm^3 or 4×10^{-9} mol/l. This concentration is less than the solubility of $PbSO_4$ by at least three orders of magnitude, so the conditions are favourable to the dissolution of Pb^{2+} ions. In these calculations, it was shown that the average pore diameter was about 200 Å and even with a very high superficial CD of 1 A/cm^2, the true CD throughout the active mass would be as low as 2×10^{-5} A/cm^2, making conditions for a dissolution/precipitation mechanism even more favourable.

4.5.4.2 The PbO_2 cathode: Agreement on the nature of the discharge/charge processes was not quite so general. Feitknecht and Gaumann[55] considered that the first stage of the discharge reaction was the release into the electrolyte of tetravalent Pb^{4+} ions. These were then reduced to Pb^{2+} ions and precipitated as $PbSO_4$. The existence of tetravalent lead ions has, however, not been confirmed, although Ruetschi[30] has referred to the possible formation during charge of intermediate complexes with OH^- or SO_4^{2-} ions. Thirsk and Wynne Jones,[56] on the other hand, suggested that the first step involved an electron transfer directly to the tetravalent lead in the solid state, i.e.

$$PbO_2 + 4H^+ + 2e \rightleftharpoons Pb^{2+} + 2H_2O \qquad (4.35)$$

with immediate precipitation of $PbSO_4$. Vetter[54] also favours this reaction, which leads to the same sort of dissolution/precipitation mechanism proposed for the lead anode. Fig. 4.19A shows schematically how the discharge reaction might take place in the boundary phases. According to Vetter, the complete conversion by a solid state reaction must be ruled out, since protons were the only ions which could diffuse into the solid matrix. In view of the high electrical resistivity of $PbSO_4$, only very thin films about 1 μm, if continuous in thickness, would be required to polarise the material completely. It is obvious, therefore, that these films of crystalline sulphate grow around the particles of PbO_2 with many cracks and imperfections. This was confirmed by Kabanov *et al.*[57], who showed by capacitance measurements of the double layer that polarisation of the PbO_2 electrode was not complete until over 90% of the surface was covered by $PbSO_4$.

During recharge, the discharge reactions take place *in reverse*, i.e. first the dissolution of $PbSO_4$ by reaction 4.33, followed by the oxidation reaction 4.35. In Fig. 4.20, Vetter indicated schematically how diffusion processes operate in the re-formation of the solid PbO_2. Simon and Caulder[58] have described SEM examinations of the structural changes which take place in lead dioxide during normal cycling in cells. The initial fibrous, dendritic, disordered form gradually changed to a more uniform crystallinity, with a brain-like, ovoid or coralloid structure and this appeared to be associated with a gradual fall in capacity. The authors attributed this to the formation of an inert form of PbO_2, possibly caused by the migration into the crystal lattice of protons or H_2O molecules. This inert form could, of course, be associated with the phase change attributed by Katz and Le Faivre[40] to the reduction in the ratio of oxygen to lead from 1·98 to 1·90

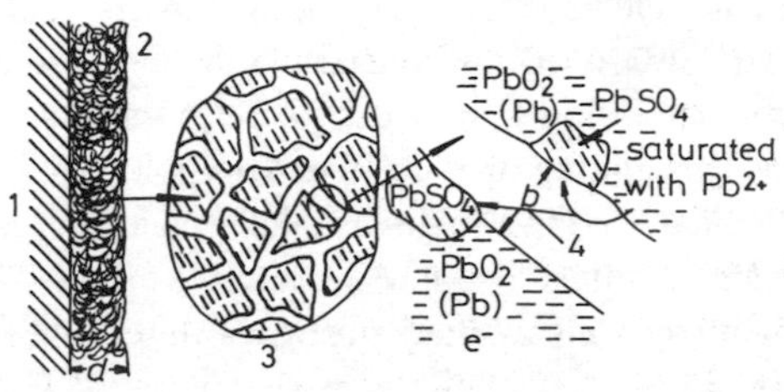

i

Fig. 4.19 *Schematic diagram of the structure of a porous electrode*
1 Metallic conductor
2 Porous electrode (thickness)
3 Idealised network (porosity about 50%, p = 0·5)
4 Average width of the diffusion layer
[Courtesy K. Vetter, *Chemie-in-Techn,* **45, 1973**]

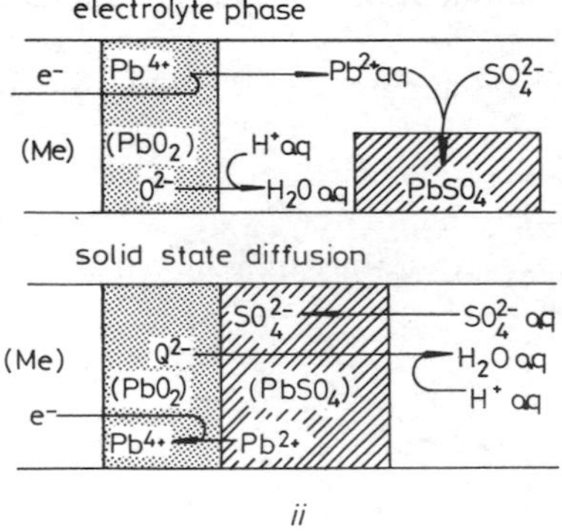

ii

Fig. 4.20 *Special phase diagram for the PbO_2/$PbSO_4$ and Pb/$PbSO_4$ electrodes*
[Courtesy K. Vetter, *Chemie-in-Techn,* **45,** 1973]

Dawson *et al*[59] have also related kinetic behaviour to morphological studies of beta PbO_2 during cycling. They concluded that, at low over-potentials, i.e. low discharge rates, a dissolution/precipitation mechanism prevailed, but at high over-potentials (high discharge rates), at least during the first few cycles, a solid state mechanism appeared to be involved in the inner recesses of the plate. This second reaction appeared to be the cause of the major change in the electrocrystallisation of the product and the development of the reticulate coralloid structures reported by Simon and Caulder.[58] Dawson reported that the kinetics of the $PbSO_4$ → beta PbO_2 reaction followed a $t^{3/2}$ type relationship, which is typical of a three dimensional nucleation and growth process. At the outset, it appeared that the PbO_2 was non-stoichiometric but electrochemically active. With continued cycling, the stoichiometry improved; in contrast to the conclusions of Katz and Le Faivre[40] it became less reactive and the cohesion of the particles decreased, causing shedding of the active material.

The dimensional changes which take place on every reversible cycle $PbO_2 \rightleftharpoons PbSO_4$ must also have an adverse effect. This amounts to +45% and this regular disruption must be a determining factor in the life of the material.

4.5.4.3 Coup de fouet or Spannungsak: This somewhat obscure phenomenon deserves a brief note. Vinal has referred to the sudden initial fall in the voltage, followed by an equally sharp upward step at the beginning of the recharge, which sometimes occurs when a cell is discharged at a constant current. The 'whip-crack' is observed within the first few seconds or minutes, depending on the rate of the discharge. Berndt and Voss[60] have made a systematic study of the effect, shown schematically in Fig. 4.21 and came to the following conclusions:

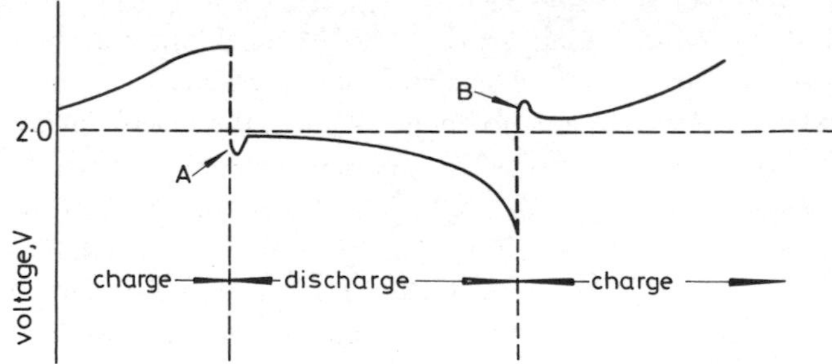

Fig. 4.21 *Discharge/charge cycle, showing cout de fouet (A) on discharge and peak voltage (B) on charge (arbitrary scales)*
[Courtesy of D. Berndt and E. Voss[60], *In Batteries 2*, COLLINS, D.H. (Ed.), Pergamon Press, 1965]

(*a*) the effect was confined to the positive plate and to the β PbO_2 polymorph: it did not occur on the negative plate

(*b*) the dip reached a maximum of 30 mV at a CD of 30 mA/cm^2

(*c*) at 19 mA/cm², it occurred after 60 s and at 2·5 mA/cm², after 2 min
(*d*) On discharge, the effect was due to supersaturation of the electrolyte by Pb^{2+} ions, which were ultimately removed at the lowest point by nucleation and crystallisation
(*e*) on charge, the voltage spike was due to the resistance of films of $PbSO_4$, effective in thicknesses as low as 1 μm, which were not removed until the dissolution process $PbSO_4 = Pb^{2+} + SO_4^{2-}$ had reached an equilibrium value.

In view of the small effects on the cell voltage, the phenomenon has little practical significance for most applications.

4.6 Manufacturing processes

4.6.1 *The Planté process*

The early history has been described in Section 1.5.3. To overcome the tedium of repeated cycling and to make the process competitive with the pasted-plate process, Planté had experimented with a preliminary attack stage using dilute nitric acid to accelerate the corrosion of the pure lead plates. This method has been perfected in the modern process. in which other lead-corroding acids, chloric or perchloric may be used as well as nitric, perchloric acid being the most popular.

To extend the surface area to about 12 times the superficial area of one side, the grids are cast with large numbers of closely-pitched vertical lamelles, held in position by robust horizontal cross members. The traditional Planté positive was 12mm in overall thickness, but this has largely given way to the 'high performance Plante' (HPP) plate, which is only 8 mm in thickness. These plates are used only as positives and are assembled with pasted negative plates, usually of the conventional open type, and, sometimes, of the 'box' type, in which the paste is retained within perforated thin lead sheets, welded to each side of the grid.

The modern Planté process is carried out in two stages. In the 'attack' stage, double rows of plates, assembled in parallel in large tanks containing sulphuric acid of 1·075 SG, to which is added 8g/l of potassium perchlorate, are charged at a CD of 10·6 ma/cm² of the superficial area for 22 h. The amount of charge is determined by the 'attack factor', the total charge being usually $4{\cdot}25 \times C_{10}$, where C_{10} is the capacity of the plate at the 10 h rate of discharge. After a short stand, the current is reversed for a similar period when the reduced plates, now covered with a thick layer of spongy lead, are replaced by another set of unformed plates and the cycling schedule repeated. The reduced plates are thoroughly washed in running water to remove all traces of the corroding acid and are then submitted to the 'formation' stage. In this the plates are again assembled in parallel, this time in pure dilute sulphuric acid, and charged anodically against pure lead sheets for about 17 h, with occasional rests towards the end of the formation. The formed plates, now dark brown or black in colour, are again washed in running water and finally dried. Where strict capacity specifications are laid down, for example by the British Post Office, sample plates are generally removed for test and the formation can be adjusted to suit.

In another design of Planté type plate, known as the 'rosette' or 'Manchester' plate, heavily-profiled pure lead tapes were coiled into rosettes and pressed into holes in a thick grid cast in lead antimony alloy. The rosettes were submitted to a Planté type formation with nitric acid as the corrodant. Rosette plates, now obsolete, were widely used for many years for standby power applications in the USA and for rail-car services in India and South America.

4.6.2 *The pasted plate*

Pasted plates are made by the process generally attributed to Fauré and Brush, in which the active materials are prepared in the form of a stiff paste by mixing certain lead oxides with water and sulphuric acid and pressed into die-cast lattice type grids, usually by an automatic pasting machine. For small production runs the paste may be applied by hand, using tools and techniques similar to those used in the application of cement or plaster of Paris. The process took a great leap forward during the two decades between the two World Wars, following the development of the automotive starter battery. The two main criteria for this service were that the battery should be as light and small as possible while still able to supply the instant power of around 2kW required to start the internal combustion engine. Because of the mounting global demand for automobiles, mass production methods were necessary. Pasted plates are, however, used for many other applications in both the traction and stationary fields.

4.6.2.1 Design of die-cast grids for pasted plates: The grid lattice not only supports the active materials physically, but also serves as the current collector to transfer electrons from the chemical reactions to the cell busbars. The grid pattern has evolved over many years of intensive testing, including attempts to map the resistance contours of grids and finished plates at various rates of discharge. The early developments were described by Wade in 1905,[61] and, although a number of refinements were introduced in the interim, many of the basic designs were still listed by Vinal[11] some 50 years later. Some current designs are shown in Fig. 4.22. Generally these take a rectilinear form, with a fairly robust outer frame, having the current take-off lug at one corner. The vertical ribs generally have the same thickness as the plate. Horizontal members are about half this thickness and staggered to provide a continuous ribbon of paste. As a further aid to keying of the active material, both vertical and horizontal members are cast with re-entrant angles, a design feature first introduced by Sellon in 1882. Similar designs are used for both positive and negative plates, but the

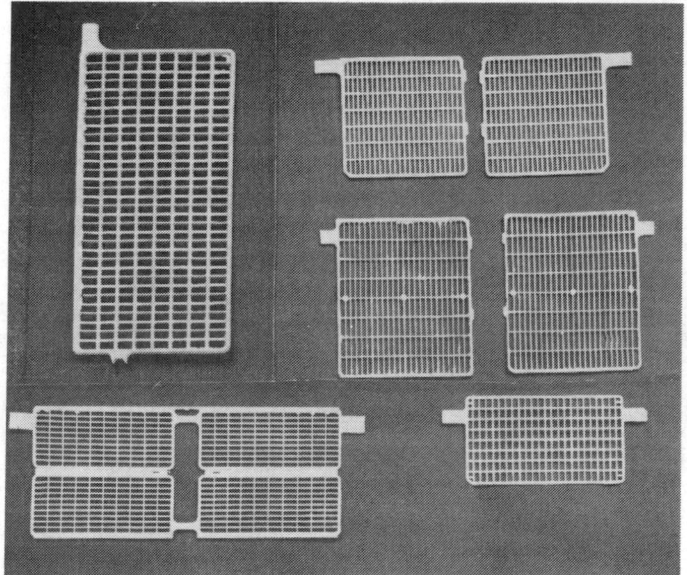

Fig. 4.22 *Die cast grids for pasted plates*

a Negative grid for a flat plate traction cell
b Standard grid for SLI automotive batteries
c Lightweight grid for SLI automotive batteries
d Grids for pasted plates for miners' cap-lamp batteries
e Grids for plates for aircraft batteries
[Courtesy of Chloride Industrial Batteries Ltd.]

former are somewhat thicker and more robust. This is done partly because the positive grid is subjected to continuous anodic attack during the charging phase of each cycle and partly because the negative plate group in each cell generally contains one more plate than the positive group. This ensures that all of the surfaces of the positive plates are fully worked and that the current density is thereby kept to a minimum. Designs with equal numbers of positive and negative plates have occasionally been proposed and the ratio of positive to negative active material is sometimes adjusted in a 'reverse assembly', having one excess positive plate.

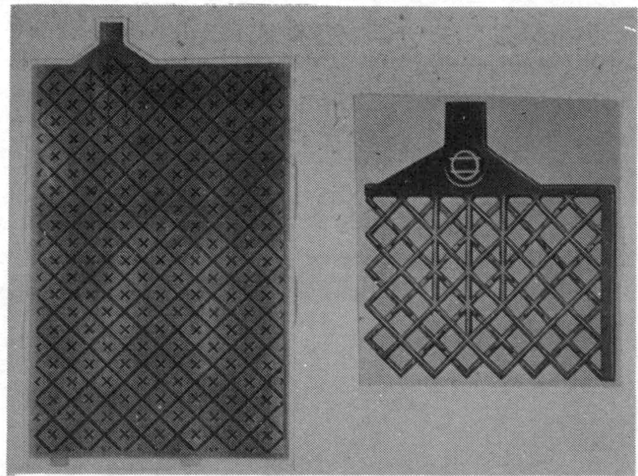

Fig. 4.23 *Positive plate with diamond patterned grid* [Courtesy Varta Batterie AG]

The amount of metal in the grid and its distribution throughout the active material are critical factors. Although *IR* voltage losses in the grid must be kept to a minimum, particularly at high rates of discharge, it is important to include as much active material as is practicable. For the best all-round performance, the ratio of the weight of active material to grid should not fall much below 1·4 to 1. As indicated in Section 4.4.4, the polarisation of the plate is related to the current density. To secure high discharge voltages, plate thicknesses have been steadily reduced, making it possible to increase the number of plates in each cell. Whereas about 20 years ago the thicknesses of positive and negative plates in standard SLI batteries were 2·4mm and 2·0mm respectively, nowadays they are nearer 1·75mm and 1·4mm. In 'heavy duty' SLI batteries for commercial vehicles, buses, trucks etc, where weight is not so critical and service conditions more severe, grid thicknesses are higher, 2·5 to 3mm, and in traction batteries, where durability and long life under arduous conditions of regular deep discharges are essential, plate thicknesses may be as high as 6mm.

In all cases, both horizontal and vertical members of the grids have sharply-angled profiles to ensure that the maximum surface of the active materials is exposed to the electrolyte, and, to secure the highest coefficient of use, they may be submerged below the surface of the active materials. In another grid form, shown in Fig. 4.23 and used in some European traction batteries, the ribs are arranged in a diamond pattern. Although this design appears to show no significant gain in conductivity, the flow of molten metal during die-casting is claimed to be more uniform, causing less hot-tearing and cracking at rib intersections.

On the assumption that the current is gathering in intensity towards the take-off lugs, designs have been proposed with the main vertical conductors running fan-wise up the grid, as shown in Fig. 4.24., a computer-assisted design, proposed by the Globe Union Co., USA. The importance of grid conductivity, particularly at high rates of discharge, is obvious, but the distribution of the grid members throughout the active material also affects the performance. During a discharge, the flux of electrons is highest in the vicinity of the conductors, where layers of lead sulphate may isolate the pellets of active material.

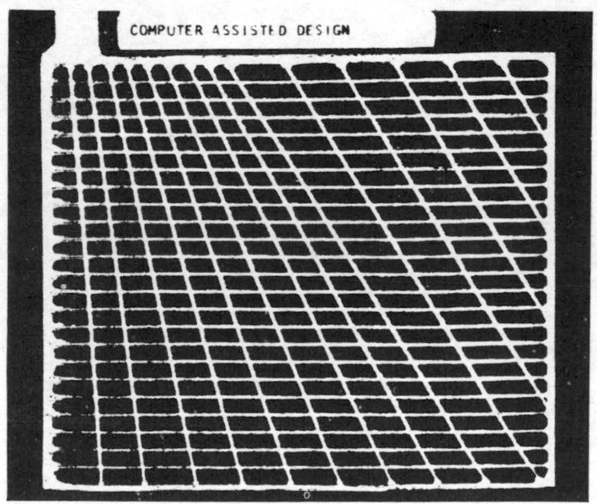

Fig. 4.24 *Computer assisted design of Grid*
[Courtesy Globe-Union Co., USA]

Faber[62] has indicated (Fig. 4.25) how the coefficient of use of the positive active material at the 5h rate fell from an estimated 92% to about 30% as the average distance from the nearest conductor increased from about 0·1mm to 6mm. The first figure was obtained by using a felted mat of fine titanium fibres as the grid. Claims for a grid made of felted lead fibres with a cast-on frame were made about 90 years ago[61] Unfortunately, in the highly corrosive environment of the positive plate the life-time of the fibres was relatively short and the principle was not pursued. Further evidence of the benefits gained, particularly at high rates of discharge by improving the conductivity of the grids, has been produced by experiments with tall cells, having plates 1000 mm in height, described by Bagshaw *et al.*[63] and Brinkman,[64] referred to in Section 4.13 below.

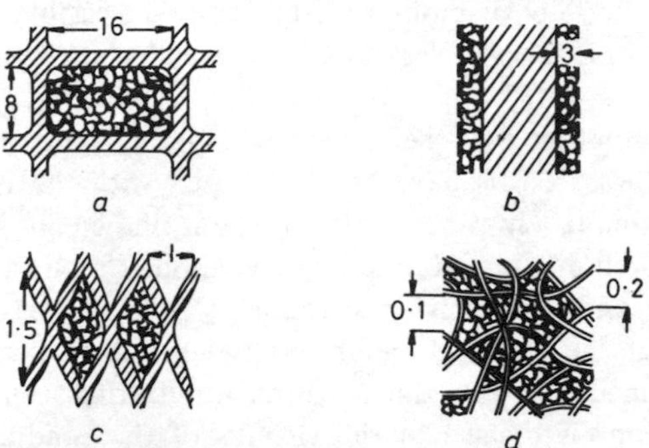

Fig. 4.25A ***Lead/acid battery titanium grid structures***
a Conventional
b Tubular type
c Expanded metal
d Fibre
[Courtesy of P. Faber,[62] *In Power sources 4*, COLLINS, D.H. (Ed.), Oriel Press, 1973]

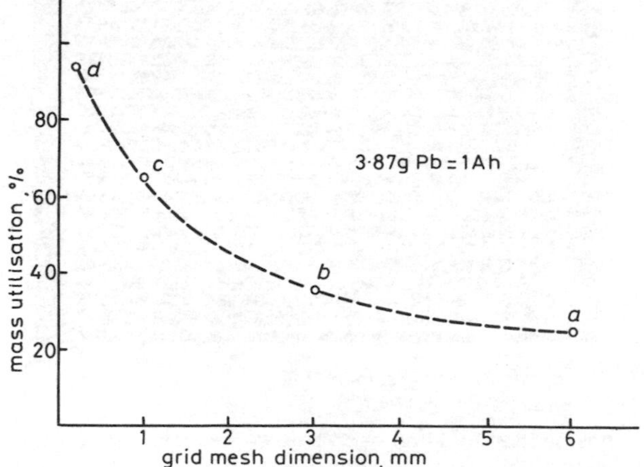

Fig. 4.25B ***Mass utilisation and local grid geometry***
[Courtesy of P. Faber[62], *In Power sources 4*, COLLINS, D.H. (Ed.), Oriel Press, 1973]

It has been traditional practice to cast grids with the current lugs at one of the top corners. The busbars to which the plates are welded in parallel are then spaced as widely as possible in the cells, minimising risks of short-circuits at the tops. Recently the Delco-Remy Division of the General Motors Corporation of the USA announced a change in their 'Freedom' SLI battery, in which the take-off lugs are cast nearer to the centre of the grids. This is claimed to give more uniform current density over the plate surface, with a corresponding gain in the coefficient of use of the active materials.

4.6.2.2 Wrought metal grids: Several attempts have been made to use wrought metal technology for grid production, thereby eliminating the energy-consuming melting stage, needed in die-casting. In 1921, the Hazelett Storage Battery Co.[65] described a continuous process in which lead sheet, produced by rolling billets, was punched to the required

Fig. 4.26A *Wrought lead/calcium grid*
[Courtesy of Delco-Remy Division of General Motors Corp., USA]

porosity, passed under a pasting hopper and finally into a sharp-setting oven, after which the plates were cut to size and stacked in heaps in a seasoning oven. The process had a number of drawbacks. The porosity of the grid was low, giving an unfavourable weight ratio of paste to grid. Keying of the active material was poor and the work-hardened metal was subject to corrosion and growth.

Recent experiments with lead-calcium alloy and the expanded metal technique appear to have overcome many of the earlier difficulties. Helms, Coyner and Hill[66] of the Delco-Remy Co. have described the development of their Freedom battery, assembled with plates made by a continuous process of this type, the lead sheet being cast between stainless steel belts and sized for thickness before being passed to the expanded-metal operation. The use of wrought lead-calcium alloy is claimed to offer numerous benefits over cast metal, including work-hardened surface strength, uniform, small grain structure and resistance

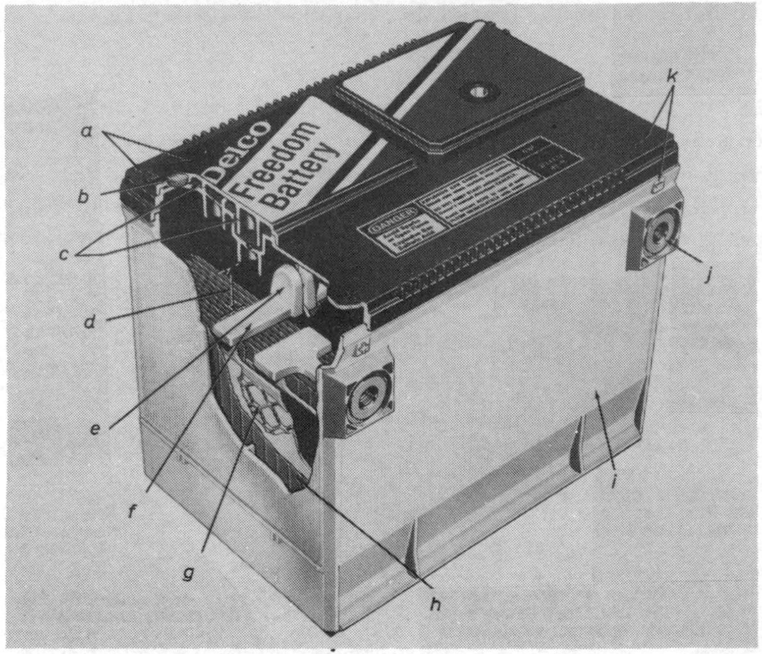

Fig. 4.26B *Delco Freedom Battery, showing wrought expanded metal grid*

a Heat-sealed covers prevent electrolyte contamination and increase case strength
b Small gas vents have built-in flame arresters
c Special liquid-gas separator returns any liquid to reservoir
d Generous electrolyte reservoir prolongs battery life
e Extrusion-fusion intercell connections provide increased performance and reliability
f Centre plate straps are highly resistant to damaging vibration
g Wrought lead/calcium grids are fine grained, strong and highly corrosion resistant
h Separator envelopes prevent shorting between plates and misalignment caused by vibration
i Special ribbed polypropylene case combines light weight with high impact strength
j Maintenance-free sealed terminal connections need no periodic tightening or cleaning
k Moulded symbols permanently identify terminal polarity

to grain-boundary attack. Fig. 4.26A shows a wrought, expanded metal grid in lead-calcium alloy. and Fig. 4.26B shows a Delco Freedom battery, assembled with grids of this kind. Fig. 4.26C indicates that fine-grained wrought metal has a greater resistance to anodic corrosion than gravity die-cast alloy of the same composition.

Fig. 4.26C ***Anodic attack on wrought metal and gravity die cast alloy of the same composition***
a Delco's wrought grid
b Competitive cast grid
[Courtesy of Delco-Remy Deivision of General Motors Corp., USA]

4.6.2.3 Lead antimony alloys for grids: Thin, open-mesh grids in pure lead are too fragile to withstand the rapid handling through the automatic factory processes and also to the adverse working conditions which apply to many portable applications. To improve the hardness and strength, lead is alloyed with antimony in varying amounts up to the eutectic composition of 11·4%, at which the melting point is reduced from 327°C to 247°C. The addition of antimony improves the fluidity of the metal, reduces the volume change in setting and the setting temperature range and thereby improves the sharpness of the grid profile. Depending on the antimony content, these alloys show in varying degree a characteristic fern-like, dendritic structure, with an average grain size much less than that of pure lead. Because of these properties, the alloys can be used in high-speed gravity die-casting equipment without cracks and hot tears. The density of the eutectic alloy is about 8% less than that of pure lead and the electrical resistivity about 35% higher. For a detailed list of mechanical and electrical properties, the reader should refer to Vinal.[11]

As indicated in Section 4.6.2.10 antimony is believed to have a subtle, beneficial effect on the structure and life of the positive active material. However, it also has a prime electrochemical disadvantage. Under the anodic attack on the grid during the overcharge of each cycle, antimony passes into solution in the electrolyte in the form of the Sb^{5+} ion. As Ruetschi and Angstadt[67] have shown, some of this is adsorbed by the positive active material and may contribute to its self-discharge, but the bulk is reduced to Sb^{3+} at the negative plate, where, because of the difference in the electrode potentials of lead and antimony, the latter is deposited on the surface. The hydrogen over-potential of Sb is lower than that of Pb, and, as Crennell and Milligan[68] and Vinal, Craig and Snyder[69] have shown, this causes self-discharge of the spongy lead with the evolution of hydrogen. According to Dawson *et al.*,[70] the reaction may be represented as

$$2SbO^+ + 3Pb + 3H_2SO_4 = 2Sb + 3PbSO_4 + 2H_3O^+ \quad (4.36)$$

During the succeeding gassing overcharge, some of the electrodeposited Sb may be released as stibine, by a reaction of the type:

$$Sb + 3H^+ + 3e = SbH_3 \quad (4.37)$$

first recorded by Haring and Compton in 1935. Holland[71] has examined in more detail the conditions for the formation of stibine and concluded that (*a*) this only takes place when the cell voltage reaches 2·6V or higher, and (*b*) the amount evolved rapidly reaches a peak within the first hour or so of the overcharge and then falls sharply, suggesting that only the surface layers of Sb are converted to SbH_3 on each overcharge. The bulk of the antimony stays in the negative active material, and, with the highest concentrations of antimony in the grid alloy, this may accumulate to the point where it is practically impossible to charge the negative plates completely, signalling the end of the useful life of the battery. To overcome these 'standing losses', extra overcharge is needed, leading to increased electrolysis and loss of water and accelerated anodic attack on the grid, through a reaction of this kind

$$2SO_4^{2-} + 2H_2O = 2H_2SO_4 + O_2 + 4e \quad (4.38)$$

Some measurements of water losses in cells with grids having different antimony contents are recorded in Section 4.6.2.8. below.

4.6.2.4 Alternative hardeners: Apart from the technical disadvantages, the rising cost and the vagaries in the supplies of antimony have stimulated the search for alternative hardeners. Many claims have been made for improved alloys, containing particularly arsenic, with reduced amounts of antimony and one or more of a number of other elements, such as tin, copper, silver, selenium, tellurium, sulphur, cadmium etc. Perkins and Edwards[72] have described the effects of these elements on the properties of the alloys concerned and emphasised the need for microstructural control in the use of alloys for battery service. Two of the major factors are the grain size and the composition of the grain boundary phases. Corrosive attack may be concentrated on the latter or it may take place over the whole of the exposed surface of the grains. In general, the former is more damaging than the latter, since it causes growth and premature breakdown of the structure of the grid. Although they may not have superior mechanical properties, therefore, fine-grained structures are to be preferred, since grain boundaries are thereby greatly extended and the thicknesses of grain boundary phases correspondingly reduced.

Table 4.5 *Dimensional changes in active material and grid metal*

	Pb	PbO_2	$PbSO_4$
Weight, gram atom or gram mole	107	239	303
Density, g/ml	11·2	9·2	6·4
Volume, gA or gM (ml)	18·5	26	47
Volume changes, %			
PbO_2 to $PbSO_4$		← 181 →	
Pb to PbO_2	←141→		
Pb to $PbSO_4$	←	254	→

Corrosion at grain boundaries and lattice defects is accentuated by the stresses to which positive grids are subjected during cycling, caused by volume changes in both the active materials and in the surface layers of the metal itself. As Table 4.5 shows, the former may amount to 180% and the latter to 250%.

The difficulties have been increased by the demand for batteries with higher energy and power densities, particularly at low temperatures, which can be met only by using greater numbers of thinner plates. This has evoked great ingenuity in the design and operation of both casting machines and moulds for die-cast grids. The almost endless variety of possible compositions has encouraged metallurgists to find a practical castability test which can be carried out in the laboratory without the need to use full-scale factory casting plant. The pine-tree shape mould, described by Mao and Larsen[73] and White and Rogers[74] provides one of the more reliable screening tests of this kind.

Torralba[75] has reviewed present trends in the use of alloys and Heubner and Sandig[76] have listed 8 different alloys used in Germany (1971), shown in Table 4.6, and also described their mechanical properties. Both of these papers contain useful lists of references to other work on this aspect. The Indian Lead Zinc Information Centre[77] has also published a list of ten manufacturers' typical grid metal specifications, mainly from American sources.

4.6.2.5 Effect of additives to lead antimony alloys:

Arsenic, added in amounts up to 0·5%, but generally around 0·2%, refines microstructure, enhancing castability and giving greatly improved resistance to anodic attack and growth. It also increases the rate and the degree of age-hardening, when grids are cooled after casting. Rapid quenching also causes grain refinement, although this may revert after long periods of ageing.

Tin is claimed to add a favourable solid solution strengthening effect. It also alters the surface tension of the molten metal, thereby improving castability. The high cost of the metal, about 10 times that of lead, limits its use. Higher tin contents are more popular in the USA, where a typical modern alloy has the following composition: Pb, 4·5Sb, 0·5Sn, 0·2As. As Heubner and Sandig have indicated, European companies generally specify low concentrations of tin. Intermediate levels are generally used in the UK. With additions of from 0·2% to 0·5% of arsenic and up to 0·5% of tin, the antimony content can be reduced to 6% without significant loss in mechanical strength, tensile strength, creep, Brinell hardness etc. During the past 15 years or so, efforts have been made to reduce the amount of antimony still further to 4% and even to 2·5%. Tin is also believed to confer some electrochemical advantages, reducing the insulating effects caused by layers of lead sulphate at the grid-active material interface.

Silver: The addition of silver is claimed to reduce the rate of anodic corrosion. From tests with an alloy composed of Pb, 8%Sb, 0·3%Ag, Dasoyan[78] attributed the effect to changes in the microstructure of the alloy. Other workers concluded that the beneficial effects were mainly due to the catalytic activity of silver in decomposing persulphate as it was formed, giving more compact layers of protective PbO_2 on the surface of the grid metal. Mao and Rao[79] have described tests with alloys containing 4·5% Sb, with and without Ag in the range 0·02% to 0·2%. Grain boundary phases were monitored with an electron beam microprobe. The addition of Ag caused a significant fall in both the

Table 4.6 *Examples of lead alloys used for SLI and sealed batteries in Western Germany, 1971*

Element or property	Grid alloy number Conventional SLI batteries							Sealed maintenance free
	1	2	3	4	5	6	7	8
*Sb weight, %	8	7	7	6	6	6	5	0·0015
As weight, %	0·007	0·12 0·14	0·20	0·12	0·15	0·25	0·10	nd
Cu weight, %	0·01	0·043	0·048	0·05	0·04	0·012	0·044	<0·001
Sn weight, %	⩽0·001	0·008	0·005	0·004	<0·001	<0·001	⩽0·005	<0·001
Ag weight, %	0·006	0·005	0·003	0·008	0·004	0·003	0·007	0·001
Bi weight, %	0·022	0·016	0·014	0·02	0·02	0·023	0·02	0·001
Tl weight %	0·0005 0·001	⩽0·0004	⩽0·0004 0·0006	0·004	<0·0004	<0·0004	0·005 0·015	0·0005
Co weight, %	-	-	-	-	-	-	-	0·09 0·07
Brinell hardness, kg/mm²	9·6	10·59	12·8	-	11·2	11·1	11·8	9·25
Tensile strength, kg/mm²	2·3	4·05	1·9	-	3·1	3·1	4·3	

*Where two figures are given the upper applies to the positive grid

nd = not determined   Tl = thallium

These alloys show a compromise between good castability, favourable mechanical properties, low self-discharge and cheapness

Since this information was published, efforts to reduce the antimony content of the grid metal have continued. Many manufacturers are now working in the 3·5 to 4% level, with the other constituents more or less in the proportions shown in the Table.

[Chemical analyses by Metallgesellschaft A.G. From Tables 1 and 2 of Heubner and Sandig paper, Hamburg, 1971]

rate and the degree of corrosion. With the Ag-free alloys, the attack appeared to be concentrated on the antimony-rich phase, while in the case of the silver-bearing alloys, corrosion took place mainly by oxidation of the lead-rich phase.

Selenium: Small additions of selenium, 0•02% to 0•05%, to alloys containing antimony around 3% cause marked grain refinement, an observation first recorded by Waterhouse and Willows.[80] The patented alloy, Pb, 3% Sb, 1•5% Sn, 0•05% Se, sometimes referred to as 'Admiralty B' alloy, has been used for many years for the large grids in UK submarine batteries. A low antimony content is necessary to keep the rate of evolution of hydrogen gas on open-circuit within the specified low level. Arsenic cannot be used as a hardener because of the possible risk of arsine evolution during over-charge. The grain refinement caused by selenium inhibits the formation of hot tears during casting and reduces the corrosion of grain boundary phases.

Nolan *et al.*[81] have compared the results of tests on railway lighting and diesel engine starting batteries, having positive grids case in three alloys

(*a*) binary 8% alloy
(*b*) Pb, 3/Sb, 1•3% sn, 0•05% Se
(*c*) Pb, 3% Sb, 0•05% As, 0•1% Ag.

Special mention was made of the ease in casting (*b*). After 5 years service, the cells with alloys (*b*) and (*c*) were in better condition than those with (*a*). Retained capacity was higher, standing losses were lower and there was less grid corrosion and fewer fractures by grain boundary attack.

Borchers *et al.*[82] have described tests for SLI battery grids with alloys containing 1•6 to 3•5% Sb, with and without additions of 0•02% Se, 0•05% As and 0•02% Sn and compared their behaviour with that of alloys containing 5 to 8% Sb. The decrease in mechanical strength caused by the reduction in Sb content would be offset to some extent by heat-treatment, quenching and ageing. The addition of Se caused marked grain refinement, and, even with the lowest amounts recorded, there were fewer casting faults than with the larger grained binary alloys. The tests also confirmed the improved castability with additions of tin.

Borchers and Nijhawan[83] have also examined the structural changes and the hardening effects on 2% Sb alloy, caused by additions of 0•02% Se, 0•05% As and 0•01% Ag, after heat treatment, quenching and ageing. Although these operations themselves raise the Brinell hardness to within 80-90% of the specified level, additions of As and Ag increase both the rate and the extent of hardening. No additions of Sn were made in this case.

Tellurium: In amounts up to about 0•1%, this element modifies the crystal structure of lead and greatly improves its resistance to fatigue under sustained stress, hence its use in lead pipe. Claims have been made[84] for an alloy for SLI battery grids, containing 1 to 3•5% Sb, 0•001 to 0•2% Te, 0•01 to 0•2% As to improve the mechanical properties, 0•005 to 0•05% Sn to improve castability and 0•02 to 0•3% Ag to claimed to have high corrosion resistance and cause low standing losses, termed 'Astag', containing 99•9% Pb, 0•07% Te, 0•01%Ag and 0•01% As. Under special circumstances, Sb may be added in such low concentrations that Ag crystallises before As from the solidifying mass. This alloy has been tested in spine castings for tubular positive plates as well as in SLI battery grids. Tellurium has an objectional electrochemical property. Its hydrogen over-potential is lower than that of lead, and, like antimony, therefore, if deposited on the negative plate it increases standing losses and the evolution of hydrogen on open-circuit. The amount used in the positive grid alloy must therefore be strictly controlled.

Sulphur: Dasoyan[78] has pointed out that dispersal of the structure of metals increases their hardness and resistance to fracture and improves their mechanical properties generally. In casting metals, therefore, the aim is to get fine structures by special additives, referred to as crystallisation regulators. It was found, for example, that the addition of 0•1% of very finely powdered ebonite to a lead-antimony alloy considerably improved its castability and practically eliminated casting defects. The effect was finally traced to the sulphur in the ebonite. As previously indicated, by reducing the grain size and thereby increasing the total length of the grain boundaries, dispersives of this kind improved the resistance to corrosion.

Heubner and Uberschair[87] have reported systematic investigations of the grain refinement of Pb-Sb alloys by additions of copper and sulphur. With a 2•5% Sb alloy, additions of copper up to 0•04% had a marked grain refining effect. Additions of only 0•01% of sulphur had an even more pronounced effect, but the greatest improvement was achieved by a combination of 0•05% of copper and 0•006% of sulphur. It was concluded that the compound formed with copper was Cu_2Sb and with sulphur, PbS, and, when both elements were present, $CuSbS_2$ and Cu_2S, all of which may act as nucleants during freezing. It was possible, however, that the effect may have been the result of the retardation of growth rather than grain refinement.

4.6.2.6 Dispersion-strengthened lead (DSL): Lead monoxide (PbO) is another dispersant which can have a favourable effect on the mechanical properties of pure lead. As Roberts *et al.*[88] have pointed out, DSL is produced by rolling and/or extrusion of a powdered metal compact containing 1 to 1·5% of a very fine uniform dispersion of PbO in the pure lead powder. The presence of this dispersion inhibits grain-growth and gives a material with a fine-grained structure and high strength and mechanical properties, especially creep resistance, higher than those of Pb/6% Sb alloy. Figs. 4.27 and 4.28 show typical microstructures of pure lead and DSL and Fig. 4.29 comparative tensile properties of DSL,

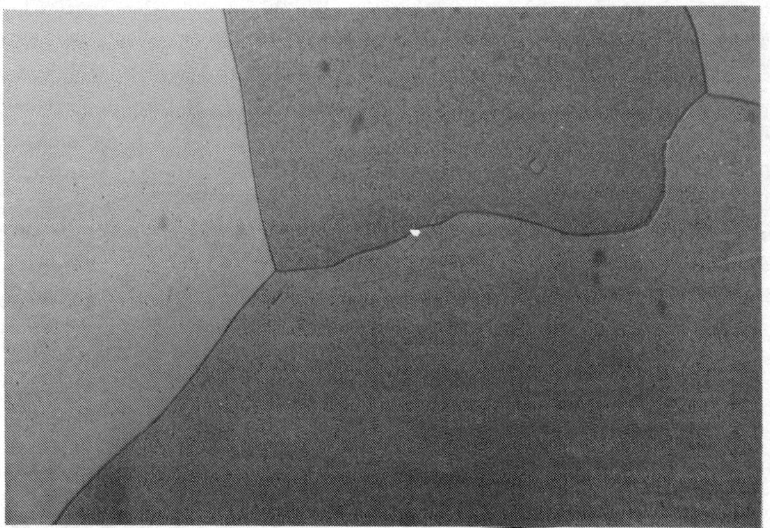

Fig. 4.27 *Microstructure of pure lead (fine grained) x 630* [Courtesy St. Joe Minerals Corporation, USA]

Pb and Pb/6% Sb alloy. Bagshaw and Hughes[89] have reported results of tests with cells having positive grids machined from DSL sheet as near as possible to the standard pattern. PbO content was in the range of 3 to 4%, giving a metal with similar mechanical properties to an 8% Sb alloy. It was noted that, if the PbO exceeded about 7%, there was a marked decrease in ductility. The cells performed satisfactorily on over-charge tests, with good retention of top-of-charge voltage, but they gave poor results on cycling tests, owing partly to shedding of active material and partly to failure of the welds between plate lugs and busbars.

The effect of the dispersive was completely destroyed when the material was melted, so casting or welding processes could not be used. Bird *et al.*[90] have also reported results of laboratory and service tests

Fig. 4.28 *Microstructure of DSL x 630*

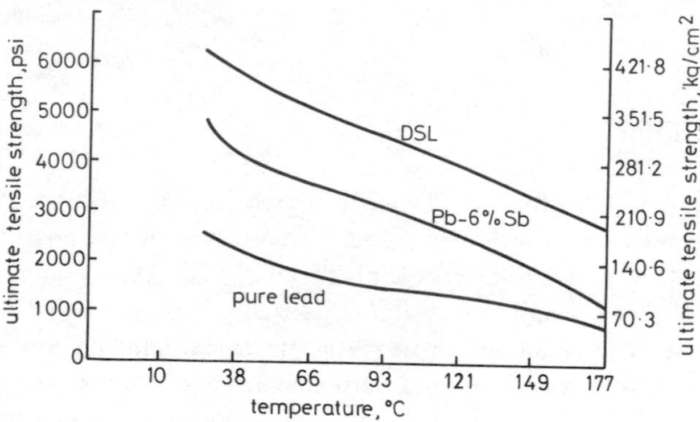

Fig. 4.29 *Comparison of the tensile properties of DSL with lead-antimony and pure lead at elevated temperatures*
[Courtesy St. Joe Minerals Corporation, USA]

with cells of 250 Ah capacity, used in train-lighting service. Some cells failed after only 11 months, owing to breakdown of the joints between plate lugs and busbars. In these cells the busbars were cast in a low antimony alloy, Pb, 3% Sb, 0·5% As, 0·1% Ag, and the plates were attached by brazing with similar alloy or soldering with a high tin-lead solder. The normal life expectancy of standard cells on this service is 6 years.

4.6.2.7 Miscellaneous metals for grids: Various attempts have been made to use metals of lower density, higher electrical conductivity and greater mechanical strength than lead, e.g. aluminium, copper and titanium. As indicated in Table 4.7, the relevant properties of aluminium and copper are particularly favourable, and although the resistivity of titanium is over 2·5 times that of lead, its density is less than 50% and mechanical strength over 7 times as high.

Aluminium and Copper: Both metals are attacked by sulphuric acid, especially under the highly anodic environment of the positive plates. Grids in these metals made from thick wire mesh or by the expanded metal process have been coated with a protective layer of lead by electroplating, or, in the case of copper, by hot-dipping in molten lead-tin alloy.[91] In neither case was it possible to ensure a pore-free coating, even with reverse-current plating schedules and successive hot-dipping stages. Although grids coated in this way might be adequate for negative plates, they were not sufficiently durable for use as positives.

Table 4.7 *Properties of possible metals for grids*

	Pb	Al	Cu	Ti
Density, g/ml	11·3	2·7	8·9	4·5
Density, %	100	24	78	40
Resistivity, Ω cm x 10^{-6} at 0°C	19	2·45	1·56	50
Resistivity, %	100	13	8	263
Young's modulus, x10^{11} dynes/cm^2 at 18°C	1·62	7·05	12·98	11·60
Young's modulus, %	100	435	800	716

Titanium: As Cotton and Dugdale[92] pointed out, this metal has good resistance to corrosion in battery electrolyte, if the potential is raised above the standard reversible potential of −O·42V. At a potential of +0·25V, for example, corrosion is virtually stopped by the high resistance layer of oxide which forms on the surface and passivates the metal. To make it suitable for a battery electrode, it must be coated with a thin conducting layer, for the positive, platinum or lead dioxide or both and for the negative, lead. Tests were described with 5-plate cells, having plates made by the expanded metal process and coated as above, with strips of titanium attached to serve as current collectors. Standard active materials were used and the cells were cycled at the 5h and 1·5h rates. At the 5h rate, the coefficient of use of the active materials in the titanium cells matched that in the lead cells for the 12 cycles recorded, but at the higher rate there was a marked reduction in efficiency after only 2 or 3 cycles and in some cells evidence of dissolution of titanium in the negative grids. It was estimated that a reduction of about 10% in the weight of an aircraft battery, weighing 21·6 kg, could be achieved by replacing the lead grids by titanium. The high cost of platinum, coupled with the complex processes of manufacture and risks of chemical attack on the negative grids would, however, seem to exclude titanium for most commercial applications.

Faber[62] has also described tests with positive plates made with titanium grids, fabricated in a number of ways, e.g. by the expanded metal procedure, by sintering a mat of felted fibres and by welding thin rods to a top-bar to give a spine assembly suitable for a tubular plate. In all cases the metal grids were coated with a thin layer of PbO_2 and then filled with positive active materials of conventional type. Single positive plates were assembled between two negative plates in an excess of electrolyte of 1·21 SG, and, after a series of discharges at rates from the 5h to the 0·5h rate, the plates were cycled at the 5h rate. As indicated in Fig. 4.25B, Section 4.6.2.1, much higher coefficients of use of the active material were obtained at all rates with the expanded metal grids and particularly with the grids made of felted fibres than with standard pasted and tubular plates. The former were, however, somewhat exaggerated by the large excess of electrolyte. It was claimed that the titanium plates had given 300 to 500 cycles without significant loss in capacity.

4.6.2.8 *Plastics for grids:* The use for the grid framework of plastics which are inert to sulphuric acid, such as polyvinyl chloride, polyethylene, polystyrene, polypropylene, to give rigidity and strength was obvious and experiments have followed several lines. In one case, moulded plastic grids, after a flash-coating with silver, were electroplated with a thin layer of lead to give the necessary conductivity. However, the thin lead coating was not sufficiently durable for a positive plate and the process would have been too costly. In another method, described by Sladin,[92a] metallic conductors, arranged like the spines in tubular plates were inserted into the plastic frame after moulding. Alternatively, as described by Robson,[92b] the metal conductors, made of pure lead or any chosen alloy, were set in the plastic frame during moulding or clamped between two half-sections of the plastic frame, which were then rigidly sealed by heat or a suitable sealant. Booth[93] has described small cells of 20Ah capacity assembled with grids moulded in polystyrene with pure lead conductors inserted in this way. The cells gave over 200 cycles without loss of capacity and standing losses were only 35% after a stand of 6 months at ambient temperatures. The saving in weight probably amounted to about 10%. These processes were used in a 12V/90 Ah low-loss battery for army signals service, the construction and performance of which have been described by Thomas.[94] The battery was 25% lighter than the standard model, but weight savings had been made in other components apart from the plate grids, e.g. in the battery container. Standing losses were only about 33% of standard and there was no significant loss in performance after 400 cycles on a 4h discharge 8h charge regime.

Faber[93a] has also described tests with a metal/plastic composite grid, in which the thin metallic conducting grid is sandwiched between two net-like layers of plastic material, which serve mainly as supports for the active materials. As a further aid to the retention of the active materials, the surfaces of the plates can be covered by welding a fine non-woven layer of a suitable plastic material to the underlying framework. Claims were made for an energy density of 36 Wh/kg at C/5 for a traction type cell assembled with plates of this kind.

4.6.2.9 Alloys for maintenance-free batteries: Provided the terminal posts are kept free from high resistance films caused by corrosion, and the top of the battery clean and dry to avoid tracking leakages, the only maintenance required should be regular topping-up with distilled water, lost by evaporation or minimal gassing overcharge. It is assumed, of course, that the charging equipment is in good order. To the automobile manufacturer, and many users, even this limited maintenance is objectionable, and some years ago, two of the largest international manufacturers, the Ford Company and General Motors Corporation of the USA, issued a specification for a maintenance-free (MF) battery, which should operate for 30 000 miles or 3 years without maintenance. As mentioned in Section 4.6.2.3, the chief need for maintenance lies in the deleterious effect which antimony has on the negative plates and manufacturers world-wide, stimulated also by the rising cost, have been making determined efforts to eliminate antimony altogether or at least to reduce the content in the positive grid alloy.

Lead-calcium alloys: In the USA particularly, interest has been concentrated on alloys with 0·06 to 0·085% of calcium, first described by Schumacher and Bouton[95] and Haring and Thomas[96] of the Bell Laboratories in the USA over 40 years ago and used for many years in stationary batteries for telephone and stand-by power services. Rose and Young[97] have shown that age-hardening of the binary alloys was due to precipitation from the super-saturated solid solution of fine particles of Pb_3Ca, which produced lattice distortion of the single phase matrix. With up to about 0·07% of calcium, the increase in strength was roughly linear, but at higher contents, 0·08 to 0·1%, large crystallites of Pb_3Ca formed at the grain boundaries, reducing mechanical strength and making the alloy more vulnerable to anodic attack. Additions of 0·5 to 1·0% of tin improved both characteristics, apparently through the formation of another fine grain boundary phase, Sn_3Ca, causing even more lattice distortion. The structure and strength of die-cast alloys were sensitive to both the rate of cooling and the composition. SLI battery grids could, however, be cast at the rate of 14 double castings per minute with metal temperature in the range 475 to 540°C and mould temperature, 155 to 215°C.

Cold working, as in wrought metal technology, improved the fine-grained structure and mechanical properties, which remained stable at room temperatures for periods of 4 years. The rate of corrosion was also related to grain size and was generally more severe on grain boundary phases than on the grains themselves. For both reasons, it was obviously an advantage to use a fine-grained structure as indicated by Fig. 4.30.[97]

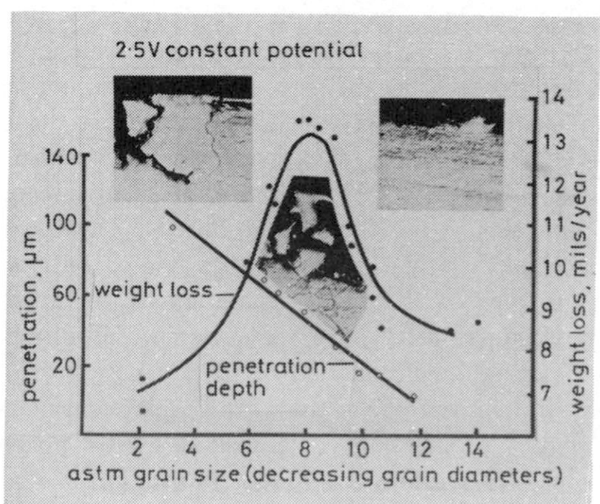

Fig. 4.30 *Lead-calcium alloys; effect of grain size on rate and depth of anodic attack*
[Courtesy St. Joe Minerals Corporation, USA]

In gravity die-casting the emphasis was slightly different. Provided there was a minimum segregation of grain boundary phases and these were relatively fine, freedom from defects such as hot-tears, shrinkage cracks, gas cavities etc. was more important than achieving the finest grain size. In die-casting, care must be taken to keep the calcium content within the specified narrow limits. Oxidation of calcium in the metal pot can be largely prevented by covering the surface of the molten metal with a layer of carbon powder or a steel plate or a shroud enclosing an atmosphere of inert gas. The addition to the melt of small amounts of aluminium has been found to reduce the drossing of calcium.[98] During production the calcium content of the molten metal must be regularly monitored. This can be done spectrographically or on the shop floor, by a simple method of 'antimony titration', first described by Bouton and Phipps[99] and later developed in the ternary bloom test by the St. Joe Minerals Corporation of the USA.[100]

Although the general behaviour of lead/calcium alloys is more or less equal to that of lead/antimony alloys, two major problems have been encountered. The first concerns the adherence of the active material to the grids during cycling. Shedding of the active material has sometimes led to a premature fall in capacity. In other cases, the formation of what appears to be high resistance films at the grid-active material interface has retarded normal charge acceptance. The latter effect appears to be associated with the high corrosion resistance of lead/calcium alloys.

Lead/strontium alloys: Weinlein and Pierson[101] of Globe Union Inc., USA, have reported the results of tests with an alloy containing strontium, tin and aluminium, claimed to have comparable electrochemical properties to lead/calcium/tin alloys, better mechanical properties and greater stability to drossing in the molten state. No concentrations of the alloying elements were given. Maintenance free characteristics of batteries with grids in this alloy were claimed to match those of batteries with calcium/lead alloys and Globe Union announced that they were proposing to use the alloy in their MF batteries. Bagshaw[102] has also described metallurgical tests on lead/strontium alloys and cycling tests on cells with grids in these alloys. He concluded that a binary alloy, containing 0·1% of strontium, would probably be satisfactory for floating standby service, a ternary alloy, with 0·5% of tin in addition, for automotive duties and a quaternary alloy with an additional 0·25% of silver for deep cycling motive power service. A small addition of aluminium to the molten metal protected all of these alloys from oxidation of the constituents.

Low antimony alloys: Although lead-calcium alloys have generally been accepted as the standard for MF batteries in the USA, manufacturers in Europe and Japan have concentrated on the use of alloys with antimony in the range 2·5% to 3·0%, arsenic up to 0·2% and small additions of tin. These alloys call for the least changes in casting techniques and casting rates and thereby avoid undue increase in cost. Haworth, Peters and Throw[103] have reported tests on simulated service routines with SLI batteries of 12V/39 Ah capacity, assembled with grids cast in the alloys as shown in Table 4.8.

The basis of the test programme was the shallow-cycle overcharge regime, specified by the International Electro-Technical Commission,

Table 4.8 *Composition of grid alloys and their associated water losses*

	Alloy		
	A	B	C
Antimony, %	5·8 - 6·2	2·8 - 3·2	0·005
Arsenic, %	0·15 - 0·20	0·15 - 0·20	0·001
Tin, %	0·03 - 0·06	0·03 - 0·06	0·55 - 0·65
Calcium, %	0·001	0·001	0·06 - 0·08
Water losses:			
at 20°C, ml	96·8	50·8	29·8
at 40°C, ml	204·2	132·4	75·7

1972, with an extended charge by an uncompensated alternator after each test unit. Estimated water losses after 3 years and the equivalent of 30 000 miles of service (1500 h at 20 miles/h and 24 700 h O/C) at 20°C and 40°C were as shown in Table 4.8.

Water losses for the 3% alloy were nearly half those for the 6% alloy and twice those for the lead/calcium alloy. It was concluded that at ambient temperatures around 20°C, both alloys B and C would be satisfactory, but at 40°C, some temperature compensation would be necessary, even for the battery with lead/calcium alloy grids.

To relieve casting as well as technical problems, hybrid assemblies have been proposed, with the lead/calcium alloy in either the positive or negative grids. A Japanese battery, known as 'carec', includes the latter combination, presumably to keep a high hydrogen over-potential on the negative group. The alloy astag[85] referred to in Section 4.6.2.5 has also been recommended for MF batteries.

Lead/antimony/cadmium alloys: Early claims for 'low maintenance' alloys for positive grids, based on additions of cadmium and tin to low antimony alloys were made in 1948 by Couch *et al.*[104] The proposed alloy, containing 2% Sb, 1·5% Cd, with or without 0·5% Sn, was claimed to give reduced negative poisoning, improved resistance to corrosion and growth and a lower rate of evolution of hydrogen on open-circuit, as compared with the then standard lead-antimony alloys, making it suitable for the large die-cast grids used in submarine batteries. The alloy was also suitable for wrought metal grids, made from rolled sheet, which could be further improved by heat-treatment, quenching and ageing.

Bagshaw[105] has also reported tests on batteries with grid alloys containing equal amounts of cadmium and antimony in the ranges 1·5%, 2·0% and 2·5% with and without silver, in the range 0·05% to 0·25%. It was claimed that the formation of an inter-metallic compound, Sb-Cd, improved the hardness, resistance to creep and stress corrosion, which was further improved by additions of silver. Batteries with grids in the alloy Pb, 2% Sb, 2% Cd, 0·1% Ag gave satisfactory performances on cycling tests, 40 to 100% longer lifetimes on overcharge tests than batteries with standard 3% Sb alloy and an increase in charge retention of about 100% at 40°C. There was a slight increase in drossing of both lead and cadmium in the metal pot and metal and mould temperatures had to be kept at somewhat higher levels than for standard alloys. The high cost of cadmium would also impose restrictions for normal commercial service.

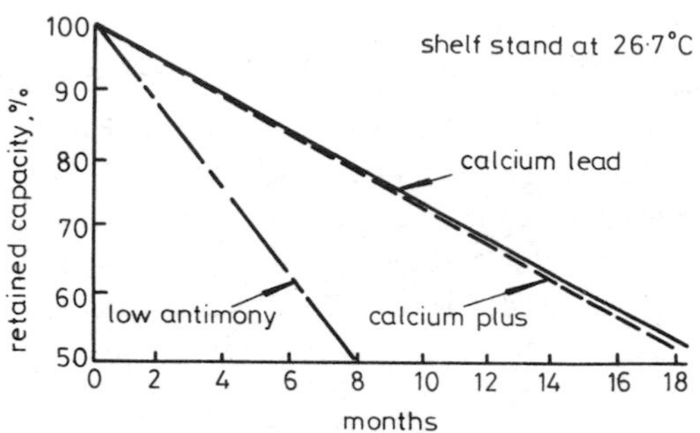

Fig. 4.31 *Retention of charge during storage*
[Courtesy A. Sabatino, *In Lead and zinc into the 80's*, Lead Development Association, 1977]

Abdul Azim *et al.*[106] have confirmed that the corrosion rate of a 4% Sb alloy was reduced by about 50% by the addition of 0·1% of cadmium, owing to increased nucleation in the formation of lead sulphate, which blocked the pores of the surface film and protected the underlying metal.

Sabatino[107] has described tests with batteries assembled with grids in the alloy Pb, 1·5% Sb, 1·5% Cd, in comparison with lead-calcium alloys and low antimony alloys. The concentration of the latter was not specified, but was thought to be not more than about 50% of the content of standard alloys, containing 4·5 to 6%. The cadmium-bearing alloy was claimed to be better in a number of respects than the calcium alloy and was therefore termed 'calcium-plus'. To contain the high cost of cadmium, a hybrid assembly was recommended, with the cadmium alloy for the positive grids and the calcium alloy for the negatives.

Figs. 4.31 and 4.32 show comparative performances, relevant to maintenance-free assemblies for batteries with the three alloys. Retention of charge during storage at 26·7°C was similar for calcium and cadmium alloys, with 80% retention after 6 months and 50% after 18 months, compared with only 3 and 8 months, respectively, for the low

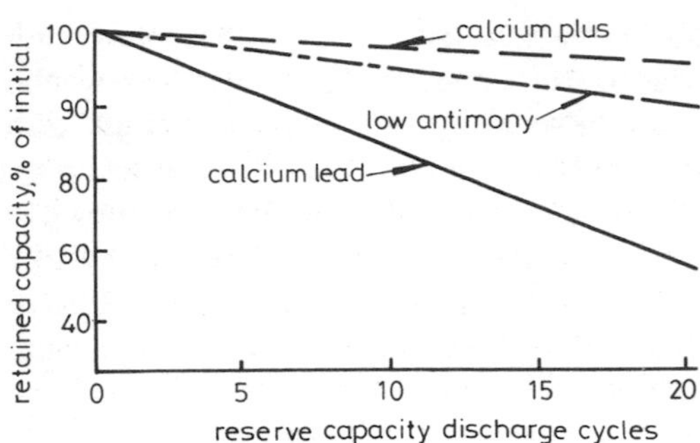

Fig. 4.32 *Retention of capacity after cycling*
[Courtesy A. Sabatino, *In Lead and zinc into the 80's,* Lead Development Association, 1977]

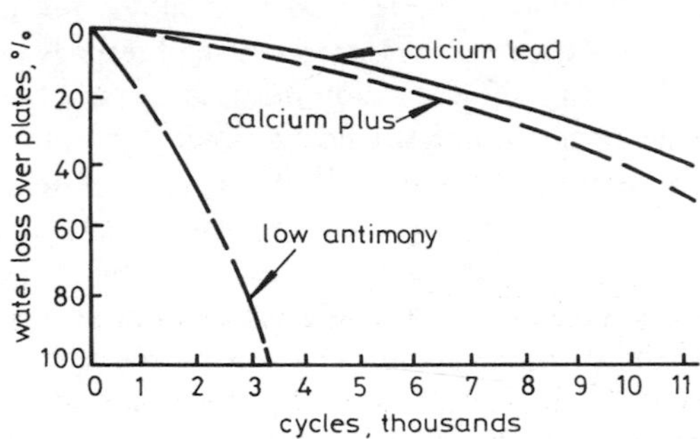

Fig. 4.33 *Water loss in SAE J240 cycle testing*
[Courtesy A. Sabatino, *In Lead and zinc into the 80's,* Lead Development Association, 1977]

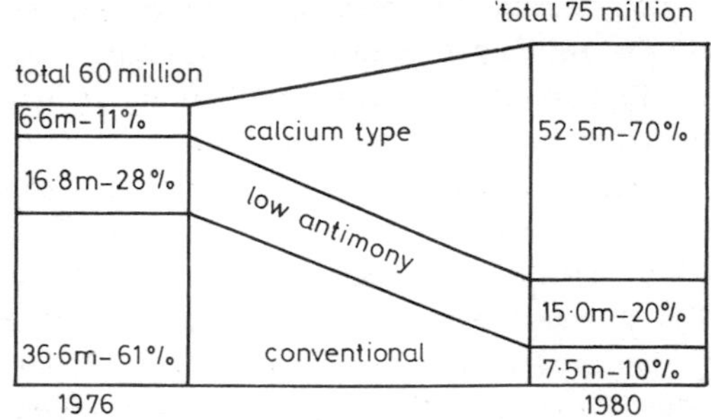

Fig. 4.34 *Gould's own forecast of projected sales mix in the USA*
[Courtesy A. Sabatino, *In Lead and zinc into the 80's,* Lead Development Association, 1977]

antimony alloy. On the test for retention of capacity during cycling, batteries with lead calcium alloy lost about 50% after only 20 cycles, those with low antimony alloy about 10% and those with cadmium alloy only 5%. On the SAE J240 cycling test, water losses were appreciably lower for the calcium and cadmium alloys, as shown in Fig. 4.33. Fig. 4.34 indicates the forecast by the Gould Company of the USA of the trend in the use of MF alloys in 1980. It was further thought that calcium alloys would be superseded by strontium and cadmium alloys of the type described.

4.6.2.10 The effect of antimony on the positive active material: The way in which antimony is leached out of the positive grid metal and adsorbed by the positive active material was briefly described in Section 4.6.2.3. In the early work with lead-calcium alloys, several authors recorded that the performances and deep cycling lifetimes of batteries with positive grids in this alloy fell away much more rapidly than those with grids in lead-antimony alloys. Using X-ray diffraction techniques and the electron microscope, Burbank[119,120] made a systematic study of the behaviour and morphology of the positive active material in cells with alloys of both types, submitted to a shallow cycling regime. The material in the antimony-free cells became progressively softer and the capacity began to fall away almost from the outset, whereas the cells with antimonial alloy maintained their capacity and the active material remained firm throughout the tests. Intensive examination showed that the active material in the former cells had assumed a nodular crystalline structure, whereas in the cells with antimonial alloy, the PbO_2 took the form of massive prismatic conglomerates, with clusters of inter-locking crystals, which appeared to impart mechanical strength giving good cycling lifetimes. It was also recorded that the active material in the antimonial cell contained a higher ratio of alpha to beta PbO_2, a point also observed by Sharpe.[120a]

In seeking an explanation of these phenomena, several authors have described their investigations into the conditions of formation and the morphology of corrosion layers on pure lead and lead-antimony alloys. Using an ingenious method of precipitating $PbSO_4$ in thin cellophane film, Ruetschi[120b] showed that thin layers of $PbSO_4$, in thicknesses above 1 μm were perm-selective, being permeable to H^+ ions and H_2O, but non-permeable to Pb^{2+}, SO_4^{2-} and HSO_4^- ions. The diffusion potential of such membranes exceeded several hundreds of millivolts. Under SEM examination, anodic corrosion layers on pure lead exhibited characteristic multi-phase structures, the exterior layer at the electrolyte interface being $PbSO_4$, the innermost layer against the metal being tretagonal PbO with some alpha PbO_2 and the intermediate layers being a progression of basic sulphates - $PbO.PbSO_4$, $3PbO.PbSO_4.H_2O$ and 5 $PbO.PbSO_4.2H_2O$, in agreement with thermodynamic predictions, based on the levels of the PD and the pH. During the anodic phase of the cycle, the surface layers of the $PbSO_4$ in direct contact with the electrolyte were converted to beta PbO_2, but at the metal interface, where the pH may have been as high as 9·34 and conditions highly alkaline, the PbO formed from the OH^- ions present would be converted to alpha PbO_2.

The specific effect of antimony on the formation of the beta polymorph has been described by Hampson *et al.*,[120c] who have reported an examination by SEM and X-ray diffraction techniques of the anodic oxidation in 5M H_2SO_4 of the corrosion layers on pure lead and 5% antimonial lead electrodes. They concluded that, in areas where there was a depletion of acid, a primary layer of mixtures of alpha and beta PbO_2 was formed with a strong preponderance of alpha on the pure lead and beta on the antimonial alloy. The grain size of the surface layer of the beta PbO_2 appeared to be finer than that of the alpha form and there were fewer massive residual crystals of $PbSO_4$. On further oxidation, a duplex layer of beta PbO_2 was formed on the antimonial alloy, and, in agreement with Burbank and other authors, it was concluded that the antimony present had set up nucleation centres for the formation of beta PbO_2.

Fig. 4.35A shows typical SE micrograms of these structures. The mixture of alpha and beta PbO_2 appeared to be firmly attached to the metal surface by a micro-crystalline coating and it could be inferred that, provided strong bonds were formed by inter-crystalline growth between the primary and secondary layers of beta PbO_2, practical plates would have satisfactory lifetimes. Hampson observed, however, that in some cases the secondary layer of beta PbO_2 was composed of much larger crystals than those in the sponge-like primary layer. The bonds appeared to be weaker, voids began to develop on further charging and these led finally to the complete disruption of the corrosion layer.

Attempts have been made to realise the beneficial effects of antimony by the addition of small amounts of antimony oxide to the positive active material at the paste-making stage. Unfortunately, antimony added in this way appears to find its way very rapidly to the negative plate and any structural improvement to the positive material was more than offset by poisoning of the negative material to a degree which would be intolerable in a maintenance-free assembly.

Fig. 4.35A *SE micrograms of anodic corrosion layers*
a Pure lead electrode showing primary layer of fine grained alpha PbO_2, with massive residual plate-like crystals of $PbSO_4$
(i) x 6000
(ii) x 20 000

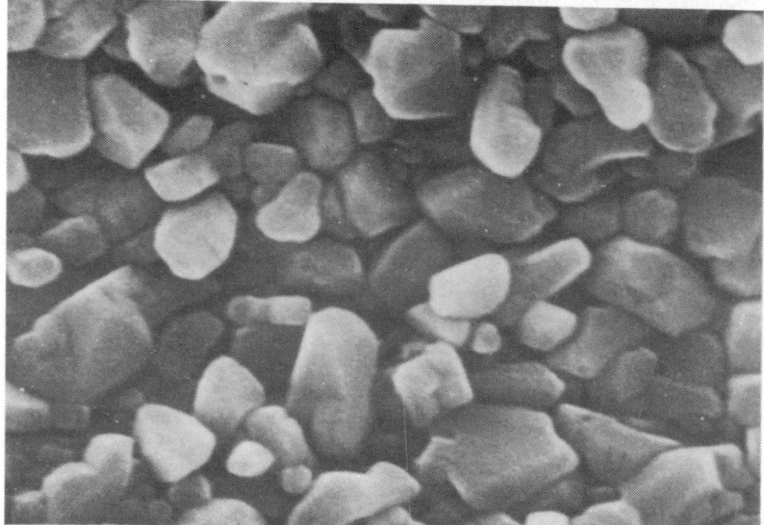

Fig. 4.35B *SE micrograms of anodic corrosion layers*

b Lead - 5% Sb electrode, showing duplex layer of beta PbO_2 crystals at a similar stage of oxidation with a minimum of $PbSO_4$ crystals
(i) x 6000
(ii) x 20 000
[Courtesy Hampson *et al.*[52a], *J. Appl. Electrochem.*, 1980, Chapman & Hall Ltd.]

Some claims have been made that softening of the positive active material of the type sometimes observed with maintenance-free batteries, having grids in lead-calcium alloy could be allayed by additions of small amounts of phosphoric acid, either during paste-mixing or to the electrolyte during the first charge. However, a reduction in the cell capacity of from 5 to 10% was noted and the beneficial effect was apparently not sustained during deep cycling.

4.6.3 *Grid casting machines*

For convenience in handling through the factory processes, grids for SLI automotive batteries are generally made in twin castings, as shown in Fig. 4.22. They are joined along the bottom edges with a minimum of metal so that, when fully processed, the two plates can be readily parted. The double grids are cast in the horizontal position in the mould, with the reservoir of metal or 'gate' running along the whole of one side of the casting. A typical grid casting machine has been described by Williams.[108] In brief, the two most important points are the control of the temperatures of both the metal and the mould and the application to the surfaces of the mould of a thin insulating dressing, which regulates the heat transfer between the molten metal and the mould. To reduce drossing, the temperature of the metal in the melting pot is kept just above its melting point, about 350°C for Sb alloys of 4 to 8%, and is raised by about 100°C as the metal is pumped to the pouring orifice plate just above the mould. The temperature of the mould is regulated by both heating and cooling coils in both half sections and for these alloys generally runs between 150°C–180°C. The mould dressing is based on a suspension of very fine cork powder in a dilute solution of sodium silicate and is applied by spraying, usually twice per shift.

For common types of automotive castings, the rate of production may be from 12 to 15 double castings per minute, giving a daily (8h) production rate of 5000 to 6000 castings. Similar automatic machines are used to cast grids for industrial batteries, but with the thicker and larger sizes of plates these are generally cast as single plates and at a slower rate.

4.6.4 *Hand casting*

Manually-operated equipment is available for small manufacturers or where only short runs of a variety of castings may be required. In this case the molten metal is man-handled in a large ladle. The temperature of the mould and the metal at the pouring point are kept closely in line with those in the automatic casting machines and the procedures for applying the mould dressing are similar.

4.6.5 *Oxides for pastes: methods of manufacture*

4.6.5.1 Grey Oxide: For many years litharge and red lead were the basic ingredients, and, as previouly mentioned, they are still used in certain circumstances. Nowadays the material most widely used is known as 'grey oxide'. This generally contains 60 to 70% of lead monoxide and 30 to 40% of fine lead powder, so named because of its greyish green colour, and it is the product of a ball-milling process, first developed by Shimadzu in Japan in 1924.[109] Many modifications have been made to the original equipment, but the basic principles have been retained. Pure lead balls, cast in varying diameters from 13mm to 125mm, are tumbled in a steel drum, rotating with its central axis in the horizontal plane. Considerable heat is caused by the friction between balls and the temperature rises rapidly. An air stream is blown through the drum and the particles of lead which are rubbed off by the abrasion are partially oxidised and then subjected to further ball-milling. Water is usually injected into the air stream to assist both the oxidation process and in the control of the temperature. In modern equipment, sections cut from complete lead ingots, or, in the largest units, even the ingots themselves may replace the balls, thereby saving the cost of the casting plant and labour to make the balls. There are two main forms of milling equipment. In the original Japanese mill and later modifications, sometimes referred to as *Tudor Mills*, the powder falls continuously through small holes in the periphery of the drum onto a concentrically-placed cylindrical outer casing. From here the air stream carries it through into a series of cyclone separators, where the fine particles are aerodynamically separated from the coarser, heavier particles. The dust-laden air stream finally passes through bag collectors, where the finest particles are collected, the purged air then being vented to the atmosphere. The average temperature in the mill runs around 90°C. In the other type, known as the *Hardinge Mill*, the drum is not perforated, but is cone-shaped at one end, from which the powdered oxide spills out as it is made. A typical construction has been described by Williams.[108] Here also the oxidation process is accelerated by a stream of humidified air, which carries the dust through the sizing apparatus and the bag collectors and provision is sometimes made for the return to the mill of the coarsest particles precipitated in the first cyclone separator. The inner temperature of this mill usually runs around 170-180°C. Both types are fitted with an automatic feed of balls or ingots and the final product is automatically collected in weighed drums.

4.6.5.2 The Barton or Linklater pot process: Molten lead is held at the specified temperature in a large crucible and agitated by a rotating paddle. A stream of air, blown over the surface of the molten metal, carries the oxide powder into a series of size-classification cyclones, similar to those described for the ball mills. Here also, the coarser particles may be returned to the pot for further comminution. By careful control of the metal temperature and the speed of the air stream, a product containing about 30% of fine metallic lead can be readily obtained. Brachet[110] has described a typical installation. In both of these processes, the gain in weight caused by the oxidation of the lead varies from about 7·5% for 60% of PbO to 8·8% for 70% of PbO, and, in considering the economics of the processes, this gratuitous gain can be offset against the cost of running the plant.

4.6.5.3 Fume oxide: Fume oxide represents almost the ultimate in particle fineness. Molten lead is allowed to flow through a narrow exit and is atomised by a blast of air, blown at right angles to the direction of flow, like an aerosol spray. Particle size is similar to that of the fine dust collected in the bag chambers of the oxide mills, but the lead content may be lower than is desirable and the material is costly to produce. In this connection, it has been reported that the overall costs, capital and running costs, of the Barton Pot process are marginally lower than those for ball mills.

4.6.6 Properties of oxides

Several properties have a bearing on the performance and life of the active material.

4.6.6.1 Polymorphism: Extensive research and development studies have confirmed that the red tetragonal beta PbO is the preferred oxide for the paste-making process, so the running conditions of the oxide mills are adjusted to ensure that the bulk of the product is in that form. As previously mentioned, the transition temperature at which the red modification changes to the yellow ortho-rhombic variant is 468°C.[38] The internal temperature of the ball mills is well below this level, but this may not be the case on the surface of the molten metal in the Barton-Linklater pot. However, tests have shown that no deleterious effects on performance and life of the finished positive plate may be expected if the concentration of the yellow polymorph does not exceed about 15%, and this is the limit generally specified for this process.

4.6.6.2 Particle size: This is one of the two main factors which determine the internal surface area of the active material, the other being the porosity, and both have a direct influence on the polarisation during discharge. As might be expected from the methods of preparation, the oxides are formed with a wide range of particle sizes, but it is an advantage to keep the mean size to a practical minimum. Some operators have found that the most convenient way to produce a uniformly fine product is to pass the whole of the output from a Tudor or a Hardinge mill through a second grinding mill of the Impax type, in which the powder is subjected to blows from rapidly rotating hammers. Oxide ground in this way is sometimes called 'GOX'. As might be expected from the abrasive method of production, Tudor oxide generally takes the form of minute flakes of lead, coated with a thin layer of oxide. Particles of Hardinge and Barton oxide are more nearly spheroidal, and this applies also to GOX and to oxide collected in the bag chambers on the mills.

Various methods have been used for the determintion of particle size, inner surface area and porosity of the active material and so on. For a detailed description of these, the reader is advised to refer to an excellent book by Allen.[111] Brief accounts of some of these methods follow.

Visual examination: The optical microscope is used for the larger particles in the range above about 5μm and the transmission and scanning electrons (TEM and SEM) in the size ranges below 5μm. These are the only methods which reveal the true shape of the particles. By using graticles the particles can be sized and counted, but the method is tedious and many fields must be examined to get representative values.

Sedimentation processes: The sizes of particles can be calculated from their terminal velocities, when falling under the force of gravitation in fluids of various types, air or nitrogen, water, xylene etc. Special sedimentometers are available that record continuously the weight of powder falling onto a pan at the base of the column of fluid. Alternatively, samples of the fluid with the falling particles can be withdrawn at intervals and at different depths and the weight of the particles determined (Andreasen's method). Unfortunately these processes depend on Stokes' law

$$6\pi\eta r v = (m - m_0) g \tag{4.39}$$

where v is the terminal velocity, r is the radius of the particle, η is the viscosity of the fluid, m is the mass of the particle, m_0 is the mass of fluid displaced by the particle, g is the force of gravity, and this assumes that the particles are uniformly spheroidal and that there is no turbulence to disturb the laminar flow. Stokes' law is held not to apply if the Reynolds number exceeds 0·2.

Optical turbidimetry: In this method, a comparison is made of the obscuration of a beam of light, passing at right angles through a vertical column of fluid containing the falling particles with a second beam from a similar source passing through the clear fluid. The difference in the intensity of the two beams as recorded by photocells, is a measure of the cross-sectional area of particles passing through the beam and this can be used to calculate the size and numbers of the particles.

Electrical sensing zone method, developed by Coulter: An electrolyte, such as a solution of 0·9% W/V of sodium chloride, containing a weighed sample of the powder is forced through a small orifice in a glass tube, having electrodes on either side, by which the resistance of the column passing through the orifice can be measured. The change in the resistance, due to the particles generates voltage pulses, which are proportional to the volume of the particles. Calibration is first carried out for each tube and electrolyte with narrowly classified powders of known diameter. It is claimed that analyses by semi-skilled operators can be carried out with good reproducibility.

Sieving: In view of the difficulties mentioned, many operators rely on a straightforward screening method, using standard sieves with wire meshes, specified by British Standard 410. The nominal aperture sizes of a family of these sieves are shown in Table 4.9. Some dimensions, calibrated under a microscope, are also included for some of the finer sizes.

One procedure is to mount the nest of sieves, with apertures in descending order of size on an electro-magnetic vibrator. A suspension of the power in acetone, 2•5g per 60 ml for sieves 6″ in diameter or 1g per

Table 4.9 *Standard sieves; BS No. 410*

Meshes/in²	Size of aperture nominal mm x 10^{-3}	actual mm x 10^{-3}
85	180	-
100	150	150
150	105	-
200	75	-
300	53	55-54
Not stated	30	27-34
Not stated	10	6-11

20 ml for smaller sizes, is poured onto the top sieve, and, with the nest on continuous vibration, jets of fluid are sprayed on at intervals. After about 45 min, the sieves are removed, dried at about 80°C, cooled and weighed. The fraction retained on each sieve is then estimated.

Systematic size-distribution measurements of this kind have been made with oxides produced in a Barton type reaction pot, A Tudor mill and large and small Hardinge mills with outputs of 200 and 40 tonnes per week. Average values for a large number of tests are given in Table 4.10, in which the PbO content of the oxide is shown and also the water absorption figure, determined as described in Section 4.6.6.6. In these tests, Barton pot oxide had the greatest amount of fine material, with 75% below 10 μm and 95% below 53 μm. Oxide from the Tudor mill had the coarsest range, with 59% below 10 μm and only 50% below 53 μm.

Table 4.10 *Grey oxides: particle size ranges*

	Percentage of oxide under-size			
Size (μm)	Barton Pt.	Hardinge (200)	Hardinge (40	Tudor
53	95	95	88	50
32	90	75	55	25
10	75	30	25	15
PbO (%)	70	63	54	59
Water absn. ml/100g (ref. 4.6.6.6)	12•3	10•2	11•8	13•7

4.6.6.3 Surface area: The gas adsorption process, developed by Brunauer, Emmett and Teller, the BET process,[112] has been universally adopted as one of the most effective ways of monitoring surface areas of this kind. The principle is to measure the volume of gas physically adsorbed as a mono-layer on the surface, and, with the aid of an adsorption isotherm of the Langmuir type and knowledge of the area occupied by one adsorbate molecule, it is possible to calculate the total surface area involved. Nitrogen gas is generally used as the adsorbate, but other gases, such as ethylene and krypton, have also been used. The quantity of adsorbed gas increases with decreasing temperature and tests are therefore made at very low temperatures. In the case of nitrogen, temperatures around the liquefaction point, $-195{\cdot}3°C$, are used. The sample is first thoroughly degassed by heating for some hours under vacuum at temperatures between 100 and 400°C. The temperature is then reduced to the critical point before admission of gas to the sample and the adsorption monitored by changes in the pressure. The BET equation takes the form

$$\frac{P}{V(P_0 - P)} = \frac{1}{V_m c} + \frac{c-1}{V_m c}\,\frac{P}{P_0} \tag{4.40}$$

Where V_m is the volume required to form a mono-layer, P is the pressure, P_0 is the pressure at saturation vapour pressure, V is the volume of gas adsorbed and c is a constant. The surface area occupied by one absorbate molecule of nitrogen is $16{\cdot}2 \times 10^{-20}$ m^2.

Details of standard volumetric gas adsorption apparatus are given in British Standards BS 4359 Part 1 (1969) and Part 3 (1971). The BET process can also be used to determine the total pore volume and pore size distribution of a porous body, but errors will arise if micropores of diameters 10-20 Å and less are present.

4.6.6.4 Porosity: The mercury porosimeter is also used for measurements of pore size and distribution. Few materials are wetted by mercury, which has a high contact angle of 140°, and considerable pressure must be used to force the liquid metal into the capilliary paths. To make the measurement, the sample is held in a small pressure vessel filled with mercury and volume changes noted at increasing and decreasing pressure steps. Young and Laplace have derived the following relationship between the pressure p and the radius r of a cylindrical pore:

$$p = \frac{2\gamma\cos\theta}{r} \tag{4.41}$$

where γ is the surface tension, 480 dyne/cm, and θ is the contact angle.

If p is expressed as lb/in^2 and r as micrometres, the expression reduces to $r = 106/p$. Mercury cannot be used with metals with which it forms an amalgam, and lead is one of these, although claims have been made that this problem can be overcome by sulphiding the surface of the porous lead sample. Glycerine has been used in place of mercury with some measure of success.

4.6.6.5 Apparent density: As indicated by Vinal,[11] the Scott volumeter has long been used as an approximate production control device to check the uniformity of oxide powders, although this instrument does not give any quantitative information about the proportions of coarse and fine particles. The powder is allowed to fall gently down a vertical tube of rectangular cross section, with baffles arranged obliquely at intervals on either side, into a cube-shaped box having a volume of 1 in 3. The surface of the powder is levelled off flush with the outer rim and the 'cube weight' determined. Each operator sets up his own uniformity standards.

4.6.6.6. Water (or acid) absorption: This test is also somewhat arbitrary, but it gives useful practical information about the paste-making properties of the oxide. It is usually carried out with water, but dilute sulphuric acid of SG about 1•100 is sometimes used. 100g of the oxide are placed in a 250ml beaker and water or acid is added with continual mixing and rubbing of the paste until it forms into a small ball. When this reaches the required paste consistency, the volume of liquid used is taken as the measure of the absorption. Typical figures are shown in Table 4.10.

4.6.6.7 Purity standards: For obvious reasons, much lower limits are specified for impurities in metal used to make oxide than in metal used for grids. Maximum limits generally accepted for oxides are shown in Table 4.11.

Table 4.11 *Limits for impurities in lead for the manufacture of oxide*

		Limits %
Antimony	Sb	0•005
Arsenic	As	0•001
Bismuth	Bi	0•03

Copper	Cu	0·003
Iron	Fe	0·001
Manganese	Mn	0·0005
Nickel	Ni	0·001
Selenium	Se	0·0005
Silver	Ag	0·005
Tellurium	Te	0·0005
Tin	Sn	0·001
Zinc	Zn	0·001
Cadmium	Cd	0·003
Thallium	Tl	0·01

Lead for alloy blending only may have impurity limits exceeding those specified above, but not exceeding those specified for the alloy

4.6.7 *The paste-making process*

The principle of this process is to provide a porous mass with sufficient porosity to give the required coefficient of use of the formed active material and adequate rigidity and cohesion to withstand vibration in service, as well as the dimensional changes involved in each discharge-charge cycle, for the specified lifetime of the battery. Water is used as the pore-forming agent and sulphuric acid is added to provide the basic sulphates, which cause the necessary cementation. Paste mixing may be done in a variety of machines, some even resembling cement mixers. For many years, machines similar to those used for mixing dough in bread-making, with counter rotating Z-shaped blades, were used. In recent years these have been superseded by muller type mixers, fitted with rotating milling rollers, as well as paddles and scrapers to assist in the mixing. Mixing may be done in three stages. If more than one oxide is used, e.g. grey oxide and red lead, these are first mixed in the dry state in a rotating drum type of machine. Any dry additives, such as fibrous materials sometimes added to the positive mix or expanders for the negative paste, may also be added at this stage. The dry mixture is then dumped into the wet mixer, to which the required amount of water has already been added. If any of the negative expanders are added as liquids, e.g. barium hydrate to produce barium sulphate or ligno-sulphonate solution, they would also be added at this stage. After mixing for a few minutes, sulphuric acid, generally of 1·400SG, is slowly added as a fine spray over a period up to about 30 min. Some manufacturers simplify the procedure by carrying out dry and wet mixing in the same machine, adding the water and acid in one operation as dilute acid of 1·100 SG. Advantages claimed for the dual process are that there is some solution of the PbO in the water, giving more uniform reaction with the acid and also that the water causes some oxidation of the lead powder in the grey oxide.

The presence of a proportion of very fine particles gives the paste thixotropic properties and the mixing time must be adjusted to give a suitable consistency for pasting. If mixing is unduly prolonged, the paste may become too soft. A complex mixture of compounds is formed in the mixing process. From thermodynamic and other data, listed by Bode and Voss,[113] Barnes and Mathieson[114] have constructed a typical potential-pH diagram, Fig. 4.35C, showing the conditions under which the various compound form. Using X-ray diffraction procedures, Lander[115] reported the presence of the monobasic sulphate, $PbO.PbSO_4$, and the hydrated tri-basic sulphate, $3PbO.PbSO_4.H_2O$. With similar techniques, Mrgudich[116] detected the treta-basic sulphate,

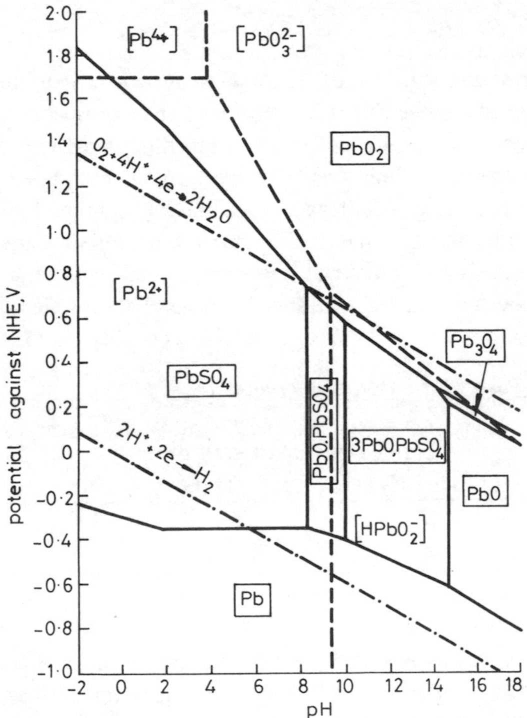

Fig. 4.35C *Potential pH diagram of lead in the presence of sulphate ions at unit activity and 25°C*
[Courtesy S.C. Barnes and R.T. Mathieson), *Batteries 2*, D.H. Collins (Ed.), Pergamon Press Ltd., 1965]

4PbO.$PbSO_4$, a compound formed only at temperatures well above ambient.

The crystal habits of these compounds are important, since they appear to determine the rigidity and strength of the active materials. The chemical reactions are exothermic and the temperature rises as the acid is added. This causes evaporation and some manufacturers cool the mixing machine to keep the temperature below about 50°C. Control of the temperature also promotes the formation of the tribasic sulphate,[117] which, because of its monoclinic crystalline form, provides the interlocking fibrous type of structure, on which the strength and lifetime of the active materials depend. At the end of the mixing operation, the pH of the paste is around 9 to 10 and the metallic lead content is reduced from 30 – 40% to 10 – 15%. Although the starting mixtures and the additives may be different, the same procedures are used for both positive and negative pastes.

4.6.7.1 Paste density: One of the main criteria of acceptability of the paste is its density, since this is a measure of the porosity and hence the ultimate capacity of the formed active material. Density is determined in the usual way by filling a cylindrical cup of known volume, with gentle tamping to release entrapped air, and weighing. Typical values for positive and negative pastes for automotive service are shown in Table 4.12. Pastes for industrial batteries, which may be subjected to regular deep cycles as in traction service, generally have higher densities, around 4·5g/ml

Table 4.12 *Pastes for automotive plates*

	Liquids (ml/kg of grey oxide)			Density
	1·40 SG H_2SO_4	H_2O	Total	g/ml
Positive	29	54-56	84-85	4·1
Negative	27	53-55	80-82	4·3

4.6.7.2 Additives to negative paste: The spongy lead of the negative plate, particularly under the influence of charging tends to agglomerate and lose porosity. To allay this tendency, 'expanders' are added to the paste and these take several forms. Finely divided gas carbon black was originally added, largely to improve the conductivity of the active material as the concentration of highly resistant lead sulphate increased during discharge. Carbon has a lower hydrogen over-potential than lead and during each recharge hydrogen gas accumulates around the carbon particles, causing mild expansion of the spongy lead. Finely divided barium sulphate or 'blanc fixe' was also found to be effective for another reason. This compound is isomorphic with lead sulphate, having a similar face-centred cubic crystal form. As Willihnganz[118] has pointed out, the fine particles of barium sulphate act as foci for the formation of lead sulphate crystals during the discharge reaction, ensuring uniform distribution of the reaction product throughout the active mass. For some years, some manufacturers added this expander as barium hydrate solution, in the belief that this gave the most uniform distribution, but, with the effective mixing provided by modern machines, this has been abandoned in favour of the use of the powdered sulphate.

The third substance commonly added for this purpose belongs to the class of compounds known as ligno-sulphonates, formed as a by-product during the bi-sulphite treatment of wood pulp in the paper industry. The use of this material also started impirically. When wood separators were in common use, up to about the end of the Second World War, it had been observed that negative plates retained their activity longer and batteries assembled with wood separators gave a better high rate performance at low temperatures than batteries having artificial or synthetic separators, which were ultimately to replace wood. This led to the use as an expander of wood sawdust, usually first treated with dilute sulphuric acid to leach out lead-corrosive acids of the acetic acid family and dried. This brown 'lignin', as it was known, was added in amounts up to about 1% of the weight of oxide in the mix. Ligno-sulphonates were later found to give an improved performance. The reasons for this are still subject to some conjecture. Because of their sulphonated structures, these substances have surfactant properties. They are adsorbed on the surface of the spongy lead and also on the surface of the $PbSO_4$ crystals, as they form during the discharge. In both cases they inhibit crystal growth during the dissolution-precipitation processes and thereby cause the electrochemical discharge-charge reactions to proceed more smoothly. A typical negative paste mix contains 0·34% of barium sulphate, 0·17% of carbon black powder and 0·1 to 0·2% of ligno-sulphonate.

4.6.7.3 Additives to positive paste: As mentioned earlier, the capacity of the positive active material can be improved by increasing the porosity, but there is a practical limit to this, beyond which the lifetime may be seriously reduced by premature disruption and shedding. Attempts have been made, with some success, to overcome this problem by additions of either powdered silica gel or of materials composed of short fibres. The former provided support and rigidity to the highly porous matrix and the latter reinforced the fibrous condition, set up by the basic sulphates, formed in the cementation process. In the former case, fine sand has been used, first cleansed of impurities by washing with sulphuric acid, and, in the latter case, solka floc, a short-fibred product of the wood pulp industry or similar fibrous materials made of synthetic plastics, polythene, polypropylene etc. were found to be fairly effective. In both cases the additions amounted to about 0·5% of the weight of oxide in the mix.

One substance which should not be used is barium sulphate, which has a deleterious effect on the life of the positive active material, for the very reason that it improves the performance of the negative plate, by breaking down the crystallite structure during cycling. It must be remembered that SLI batteries are generally subjected only to shallow discharges. Such additives may cause other problems in industrial batteries, which are submitted to regular deep discharges.

4.6.8 The pasting process

4.6.8.1 Machine pasting: Where long runs of plates are required, as for SLI batteries, automatic machines are used. In the horizontal type of machine, the double-grid castings are carried forward on a moving belt that passes under the hopper holding the paste. In this a rotating twin-bladed propellor forces the paste into the grids, which then pass under diagonally fixed smoothing knives and rubber rollers. When the pasted castings reach the end of the continuous belt, they drop into a vertical position, the take-off lugs at either end being picked up at close intervals by a moving chain, which carries them through a long tunnel oven, heated by electric heaters or gas flames. Passage through this oven takes only a few minutes, long enough to dry the surfaces so that the castings can be stacked without risk of sticking together. The stacks of castings are placed in enclosed cubicles to allow the setting process to develop.

The horizontal type of machine has one prime disadvantage; namely, the paste is applied to one surface only and although the smoothing knives are devised to ensure uniform filling of the whole casting, it is difficult to avoid some irregularity, which can lead to buckling of the plates in the formation process.

The vertical type of machine was designed to overcome this problem. In this the castings move downwards through the machine in a vertical position and paste is simultaneously applied to both side by fluted metal rollers. This machine is more complex and considerably more expensive to construct, but it is claimed to give more uniform filling, better control of paste weight and plate thickness.

4.6.8.2 Hand pasting: For short runs, the paste may be applied by hand, usually with a broad steel or wooden spatula, the casting being supported on a board, fitted with a fluted rubber mat, covered by a tightly-fitting heavy cotton cloth. Paste is generally applied to both sides of the casting and levelled by means of a flexible rubber paddle. Undue working of the paste is avoided, as this tends to draw water to the surface, altering the porosity of the main bulk of the paste, a phenomenon also noted with cement or plaster of Paris.

4.6.8.3 Chemical processes in setting and drying: The cycling lifetime of the positive plate is largely determined by the efficiency of the setting or cementation process. As Armstrong *et al.*[121] have indicated, there are clear signs that the morphology of the crystalline divalent lead compounds, present in the cured paste, still exists after subsequent oxidation to PbO_2 in the formation or first-charge process. Many systematic studies have been made of the chemical reactions, which bear similarities to phenomena in the setting of cement. The final porosity of the paste is mainly dependent on the amount of water in the wet mix. To retain this porosity in the finished plate, shrinkage must be kept to a minimum. Conditions in the setting cubicles are, therefore, devised to stimulate the growth of certain acicular, crystalline compounds which form a rigid inter-locking fibrous type of matrix. The relative humidity in the cubicles is held at about 100% to retard the evaporation of water while these compounds are forming. The chemical reaction between lead hydrate and lead sulphate and the oxidation of the lead powder, catalysed by the water present, are both exothermic and the temperature rises appreciably. Using a variety of techniques, microscopy, wet chemical analysis and X-ray diffraction, Pierson[122] has examined the changes in crystal structure and chemical composition during the curing of paste made from Barton pot oxide. The rate of oxidation of the lead powder closely followed the fall in the amount of residual water, the bulk of both changes taking place during the first 20 h, as shown in Fig. 4.36. At medium temperatures of about 50°C, the main constituent was hydrated tribasic lead sulphate, $3PbO.PbSO_4.H_2O$, but at higher temperatures, 82°C, the finely-divided crystals of the tri-basic sulphate recrystallised as coarse tetra-basic lead sulphate, $4PbO.PbSO_4$. During the subsequent forming process, the surface layers of these crystals were more readily converted to PbO_2 than the tri-basic sulphate and there was a higher amount of the alpha PbO_2 polymorph in the product, but the hard inner layers of the tetra-basic sulphate were converted to PbO_2 only after several cycles. This behaviour was considered to be an improvement in the efficiency and strength of the positive active material during formation and cycling.

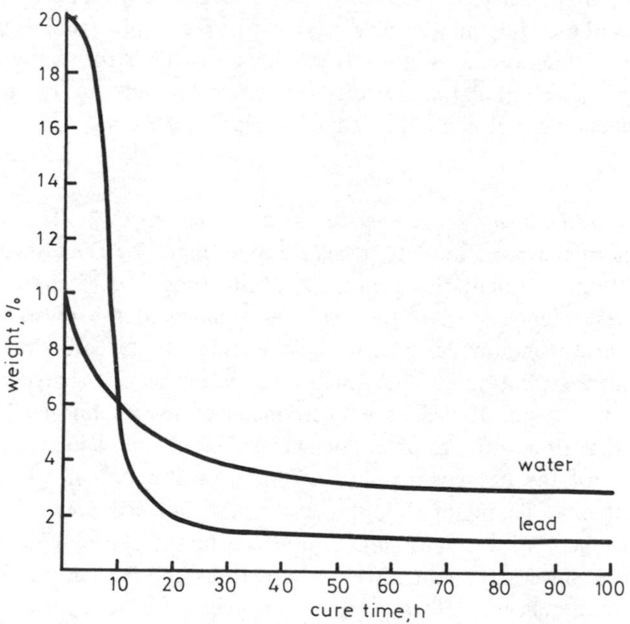

Fig. 4.36 *Metallic lead and water content against cure time*
[Courtesy of J.R. Pierson, *Power sources 2*, p. 104, Pergamon Press Ltd., 1970]

Humphreys, Taylor and Barnes[123] have also described the curing of pastes made with ball mill oxide. One of the main aims was to reduce the free lead content to below 5% and it was found that the rate of oxidation of the lead was highest when the paste contained from 7 to 8·5% of H_2O. For thicker plates, setting may be continued for periods

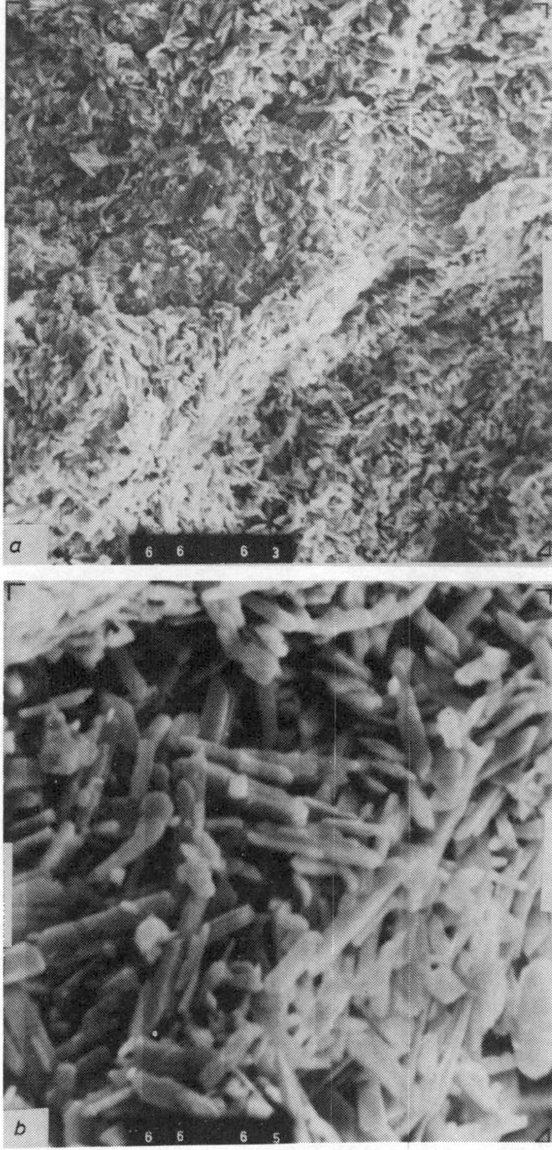

Fig. 4.37 *Microphotograph of positive and negative active materials after setting and drying, showing crystalline structures*
a Positive ×3000
b Negative ×10 000
c Negative ×3000
[Courtesy S. Hattori, Yuasa Battery Co., Japan, and ILZRO Project LE-276, sponsored by the International Lead Zinc Research Organization, 1978]

Fig. 4.37*c*

up to 3 days, with the plates stacked in heaps, covered with canvas or plastic sheets. Final drying is generally done in ovens at temperatures of about 100°C. Figs. 4.37*a-c* show typical microphotographs of the crystalline structures of active materials after setting and drying. These excellent scanning electron micrographs were made by the group at the Yuasa Battery Company, Japan, working under S. Hattori and sponsored by the International Lead Zinc research Organisation Inc., USA[123a] on their project LE-276 and covered in ILZRO Progress Report No. 2.

It was concluded that tri-basic lead sulphate was the predominant material in both positive and negative plates. Fig. 4.37*b*, at the higher magnification, clearly indicates how a rigid durable structure is produced by the random distribution of inter-locking fibre-like crystals of the tri-basic sulphate.

Peters[124] has also carried out an extensive chemical analysis and physical examination of positive and negative plates, starting from the dry oxide and working through the various processes to the completion of 150 discharge-charge cycles. Two physical characteristics of special interest, since, for reasons mentioned earlier, they both have a direct bearing on the capacity, namely, the internal BET surface area and the porosity are shown in Figs. 4.38 and 4.39. While the surface area of the *positive* active material increased fairly steadily from about 1 m^2/g to 7 m^2/g, with a point of inflection at the full formation stage, the area of the *negative* material, after an increase from 0·75 m^2/g to 1·75 m^2/g on setting, decreased rapidly to around 0·5 m^2/g during formation and subsequent cycling. This confirmed the tendency of negative active material to agglomerate on working.

The porosity figures showed much greater similarity. Starting at about 50% with the pastes, this fell to below 40% at the early formation stage, presumably due to the formation of lead sulphate, returned to 50% after the first cycle and then rose steadily during later cycling to 60% for the positives and 75% for the negatives after cycle number 150.

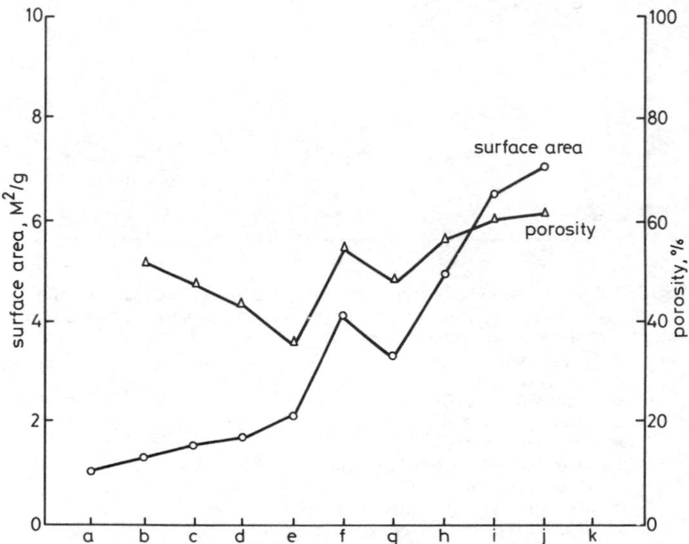

Fig. 4.38 ***Positive plates: surface area and porosity changes of the active materials at various stages of manufacture and service up to 300 discharge/charge bench cycles***

a Dry oxide
b Wet paste
c After setting (16h)
d After drying (4h)
e After formation (4h)
f After formation (completion)
g After initial discharge
h After 50 discharge/charge cycles
i After 100 discharge/charge cycles
j After 150 discharge/charge cycles
k After 300 discharge/charge cycles
[Courtesy of K. Peters]

Although many authors regard the hydrated tri-basic sulphate, made from tetragonal PbO, as the best starting compound for positive active material, Biagetti and Weeks[125] of Bell Telephone Laboratories Inc., USA, have expressed a preference for the tetra-basic sulphate, made by

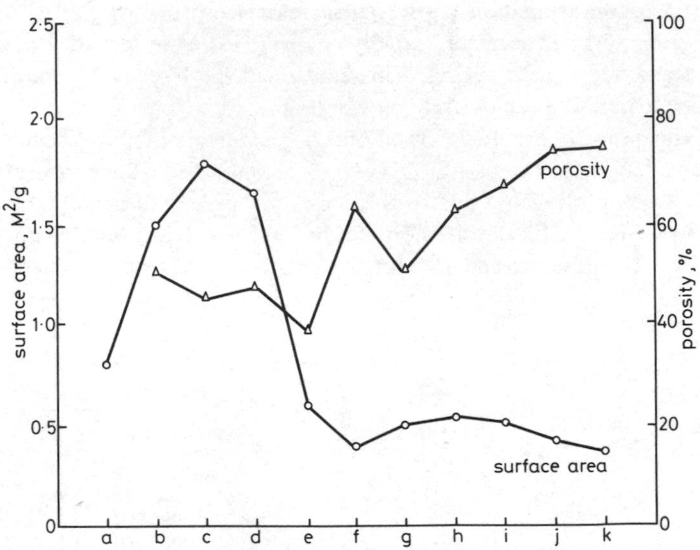

Fig. 4.39 *Negative plates: surface area and porosity charges of the active materials at various stages of manufacture and service up to 300 discharge/charge bench cycles*

a Dry oxide
b Wet paste
c After setting (16h)
d After drying (4h)
e After formation (4h)
f After formation (completion)
g After initial discharge
h After 50 discharge/charge cycles
i After 100 discharge/charge cycles
j After 150 discharge/charge cycles
k After 300 discharge/charge cycles
[Courtesy of K. Peters]

mixing stoichiometric quantities of ortho-rhombic PbO and sulphuric acid at temperatures between 80°C and 100°C. This is claimed to give increased product yield, longer lifetime and a crystal morphology more suitable to battery operation than that produced from other compounds. Yarnell[126] has claimed some advantages also for the use of tetra-basic lead sulphate for negative paste. Hattori has also examined by SEM the changes in morphology of the active materials after first formation and cycling, and these are shown in Fig. 4.40*a-f* all at magnification x 3000.

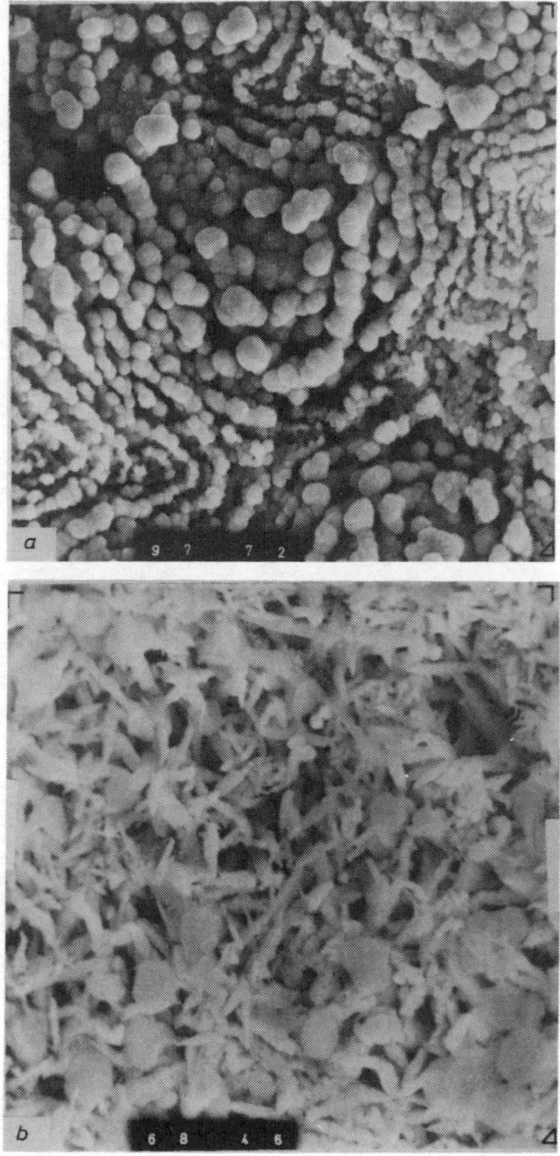

Fig. 4.40 *Microphotographs showing the changes in morphology of the active materials after first formation and cycling (magnification x 3000)*

(*a*) Positive active material, after first formation, showing typical development of crystalline PbO_2
(*b*) Negative active material after first formation, showing development of metallic lead dendrites

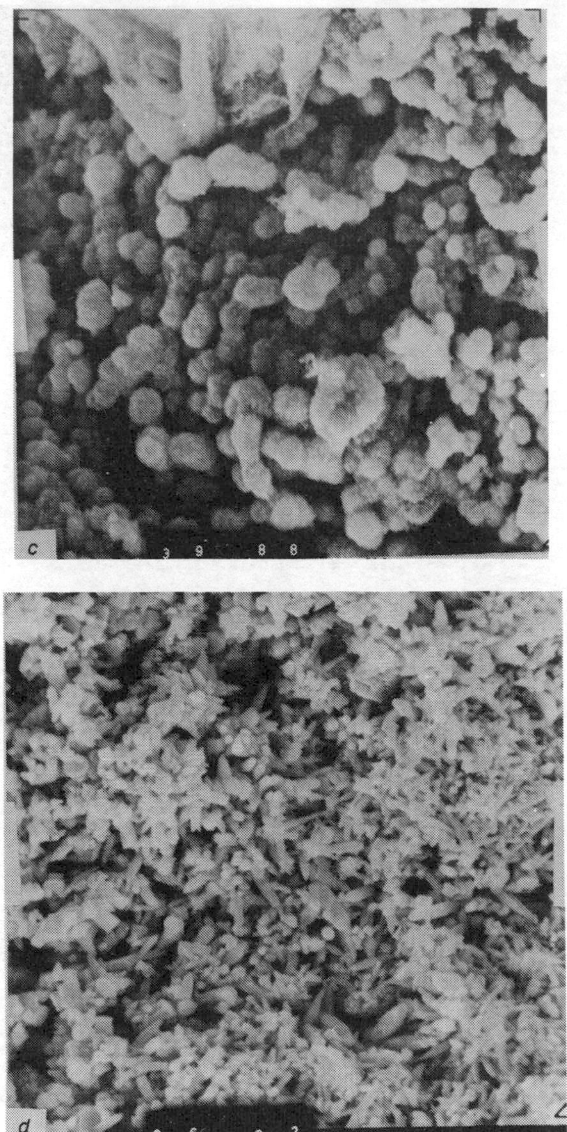

Fig. 4.40 (*c*) Positive active material after 4 discharge/charge cycles, showing increased development of PbO_2 crystals, with minute spike-like growths on each crystal, the crystals beginning to assume the coralloid structure, described by Simon and Caulder[58]
(*d*) Negative active material after 4 discharge/charge cycles showing the randomly-oriented dentrites of lead becoming less acicular and more rounded in contours

Fig. 4.40 (*e*) Positive active material after 4 cycles, but in the discharged condition, showing well-defined crystals of $PbSO_4$, confirming the dissolution-precipitation mechanism proposed by Vetter[54]
(*f*) Positive active material made from tetra-basic lead sulphate, $4PbO.PbSO_4$, prepared by curing the paste at 90°C and relative humidity of 90/, after 4 cycles, showing massive crystalline growth, suggesting a hard and rigid matrix
[Courtesy S. Hattori, Yuasa Battery Co., Japan, and ILZRO Project LE-276, 1978]

4.6.9 Formation

The next stage in the preparation of the fully-charged battery is the electrochemical conversion of the dried paste into the positive and negative active materials; lead dioxide and spongy lead. This is done in two ways.

4.6.9.1 Tank formation: Positive and negative plates or castings are assembled in parallel in large tanks, made of wood, lined with lead sheet, or ceramic or plastic material such as PVC. Thin automotive castings may be grouped in pairs or even in packs of three, held in grooved supports with one take-off lug protruding above the electrolyte. The lugs of the positive plates are welded to busbars running the length of each tank. In the case of the negative plates, since there is no growth of resistive oxide-sulphate layers at points of contact, the lugs at the bottom of the castings simply rest on lead busbars placed at the bottom of the tanks and brought to the surface at each end. The tanks are filled with a liberal excess of sulphuric acid, generally in the SG range 1•050 1•150. Several tanks are connected in series to the DC power supply, which may be provided from the AC mains source through a rectifier of the solid state or mercury arc types, or, alternatively, by an MG set. A suitable ballast resistance is included in each circuit to cover the rise in voltage at the end of charge. Charging is continued to the gassing state, when the voltage of each cell reaches about 2•60V. 40 tanks can therefore be accommodated on a DC line of 110V.

Simple charging regimes are used, e.g. continuous charge for 20 h at a current of 16•5A/kg of dried paste, corresponding to an input of 330 Ah/kg. When plates are assembled in pairs or threes, the current is simply increased *pro rata.* Gassing generally starts about 3 h before the end of the schedule. Complete formation is indicated by the colour of the plates, dark maroon for the positives and light grey with a metallic texture for the negatives.

The gas evolution towards the end of formation can be a nuisance. The gas bubbles, particularly the hydrogen bubbles, carry acid spray into the atmosphere, and this is harmful to the operatives and to the structure of the building. Various methods are used to contain the spray, by, for example, covering the tanks with sheets of plastics, inert to sulphuric acid, such as PVC. Alternatively, the spray can be substantially reduced by the addition to the acid of small amounts of a wetting agent or surfactant, usually of the sulphonated type. These reduce the surface tension of the acid, the gas bubbles become much smaller and remain on the surface of the liquid as a fine foam. The surfactants are partially decomposed in the charging process and small additions are needed after each formation run. Care must be taken, however, to avoid excess, as these compounds also affect the over-potential of the negative plate, causing premature evolution of hydrogen before the material has been completely reduced.

During formation, sulphate ions are released by both plates and the concentration of the acid rises. At the end of each run the SG is adjusted by the addition of water. After a number of runs, the levels of impurities leached out of the pastes increase, and, when these reach intolerable limits, adjustments are made with pure acid.

4.6.9.2 Jar formation: Tank formation has many obvious disadvantages and in recent years most manufacturers have turned to the simpler jar formation process. In this, the 'green' plates are assembled in their final container, the electrolyte is added, and, after a short stand to enable the acid to diffuse into the elements, the unit is put on charge with a number of other batteries in series, depending on the voltage of the DC power line. The amount of charge given is generally equal to $4{\cdot}5 \times C_{20}$ Ah (where C_{20} is the nominal capacity of the battery in Ah at the 20h rate) over a period of 18 to 24 h. Temperature should not exceed 60°C, otherwise the batteries must be cooled. Since the pastes contain a certain amount of sulphate, the concentration of the filling acid is chosen to give the specified final fully-charged SG of 1·270–1·285, or, in tropical climates, 1·240–1·250. Final adjustments are made by the addition or removal of acid.

For some years a 2-stage process was common, in which the whole of the acid was dumped after the first formation charge. The battery was then re-filled with pure acid of the required SG and given a short mixing charge. This served the double purpose of getting rid of unwanted impurities and of adjustment of the electrolyte. The additional handling operations were, however, objectionable and costly, and nowadays the single stage process is generally used.

4.6.10 Drying processes

4.6.10.1 Tank-formed plates: Plates are left in the formation circuits for 1 to 2 h after the schedule is completed. They are then lifted, placed in racks on skids, allowed to drain for a few minutes and dried in steam-heated ovens at temperatures just above 100°C. The dried positive material contains at least 75% of PbO_2, with less than 5% of $PbSO_4$. The highly reactive negative active material is largely oxidised in the drying process, and, since the PbO is converted to $PbSO_4$ when the assembled battery is primed with electrolyte, plates dried in air in this way require a full first charge after assembly.

4.6.10.2 The dry-charging process: Dry-charged batteries can be put straight into service after filling with electrolyte and a short stand to allow the acid to soak into the elements. Such batteries are expected to give at least 75% of their nominal capacity, after a storage period of not less than 1 year at ambient temperatures. Dry-charging therefore eliminates a second formation or first charge, although, to get the full nominal capacity on the first discharge, a short boost charge may be necessary.

The essential requirement is to restrict the oxidation of the fully-formed negative active material. Immediately after formation the plates are immersed in water before being moved to the drying ovens. These take several forms. Sealed vacuum ovens have been used but they are expensive. In another type, an inert atmosphere is produced by using a naked gas flame to consume the bulk of the oxygen. In a third process for limited production, the wet plates are piled in stacks of about 10 castings and tightly held between two heated platens, kept at a temperature well above 100°C. In a third process, the 'oil-frying' process,[127] the wet plates are immersed in hot kerosene held above 100°C until the water is driven off. After lifting from the hot oil, the plates are held in a stream of hot air to evaporate any residual kerosene.

As an additional precaution and to prevent a loss of capacity during long periods of storage, substances known to inhibit the oxidation of the highly active spongy lead, e.g. lanoline,[128] stearic acid[129], mineral oils,[130] may be added to the paste at the mixing stage, generally in amounts up to about 0·2% of the weight of oxide. Claims have also been made for the use of boric acid as an oxidation inhibitor,[131] added either to the paste as with the other materials, or by dipping the plates in a dilute solution before the final drying in one of the special ovens. For special cases the double process may be used. Plates treated by these processes acquire a certain degree of hydrophobicity. If they are made too hydrophobic, however, the delay in activation when the battery is primed may be unacceptable.

Positive plates for dry-charged batteries are given the normal tank formation and drying processes. They may be held in dry storage for long periods without significant loss of charge. Dry-charging of jar-formed batteries is more complicated. After formation, the electrolyte is drained off, the battery may be flushed out with water once or twice and then moved into a drying oven kept at an appropriate temperature for the minimum time required to remove the water in the elements, sometimes assisted with a stream of hot air. In another process, developed by the Globe Union Company of the USA,[132] batteries held on their sides are spun in a type of centrifuge, in which the free electrolyte is ejected. One minor problem concerns some separators used in jar-formed dry-charged batteries. Some of the basic materials may be hydrophobic, e.g. PVC and wetting agents are added during the manufacturing process. These may be washed out during the dry-charging process, and, unless the wetting properties are re-established, air entrainment in the separators may cause high internal resistance. For separators based on PVC, this problem has been overcome by the use of a co-polymer, one component of which has hydrophilic properties.

4.6.11 Separators and 'retainers'

In most storage batteries, the positive and negative plates are welded to the busbars of 'equaliser-bars' in parallel groups, generally with one extra negative plate and the groups are inter-leaved with insulating separators. The main technical requirements for separators are

(*a*) adequate mechanical strength to withstand handling through assembly and the stresses likely to be met in service

(*b*) resistance to oxidation under the severe conditions in service, i.e. contact with PbO_2 in 30% sulphuric acid at temperatures up to 60°C, and with nascent oxygen evolved during overcharge

(*c*) a micro-porous structure, with an overall porosity of at least 50%, to permit free diffusion of electrolyte and migration of ions, giving a low electrical resistance. Values for the latter generally range from 0·030 to 0·045 Ω/in^2 ($0{\cdot}19 \times 10^{-4}$ to $0{\cdot}30 \times 10^{-4}$ Ω/m^2) of the superficial surface area. A small pore size is also an advantage, since this restricts the penetration of particulate matter, sulphate crystals etc., which leads to short-circuits, sometimes referred to as 'leading through'.

For many years, wood was the prime material for this purpose, principally Port Orford Cedar and Douglas Fir from the forests of North-West

America. During the Second World War, supplies neared exhaustion and various artificial separator materials began to appear. Felt made from glass fibres was a popular starting material, sometimes bound together by latex made from rubber or a plastic resin, mixed with fine siliceous powder or kieselguhr. During the 1930s several types of microporous rubber had been developed and some of these techniques were applied to other plastics then coming onto the market, such as polyvinyl chloride (PVC) and polyethylene. Two main routes were followed. In the wet process, the plastic which formed the continuous phase was mixed with a pore-forming agent, such as starch or a water soluble salt, such as ammonium sulphate and a plasticiser, the mixture being warmed to form a pliable dough. This was forced through a profile-forming extruder and the plasticiser evaporated by passing the extruded sheet through a heated oven. The dried sheet was then treated in hot water to remove the pore-forming dispersed phase, and, after a final wash in a solution of a wetting agent, dried again before being passed to a guillotine, where the separators were automatically cut to the required size.

Porvic 1 is a typical separator made by this process, the continuous phase being PVC. The pore size is very uniform, being similar to that of the starch particles and the maximum dimension is about 5×10^{-3} mm. The final product is therefore highly micro-porous, with an overall porosity exceeding 80%.

Porvic 2 is a PVC-based separator made by a dry process, in which the fine powder, spread on a continuous steel belt, passes under a 'doctor knife', which fixes the profile and the thickness and then into a sintering oven, where the powder is sintered into a rigid form. This material has a much wider range of pore sizes, the largest being nearly 10 times that of the pores in Porvic 1 and the overall porosity is nearer 50%. For any given size of separator, therefore, Porvic 2 contains nearly twice as much plastic as Porvic 1. The material cost of the former is higher, but the plant and the process of manufacture are much less costly.

Paper pulp is a third basic raw material, now widely used for SLI separators. To obtain the required mechanical strength, rigidity and resistance to oxidation in the battery, the pulp is impregnated with a resin, usually of the phenol-formaldehyde type and oven cured. The appropriate profile is made during curing or by applying ribs of powdered PVC which is sintered in the curing oven.

Within the past few years, at least three other microporous separator materials have come into service; Darak 5000 and Daramic, both developed by W.R. Grace and Company of the USA, and Yumicron, by the Yuasa Battery Company of Japan. These have a plastic polymer, based on phenol or resorcinol resin as the continuous phase, supporting powdered silica gel filler with an extractable oil or water as the pore-forming agent. Alternatively, the microporous plastic is retained on a thin fleece of acid resistant fibres, such as polyethylene, terylene or polypropylene. It is claimed that the micro-pores in Darak range around 1 μm (10^{-3} mm) in size, while the bulk of those in Daramic are as low as 0·1 μm (10^{-4} mm).

Table 4.13 lists some of these separators, the basic materials from which they are made and some of their physical properties. Apart from their function in preventing short circuits between plates, some of these micro-porous separators also inhibit poisoning of the negative active material by antimony. This can hardly be classed as filtration, since the smallest micropores are appreciably larger than the diameter of the ionic antimony species, even if surrounded by a large solvent sheath. The microporous materials, however, have a high tortuosity factor and the blocking effect may be due to electrophoretic forces, similar in some respects to those which develop in ion-exchange membranes.

Table 4.13 *Separators*

Separator	Manufacturer	Basic materials	Bulk Porosity %	Pore Size μm	Electrical resistivity Ω/cm^2
Acesil	American Hard Rubber Co.	vulcanised rubber lates; silica gel	61	1·0 to 2·5	0·30
Amersil	A.B. Tudor Co. Sweden and Amerace Corpn.	PVC lates with silica powder	75	0·01 to 0·5	0·125
Darak 5000	W.R. Grace & Co., Germany	P/F resin polymer on non-woven plastic fleece	55	0·5 to 0·8	0·20
Daramic	W.R. Grace & Co., USA	polyethylene with silica powder and mineral oil	63	0·01 to 0·10	0·15
Mipor	Varta A.G., Germany	vulcanised rubber latex; silica gel	65	0·1 to 5·0	0·35/·40
Paper pulp	Various	wood pulp, stiffened with P/F resin	60-70	<30	0·25
Porvic 1	Chloride Group	PVC with starch or ammonium sulphate and plasticiser.	80	<5	0·40
Porvic 2	Chloride Group	sintered PVC	48	<50	0·30
PVC Industrial	Leopold Jungfer, Austria	sintered PVC	40-45	<35	0·30
Permalife	Permalife Corpn., USA	latex with kieselguhr on glass-wool mat	70	5·0 to 12·5	0·35
Polymion	Japan Storage Battery Co.	polyolefin resin	60	0·1 to 1·0	very low
Ymicron	Yuasa Battery Co., Japan	vinyl chlordie resin	60	0·1 to 1·0	very low

The values of the resisitivity must be regarded as approximate only. The measurements were made on actual separators, the thicknesses of which were not entirely uniform. The figures serve, however, as a guide to the differences between the different types of material.

Separators are generally made with vertical ribs on the side in contact with the positive plate. These serve two purposes. They give firm support to the closely packed element and they also provide a good reservoir of free acid for the positive active material. As indicated in eqn. 4.13, during the discharge, water is produced at the positive plate and this reservoir helps to equalise the concentration of the electrolyte.

Sometimes dual separation is used, with a felted glass fibre 'retainer', held against the positive plate, to retain the active material. The introduction of artificial separators, made of acid-resistant plastics, was a crucial factor in improving the performance and the lifetime of SLI batteries. They had greater mechanical strength and durability and were free from the contaminating impurities in the wood separators which they have largely replaced. Even greater durability is required in traction battery separators, since they are required to survive at least 1500 deep cycles over a working lifetime of about 6 years.

4.6.12 Containers and covers

SLI batteries are generally assembled in 3-cell (6V) or 6-cell (12V) monoblock mouldings, which must be tough enough to withstand rough handling and vibration on the vehicle, as well as the thermal cycles of heat and cold that they meet in normal service. Originally these containers were made in a heavily-loaded bituminous pitch ('Milam', 'Plastok' and 'Dagenite' were typical examples) with thick walls and separate lids for each cell, sealed in position with bituminous compound.

The use of hard rubber, first loaded with clay and later with finely-powdered coal dust, brought substantial reductions in wall thickness and weight and a significant increase in impact strength. The introduction of resin rubbers, mixtures of natural or synthetic rubber with styrene or phenolic resins, of which 'Fortex' is an example, led to further reductions in wall thickness and weight, with impact strength three to four times that of the older types of hard rubber and curing times as low as three to four minutes. Single piece covers fitting over the complete container have generally replaced the multi-lid assemblies, resulting in economies in assembly and inventory. The advent of new plastic resins; polystyrene, polyethylene, acrilo-nitrile-butadiene-styrene (ABS) and polypropylene, has revolutionised the battery industry. ABS and polypropylene have now come into service on a steadily growing scale. The relatively high costs of these materials, all products of petroleum feedstocks, and the moulds and moulding presses are offset by a number of distinct advantages. The weight has been reduced from about 2·73 kg (6 lb) for a typical container in hard rubber to 0·82 kg (1·8 lb) for the equivalent, thin-walled polypropylene model. The use of translucent or brightly-coloured containers and covers in the same material has also made it possible to break away from the traditional 'black box', which has become the rather disparaging image of the SLI storage battery and also made it possible to attach the one-piece cover by a heat-welding technique instead of the more complex and costly process of pouring bitumastic or plastic seals.

4.6.13 Assembly of cells and batteries

Many ingenious methods have been devised to simplify assembly and to reduce both the labour and the material contents. In the older process, the plates were dropped in the 'burning boxes' by hand and the plate lugs were welded to the equaliser-bars with terminal posts at each end. Positive and negative groups were fitted together, and, after insertion of the separators by hand, the elements were pressed into the cell compartments. Individual lids were fitted over the elements in each cell with the terminal posts protruding through moulded-in metal inserts in the lids, to which they were then welded. Finally, the lids were sealed in position with bitumastic compound, and the cells connected in series with heavy connectors, welded to the terminal posts on top of the lids.

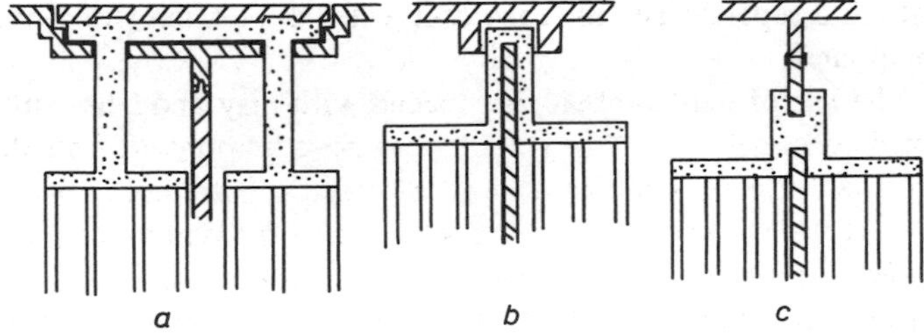

Fig. 4.41 *Three kinds of intercell connection*

a Conventional post link construction *R*-0·0016Ω
b Over partition construction *R*=0·003Ω
c Through partition construction, *R*=0·0003Ω
[Courtesy of Muto and Uno,[135] Japan Storage Battery Co.]

Later, automatic stacking machines were developed, in which complete elements of plates and separators were stacked automatically, before being handled into the burning boxes for the final welding operation. It had been recognised that this type of assembly carried a good deal of surplus 'top lead', but care had to be taken to avoid leakage of electrolyte between cells. The next step was to cast the inter-cell connector as a small metal saddle, which was fitted over the cell partitions beneath a single unit cover over the whole battery.

The development of thin-walled plastic containers opened the way for another advance, described by Halsall and Sabatino,[133] of reducing the top lead still further, and, at the same time, saving valuable voltage losses, by welding the inter-cell connectors through the cell partitions. The different stages of these developments are shown in Fig. 4.41.

4.6.14 *Weight analysis of typical SLI 12V batteries*

This is a useful exercise for engineers who may be seeking ways of reducing the weights of various components in an effort to improve the energy density of the complete unit. Fig. 4.42A shows such an analysis of a typical modern 40 Ah UK battery, having an energy density at the 20h rate between 35-40 Wh/kg[134] and Fig. 4.42B, a similar breakdown of a standard Japanese battery having an ED of 37 Wh/kg.[135]

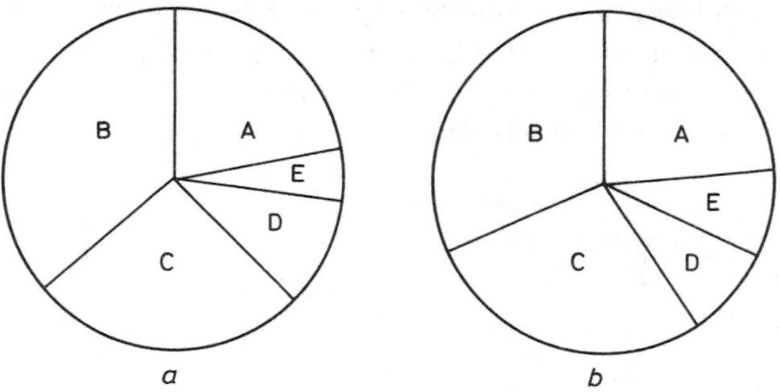

Fig. 4.42 *Weight analysis of batteries*
a 40 AH UK battery with energy density (ED) at the 20h rate between 35-40 Wh/kg
b Standard Japanese battery with energy density of 37Wh/kg

	UK	*Japanese*
Grids	A 21•5	23•8
Active materials	B 36•0	31•7
Electrolye	C 27•5	29•6
Containers, lids separators	D10•3	7•4
Top lead	E 4•7	7•5
	100 %	100 %

The weights of the components are expressed as percentages of the total weight. One noteworthy point of interest is that, in both cases, the active components which contribute directly to the performance; the grids, active materials and the electrolyte, make up about 85% of the total weight, with slight differences in emphasis in the two models.

4.7 Performance of SLI batteries

4.7.1 General characteristics

Fig. 4.43 shows typical families of discharge voltage curves for a 12V, 100 Ah battery at various rates of discharge from the 20-hr rate to the 1-min. rate at a temperature of 25°C (77°F). Bench tests of this kind are generally carried out at constant current. In assessing the capacity at any rate, the choice of the final voltage is important. As mentioned in Section 4.4.4.1, the 'knee' of the curve, particularly at high (i.e. fast) rates is less pronounced for a 12V battery than for a single cell. In this case, the final voltages were as shown in Table 4.14. Mean voltages during the discharges are indicated by the dotted lines. Table 4.15 lists the values in the curves and includes figures for the energy, expressed as watt-hours and the power, as watts, in both cases using the values shown for the mean voltages. The figures for energy are also

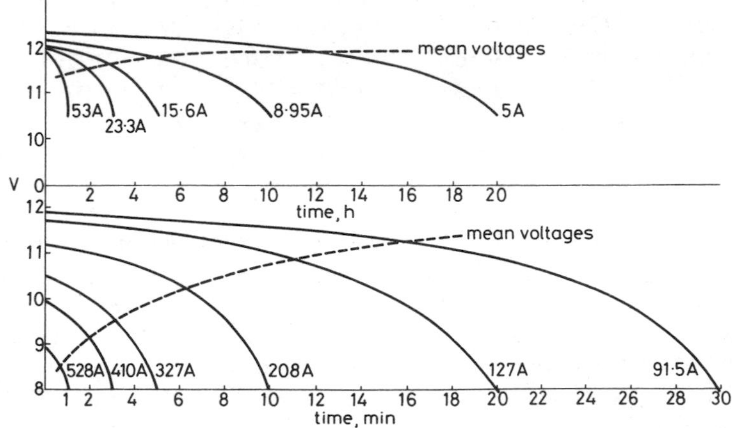

Fig. 4.43 *Discharge voltage curves for a 12V, 100Ah (20h) SLI battery at 25°C (77°F)*

Table 4.14 *Final voltages for different rates*

Rate	Final Voltage Cell	Battery
From 20 h to 1 h	1·75	10·50
30 min to 1 min	1·33	8·00

expressed as percentages of the 20h value. In the case of power, interest lies mainly with the highest rates and the values for rates down to the 30 min rate shown in reverse as percentages of the figure for the 1 min rate.

Table 4.15 *Performance of a 12V SLI battery*

Rate of discharge	Current	Capacity	Mean voltage	Energy		Power	
h	A	Ah	V	Wh	%	W	%
20	5·0	100	11·85	1185	100	59	
10	8·95	89·5	11·75	1052	89	105	
5	15·6	78	11·55	901	76	180	
3	23·3	70	11·48	804	68	267	
1	53·0	53	11·40	604	51	604	
min							
50	63·0	52·5	11·35	596	50	715	
40	75	50	11·30	565	48	848	
30	91·5	45·8	11·20	513	43	1025	23
20	127·0	42·3	10·85	459	39	1378	31
10	208	34·6	10·20	360	30	2122	48
5	327	27·2	9·50	261	22	3107	70
3	410	20·5	9·20	189	16	3772	84
1	528	8·8	8·50	75	6	4488	100

Fig. 4.44 shows that the relationship between log I and log t, where I is the current density and t is the duration or rate of discharge, is linear down to about the 10 min rate, in accordance with Peukert's expression, expr. 4.20. As indicated in the Figure however, divergences appeared at rates below the 10 min rate, probably due to errors in measuring the parameters accurately over these very short periods. Some workers have, however, found Peukert's equation useful for predicting performance in the 1 to 3 min range by interpolation of results not too far apart. In this case, the value of n representing the slope of the line

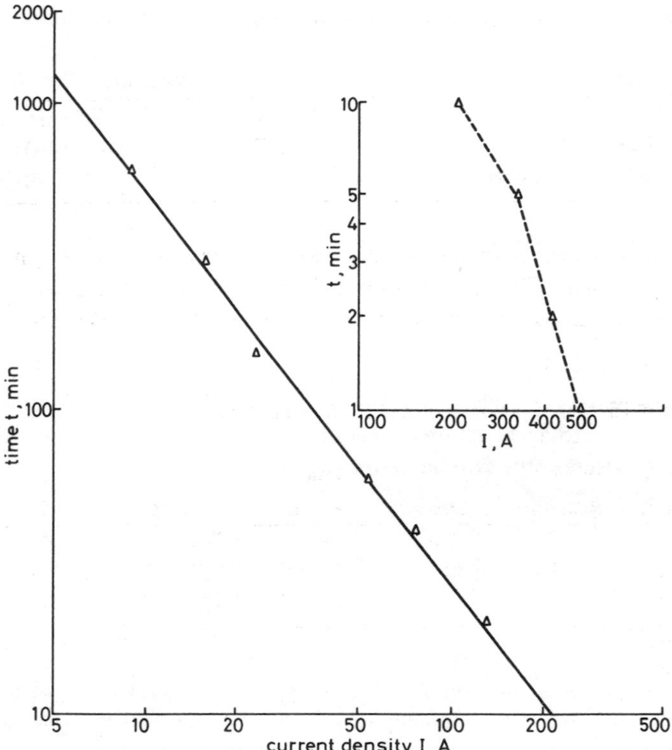

Fig. 4.44 *Relationship between log I and log t*

was about 1·3, in reasonable agreement with the value of 1·36 obtained by Vinal[11, p.217] and 1·39 by Gillibrand and Lomax[31] by tests on single positive and negative plates.

4.7.2 *Specifications for SLI batteries*

Battery performance is, of course, largely determined by the standards laid down in official specifications. Unfortunately, there is no universal specification setting out standards of performance on an international basis. The standards most widely used are those drawn up some years ago, and modified from time to time to meet changing needs, by the US Society of Automotive Engineers (SAE) and recently adopted by the Battery Council International (BCI) with headquarters also in the USA. In Europe and Japan, standards have been drawn up by the various national standards Institutions, such as the DIN standards in

Germany, and, in some cases, by the large automobile manufacturers themselves.

In recent years, attempts have been made throughout Europe to unify the requirements through a battery section of the International Electrotechnical Commission (IEC). Using SAE specifications J537 (latest revision J) and J240a and British Standard BS 3911-65 as a basis, a draft, stating the methods of test and the minimum requirements is under consideration by the IEC and likely to be published in 1981.

In Japan, the National Standards Institution has prepared a series of specifications, of which JIS D5301 (1973) relates to SLI batteries, but the Japanese Standards Committee have indicated their willingness to adopt the SAE methods of test. In the short space available here, it is not possible to give detailed accounts of these specifications. Outstanding features only will be described.

4.7.2.1 Cranking current in amps (CCA): As indicated by SAE/BCI, the primary function of the SLI battery is to provide power to crank the engine during starting. With the increase in the compression ratio of internal combustion engines, this duty has become more and more onerous and added difficulty arises at sub-zero temperatures, owing to the increased viscosity of the engine oil. In 1978, the SAE/BCI standard was declared as the current, in amperes, delivered at a temperature of −17·8°C (0°F) for not less than 30 s to a voltage, not less than 1·2V per cell or 7·2V for a 12V battery.

The corresponding IEC requirements, also at −18°C, is a minimum duration of 1 min to not less than 1·4V per cell or 8·4V for a 12V battery. The current delivered under SAE conditions is greater than that specified by IEC by a factor of from 1·6 to 1·8. To cater for the lowest temperatures met in America and elsewhere, the SAE Specification includes a second low temperature requirement, namely, the current in amperes, delivered at −29°C (−20°F) for not less than 90 s to a voltage, not less than 1·0V per cell or 6·0V for a 12V battery. The IEC has also adopted the conditions: −29°C, 1 min to not less than 8·4V for a 12V battery as a high rate (CCA) test for extremely low temperatures.

The Japanese Standard differs considerably. JIS D5301 specifies a temperature of −15°C and a current of 150, 300 or 500 A, to a voltage of 1·0V per cell or 6·0V for a 12V battery for a period between 1·4 and 4·3 min, depending on the size of battery. Certain minimum voltage levels are also specified; 7·1 to 8·9V after 5 s and 7·6 to 8·2V after 30 s, again depending on the size of the battery.

4.7.2.2 Reserve capacity: The battery must also provide emergency power for the ignition, lights and ancilliary services, such as the radio, fans for heating and air-conditioning, window-winding equipment etc. The reserve capacity specified by SAE and IEC, but not by JIS D5301, is defined as the number of minutes a fully-charged battery can maintain a current of 25A to a voltage, not less than 1·75V per cell or 10·5V for a 12V battery at a temperature of 26·7°C (80°F) (25°C for IEC).

4.7.2.3 Capacity at the 20h rate at ambient temperatures - 23°C (77°F): This low-rate test was originally devised to give some measure of the amounts of active materials used. In modern batteries with thin, closely-pitched plates, the 20h capacity is often limited by the amount of electrolyte and has little relevance to the cranking power, which is largely determined by the surface area of the plates. However, the 20h capacity to a final voltage of 1·75V per cell has been retained in some specifications as an identifying link with the older standards and particularly where deep discharges are required, e.g. on buses, fire engines etc.

4.7.2.4 Miscellaneous tests: The following may also be specified:

charge rate acceptance test
overcharge life test
cold activation test for dry-charged batteries
vibration test.

For details, the reader should refer to the specifications listed.

As noted, there are some radical differences in the methods of testing. Reserve capacities are not included in the JIS Specification and fully-charged weights are not given in the latest SAE/BCI standards. It is difficult, therefore, to make quantitative comparisons of batteries covered by the three different groups. However, Table 4.16 lists some of the specified characteristics of typical batteries, taken at random from the documents noted. Groups 1 and 2 represent European standards, as laid down by the IEC, Group 3, the SAE American standards, and Group 4, the Japanese JIS standards. Groups 5 and 6 represent some of the latest standards for maintenance-free batteries.

Table 4.16 *SLI battery specifications*

Group	Code or catalogue number	20h capacity at 25°C	Dimensions* L/WX	Volume	Weight	CCA at −17·8°C	Reserve capacity mins at 25°C and 25A	
		Ah	mm	dm³	kg	A	Ah	Ah
1	383	35	207/175/175	6·3	12	140	50	21
	368	68	267/173/206	9·5	20	330	120	50
	298	75	366/167/217	13·4	26	210	190	45
	361	80	318/175/277	15·4	32	300	120	50
	222	120	508/208/207	21·9	43	450	205	85
	By courtesy of the Chloride Group: Catalogue dated February 1979							
2	141	40	218/133/205	5·9	12	170	58	24
	274	60	305/172/219	11·5	23	170	80	33
	389	68	255/174/204	9·1	21	320	120	50
	327	110	513/189/225	21·8	40	450	210	88
	By courtesy of Joseph Lucas (Batteries) Ltd.: Catalogue dated 1978							
3	24-255	-	260/173/225	10·1	-	255	60	25
	24-285	-	260/173/225	10·1	-	286	75	31
	24-375	-	260/173/225	10·1	-	375	86	36
	27-360	-	306/173/205	11·9	-	360	110	46
	27-440	-	304/171/222	11·5	-	440	102	43
	SAE Specification J537: revision j, June 1978							
4	NS4OZ	35	197/129/227	5·7	12	150,	3·5 min to 6·0V	
	N50	50	260/173/225	10·7	20	150,	3·6 min to 6·0V	
	NS60	40	238/129/227	7·0	14	300,	1·4 min to 6·0V	
	N100Z	100	395/176/233	16·2	35	300,	4·1 min to 6·0V	
	JIS Specification J5301; courtesy of Kozawa and Takagaki†							
5	071-85 MF1	41	209/178/214	8·0	15	275	60	25
	071-85 MF2	48	214/178/214	8·2	15	350	80	33
	071-87 MF3	60	229/178/214	8·7	17	430	100	42
	072-1200	78	331/173/239	13·7	24	475	130	54
	072-1250	108	331/173/239	13·7	27	575	180	75
	By courtesy of AC Delco: MF Freedom Battery leaflet 07-2, Jan. 3rd 1978							
6	MF-22F	-	241/174/210	8·8	16	375	82	34
	MF-24	-	260/174/222	9·9	21	470	120	50
	MF-27	-	305/171/222	11·6	24	550	140	58
	By courtesy of Gould Inc., Automotive Battery Division: leaflet PS04892							

* Height includes terminal posts and top hamper over vents, if used, equivalent to about 22 mm or 10% of the total height

† Kozawa, A., and Takagaki, T., Japanese Lead-acid Battery Industry, published by the US Office of the Electrochemical Society of Japan, Cleveland, Ohio, USA, 1978

Models 072-1200 and 972-1250 are the latest AC Delco Freedom M-F batteries, with thin wrought-metal grids. In a number of instances, a common container is used for two or three types with a different combination of plates and separators and therefore giving different performances.

4.7.2.5 High rate performance at −17·8°C: The importance of the high rate performance, particularly at low temperatures, is clearly emphasised in all of the SLI specifications. Fig. 4.45 represents graphically the results of a series of tests on battery, Code No. 389 in Group 2 of Table 4.16, showing the duration in minutes to a final voltage of 6·0V with currents from 210 to 600A at −17·8°C. Mean voltages are shown by the dotted line and in Table 4.17 values have been estimated for the power densities by weight (W/kg) and volume (W/dm³). As indicated in Table 4.16, the weight of this battery is listed as 21 kg and the volume, 9·1 dm³. The capacity of this battery at the 20-hr rate at 25°C (77°F) is listed as 68 Ah and the mean voltage for a discharge at this rate is 11·85V, from which the following *energy densities* are derived:

by weight 38 Wh/kg

by volume 87 Wh/dm^3

These values fall among the upper levels for batteries listed in Table 4.16.

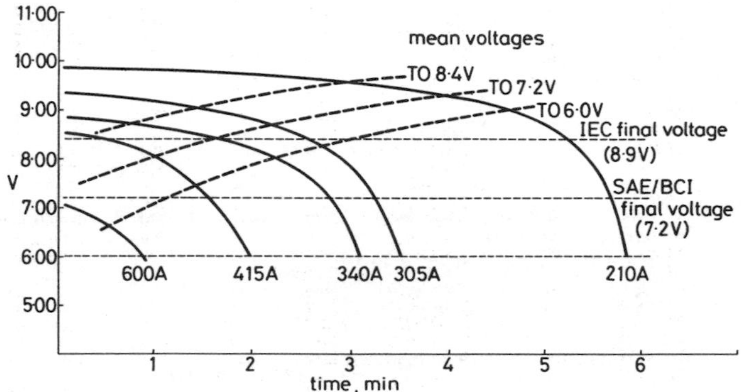

Fig. 4.45 *High rate performance at −17·8°C, battery code no. 389. 68Ah (20h rate)*
[Courtesy Lucas Batteries Ltd.]

Table 4.17 *Power densities at high rates at −17·8°C*

Current	Duration	Mean Voltage	Power	Power densities	
				by weight	by volume
A	min	V	W	W/kg	W/dm^3
210	5·9	8·8	1878	89	206
305	3·5	8·1	2471	118	272
340	3·1	7·8	2652	128	295
415	2·0	7·4	3071	146	337
600	0·9	6·5	3900	186	429

4.7.2.6 Effect of temperature on high rate performance: Fig. 4.46 shows the results of discharge tests, carried out on battery number 389 at a constant current of 340 A at temperatures from 25°C to −30°C, to a cut-off voltage of 6·0V. In Table 4.18, the capacities (Ah) are also expressed as percentages of the value at 25°C.

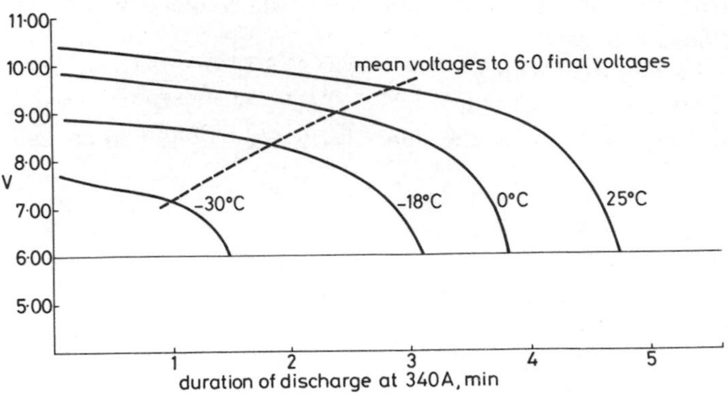

Fig. 4.46 *Battery code no. 389. 60Ah (20h rate): Performance at avious temperatures at 340A*
[Courtesy Lucas Batteries Ltd.]

Table 4.18 *Effect of temperature on high rate performance*

Temperature	Duration to 6·0V	Capacity		Mean voltage	Energy		Temperature coefficient
°C	min	Ah	%	V	Wh	%	Ah/1°C x 10^{-2}
25	4·7	26·6	100	9·0	239	100	-
0	3·8	21·5	81	8·6	185	77	0·77
-18	3·1	17·6	66	8·1	143	60	0·79
-30	1·5	8·5	32	7·0	60	25	1·23

The temperature coefficient of capacity (α) was calculated from the expression

$$C_2 = C_{25} \, [1 - \alpha(25 - \theta_2)] \tag{4.42}$$

where C_{25} is the capacity (Ah) at 25°C and C_2 is the capacity at temperature θ_2.
Smith[136] has quoted a figure of $0{\cdot}7 \times 10^{-2}$ Ah per 1°F for the temperature coefficient of an SLI battery at the 5min rate. This is equivalent to $1{\cdot}26 \times 10^{-2}$ Ah per 1°C, in close agreement with the higher figure in Table 4.18. Smith also states that the coefficient increases as the rate increases and as the temperature decreases.

4.7.2.7 Charging equipment for SLI batteries: After a discharge, the electrochemical energy in the active materials is replenished by recharging from a DC source, the current being passed into the battery in the reverse direction to that of the current on discharge. The voltage of this source must obviously be set at a level somewhat higher than the back EMF of the battery under the appropriate charging current. The battery voltage is inversely related to the temperature: it falls as the temperature rises and increases as the temperature falls, and relays are generally included in the charging circuit to make the required adjustments. On the vehicle, the SLI battery is connected to the starter motor in a separate circuit and in series with the other electrical equipment, the lights, the ignition and the generator. For many years this generator took the form of a dynamo, but there were several objections to this mode of charging. The DC circuit included a cutout and a regulator to adjust the voltage.

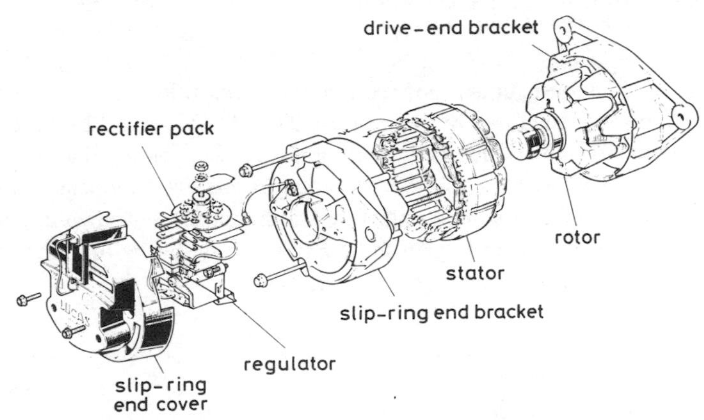

Fig. 4.47A *Exploded view of a typical ACR alternator*
[Lucas Technical Series, Fig. 25, p. 16]

As the electrical loading of modern autmobiles increased, the design of DC generators capable of the additional output raised problems of size, weight and commutation, particularly at high speeds. This has led to the almost universal replacement of the dynamo by an alternator or AC generator, which can supply twice the current of a DC machine of similar size. It can also produce a useful output at idling speeds and will operate satisfactorily at high running speeds.

The development of solid state semiconductor devices, such as diodes and transistors, has made several refinements possible. The AC is converted to full-wave rectified DC by solid state diodes mounted in the frame of the device and connected to the bridge circuit. No cutout is needed and the transistorised voltage controller regulates the output to suit the electrical load and the state of charge of the battery. The

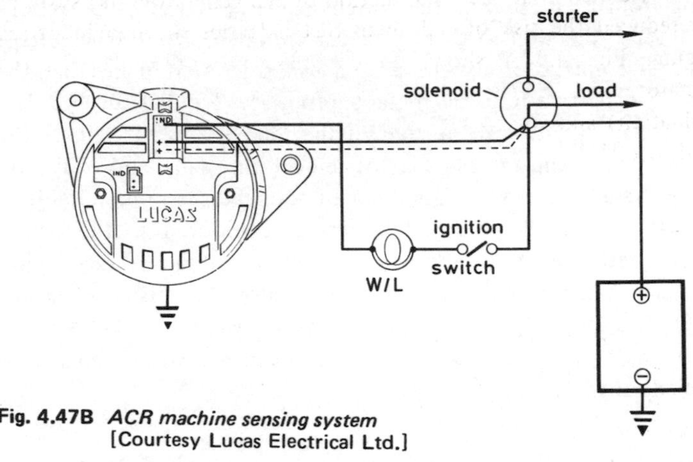

Fig. 4.47B *ACR machine sensing system*
[Courtesy Lucas Electrical Ltd.]

battery is charged under constant voltage conditions, at a floating voltage on the generator busbars of 14·2V ± 0·2V for a 12V battery, equivalent to 2·37V per cell, with a drop of no more than 0·25V from 20% to 80% of the full load. Regulators are level compensated for temperatures, but on some applications temperature sensing is applied, using a thermistor placed against the battery case, but with limitations of the upper and lower limits of the voltages.

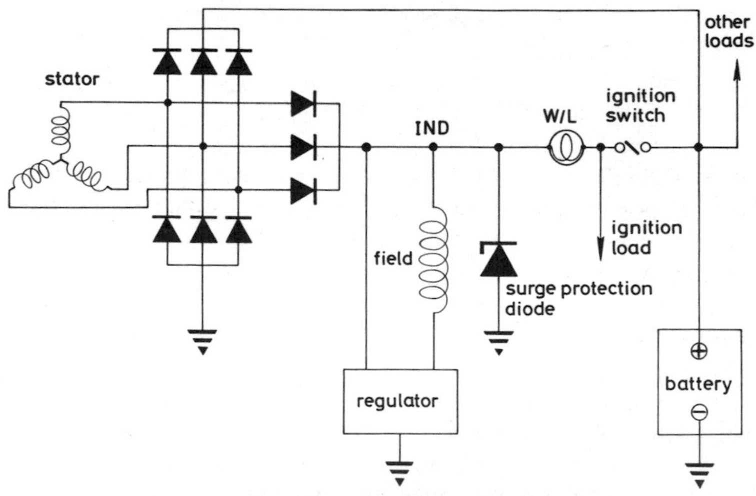

Fig. 4.47C *Alternator surge protection device*
[Courtesy Lucas Electrical Ltd.]

It is claimed that the introduction of AC generators has systematically reduced the risk of failure of SLI batteries through inadequate charging. Fig. 4.47A shows an exploded view of a typical ACR alternator, Fig. 4.47B, an ACR machine sensing system (European termination) and Fig. 4.47C, the circuit of an alternator surge protection device.[137]

4.8 Batteries for electric traction

The concept of battery-powered vehicles is not new. To quote an oft-repeated tale, in 1895 a battery-driven racing car, the 'Jamais Content', driven by the Belgian, Count Jenatzy, briefly held the world land speed record of 68•5 mile/h, and, at the turn of the 19th century, electric passenger vehicles were a cosiderable vogue in the USA and throughout Europe. In the USA, this development was actively promoted by the electricity utility companies, who recognised the commercial advantages of supplying off-peak power for over-night charging.

The discovery and meteoric rise of the Otto internal combustion (IC) engine during the first two decades of this century quickly changed the scene. IC-engined vehicles had a vastly superior speed and range, and these advantages, allied to the ease of re-fuelling, outweighed almost everything battery power had to offer. Everything, that is, except pollution. The recent universal pre-occupation over the levels of toxic gases emitted from IC engines has revived interest in electric traction. This interest has been stimulated by the rising costs of hydrocarbon fuels, the projections of their exhaustion and the growing awareness of the need to preserve these vanishing feedstocks for the production of plastics, pharmaceuticals, pesticides and other products of the organic chemical industry, rather than to burn them to waste in IC engines. As mentioned in Section 4.2.2, Great Britain has been in the forefront of the development and use of battery-powered electric vehicles, primarily industrial trucks and commercial road vehicles of various types. The service conditions for both applications are severe. The batteries are required to give their full 5h output at least 5 days per week in hot as well as temperate climates, and, to make their use economical, they must have lifetimes of about 6 years, equal to not less than 1500 discharge-charge cycles.

Many documents have been published showing the economic advantages of electric traction for these uses.[138–140] Although the capital costs of electric vehicles are appreciably higher than those of their IC-engined diesel or petrol-driven counterparts, the running and maintenance costs of the electrics are demonstrably lower. Also, the growing sanctions against pollution ensure an increasing demand for electric vehicles in enclosed areas, such as warehouses, factories, supermarkets, railway stations etc. In the case of industrial trucks battery weight may not be such a severe handicap, since this may help to stabilise the vehicle where high lifts are required in narrow storage aisles.

During the past few years, considerable sums of money have been spent in efforts to extend the uses of battery-powered vehicles, particularly as goods transporters, but also as personnel carriers, buses and automobiles. International developments in these fields have been described in the quarterly issues of *Electric Vehicle News.*[141] The quarterly publication, *Electric Vehicle Developments,*[142] performs a similar function in Europe. Barak[143] has briefly reviewed some of the more noteworthy developments in Europe up to 1975. Gallot[144] has described the progress of tests in France. Sahashi and Hattori[145] have reported the results of service trials on twelve battery-powered buses in Japan.*

In these trials, although other systems have been included, the lead/acid battery has provided the datum line. As indicated later, the energy density of traction batteries has been increased to about 35 Wh/kg at the 5h rate, with no significant reduction in cycling lifetime, but, even at this level, vehicles have a range on one charge of only about 80 km at a speed of 70 km/h. Surveys carried out in Germany and the UK of the daily regime of goods transporters, carrying payloads of around 1 tonne have shown that 60% of these vehicles travel no more than about 70 km per day. This is well within the compass of modern battery-powered vehicles, but domestic users, accustomed to instant re-fuelling of their automobiles, would regard these capabilities as intolerable for personal travel. In the latest developments in Japan, energy densities as high as 50 Wh/kg have been claimed with lifetimes reduced to about 500 cycles, but commercial production of batteries of this capability is still awaited.

Various methods of conserving the energy of the battery and thereby extending the range of the vehicle are being explored. Regenerative braking is one. In this, the energy needed to retard the

* Readers interested in a detailed account of the application of electric road vehicles are advised to refer to an excellent list of references published by the Birmingham Public Libraries, UK.

vehicle is used to drive a small motor generator, which in turn recharges the battery. This complicates the circuitry somewhat and adds the cost of an additional unit, but claims have been made for energy savings of between 10 and 20% by this process. Another method of increasing the range of the vehicle is the use of a hybrid system, in which an IC-engined motor generator set is coupled to the battery-powered drive train. The IC engine is designed to supply mechanical power to the wheels in suburban areas and electrical energy through the generator to top-up or recharge the storage battery. Müller[146] has described some developments of this type now in progress in West Germany.

Another traction application of interest is the use of battery-powered rail cars in West Germany.[147] For some years, the German Federal Railways have used over 200 cars of this type on relatively short distance stopping services. The battery weight varies from 17 to 22 tons and the capacity between 450-600 kWh, giving the units a range on each charge of about 120 km. With suitably spaced charging stations, the daily range can be extended to about 500 km.

4.8.1 Flat-plate, glass wool traction assemblies

In this battery, sometimes referred to as the 'Kathanode' type, the pasted plates are made on similar lines to their SLI counterparts. The grids are, however, thicker and more robust and the pastes generally somewhat denser to ensure the required durability and lifetime. The grids in general use have thicknesses varying from 3 to 6 mm. To simplify tooling for processing and assembly, the grids are cast in a limited number of widths, but their length may vary considerably and also the number of plates in each cell to provide the required capacity.

4.8.1.1 Alloys for grids: Alloys in general service are similar to those used in SLI batteries, with antimony in the 4 to 8% range, arsenic about 0·2%, sometimes with additions of tin and copper in amounts shown in Table 4.6. These grids are also cast on their sides, by hand or in automatic machines, similar in principle to those used for SLI castings. Because larger amounts of metal have to be accommodated in the mould, however, the temperatures of the metal and the mould and the rates of casting are more critical. If cooling of the metal is too fast, shrinkage cavities or 'hot-tears' may develop, particularly at rib intersections where turbulence may occur. The diamond pattern, shown in

Fig. 4.23 was devised to ease this problem, on the premise that there would be less turbulence with metal flowing diagonally down the rib channels. The problem of hot tears becomes more acute when the concentration of antimony is reduced, since the setting range of the alloy increases and shrinkage becomes more prevalent. As indicated in Section 4.6.2.5, this can be allayed to some extent by the addition of grain refiners, such as selenium and sulphur. This problem has come more and more into prominence lately with the advent of maintenance-free SLI batteries. Batteries with low antimony alloys and requiring restricted maintenance only, if not completely maintenance-free, would have advantages for traction service.

4.8.1.2 Positive and negative pastes: These follow conventional SLI processes with grey oxide as the main constituent, sometimes with additions of 15 to 20% of red lead in the positive mix. As mentioned in Section 4.5.2, when treated with sulphuric acid, red lead reacts as a mixture of PbO and PbO_2 and the latter assists in the electrochemical formation process. Paste densities generally come in the ranges of positive, 4·5 g/ml, and negative, 4·2 g/ml.

Setting and *drying* processes are similar to those used for SLI plates, but because of the larger amounts of materials involved they take longer and require more careful control; e.g. the period for seasoning of the positive plates in a highly humid atmosphere before drying may extend to 3 to 4 days.

4.8.1.3 Separators, assembly and containers: Double, and, in some cases, treble separation is used for this arduous service. A thick, well-matted glass wool 'retainer' is held tightly against the positive plate, sometimes supported by a perforated, flat sheet of plastic material. A third separator, generally ribbed on one side and made of microporous PVC, rubber or related material, as described in Section 4.6.11 is fitted with the flat side against the negative plate. In some assemblies the perforated sheet is omitted and the ribbed microporous separator is reversed, with the ribs resting against the negative plate and the flat back against the glass wool retainer.

Alternatively, the negative plates may be completely enclosed in an envelope made of one of the microporous materials. This provides the best insulation of both the side and the bottom of the element against short-circuits caused by particles of active materials, but it may restrict free diffusion of the electrolyte. The microporous separators should not have pores larger than about 10 μm and their tortuosity factor should be high to prevent the growth of lead bridges through the pores. The complete element (plates and separators) is built up in the form of a stack and tightly clamped while the equaliser bars are welded to the lugs of the plates in each group. The element is then pressed into the container and spacing pieces may be pressed in at each end to ensure that the whole unit is tightly packed.

Cell containers are generally made of high quality hard rubber, but with the introduction of new plastics these are now being moulded in the co-polymer, acrilo-nitrile-butadiene (ABS), polyethylene and polypropylene, the two latter by the economical 'blow-moulding' technique. Thermoplastic materials have the added advantage that the lids can be attached by a hot-welding operation.

4.8.1.4 Formation: Plates may be formed by the conventional tank formation process and dried before assembly, but, because of the economies in handling and in charging current, it is now more usual to assemble the unformed elements in their own containers and charge them by a jar-formation schedule. For this, cells are generally filled with acid of 1·250 SG and allowed to stand for about 2 h before being put on charge. Temperatures should not exceed 55°C, and, to prevent over-heating, the cells are held in shallow tanks of water and sometimes sprayed with cold water. The rate of charge is generally around $C/20$ A, where C is the nominal capacity at the 5h rate of the cell and the total charge given is approximately equivalent to 250 Ah/kg of positive paste.

4.8.2 Tubular plate assemblies

4.8.2.1 The Ironclad plate: Wade[61] has described early developments of tubular positive plates by Woodward in 1890 with 'perforated india-rubber insulating material, containing a central conducting core and active material around it' and by Currie, who, in 1891, 'perfected an electrode in which tubes of woven asbestos formed the protective covering'. Woodward's plate was the progenitor of the 'Ironclad' plate, which gave prodigious service in submarine batteries during the First World War, vindicating their name for hard work and endurance. Their growing use in traction and rail car service between the wars naturally followed. In this plate, the thin-walled hard rubber tubes, having an internal diameter of 7·7mm (0·303 in), were made porous by a number of narrow parallel slots, cut at close intervals in the periphery. The overall porosity of these tubes, based on the material cut away, was about 20%. Two ribs, running down each side, served to space the tubes suitably between the separators. The current collectors or grids consisted on a number of spines of diameter 3·2m (0·125 in), cast with a common busbar, having a current take-off lug at one side. These spines had occasional protruding fins to locate them co-axially at the centre of the tubes and were generally cast in pressure die-casting machines, the molten metal being blown into the mould by compressed air.

To complete the plate, the tubes were threaded onto the spines and the casting was held in an inverted position vertically on a vibrating anvil, while the dry oxide powder was released from a hopper immediately above the open ends of the tubes. Tamping was continued for several seconds until the filled weight of the plate reached the specified norm, when the ends of the tubes were closed by casting a metal bottom bar to the exposed ends of the spines.

4.8.2.2 New plastic materials for tubes: During the past 25 years or so, the development of acid-resistant plastics, which can be woven, braided or felted, has opened the way for spectacular advances in this type of assembly. At first glass fibres were used, stiffened by impregnation in phenolic resin and curing. Another variant, developed by the A.B. Tudor Company of Sweden and widely used in Europe, known as the 'PG' or panzer-glas type, has an inner sleeve made of woven glass fibres held in an outer tube of thin-walled perforated PVC. These types are constructed with single tubes.

In the 'gauntlet' type, developed by the Chloride Group of the UK, composite multi-tube sections are made by stitching at appropriate intervals a double layer of woven or felted, non-woven cloth made of terylene. To shape the tubes, the cloth is impregnated with a stiffening thermo-setting resin, metallic mandrels are inserted in the appropriate spaces and the resin is cured in an oven prior to removal of the mandrels. Other manufacturers use a PVC-acrilo-nitrile co-polymer for the tubular material. Figs. 4.48*a* and *b* shows typical examples of PG and Gauntlet type assemblies with the conducting spines. These materials have a much higher porosity than the original slotted hard rubber, 40 to 50%, giving much more favourable mass transport of electrolyte to and away from the active material with correspondingly higher coefficients of use. In the USA and the UK, the standard plate has 15

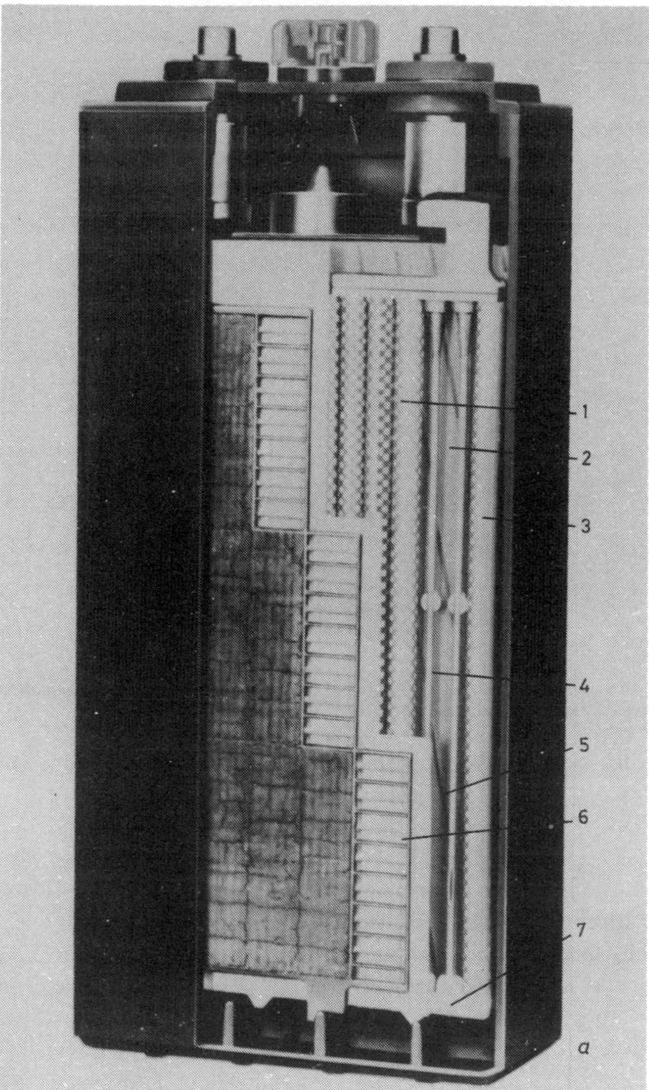

Fig. 4.48 *PG and Gauntlet type tubular plates*
a PG assembly
1 Perforated outer sleeve
2 Braided glass fibre sleeve
3 Half-perforated outer sleeve
4 Pb-Sb alloy spine
5 Positive active material
6 Negative grid
7 High-density polythene foot
[Courtesy Oldham & Son Ltd.]

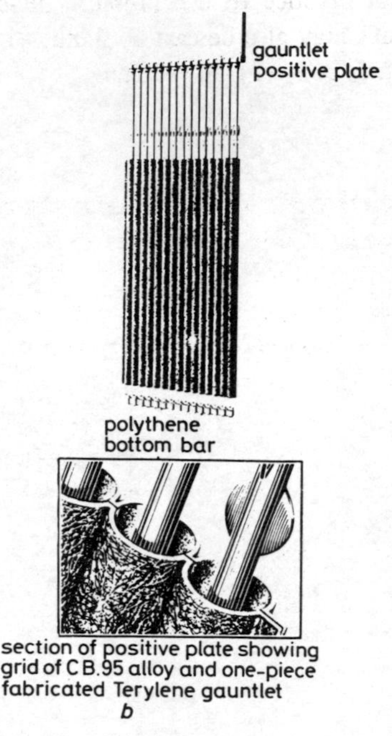

b Gauntlet with non-woven Terylene
[Courtesy Chloride Industrial Batteries Ltd.]

tubes. In some other countries, wider plates with 19, and, in some cases, up to 23 tubes, are used.

4.8.2.3 Alloys and spine-casting procedures: Alloys for this service follow to some extent those used in SLI batteries. For many years the antimony content was around the eutectic level, just over 11%, but latterly this has been reduced to 8%, 6% or even 4%, with additions of arsenic up to about 0·2% and sometimes small amounts of tin and copper. Interest in low maintenance batteries has encouraged manufacturers to explore the use of alloys with a low antimony content or no antimony at all and claims have recently been made that Astag alloy,[85] containing 0·07% Te 0·01% Ag 0·01% As and the remainder lead, offers advantages in this respect. In the pressure die-casting operation, the metal is held at a temperature of about 480°C and the mould just above ambient. Rapid chilling of the metal ensures a fine-grained structure. To eliminate shrinkage cavities, a small gate is retained at the top of the casting and removed when the grid is trimmed to length.

Although it is general practice to use pressure die-casting equipment for this purpose, spines may also be cast by hand, when the slower rate of cooling gives a slightly coarser grain pattern.

4.8.2.4 Tube-filling material: For many years, finely-divided red lead was the only oxide used for tubular plates, but in recent years, mainly to reduce material costs, blends with about 50% of grey oxide and higher have come into practice. As with the active material in pasted plates, the coefficient of use and its cycling lifetime are largely determined by the porosity and this is in turn directly related to the packed density. A characteristic value for the standard blend is 4·2g/ml. The process of tube filling is dusty and therefore hazardous to the operatives and the capital costs of the filling and ventilating plants are high. Also, variations in the packing factor of the dry oxides sometimes make it difficult to keep within the specified weight tolerance for the plates. Alternative methods of treating the oxides have therefore been examined. In one of these, the oxide was granulated by tumbling in a rotating drum, under a fine spray of dilute sulphuric acid, after which it was dried in a stream of hot air. This treatment greatly reduced the dust hazard and improved the flow of the material, but irregular capacities were observed and the additional process increased the manufacturing costs. In a second novel process, the filling material was made in the form of a paste, similar to that used with SLI plates but less viscous and with the addition of a thixotropic agent,[148] which improved the rheological properties. The paste was forced into the tubes by an extruder, but it was difficult to ensure uniform filling weights.

A third process uses a thick slush, also made with dilute sulphuric acid, and this is poured vertically down the tubes. The fluids pass through the porous sheath and the solids are retained in the tubes. Encouraging results have been obtained on tests.

In another move to reduce weight and to improve the energy density of the cell, the cast-on metal bottom bar has been replaced by a moulded plastic bar, generally made of polyethylene with suitably profiled feet, pressed or hammered over the ends of the spines.

After the standard procedure of dry filling, the plates are soaked in sulphuric acid of about 1·400 SG for 2 to 3 h and dried in a hot air oven before assembly. With this deep sulphation of the active material, the formation proceeds more uniformly and pitting corrosion of the spines, sometimes observed in plates without this treatment, during storage, is eliminated.

4.8.2.5 Cell assembly, separators and formation: Standard forms of pasted negative plates, common to flat plate, glass wool traction cells, with similar paste density and expander contents and microporous separators made of plastic materials or hard rubber are used. Since the tubular plate is contoured, the separators are not ribbed, the necessary free electrolyte for the positive plate being accommodated in the spaces around the tubes. The negative plates may be completely enclosed by the separators, as in the case of flat-plate cells.

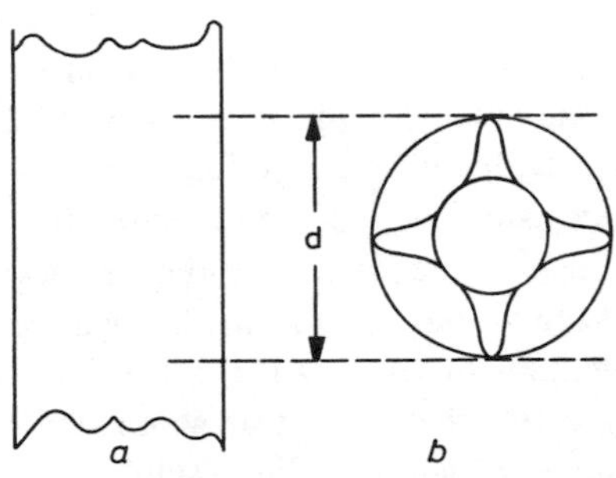

Fig. 4.49 *Surface area of a tubular plate*
a Pasted plate
b Tubular plate
Ratio of surface areas, $\pi d/2, d$ = 1·57:1

The contoured shape of the positive plate also means that the surface area exposed to the electrolyte and to the negative plate is significantly higher than that of a flat plate of similar overall dimensions, and so that at any given rate of discharge, the current density is proportionally lower. As shown in Fig. 4.49, the ratio of surface areas is 1·57: 1, but since the internal resistance increases with the length of the path between the negative plate and the periphery of the positive, the reduction in current density would not be fully realised. Cell containers, methods of assembly and jar formation procedures closely follow those described for flat-plate cells. After formation, cells are carefully checked for uniformity before assembly as a complete battery in robust crates, made of plastic-coated steel or impreganted glass fibre laminate.

The cells are connected in series to give batteries having, in the popular sizes, open-circuit voltages between 48V (14 cells) and 96V (48 cells).

4.9 Performance of traction cells and batteries

4.9.1 General aspects

Cells with both pasted and tubular plates are used in traction or motive power service, and, in general, the performances and lifetimes of the two types of cells are similar. Tubular plates, however, are claimed to have the following advantages over flat, pasted plates of similar dimensions:

(*a*) a greater surface area, giving improved diffusion and reduced polarisation, as indicated in Section 4.8.2.5
(*b*) less severe corrosion of the embedded current collectors; the spines
(*c*) reduced standing losses, since the bulk of the antimony released by corrosion of the spines is adsorbed in the surrounding pencil

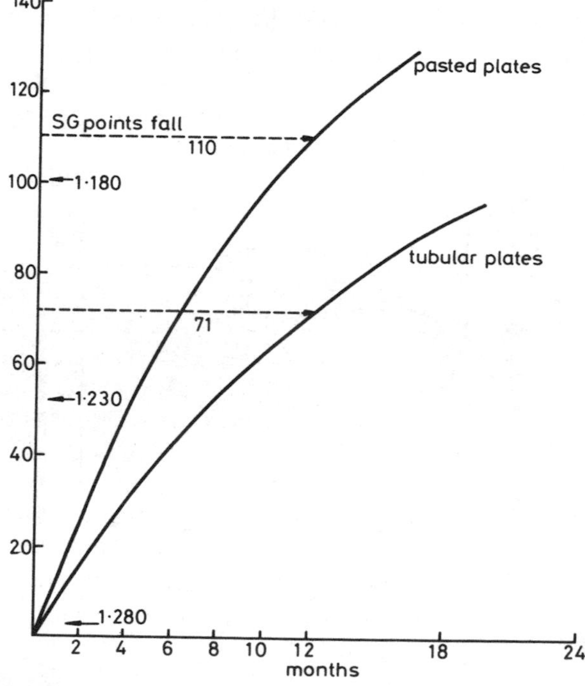

Fig. 4.50 *Pasted plate and tubular plate traction cells: standing losses at 21°C* [Courtesy G.J. Bushrod), *Lead 68*, pp. 187-190, Pergamon Press Ltd., 1969]

of PbO_2. Bushrod[149] has reported the results of storage tests at 21°C on cells with pasted and tubular positive plates with grids in 11% antimonial alloy, shown in Fig. 4.50. After 12 months, the fall in the SG of the electrolyte in the former cell was 110 points, whereas in the cell with tubular plates, it was only 72 points. Assuming that the fall in SG on a full 5h discharge would be 140 points, these figures represent capacity losses of about 80% and 50% respectively

(*d*) better retention of the positive active material by the close-fitting porous tubes.

The two main disadvantages of tubular plates are, first, the complex and costly manufacturing plant required, only justifiable with long production runs, and secondly, the difficulty in altering plate thicknesses, which involves changing many components, notably the tubes, spine castings, filling jigs etc. Because of their advantages, however, the latest trends clearly favour the tubular type of construction. It is reported that, in Germany, 90% of motive power batteries have tubular positive plates. Fig. 4.51 shows the gain in capacity of tubular cells, made by the substitution of more porous plastic materials for the older slotted hard rubber ironclad tubes. As indicated in Table 4.19, gains of about 35% in energy densities were achieved, with no reduction in lifetimes of 1500 - 1800 cycles.

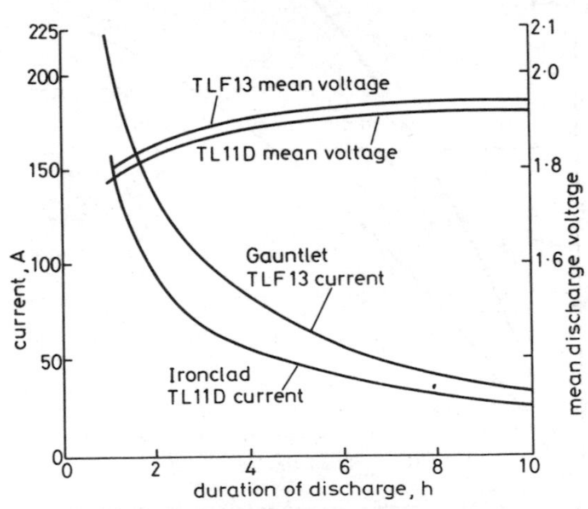

Fig. 4.51 *Discharge characteristics of Ironclad Gauntlet traction cells*

Table 4.19 *Improvements in tubular plate cells*

	Old Ironclad Type TL11	New Gauntlet Type TLF 13
Weight, kg	22·73	22·95
Volume, dm^3	8·61	8·52
Capacity at the 5h rate, Ah	240(100%)	324 (135%)
Mean voltages, V	1·91	1·92
Energy densities:		
Wh/dm^3	53·04 (100%)	73·01 (138%)
Wh/kg	20·17(100%)	27·11 (134%)

4.9.2 Discharge voltage characteristics

Fig. 4.52 shows a typical family of discharge voltage curves at rates from the 10h to the 10min rate for a tubular plate having a nominal capacity of 20 Ah at the 5h rate, with the generally-accepted terminal

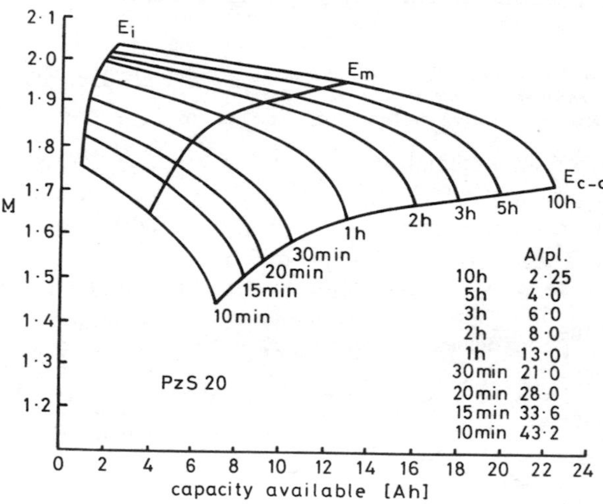

Fig. 4.52 *Discharge voltage characteristics showing capacity available from tubular plates of 20Ah capacity (5 h rate at different rates of discharge at ambient temperature)*

	A/pl		
10h	2·25	30 min	21·0
5h	4·0	20 min	28·0
3h	6·0	15 min	33·6
2h	8·0	10 min	43·2
1h	13·0		

[Courtesy Varta A.G.]

and mean cell voltages.[150] For complete cells with n positive plates, the values of the capacities should, of course, be multiplied by n. Similar families of curves can be constructed for each type of plate and corresponding cell. As with SLI batteries, the capacity is inversely related to the discharge current and therefore directly to the duration of the discharge. This is shown in Fig. 4.53, where the capacity is presented as

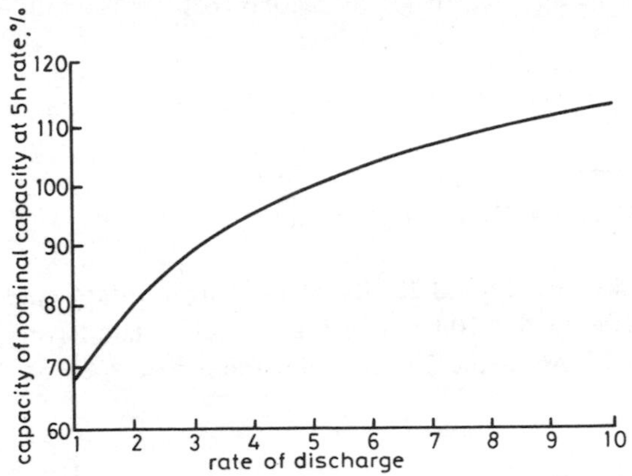

Fig. 4.53 *Traction batteries: capacity expressed as a percentage of the nominal capacity at the 5h rate*
[Courtesy of Chloride Industrial Batteries Ltd.]

a percentage of the nominal capacity at the 5 h rate. Motive power batteries are generally rated at the 5 or 6h rate, since the continuous output at this rate approximates to the capacity available on intermittent service over a normal working shift of 8 h.

In practice, batteries are frequently submitted to intermittent opera-

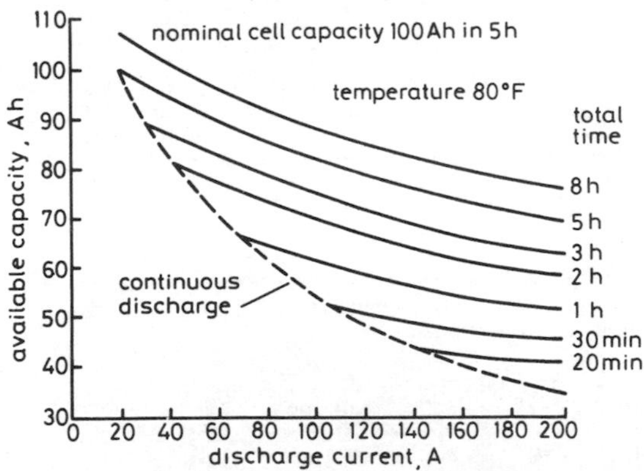

Fig. 4.54 *Traction batteries: capacity available on intermittent discharges*
[Courtesy of Chloride Industrial Batteries Ltd.]

tion. Fig. 4.54[151] shows the capacities available from a cell of 100 Ah (5 h) at different rates, when the discharges are carried out intermittently over different periods; e.g. when the cell is discharged continuously at 100 A, the available energy is 54 Ah and the duration of the discharge is 0·54 h. If the discharge is spread over 1 h, the capacity is 61 Ah and over 8 h it is 88 Ah.

4.9.3 Energy and power densities

As already indicated, the energy densities of standard assemblies at the 5h rate of discharge lie in the ranges, 25 to 30 Wh/kg and 70 to 75 Wh/dm^3. With the recent revival of interest in battery-powered electric vehicles, much work has been done to improve these factors, particularly energy density by weight, since this directly affects both the payload and the range. Power density is also dependent on the number of watts available and may be defined in terms of the duration of the discharge; e.g. a battery may deliver a certain number of watts for a period of 5 h and this value would be used to calculate the power density *at that rate*. As Takagaki, Ando and Yonesu[152] have pointed out, however, a battery also has a maximum power density, which is not related to any particular rate of discharge. In relation to traction service, the maximum power density gives a measure of the possible acceleration of the vehicle.

Seeking to improve both characteristics, battery manufacturers have tested many variants in the processes and design on the assumption that

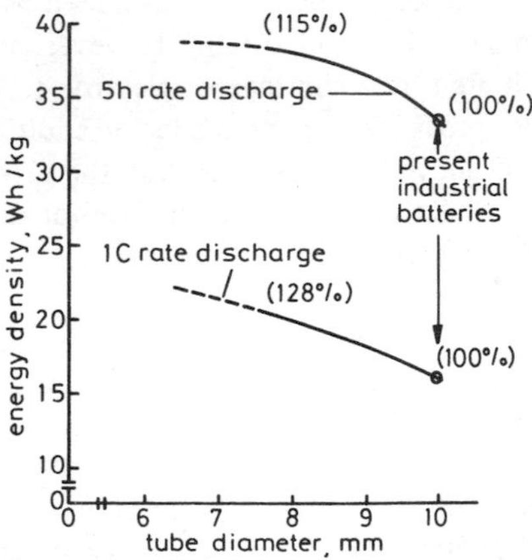

Fig. 4.55 *Effect of tube diameter on energy density at the 1h and 5h rates of discharge*
[Courtesy A. Kozawa and T. Takagaki, *Japanese lead-acid battery industry*, p. 113, Electrochemical Society of Japan, 1977]

lifetimes of 1000 cycles or 4 years would be acceptable. An obvious way to improve the performance is to reduce the thickness of the plates. Recent tests with tubular plates, having tubes of only 6mm outside diameter have shown that the energy density can be pushed up to around 40 Wh/kg, while retaining a cycle lifetime of about 1000 cycles. In reaching this goal, the weights of all of the inert components were kept to a minimum, e.g. copper or aluminium cores were used in the inter-cell connectors and light-weight glass reinforced plastic crates replaced the robust steel trays in general use.

Kozawa and Takagaki[3] have described work in Japan in an effort to produce a motive power battery with an energy density not less than 40 Wh/kg suitable for an electric bus. Tests were carried out with tubes of diameters 10, 8•5 and 7•5 mm. As indicated in Fig. 4.55, gains in energy density of 28% at the 1 h rate and 15% at the 5 h rate were achieved with a reduction in the tube diameter from 10 to 7•5 mm, but with a fall in deep cycling life from about 1500 cycles to less than 1000 cycles. This 384 V battery was built up of 64 6V units, each of 310 Ah capacity and the overall weight was 3350 kg. Sundberg[153] has also reported test results with tubes of different diameters and oval as well as circular cross-section, indicating some advantage for oval tubes at high rates of discharge.

Attempts have been made to boost the coefficient of use of the positive active material by the inclusion of small amounts of carbon black or finely-powdered silica gel, but the cycling lifetime and durability of the plates were adversely affected.

The internal resistance of the cell is contained by keeping the plate pitch to a minimum. It is important, however, to ensure that the capacity is not limited by a shortage of electrolyte. If sufficient headroom is available, provision can be made for additional electrolyte to be accommodated above the elements, but the processes of diffusion and convection are relatively slow and this cannot be tapped as quickly by the plates as the reservoir of electrolyte held in the body of the cell.

Takagaki, Ando and Yonesu[152] have reported a systematic examination of the relation between energy density and maximum power density. Fig. 4.56 shows the progress made in Japan in improving these characteristics during the decade from 1964 to 1974. Energy density is restricted by the available activity of the active materials, but power density is also dependent on the conductivity of all of the materials involved, the grids, the active materials and the electrolyte. The authors concluded that a high energy density can only be obtained at the times. The performance of practical batteries would therefore be in the expense of a high maximum power density and it is virtually impossible to have both simultaneously without seriously impairing cycling life-

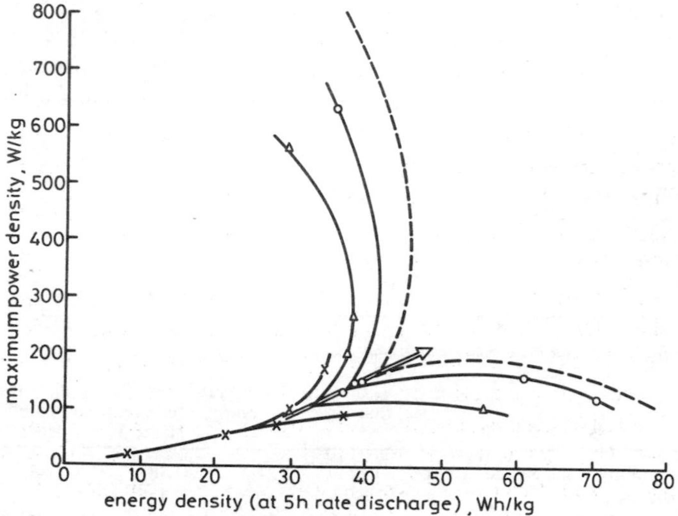

Fig. 4.56 *Progress of lead/acid battery perofrmance and future prospect*
× developed before 1964
△ developed in 1965-1972
○ developed in 1973-1974
- - - future prospect
[Courtesy A. Kozawa and T. Takagaki, *Japanese lead-acid battery industry*, 1977]

range of 38 to 40 Wh/kg and 150 W/kg, with maximum realisable values for traction service probably in the range of 45–50 Wh/kg and 150–200 W/kg.

4.9.4 *Weight analyses of traction cells*

Published data, similar to that described for SLI batteries, are available for standard UK cells with pasted and tubular plates[134] and for similar Japanese cells of standard type and also of improved design to meet the exacting demands of the electric automobile[154] Tables 4.20 and 4.21 list the weights of the different components, as percentages of the total weight of the cell in each case. Points of interest are the relatively low weights of the positive and negative grids and the liberal amounts of electrolyte in the special Japanese cells. A noteworthy feature of the Japanese cells A and B is the high proportion of the total weight taken up by the active components, the grids, active materials and the electrolyte. This is about 92%, as compared with about 85% for the standard

UK types. It is understood, however, that these Japanese cells are experimental only and have not yet been produced on a commercial scale.

Table 4.20 *Pasted plate cells: weight analysis (percentages)*

	UK Standard	Japanese Batteries Model A	Japanese Batteries Model B
Energy density at the 5h rate, Wh/kg	27	47	60
Positive grids	15·8	12·6	12·5
Positive active material	20·7	22	21·4
Negative grids	11·4	8·7	8·5
Negative active material	19·3	16·5	15·7
Electrolyte	18·5	15·1	27·3
Containers and covers	8·6	7·1	6·2
Other items	5·7	8·0	8·4
	100 %	100 %	100 %

Table 4.21 *Tubular plate cells: weight analysis (percentages)*

	UK Standard	Japanese Cells for trains	Japanese Cells for electric automobiles Model A	Japanese Cells for electric automobiles Model B
Energy density at the 5h rate, Wh/kg	27	24	38	45
Positives: grid	16·3	12·5	10·7	6·2
active material	17·5	19·6	20·3	24·0
Negatives: grid	10·4	9·6	9·7	4·8
active material	18·4	20·1	22·5	22·7
Electrolyte	21·4	24·0	29·4	34·7
Container and cover	10·6	9·6	4·8	5·7
Other items	5·4	4·6	2·6	1·9
	100 %	100 %	100 %	100 %

4.9.5 *Charging of traction batteries*

A storage battery[151] is one part of a three component system, the other members being the equipment, which uses the power supplied to do useful work, and the charger, which replenishes the energy in the discharged battery. To make the system efficient and cost-effective, there must be complete compatibility among the three components and it is particularly important to match the output of the charger with the capacity of the battery. Battery chargers may be operated from any AC mains source and include a transformer to step the voltage up or down according to the maximum voltage of the battery and a rectifier to convert the AC to DC. A ballast choke and a resistor are included in the output circuit to restrict the charging current in the early stages, when the battery is still in a state of discharge, since high currents may cause over-heating.

Most of the chargers used for motive power batteries are of the 'taper' type, in which the current falls or tapers as the charge proceeds and the battery voltage rises, in accordance with values for the two plate groups, shown in Fig. 4.16. Different battery manufacturers generally specify their own specific charging conditions. The following have been found to give a reliable and practical regime. Taper chargers fall into two classes.

4.9.5.1 Single-step chargers: These are used when a charging period of more than 10 h is available and the following empirically determined conditions apply:

(*a*) the charging current at the point where electrolytic gassing starts, i.e. 2·5V per cell should be about 1/12th of the normal battery capacity, i.e. 8·5A for a battery having a capacity of 100Ah (5h rate)

(*b*) the approximate starting current required to charge the battery in 11-12 h should be about 12½% of the 5h capacity, 12·5A for a 100 Ah battery

(*c*) the ratio of the starting current at 2·1V per cell to the current at 2·6V per cell should be 1·67 : 1. For a 100 Ah battery of, say, 24 cells, the current would be reduced to 7·5A at 2·6V per cell. The rated output of such a charger would be 12A and its designation '24/12'

(*d*) if shorter charging times, less than 10 h, are required, the ratio

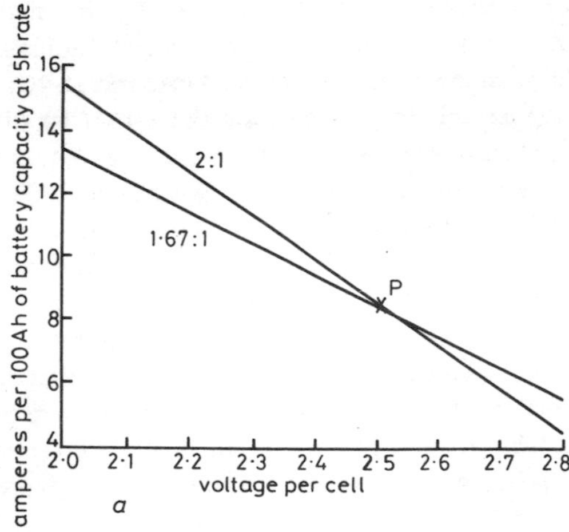

Fig. 4.57

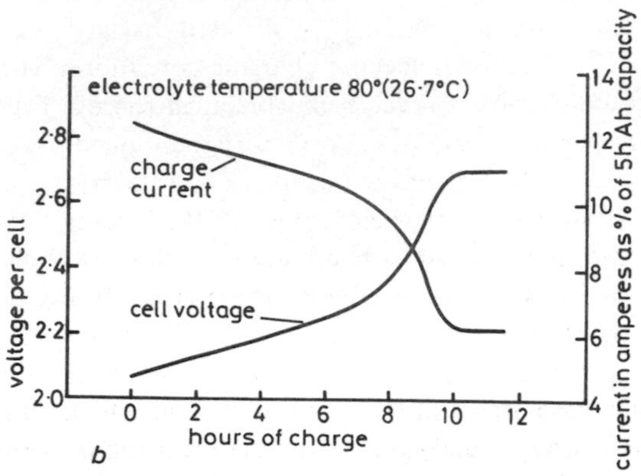

Fig. 4.57 *Single-step taper charger*
a amperes against voltage per cell
b voltage per cell against hours of charge
[Courtesy of Chloride Industrial Batteries Ltd.]

could be boosted to 2 : 1, with initial current at 2·1V of 14A and final current at 2·6V of 7A, with the current at the gassing point 8·5A as before, point P, on Fig. 4.57*a*.

Figs. 4.57*a* and *b* show voltages and currents during the recharge of a battery of 100 Ah by a single-step charger after a deep discharge at the 5 h rate. When the gassing voltage of about 2·6V is reached, a voltage-sensitive relay starts a timing device, which, after an adjustable pre-set period, ends the charge and this may also taper the current still more during the gassing period. The short gassing period of 10-15% serves to equalise the SG of the electrolyte. As shown in Fig. 4.4, under temperate conditions the SG of a fully-charged cell is near 1·280 at 15°C and at the end of a 5h discharge it falls to around 1·140, allowing a total fall of about 140 points. As already mentioned, readings of the SG provide the best indication of the state of charge, but instantaneous readings taken during charge or discharge may not be reliable and sufficient time on open-circuit must be allowed for the concentration to equalise throughout the cell. At temperatures above and below 15°C, corrections must be applied to the SG, involving the temperature coefficient, 0·0007/°C, must be applied, according to the expression:

$$S_T = S_{15} + 0{\cdot}0007\,(T - 15^\circ) \qquad (4.41)$$

Where S_T and S_{15} are the SGs at temperature $T°$ and 15°C.

4.9.5.2 Two-step chargers: These are used where the allotted time for recharging is less than 10 h. As already mentioned a storage battery can absorb a very high current during the early part of the charge, but there is a limit to the safe current after the gassing point has been reached. To make a significant reduction in the total charging time, therefore, the bulk of the charge must be given during the first step, which is generally set at about 2·4V per cell. At this point, when 75 to 80% of the charge has been given, as in the case of the single-step charger, a voltage relay starts a timing device, which runs for a predetermined period, usually about 4 h before tripping a switch, which terminates the charge.

Fig. 4.58 shows voltage and current changes during a two-step charge of this kind. In this case, the initial current at a cell voltage of 2·1V is about 20 A, tapering to 10 A at the end of the first step at 2·4V after 5 h. During the second step, the current tapers to about 6 A, after a total time on charge of about 8 h.

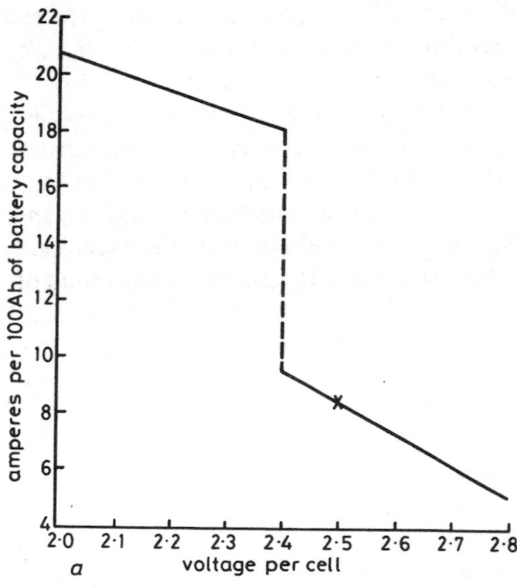

Fig. 4.58 *Two-step taper charger*
a amperes against voltage per cell

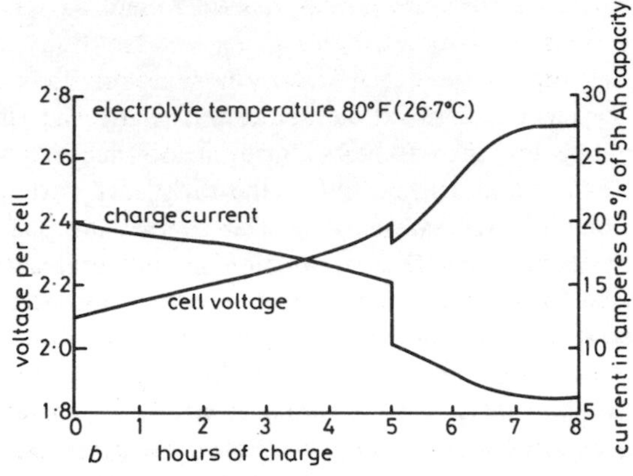

Fig. 4.58 *Two-step taper charger*
b voltage per cell against hours of charge
[Courtesy of Chloride Industrial Batteries Ltd.]

4.9.5.3 Automatic control devices: Mention has been made of the need to regard the combination of battery and charger as part of a unique system. The performance and life of the battery are, in fact closely related to the efficiency of the charging process and it is obvious from the brief description of chargers that this, in turn, depends on the accuracy and reliability of the automatic control devices, embodied in the chargers. One such device, recently developed, is the Spegel Controller.[155] This monitors the on-charge characteristics of the battery and senses the levelling off of the voltage at the top of charge, when the main charge is stopped. The battery then receives short pulses of charge, controlled by the decay in battery voltage during the intervening open circuit periods. These pulses serve initially to equalise the battery and then to maintain it in peak condition and can, if necessary, be sustained over long periods.

4.10 Stationary batteries

4.10.1 Service and battery types

Storage batteries are used for a variety of stationary applications. As in the case of portable service, the range of uses has increased year by year and the pattern of application has also changed significantly. During the first decade of this century, large batteries were used to provide lighting for country houses and other buildings remote from any local mains supply. They were also used, particularly in power stations, as standby sources of power, which might be required in relatively large amounts for short periods up to a few hours. As examples, such batteries might be called on to replace an MG set which had failed or to relieve the local mains supply of meeting peak demands when the cost of power per kilowatt was rising sharply, a function now termed 'peak-shaving' or 'load-levelling'. In these uses, the battery would be submitted to a fairly regular regime of a deep discharge at a high rate, say 1h, followed by a slow recharge and the older Planté-type batteries performed this service with great reliability. As indicated in Section 4.6.1, batteries of this type were assembled with Planté plates, 12mm in thickness, pasted box negative plates and separators made of glass rods or wood veneers, held in position by thick wooden dowells. The plate pitch was devised to allow a liberal excess of electrolyte, generally of 1·240 SG, or, in tropical climates, 1·210. The elements were assembled in containers made of glass, ceramics or wood, lined with lead sheet, the plates being supported by their lugs on ledges on the sides of the containers or on thick glass sheets. To reduce acid spray during overcharging, the cells were sometimes covered with glass sheets or a layer of heavy paraffin oil.

With the spread of the electricity mains network, the number of batteries on this deep cycling service has steadily fallen, giving way to a range of lighter standby services, such as switch operation and other similar duties in power stations, emergency lighting in all types of public buildings, cinemas, hospitals, theatres etc. The growth of the telephone service has also led to a considerable extension in the use of storage batteries. For many of these uses, the battery floats on the mains busbars, with a rectifier in the circuit to convert the input AC into DC for recharging and sometimes with an inverter to invert the output DC into AC.

4.10.2 High performance Planté batteries (type HPP)

The newer duties have been met in some countries in Europe, including the UK, by the development of a Planté cell, with positive plates only 8 or 6 mm, as compared with the older 12 mm in thickness, pasted negative plates, pitched at much closer separation and one of the newer forms of microporous separators, such as mipor or porvic. Cell containers are now generally made in one of the clear or translucent plastics (polystyrene, polyethylene, polypropylene) by the injection moulding process. Covers are usually sealed in position, either with a suitable sealant or by head-welding and fitted with spray-arresting vent plugs. This makes it possible to install cells of this type in closed rooms without risk of corrosion of associated equipment.

Fig. 4.59 shows an exploded view of a typical HPP cell and in Table 4.22 figures are given for a cell of 200 Ah (10 h rate), showing the improvements in performance over the older type of Planté cell. Improvements include a saving of space of over 70%, more than a three-fold gain in switch tripping capability and a reduction in internal resistance of over 50%. Batteries of this type generally operate at floating voltages of

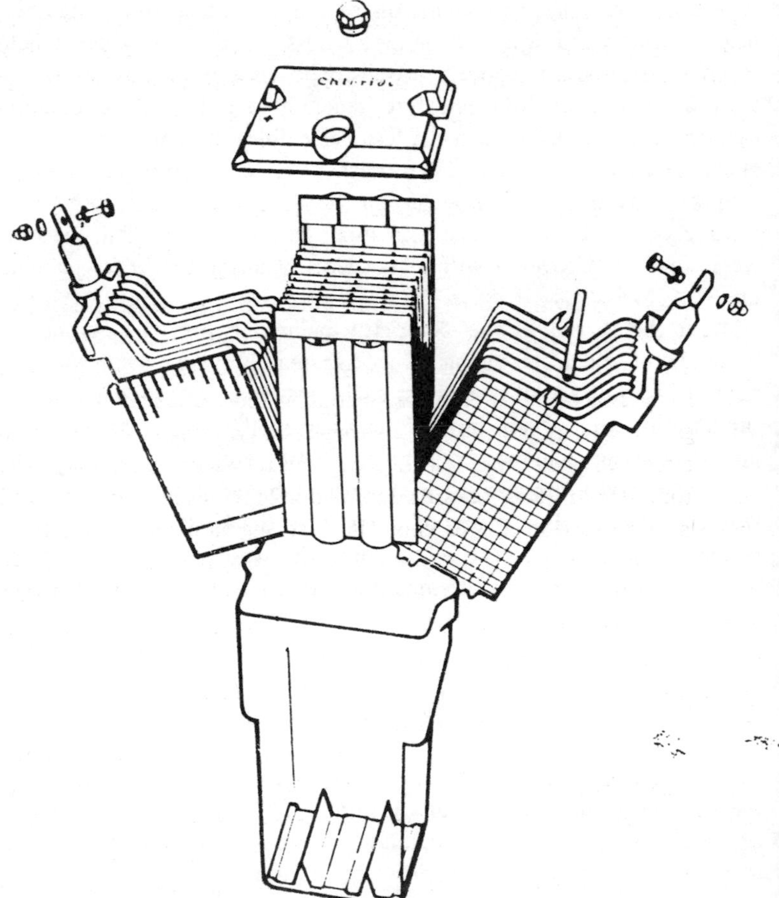

Fig. 4.59 *HPP Planté cell*

Table 4.22 *Performances of old and new Planté cells*

		Old standard DSNG8	New type YCG17
Energy density at 10h rate:	Wh/dm^3	18·2 (100%)	31·6 (173%)
	Wh/kg	7·4 (100%)	10·7 (145%)
Switch closing; 1 min rate to 1·60V/cell	A	324 (100%)	640 (198%)
	A/dm^3	15·2 (100%)	51·6 (340%)
Internal resistance	$\mu\Omega$	1130 (100%)	500 (44%)

between 2·15 and 2·25 V/cell, the trickle current supplied being just sufficient to replace charge lost by chemical reactions. With pure lead grids throughout, these standing losses are small, and, under normal working conditions, lifetimes of at least 15 years have been regularly achieved.

4.10.3 Cells with tubular and flat, pasted plates

Changes in the service and confirmation of the reliability of both types for the duties involved have encouraged users to introduce these cells for a variety of stationary applications. Ausderau[156] has reported, for example, that in Switzerland during the past 20 years or so all stationary and portable batteries, including those used for telephones, emergency lighting, electric road vehicles and mining locomotives, are of the tubular plate type. For stationary battery service, cells of capacities from 100 to 2000 Ah are assembled with plates having 19 tubes of circular cross-section of 8 to 9mm OD in strong hard-rubber containers. In train-lighting applications, gains in capacity from 90 to 150 AH, 67% were achieved by replacing Planté cells by tubular cells. For standby batteries, required to give high currents for short periods from some seconds up to 1 h, oval or rectangular tubes, 6 to 7 mm in cross-section were used, with the plate pitch reduced from 21-23 mm to13-16 mm, giving the added benefit of a reduction in internal resistance from 330-350 to about 200 $\mu\Omega$ in a cell of 1000 Ah capacity at the 10h rate of discharge. These batteries remain fully charged on a float voltage just above 2·23V per cell, and according to Ausderau, lives of 15 years can be confidently predicted. Water losses from 2000 Ah batteries have been as low as 2·6 ml/Ah/year, making topping-up necessary only twice per year.

In America, presumably because of the lower cost, the flat pasted plate cell is generally preferred to the tubular cell for stationary applications, and for most of these standby, emergency lighting type of applications the grids are cast in lead-calcium alloy. Limitations on space do not permit detailed descriptions of the different applications of stationary batteries, but the following are of special interest.

4.10.3.1 Telephone and telecommunication services: There is an interesting diversity of views about the best batteries for this service. As mentioned previously, in Switzerland and some other countries in Europe, the tubular plate battery is now the standard. In America, flat, pasted plate batteries are generally used, and in the UK Planté batteries are still the first choice. Edwards and Baxter[157] have listed the reasons why the British Post Office has retained the Planté battery. The prime reason is the proven reliability of this assembly, giving under float conditions a working life of more than 15 years. Changes in the working conditions, however, have been contributory factors. In 1960, for example, a major change was made from a large battery with a capacity reserve of 24 h to a much smaller, and therefore much cheaper, 1h battery, coupled to a prime mover AC generator which provided the main source of reserve power. With a second battery to start the prime mover, this 'dual reserve' system has now been adopted for the modern co-axial and microwave transmission links, powered by the 28V DC standard. High performance Planté cells, with plates 7 to 10 mm in thickness, are used for this service.

For the ordinary telephone exchanges, operating over the voltage range 46-52V, in the largest exchanges open-cell batteries with capacities up to 15 000 Ah and 12 mm plates are used. In the small and medium ranges up to 2200 Ah, enclosed, thin plate cells are standard. Both of these systems operate successfully on a floating voltage of 2·26V per cell.

In America, mainly to make more economical use of floor space, pasted plate cells are generally selected. The same embargo on antimony applies and this led to the development over 20 years ago by the Bell Telephone Company of lead-calcium alloys, containing 0·065 to 0·085% of calcium, described in Section 4.6.2.8. This alloy was used almost exclusively for this service for over 15 years. Some manufacturers had, however, met difficulties in handling this alloy through the manufacturing processes; it had to be carefully segregated from lead-antimony alloys, for example, and technical problems arose in service before the anticipated 15 years of lifetime had elapsed. These took two forms; a variation in the floating voltage, which was attributed to adventitious metallic impurities, particularly antimony, and a deterioration in the charge acceptance and in the capacity of the positive plates. The cause of failure of the positive plates was obscure, but two factors appeared to be involved, first, the formation of an inter-facial high resistance layer between the grid metal and the active material, leading to a steady fall in the coefficient of use of the active material, and secondly to accelerated corrosion at the grain boundaries of the grid alloy. As a result of these difficulties, the Bell Telephone Company have now specified a completely new design of cell for this service, known under the Western Electric Company Trademark as the 'Bellcell'.

The main features are that the circular lattice-type grids are cast in pure lead, cupped at an angle of about 10° and assembled in a horizontal position in a clear plastic container of circular cross-section. Koontz *et al.*[158] have described the main features of this cell, shown in Fig. 4.60. The substitution of pure lead for lead-calcium alloy was based on the claim that the growth of grids in the latter was about 8 times that of grids in the former. The circular and slightly concave form of the plates were considered to counter the effect of growth and ensure good contact of the active material and grid during the lifetime of the cell.

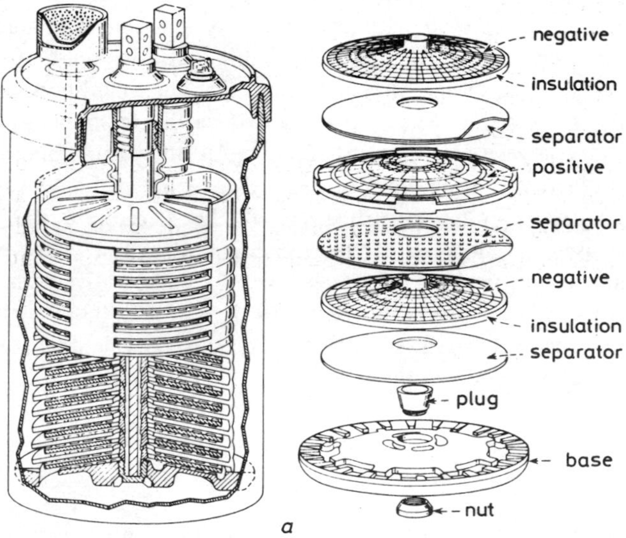

Fig. 4.60 *Bell telephone system battery*[158]
a Cutaway and exploded view
b Postivie grid
[Copyright 1970 American Telephone and Telegraph Company, reprinted by permission]

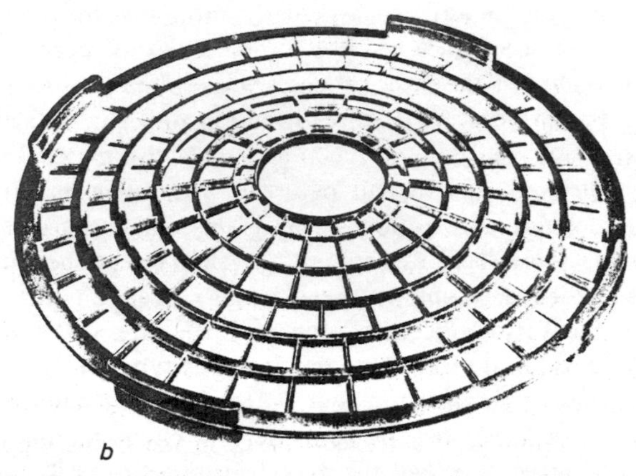

b

Another feature of this cell was the use of chemically-produced tetra-basic lead sulphate for the active materials. The interlocking rod-like crystalline particles of PbO_2 formed from this material were considered to offer greater resistance to shedding than other active materials. In spite of the wasted space in building a battery of cylindrical cells, it is claimed that energy densities approaching 21 Wh/kg at the 10h rate can be achieved with this assembly and this is appreciably higher than the most efficient HPP Planté battery could offer. In addition to the grid design, it was claimed that the long life was made possible by the development of positive grid casting technology, which produce very large grains, about 0·5 cm^3. All other lead components were manufactured using low pressure die-casting techniques. Problems may arise through stratification of the electrolyte and short circuits caused by shedding of the active materials, but the inventors have claimed that lifetimes of 30 years can be conservatively anticipated for this type of assembly. Since 1972, about 200,000 Bellcells, mostly of the 1680 Ah size have been installed and some pre-production cells are now in their 11th year of float service.

4.10.3.2 Static un-interruptible power systems (UPS): The development and mass production of solid state semi-conductor power components has opened the way to a wide range of equipment for the control and manipulation of electric power. The static inverter, which uses semi-conductor switching elements to convert a DC source to an AC output in a combination generally known as a UPS, is a good example of this new technology. Such systems include three main components:

(*a*) a battery charger, including a transformer and a rectifier, which converts the AC input to regulated DC output and keeps the battery fully charged
(*b*) a storage battery, which provides DC input power to the inverter during voltage drops or on failure of the normal AC line
(*c*) a static inverter, which converts the DC to AC, providing precisely regulated output voltage and frequency.

A schematic diagram of the basic UPS is shown in Fig. 4.61. In one such application, the battery is linked to a computer network to preserve a rigid continuity of service. A break in the mains supply, even if only momentary, could lead to chaotic disruption. In such a case a static switch with a transfer time, including sensing of not more than

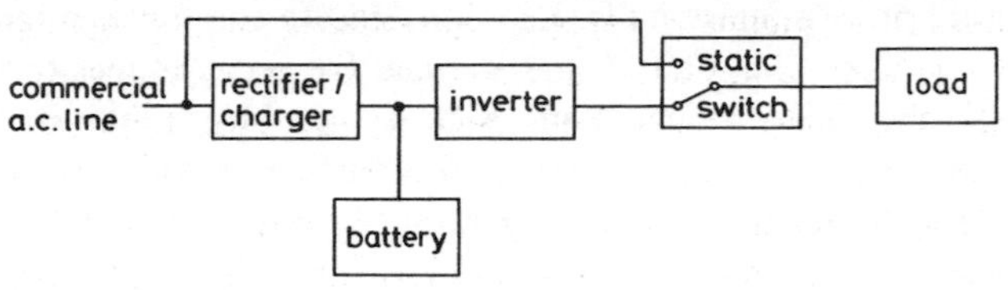

Fig. 4.61 *Uninterruptible power system*

¼ HZ would be included in the circuit. A typical installation in the reservation centre of an American air-line Company included a battery of 177 cells, each having 33 plates of the flat, pasted type, with lead-calcium alloy grids. The open-circuit voltage was 354V, the cell capacity was 2400 Ah (10h rate) and the total battery energy was 850 kWh. The weight of each cell was 182 kg and of the complete battery was 36 tonnes. The battery was maintained on a floating voltage of 2·20 – 2·25V per cell and it was claimed that equalising charges and water additions would be needed only every 3 to 5 years.

4.11 Non-spill batteries for cordless appliances

During the past decade, there has been a great increase in the use of cordless electrical appliances,[10] e.g. portable tools, emergency lights, instruments, calculators, portable radio and TV sets, toys etc. Batteries for these applications are usually of fairly low capacity, up to about 25 Ah or so, and, since they may have to function in different positions, it is an obvious advantage if they can be made 'non-spill'. Batteries with free electrolyte will fit this requirement if sufficient head-room is provided above the element to accommodate the free electrolyte in the event of inversion. Miners' cap-lamp batteries and some batteries for aircraft met this specification, the vent-plug being lengthened in the latter case to allow the entrapped gases to escape without ejection of the electrolyte. This design obviously includes a good deal of waste space and processes have been developed to entrain the free electrolyte in a powdered matrix or gel and to fill the space between the plates with highly porous separators, generally based on glass fibre mats, in which the gel is retained.

4.11.1 Gelled electrolyte

In this old process, chloride-free sodium silicate solution is mixed with dilute sulphuric acid. Vinal[11] has defined the setting times of various mixtures, the final product being a clear, translucent jelly or a fairly rigid solid. As an example, when 2 to 5 parts by volume of sulphuric acid of 1·400 SG are added to 1 part of silicate solution of 1·100 SG, the setting time ranges from 17 to 25 min. The delay in setting makes it possible to fill the cell before gelling takes place. If longer setting times are needed different ratios and concentrations may be chosen. The concentration of the acid will depend on the final SG required in the cell.

A second method is to mix very finely powdered silica or diatomaceous earth with sulphuric acid of the required final SG until the mixture forms a thick paste. Such mixtures have thixotropic properties. When thoroughly agitated, they become sufficiently fluid to be poured into the cells, where they slowly set into a jelly-like consistency.

4.11.2 Supported active material

Szper[159] has described the assembly and performance of non-spill batteries built on this principle. The key to the assembly lay in the separators, made by spraying onto a closely felted glass fibre mat or thick absorbent paper a thin suspension in water of fine siliceous powder mixed with a binding agent, such as rubber latex or other plastic emulsion and drying the resulting product. The elements of plates and separators were assembled by the normal stack-building process, the sides and bottoms of the elements being coated with a thick layer of the latex emulsion paste. Several advantages were claimed for the process:

(*a*) thin plates, down to about 1 mm in thickness and porous pastes could be used, since the tight packing prevented buckling of the plates and shedding of the active materials
(*b*) practically no mud space and very little head-room were required: plate areas could be increased to fill almost all of the available space in the container
(*c*) internal resistance was low and batteries had high energy and power densities
(*d*) the batteries could withstand a high degree of vibration and could be operated in any position.

4.11.2.1 Applications of batteries with supported active materials (SAM): The most important uses were for motor-cycles, motor scooters and for aircraft. The latter application was illustrated by a 12V, 60 Ah (10h rate) battery, with a weight of 26·5 kg (58 lb), energy densities at the 10h rate of 54 Wh/kg and 128 Wh/1 and capable of giving a current of 1000 A for 1·3 min to 1·0V per cell. An obvious difficulty with this type of assembly was the means of checking the state of charge, since this could not be done by a determination of the SG of the electrolyte. It was also found that, under deep cycling regimes, lifetimes were less than those of batteries with free electrolyte, due to segregation problems.

4.11.3 Developments in semi-sealed and sealed batteries

The recent trend towards maintenance-free assemblies has focused attention on the prospect of sealing cells and batteries for various duties. Smyth, Malloy and Ferrell[160] have described developments in semi-sealed units for floating, standby and cycling service. These batteries had grids cast in lead calcium alloy and the bulk of the free electrolyte was held in the active materials and the highly porous separators. As a further safeguard, a thick pad of absorbent material was laid on the tops of the elements. The batteries were spill-proof, if accidently overturned, but they were fitted with vents to release gas if the internal pressure rose uncontrollably. Figs. 4.62 and 4.63 show a typical 6V,

Fig. 4.62 *Exide model EMF-1 maintenance - free battery used for cycling service*
[Courtesy of ESB-Ray-O-Vac]

27 Ah (5h) battery and its discharge characteristics. This type of battery was claimed to give maintenance-free service for extended periods up to some years.

A similar construction was used in batteries for cycling service, reported to give lifetimes of 200–250 cycles, when deeply discharged to about 80% of the 5h capacity. A key to their performance lay in the special charger, supplied with each unit and fitted with a relay, which gave a sharp current cut-off at the onset of gassing. Energy densities of these 6V, 8Ah and 12V, 6Ah batteries were in the ranges 17·6–22 Wh/kg and 36–48 Wh/dm^3.

Eberts and Jache[161] and Eberts[161a] have also described developments in small sealed cells and the automatic charging equipment to be used with them. These cells are assembled with lead-calcium grids, giving maintenance-free characteristics and free or gelled electrolyte. Glass fibre mats are used as separators, the free spaces in the cells being filled with thixotropic gels containing about 6% of finely powdered

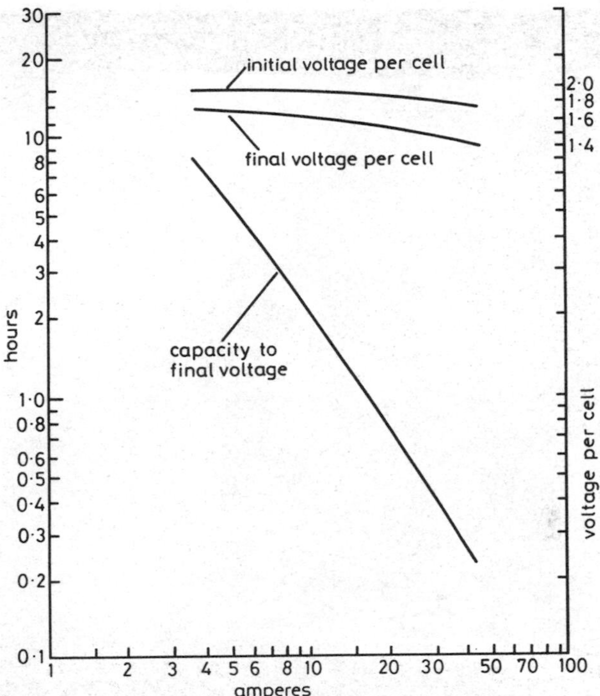

Fig. 4.63 *Discharge characteristics of Exide model 3-Lec-3 battery* [Courtesy of ESB-Ray-O-Vac]

silica, having an average particle size less than 0·1 μm. Standing losses after 1 year at 25°C amounted to only 35%. The cells are fitted with a one-way valve to relieve any deleterious internal pressure. Batteries of this type, shown in Fig. 4.64, have found a wide market and a large range of uses. An important factor in their success has been the design of the automatic charger, a typical circuit of which is shown in Fig. 4.65.

4.11.4 Gas-tight assemblies

It is obvious that gas evolution must be fully controlled if cells and batteries are to remain sealed for any length of time. Hydrogen and oxygen are the main gases evolved in storage batteries and these may be produced in a variety of ways:

Fig. 4.64 *Dry-fit Pc sealed cells and batteries*
[Courtesy Accumulatorenfabrik Sonnenschein GmbH]

Hydrogen may be evolved by

(*a*) the reaction

$$Pb + H_2SO_4 = PbSO_4 + H_2 \tag{4.43}$$

This reaction is, however, stopped fairly promptly by the formation of an insulating layer of $PbSO_4$

(*b*) the deposition on the negative plate of metals, like antimony, having a lower hydrogen over-potential than lead

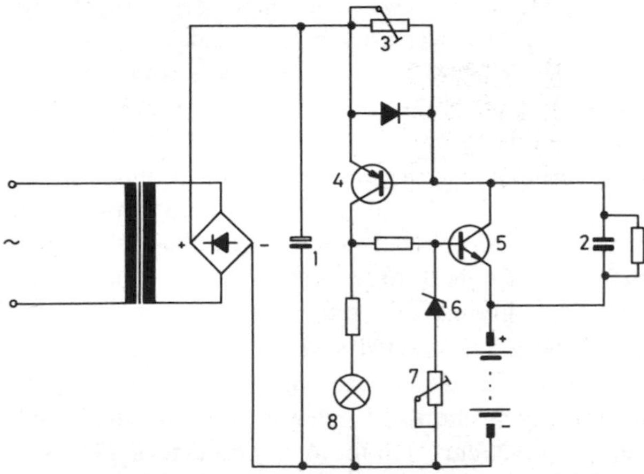

Fig. 4.65 *Dry-fit PC batteries: charging circuit with transistor switch*
1, 2 Capacitors
3, 7 Resistors
4, 5 Transistors
6 Zener diode operating at 2•47V per cell
8 Indicator lamp
[Courtesy Accukulatorenfabrik Sonnenschein GmbH]

Oxygen may be evolved by

(*c*) chemical decomposition of the PbO_2;

(*d*) *Both gases* are evolved by electrolysis of the water in the electrolyte during the latter stages of charging

(*e*) *Carbon dioxide* may be formed by oxidation of organic compounds leached out of the separators or other components in the cell.

Some of the gases may be held in solution in the electrolyte, but these amounts would be small and dissolved oxygen may be removed by chemical corrosion of metal components, particularly the grids. The use of antimony-free alloys more or less eliminates source (*b*), leaving the electrolytic gases (*d*) as the main source to be overcome. Two main routes have been explored, both depending on the re-combination of the gases.

4.11.4.1 Direct gas re-combination: In this method, the gases are allowed to form and are brought into contact with a catalyst to produce water, which is then returned to the cell.[162] The catalyst used for this purpose is generally based on platinum or palladium or their alloys, dispersed in some porous matrix, inert to sulphuric acid, held in enlarged vent-plugs, which have sometimes been termed 'catalators'. Flooding with electrolyte was thought to be a possible cause of premature failure of the catalyst. In the case of cells with antimony in the grid alloy, poisoning of the catalyst by SbH_3, released by the negative plates also reduced its activity, but claims were made that this could be prevented by including in the catalyst chamber some PbO_2, which oxidised any SbH_3 evolved. This problem would obviously not arise with antimony-free assemblies. Dyson and Sundberg[163] have described experiments on these lines to reduce the maintenance of miners' cap-lamp batteries. In this case the composite catalyst, made up of compounds of high and low activity, the composition of which was not disclosed was fully capable of causing re-combination of the charge gases. But not active enough to ignite stoichiometric mixtures at pressures up to 40 N/cm^2 (about 4 atmospheres), Fig. 4.66. On a daily cycle of 8h discharge at 0•8 A, followed by 16 h charge at the

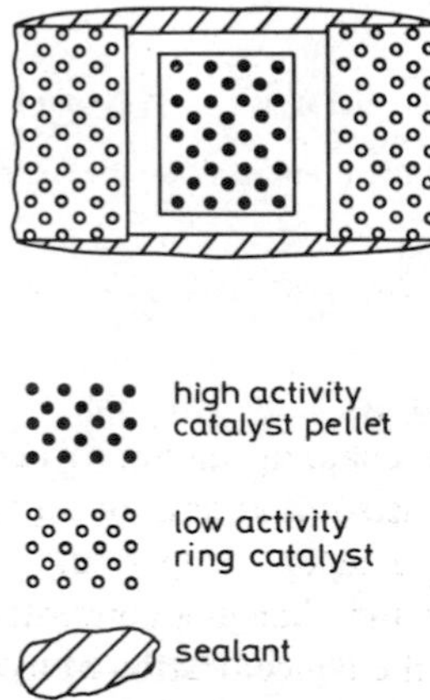

Fig. 4.66 *Diagrammatic section through a self-limiting composite catalyst* [Courtesy J. Dyson and E. Sundberg, *Power sources 4*, Oriel Press, Newcastle, 1973]

same current, the cells gave 50 cycles before the internal pressure reached 21 N/cm^2. Some degree of pressure venting was, however, recommended because excess amounts of hydrogen and carbon dioxide, which accumulated slowly, reduced the activity of the catalyst.

Watanabe and Yonesu[164] have also described catalyst plugs of this kind, introduced in Japan in 1970 and now fitted to almost all stationary batteries with capacities up to 2000 Ah in that country, making watering necessary only at 5-yearly intervals, see Fig. 4.67.

The catalytic gas re-combination reaction may cause a significant rise in temperature of the catalyst and precautions must obviously be

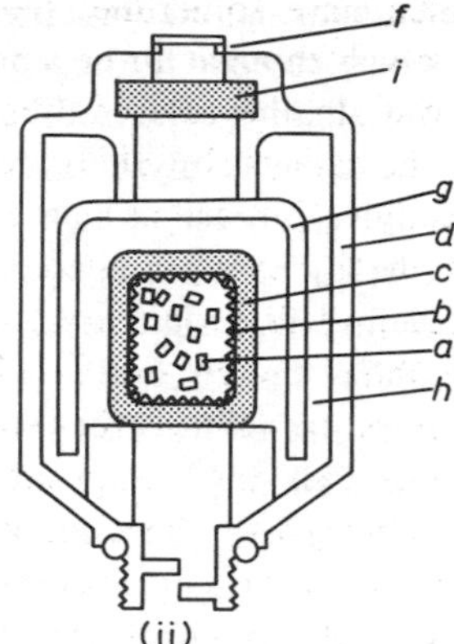

Fig. 4.67 *Japanese catalyst plug*
a Catalyst; one of the platinum group of metals—palladium, platinum or one of their alloys, in amounts from 5g (120 Ah cell) to 15g (cells up to 2200 Ah)
b Catalyst carrier, usually gamma Al_2O_3
c Sintered ceramic hydrophitic container
d Outer case
e Porous ceramic filter
f Vent
g Inner cover
h By-pass spare
[Courtesy M. Watanabe and K. Yonesu, Japan Storage Battery Co., 1978]

taken to avoid explosions. The heat can also be used to operate a thermal switch in the charging circuit, whereby the current can be automatically reduced as soon as gassing starts. One method is to use a ballast resistor having a steep temperature coefficient of resistivity. Another makes use of the changing concentration of hydrogen in the catalator chamber. As the concentration of hydrogen in the electrolytic gases increases so does the radiation of heat from a wire resistor in the chamber thereby increasing the resistivity of the resistor.

4.11.4.2 Indirect gas re-combination: A second process depends on the production of oxygen alone, steps being taken to inhibit the formation of gaseous hydrogen by keeping the potential of the negative plate group below the gassing level. Ruetschi and Ockerman[165] have reported tests with a free-acid assembly, in which a partially submerged oxygen-consuming electrode, similar to those developed for fuel cells, composed of porous carbon, impregnated with silver catalyst, was connected electrically to the negative group. During charge, any oxygen evolved from the positive plates was more or less instantly reduced at the auxiliary electrode and this raised the potential of the negative plates above the level at which gaseous hydrogen is evolved. Unfortunately, the efficiency of the auxiliary electrode deteriorated, presumably due to solution of the silver catalyst, and the method was not adopted in practice. A simpler process, which also uses the oxygen evolved to inhibit the formation of gaseous hydrogen is that successfully used in sealed nickel/cadmium alkaline cells.

The basic principles were first stated by Bureau Technique Gautrat[166, 167] and Jeannin[168] nearly 30 years ago and since confirmed by various authors. Falk and Salkind[169] have described the chemical reactions in some detail. *The salient features of the process are as follows:*

(*a*) there is an excess of capacity in the negative active material, some of which is permanently in a discharged state
(*b*) a minimum of electrolyte is used and all of this is absorbed in the active materials and the separators: there is no 'free' electrolyte
(*c*) plates should be made as thin as possible to provide a large surface area of negative active material, with correspondingly thin and highly porous separators, and diffusion paths for oxygen to reach the negative surface are then kept short
(*d*) for reasons of safety, a pressure release valve is fitted to each cell.

Because of the imbalance of the active materials, during charge, oxygen is evolved first and since the negative active material is covered with very thin layers of electrolyte, diffusion of the gas to the surface is relatively unimpeded.

Applying similar principles to the lead/acid system, the following sequence of reactions is considered to take place during charge:

At the positive plates

$$PbSO_4 + 2H_2O = PbO_2 + SO_4^{2-} + 4H^+ + 2e \qquad (4.44)$$

followed by

$$2OH^- = H_2O + \tfrac{1}{2}O_2 + 2e \qquad (4.45)$$

At the negative plates

$$PbSO_4 + 2e = Pb + SO_4^{2-} \qquad (4.46)$$

followed by the chemical reaction

$$Pb + \tfrac{1}{2}O_2 + H_2SO_4 = PbSO_4 + H_2O \tag{4.47}$$

or, in the absence of this reaction, the production of hydrogen by the electrochemical reaction

$$2H^+ + 2e = H_2 \tag{4.48}$$

If the conditions in the cell are arranged as indicated, reactions 4.45 and 4.46 prevent reaction 4.47 from taking place and there should be no accumulation of gaseous hydrogen in the cell.

Various authors have described tests on semi-sealed cells and batteries constructed on these principles. Mahato, Weisman and Laird[170] claimed to have observed the re-combination of oxygen with a lead electrode at a rate equivalent to 15 mA/cm^2 and of hydrogen on a lead dioxide electrode at more than 30 times the rate of hydrogen evolution in the self-discharge of a cell with lead-calcium grids. The latter rate would, of course, be very low, and this reaction has not been confirmed by other authors. Mahato and Laird[171] have confirmed the re-combination of oxygen in cycling tests with maintenance-free cells and 6V batteries, having capacities of 0•9, 7•5 and 47•5 Ah. In the smallest cell, the electrolyte was distributed throughout the element in the ratio of 3 parts in the separators, which were 85% saturated, and 1 part in the active materials, which were 30% saturated. In these tests the electrolyte contained 29g/l of phosphoric acid (85% W/W of H_3PO_4) but little significance was attached to this. The units were discharged at various rates from C/20 to C/5 (C = 5h rate) and recharged at rates from C/20 to C/10, with overcharge in excess of 30% of the output. Failure was defined as the point at which the capacity had fallen to 40% of the initial capacity or the internal pressure reached 30•4 N/cm^2. Cycle lifetimes were roughly related to the depth of discharge, e.g. at a current of C/10 A and depth of discharge of 0•35C, cells gave 1670 cycles, while at C/20 A, discharged to 0•55C, only 650 were obtained. The cells accepted continuous charge at C/20 A for 400C and at C/40 A for 600C. Internal pressures remained below 24 N/cm^2 and the maximum top-of-charge voltage was around 2•55V.

Using rotating ring-disc electrodes, Atkin, Bonnaterre and Laurent[172] have confirmed that a similar mechanism for oxygen re-combination occurs in MF lead/acid cells as in sealed Ni/Cd alkaline cells. They claimed that the reduction currents involved were a function of oxygen solubility. This was 5 times as high in 5M H_2SO_4 as in 7M

KOH and at an oxygen pressure of 20 N/cm^2, the re-combination rate was 15 mA/cm^2 for the lead electrode and 3 mA/cm^2 for the cadmium electrode. Major differences in behaviour arose from the instability of lead in H_2SO_4, which led to the evolution of hydrogen, when the internal pressure of oxygen was low and to corrosion of the grid metal. No similar reactions occurred in the alkaline cell, since both nickel and cadmium were thermodynamically stable in KOH electrolyte.

These authors stated that in a maintenance-free lead/acid assembly, elimination of antimony reduced the life of the positive active material, but that this weakness could be alleviated by the addition of phosphoric acid to the electrolyte, although no explanation was offered of the mechanism involved. Fig. 4.68 shows how the absorption of oxygen by a fully reduced lead negative immediately lowers the potential of the electrode by 0•4V, thereby improving its charge acceptance, without the evolution of hydrogen.

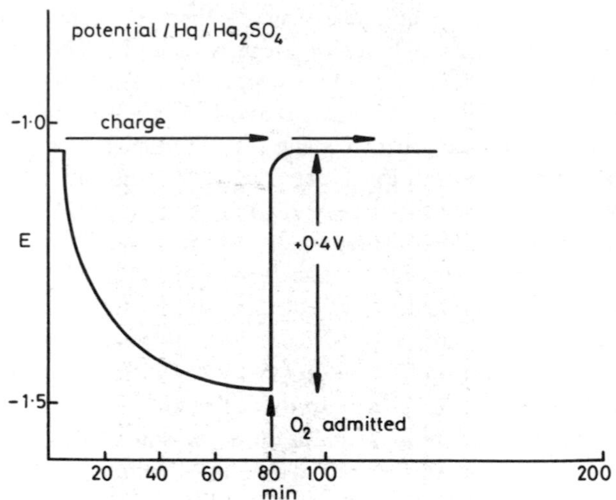

Fig. 4.68 *Change in potential of a reduced lead negative electrode on reaction with oxygen*[172]
Dimensions of Pb electrode, 4cm x 3•5cm x 0•1cm (T). SG of electrolyte 1•270, charging current (T) 200mA
[Courtesy Academic Press, 1977]

4.11.5 *Practical semi-sealed cylindrical cells*

With a better understanding of the basic principles involved, manufacturers have been encouraged to build cylindrical lead/acid cells, constructed in a similar way to the rechargeable Ni/Cd alkaline cells, now competing over the high current range with the long established primary Leclanché cells. Gates Energy Products Inc. of the USA have been pioneers in this development,[173,174] and have recently formed a joint venture with Chloride Power Ltd. of the UK, known as Chloride-Gates Energy Ltd., for the production and marketing of their Cyclon cells in Europe.

The construction and performance of these cells have been fully described in technical leaflets[175,176] and the Gates Battery Application Manual.[177] Positive and negative plates are made from thin pure lead sheet, subjected to a perforating metal process and filled with the active materials. The plates are interleaved with absorbent separator material made of felted glass fibres tightly coiled and pressed into a cylindrical plastic container, held in turn in an outer metal case.

Fig. 4.69 *Gates 'starved electrolyte' cell*
a Positive and negative plates
b Highly retentive separator
c Pure lead grids
d Metal can
e Safety van
[Courtesy Chloride-Gates Energy Ltd.]

The cells are soaked with electrolyte, given a formation charge and closed. All of the electrolyte is held in the plates and separators and the units are often described as 'starved electrolyte' cell. Figs. 4.69 and 4.70 show the construction of a typical cell, which includes a pressure release valve, operating at 40 to 60 lb/in^2. Fig. 4.71*a* shows a family of discharge voltage curves at different rates for a cell of D size (British Standard and IEC designation R20; outside diameter 33 mm; length 60 mm, volume 55 ml), Fig. 4.71*b* shows the capacities at different temperatures at a high rate, 3A or 1·2C and a low rate, 250 mA or C/10.

Hammel[178] has reported comparative performance figures for D-sized coiled electrode lead/acid and sintered plate Ni/Cd alkaline cells, shown in Table 4.23 and similar results have been described by Harrison.[179] Harrison and Peters[180] have also carried out comparative tests with continuous and interrupted discharges (Dc, 2 h, open circuit, 11 h) on primary Leclanché, lead/acid and sintered-plate Ni/Cd alkaline

Table 4.23 *Comparative performance figures for lead/acid and Ni/Cd alkaline cells*

		Lead/Acid		Ni/Cd alkaline		
Weight, g		180		147		
Volume, ml		56		49		
Open-circuit voltage, V		2·00		1·20		
Internal resistance, m Ω		9		5		
Rate of D/C						
h	A	Ah	Wh	A	Ah	Wh
1	2·00	2·00	4·00	3·50	3·50	4·2
5	0·47	2·35	4·70	0·85	4·20	5·0
10	0·25	2·50	5·00	0·44	4·40	5·2
Energy densities at the 5h rate						
Wh/L		84			102	
Wh/kg		26			34	

cells of D size. Their results were more or less in line with those of other authors. They concluded that, at loads up to about 0·1W, the Leclanché cell gave the best performance, but above 0·8W the rechargeable cells gave better outputs and under all conditions tested, the alkaline cell gave somewhat higher outputs than the lead/acid cell. Storage tests, summarised in Table 4.24 indicated that over a period of 3 months at 25°C, Leclanché cells showed the minimum loss of charge (35%), and Ni/Cd cells, the highest (81%).

In describing the charging characteristics of these sealed lead/acid cells, Hammel[181] has pointed out that the precise voltage regulation, required to prevent electrolysis in cells with free electrolyte is not necessary with this sytem. The gas recombination at the end of charge prevents any loss of water. Charging can be carried out continuously at

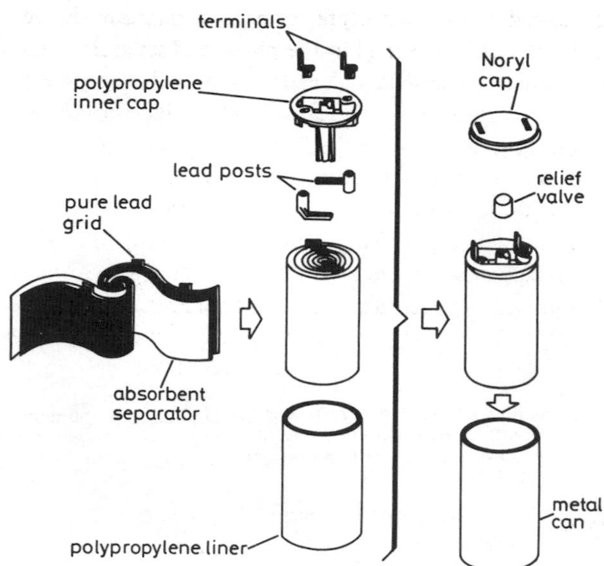

Fig. 4.70 *Gates cell: components and construction* [Courtesy Chloride-Gates Energy Ltd.]

constant current or constant voltage, provided there is no steep rise in temperature. The temperature coefficient of voltage was stated to be −2·5 mV/°C. Fig. 4.72 shows how the capacity of a D-size cell (2·50 Ah at the 5h rate) is returned during recharge after a full discharge, the recharge being carried out at a constant potential of 2·5V. Approximately 100% of the nominal capacity is returned after only 3 h. Fig. 4.73 shows the voltage characteristics of the cell, when charged at a constant current of C/15 after a deep discharge at the C/10 rate.

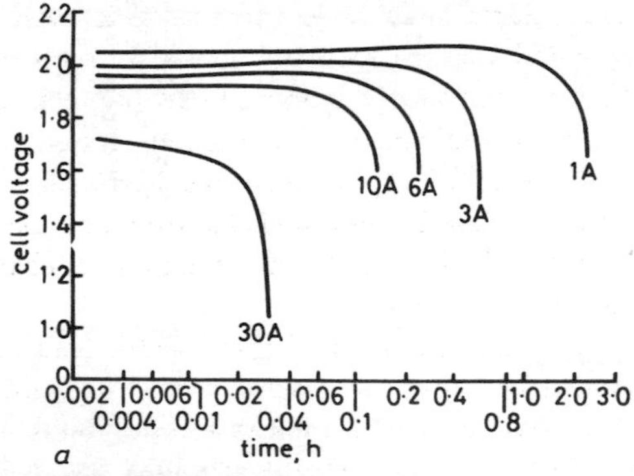

Fig. 4.71 *Gates D-size cell*
a Discharge voltage curves at various rates from 2•5h to 1•8 min

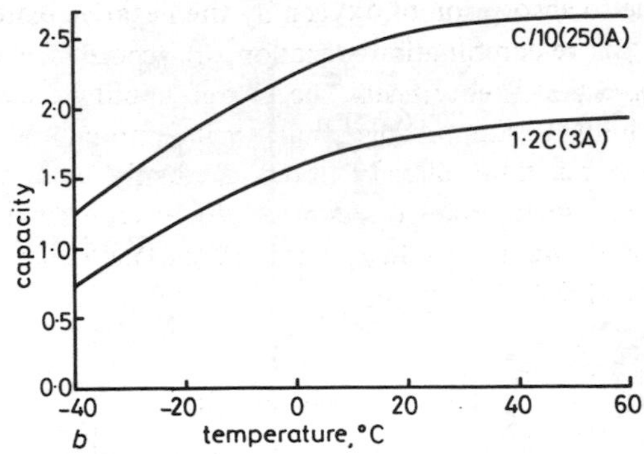

b Capacities at 0•250A (C/10) and 3A (1•2 C) over the temperature range −40°C to 60°C
[Courtesy Chloride-Gates Energy Ltd]

Table 24
O/C Losses over 3 months at 25°C.

	Exide HP2 (Le Clanche)	Exide MFR20 (Lead Acid)	Alcad W3.5 (Ni/Cd.)
Initial duration at 1•45W (hrs)	2•0	2•26	3•25
Final duration at 1•45W (hrs)	1•3	1•25	0•62
Loss of Charge (%age)	35	45	81

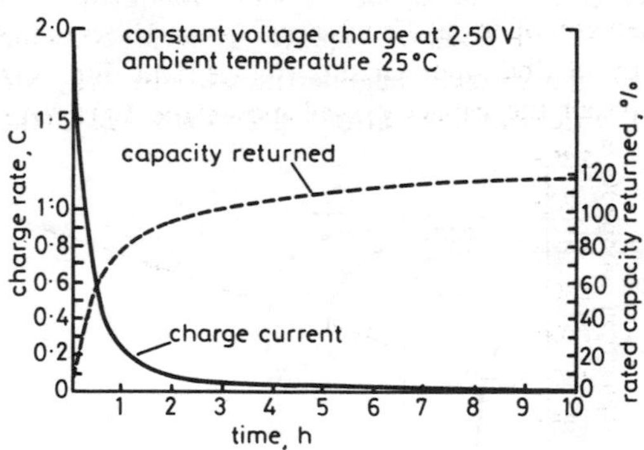

Fig. 4.72 *Capacity return against time of recharge at constant voltage of 2•50V after a full discharge at the C/10 rate*

The steep rise in voltage after about 14 h indicates that about 95% of the capacity had been returned and the cell was fully charged after about 15 h. The sharp drop in voltage to about 2•40V from the peak of

2·70V was due to absorption of oxygen by the negative plate, initiating the so-called 'gas re-combination' reaction, in accordance with values shown in Fig. 4.68. Theoretically the charge could be continued indefinitely without further change, but small amounts of hydrogen, evolved by side reactions already noted, gradually upset the steady-state conditions. According to Hammel, to ensure the longest cell lifetime, the continuous overcharge current for maintenance charging should not exceed C/400.

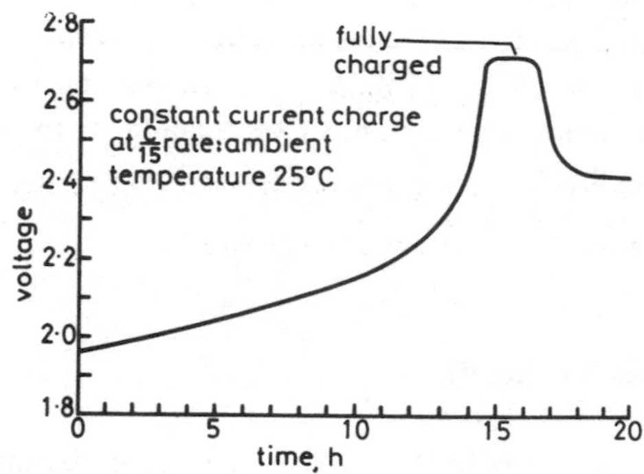

Fig. 4.73 *Cell voltage against time of recharge at a constant current of C/15 at 25°C*

One special problem with these sealed lead/acid cells, common also to Ni/Cd cells, lies in the means of determining their state of charge at any stage. This obviously cannot be done by a simple measurement of the SG, used for open lead/acid cells. Gates[177] recommend a careful measurement of the open-circuit voltage and Fig. 4.74 shows the relation between the capacity available at the C/10 rate of discharge

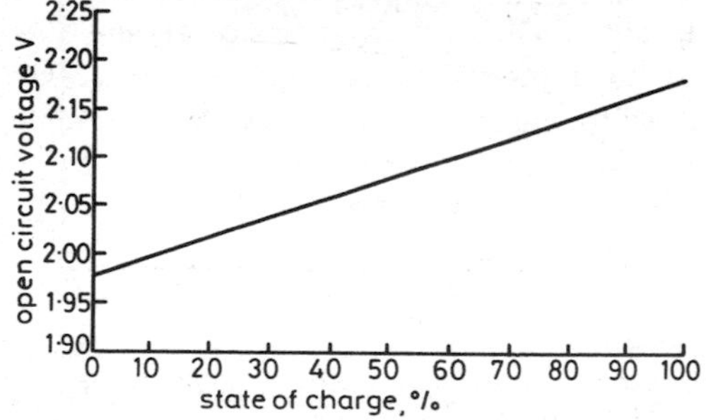

Fig. 4.74 *Available capacity against open-circuit voltage at 25°C*
The curve is accurate within 20% of the rated capacity of the cell being measured if it has not been charged or discharged within the past 24h. The accuracy increases to 5% if the cell has not been charged or discharged within the past 5 days

and the open-circuit voltage at 25°C. It is claimed that the curve is accurate to within 20% of the capacity if the cell has not been charged or discharged for 24 h and to within 5% if a period of 5 days has elapsed since the last charge or discharge. Lifetimes on cycling service depend on the depth of the discharges, but it is claimed that 200 cycles can be confidently expected, provided the discharges are of medium depth. The cells also work satisfactorily on floating service at voltages between 2·30V and 2·40V, when lifetimes at room temperatures are expected to exceed 8 years. These cylindrical cells can readily be assembled into batteries of rectilinear configuration, having nominal voltages of 6V or 12V. At present, cell capacities have been limited to a maximum of 5 Ah, but experiments are in hand with cells having capacities up to 25A.

4.12 Batteries for aircraft

As might be expected with the rapidly developing aircraft industry, the on-board power requirements have passed through several changes in the past 30 years or so. The situation in the UK is probably typical. The wholesale introduction of jet engines, for which starting currents of around 1000 A were required, imposed a heavy load on the batteries. As Robinson and Heinson[182] have indicated, this load is now provided by small turbo-generator sets, and, although batteries are needed to start these, the power required is at a much lower level. Bagshaw, Bromelow and Kirkpatrick[183] have outlined modern requirements and have described two types of batteries both having 12 cells and therefore a nominal voltage of 24V, with capacities of 18 and 34 Ah at the 10h rate. Although good high rate capabilities are necessary to supply emergency power for up to 1 h, in the event of generator failure, the main service is as stand-by power for miscellaneous equipment, as laid down by British Standard BS G205. Fig. 4.75 shows a cut-away model of the 24V, 34 Ah battery and Fig. 4.76 the performance figures at various rates and temperatures. Several special features are incorporated in these designs. The thin plates have grids cast in a low antimony alloy, 6% Sb and 0·25% As, and pastes of high porosity. The positive plates are enclosed in envelopes, made of the microporous plastic, porvic, making it possible to have a very limited mud-space in the cells. The weight of the 'top lead' and the voltage losses are both cut to the minimum by bringing the inter-cell connectors through the cell partitions and by casting the busbars and the connectors around copper

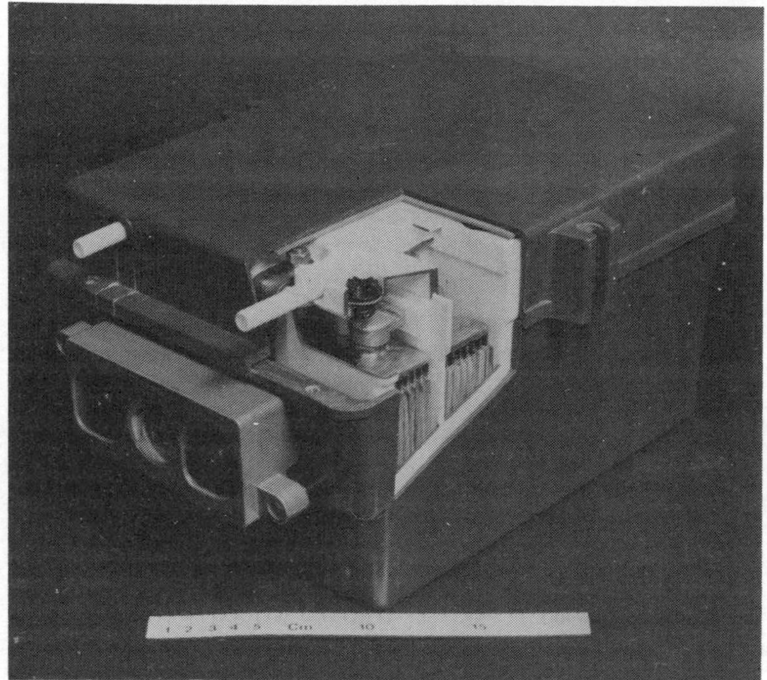

Fig. 4.75 *Cutaway view of 24V, 34Ah aircraft battery*
[Courtesy Chloride Industrial Batteries Ltd.]

inserts. Light-weight high-impact polystyrene containers and covers are used and the cells are fitted with non-spill vent-plugs of the Davis type, which ensure complete unspillability in any position during aerobatics. Special plastic manifolds are moulded into the covers to vent outboard gases evolved during cycling.

Table 4.25 shows the favourable energy and power densities of the 24V, 34 Ah battery, the weight of which, fully charged, is only 35·6 kg. On a charge retention test, at 25°C the battery lost only 10% of the 1h capacity after 28 days, and, on cycling tests of 4h discharge at the 5h rate (C), with recharges at C/8A (0·125C A) for 8 h, the battery completed 104 cycles before the 1h capacity fell to 80% of the nominal.

These batteries showed good stability when floating on the busbars at voltages as high as 30V. No thermal runaway of the type recorded by Earwicker[184] occurred with battery temperatures as high as 50°C, but at higher ambient temperatures and busbar voltages battery temperatures rose to boiling point. The loss of latent heat at this stage stopped any further rise, and, although the battery would be irreparably damaged by this treatment, the behaviour was regarded as a 'fail-safe' mechanism.

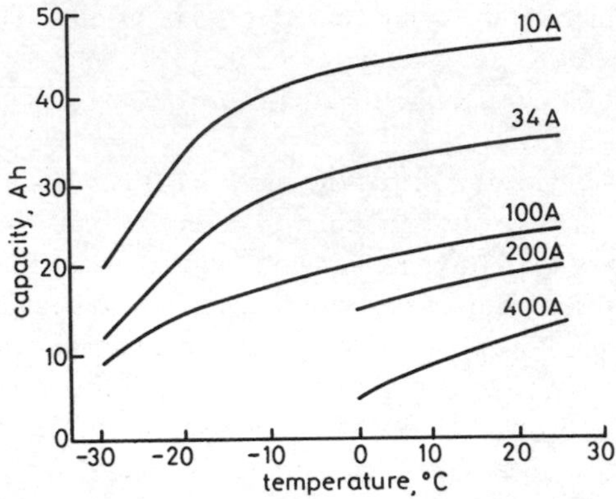

Fig. 4.76 *Capacities at various rates and temperatures*

Table 4.25 *24V - 34 Ah aircraft battery*

Rate	Current	Mean voltage	Energy density		Power density	
h min	A	V	Wh/kg	Wh/dm^3	W/kg	W/dm^3
1–00	34	23·0	22·0	48·0	22·0	48·0
0–06	200	20·9	11·7	26·1	117·4	261·3

4.13 Batteries for submarines

4.13.1 Standard types

Submarine batteries are the largest units in traction service. In the older type of ship, the storage battery was the sole means of propulsion, when the ship was fully submerged, and, in addition, supplied the 'hotel load', power for lights, instruments and other electrical equipment. The introduction of the snorkel breathing tube made it possible to use diesel engines for propulsion, but as these were noisy and detectable by sensitive listening devices, batteries were kept in reserve for emergency use. Even in modern nuclear-powered vessels, storage batteries are still used for this purpose. These may be of the flat, pasted plate or tubular positive plate types, with 5h capacities in the range 5000 to 12 000 Ah. Complete batteries may consist of 448 cells, with 4 sections each of

112 cells, which can be connected in series or parallel. The energy available may therefore be as large as 12 000 kWh and the whole battery may weigh over 300 tonnes. Batteries in nuclear-powered boats may have only 112 cells.

An over-riding criterion for this service is that the rate of evolution of hydrogen gas on open-circuit should not exceed the specified low limit. Grids are therefore cast in alloys with a low antimony content or in antimony-free alloy. The options here are somewhat limited. For example, arsenic, now one of the most popular hardeners, cannot be used for this purpose, owing to the risk of pollution by arsine, AsH_3, which may be evolved at the end of charge. In the UK, grids for pasted plates are cast in an alloy containing 3% Sb, 0•05% Se, 1•5% Sn, generally known as Admiralty B alloy, the original patent for which was filed by Waterhouse and Willows.[80] Negative grids may be cast in lead-antimony alloy, and electroplated with a thin coating of pure lead.

In the USA, it is understood that lead-calcium alloy is now generally used for this service. Spines for tubular plates are also cast in low-antimony alloys, and, in some European countries, ASTAG alloy is used for this purpose. High standards of purity are specified for the oxides used for the active materials, particularly with respect to the metals antimony, iron and nickel and tellurium.

Double separation, with felted glass fibre mats and microporous separators ensures durability, high performance and low standing losses. To resist damage through depth charges, the material for the cell containers must have a high impact strength and this is provided by plastic resin-impregnated glass fibre. As a further precaution against leakage of electrolyte, the whole element with the electrolyte is contained in a thick-walled rubber bag, fitting closely into the outer container.

These batteries are tightly packed in confined spaces in the hulls of the ships and the cells are among the tallest made, with plates as high as 1000 mm. This poses two problems. The first concerns the disposal of heat which develops during working. Most of this heat is conducted up the plates to the busbars, and to assist in its dissipation cooling water is circulated through channels cut in the copper cores, which are moulded in the group bars in each cell.

The second problem is stratification of the electrolyte, which may occur if gassing overcharges are not sufficiently vigorous. To overcome this problem, tubular air-lift pumps, reaching to the bottom of the cells, have been inserted in the free electrolyte at the side of the element.

Another interesting problem has recently come to light with tall cells of this type and this concerns the resistance of the grids. Bagshaw,

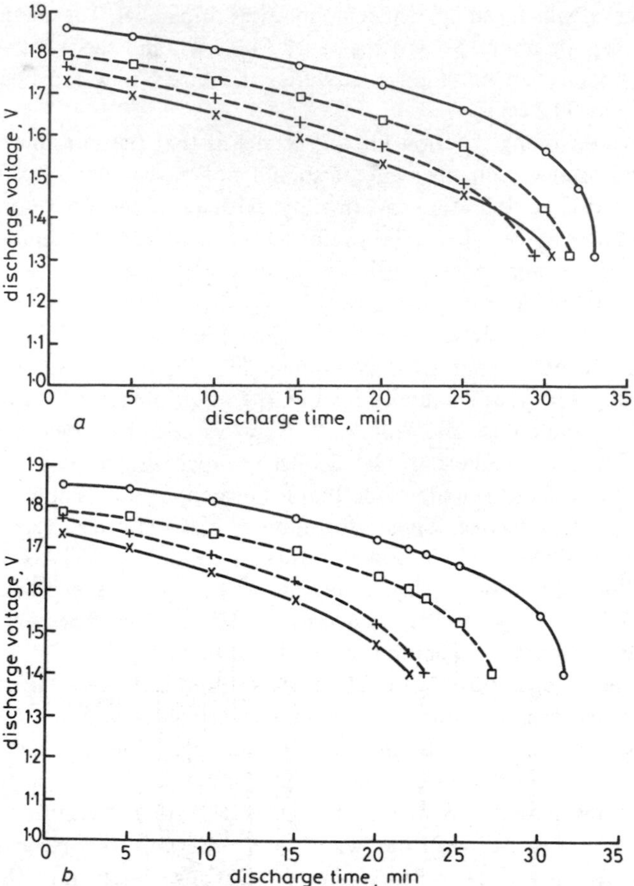

Fig. 4.77 *Tall cells: effect of improved grid conductivity on the capacity*
a At a constant current of 1244A
b At a constant power of 2•117 kW
○ experimental positive and negative electrodes
□ experimental positive/standard negative electrodes
+ experimental negative/standard positive electrodes
x standard positive and negative electrodes
[Courtesy Bagshaw, Bromelow and Eaton, Chloride Industrial Batteries Ltd.]

Bromelow and Eaton[63] have reported the results of tests, in which the conductivities of grids, 800 mm in height and thicknesses 3•9 mm (positive) and 2•9 mm (negative) were increased by a factor of 4 for the positive and 3 for the negative, by dissolving away some of the lead alloy and replacing this with electrodeposited copper, which was then lead-plated. Multi-plate cells were prepared by otherwise standard processes. Some cells were assembled with both positive and negative

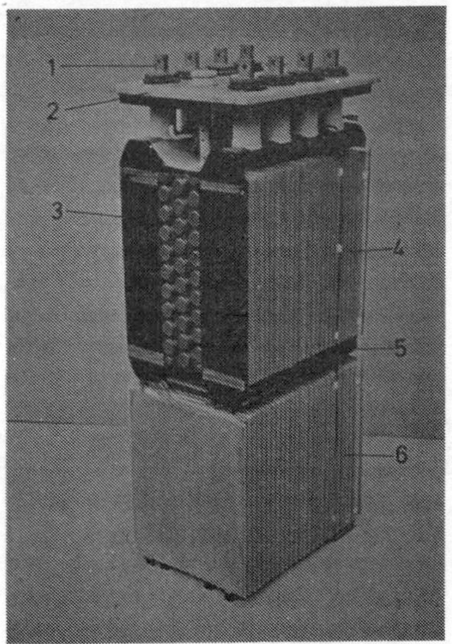

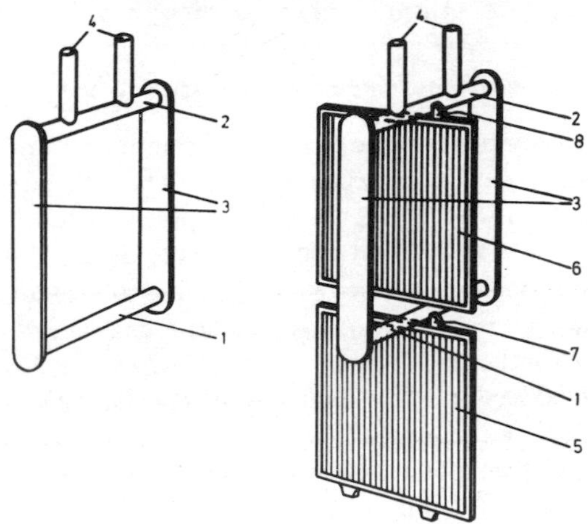

Fig. 4.78 *Tall cells*
a Double-decker construction
1 Lower polebridge
2 Upper polebridge
3 Side conductor
4 Polebolts
5 Single plate from the lower plate block
6 Single plate from the upper plate block
7, 8 Current conductor fans welded to polebridges.

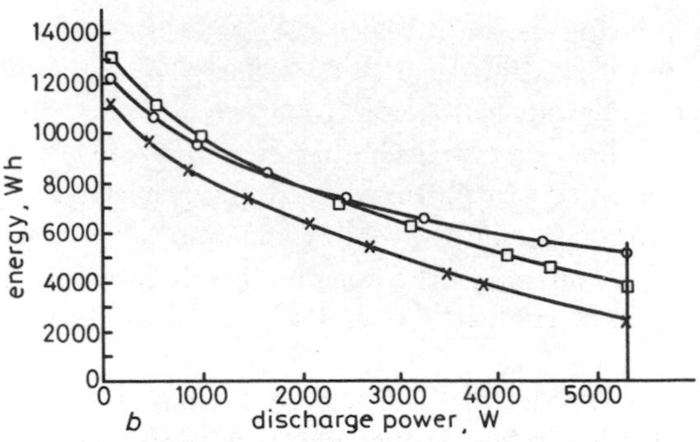

Fig. 4.78*b* ***Performance of double-decker cells***

Table: Energy of multi-electrode cells (discharge at constant power)

Cell Type	Rate of discharge				Cell	
	P=112 W		P=5306 W		Weight	Height
Tubular Plate	kWh	%	kWh	%	kg	mm
x Standard	11•170	100	2•388	100	222	1195
o Double-decker 1	12•170	109	5•041	211	253	1195
□ Double-decker 2	13•070	117	3•795	159	242	1195

[Courtesy J. Brinkman, Varta Batterie AG, also Academic Press Ltd.]

high conductivity plates and others with a combination of special plates and standard plates of opposite polarity. The cells were discharged at a constant current of 1244 A, the approximate 30 min rate, to 1•40V and 1•30V and also at constant power of 2•117 kW to 1•40V and the performances compared with those of standard cells. The results, shown in Figs. 4.77*a*/*b* and indicate that at constant current, the capacity of the cells with the special plates was 26% higher than that of standard cells and 43% higher at constant power. By far the largest portion of these gains lay in the improved conductivity of the positive grids. The effect was not so marked in the negative plates, probably because of the contribution to the overall conductivity made by the negative active material. Reference was made to the risk of copper contamination, due to the difficulty in depositing pore-free coatings of lead, which might rule out any practical application of the process.

4.13.2 Double-decker construction

Brinkmann[64] has described the construction and performance of cells built on this principle in an effort to overcome the problem of grid conductivity. In brief, the single cell element is replaced by two elements, each being one-half of the total height. These are superimposed, one above the other, in the container, with positive and negative plate groups connected in parallel by pole bridges, having copper cores, thickly coated with a protective layer of lead. Voltage losses were therefore reduced to a minimum. The construction is shown in Fig. 4.78 which also gives some comparative results for cells with tubular positive plates.

Cell number 1 was designed particularly for high rates of discharge and number 2 for low rates. At the approximate 30 min rate, constant power of 5306 W, the output of number 1 was over 100% higher than that of the standard and at the 100h rate, constant power of 112 W, the output of number 2 was 17% higher than that of standard. The cell height in this case was 1195 mm.

In addition to the improved output, other advantages were claimed. Since the plates were only half the height of standard plates, it was easier to keep specified standards in the manufacturing processes, especially in such operations as tube filling. Also the double-decker cells showed improved charge acceptance and less heat developed during charging. The danger of copper contamination from the lead-coated pole-bridges was also recognised. Since their introduction over 5 years earlier, several thousands of double-decker cells had been commissioned so far, and there was no evidence of premature failure from this cause.

Acknowledgments

The author has drawn freely from a long list of references and offers an omnibus expression of gratitude to all of those authors quoted and an apology to those whose work has been overlooked. He would like to thank the following personally for special services rendered: K. Peters, C. Hall, S. Fewster, J. Prest, N. Bagshaw and E. Baxter of the Chloride Group Ltd. for helpful advice and illustrations; E. Butler and A. Reay, Chloride Automotive Ltd. and D. Gibbons, Lucas Batteries Ltd. for information in connection with SLI battery specifications; S. Hattori, Yuasa Battery Co., Ltd and A.R. Cook, ILZRO, New York for scanning electron micrographs of active materials; D. Laight, Lucas Electrical Ltd., W. Wylam, Delco-Remi Division of General Motors Corporation, D.T. Ferrell, ESB Ray-O-Vac Inc., A. Kozawa, Union Carbide Corporation Inc., M. Rose, St. Joe Minerals Corporation, A. Sabatino, Gould Inc., J. Pierson, Globe Union Inc., G. Lander, H. Bode and J.

Brinkmann, Varta Batterie Ag, K. Eberts, Sonnenschein AG, P. Faber, RWE, Germany, P. Ruetschi, Leclanché SA, R. Hammel, Gates Energy Products, A. Hiscock, Chloride Gates Ltd, N. Hampson, Loughborough University of Technology, D. Feder, Bell Telephone Laboratories, all of whom have contributed figures or illustrations and/or helpful discussions and comments. Equally, he offers apologies for any oversights and omissions.

References

1 *Trends in lead consumption: 1968-1976* (Lead/Zinc Developmnt Association, London)

2 *The storage battery manufacturing industry year-book 1976-1977* (Battery Council International, Burlinghame, USA), pp. 95-101

3 KOZAWA, A., and TAKAGAKI, T.: *Japanese lead-acid battery industry* (US Office of the Electrochemical Society of Japan, Cleveland, USA, 1977), p. 37

4 *The storage battery manufacturing industry year-book 1971-72* (Battery Council International, Burlinghame, USA), pp. 73-82

5 *Mechanical handling equipment,* Business Monitor PQ 337 (HMSO, 1977)

6 D'ARCEY, R.L.: 'Expanding frontiers of industrial batteries'. Lead and zinc in the 80's Proceedings of the International Conference, June 1977; pp. 20-24, Lead/Zinc Development Association, London

7 KALHAMMER, F.R., BIRK, J.R., and PEPPER, J.W.: 'Storage of energy on electric power systems'. Paper presented at the World Electrotechnical Congress, Moscow, June, 1977; Electric Power Research Institute, Palo Alto, USA, p. 28

8 *50Hz and 60Hz UPS (30 to 600 KVA)* (Chloride Transipack Ltd., Kent, UK, 1977), p. 19

9 SANDERS, J.A., and LEAR, J.W.: 'Performance characteristics of sealed lead-acid cells'. Paper presented at the 27th Annual Power Sources Conference, June 1976, USA, p.4

10 BARAK, M.: 'Cordless power applications', *J. Royal Soc. Arts*, 1968, pp. 815-837

11 VINAL, G.W.: *Storage batteries* (John Wiley & Sons Inc., New York, 1955, 4th edn.), pp. 131 *et seq*

12 BODE, H.: *Lead acid batteries* (John Wiley & Sons Inc., New York, 1977), pp. 40 *et seq* (translated by BRODD, R.J., and KORDESCH, K.V.)

13 BERNDT, D.: 'The effect of temperature and current density on the utilisation of lead and lead oxide electrodes'. *In* COLLINS, D.H. (Ed) *Power sources 2* (Pergamon Press, 1970, pp. 17-31

14 FLEMING, A.N., HARRISON, J.A., and THOMPSON, T.: 'The kinetics of anodic lead dissolution in H_2SO_4'. *In* COLLINS, D.H. (Ed.) *Power sources 5* (Academic Press Inc., London, 1975), pp. 1-14

15 CRAIG, D.N., and VINAL, G.W.: *J. Res. Natl. Bur. Standards*, 1940, **24**, p. 475

16 BECK, W.H., LIND, R., and WYNNE JONES, W.F.K.: 'The behaviour of the lead dioxide electrode. Pt. 2', *Trans. Faraday Soc.*, 1954, **50**, pp. 147-152

17 KABANOV, V., FILLIPPOV, S., VANYUKOVA, L., IOFA, Z., and PROKOF'EVA, A.: 'Hydrogen overvoltage on lead', *Zhurnal Fiz. Khim.*, 1938, **XIII**, 3, p.11

18 GILLIBRAND, M.I.G., and LOMAX, G.R.: 'Discharge characteristics of the lead/lead sulphate electrode', *Electrochim. Acta,* 1966, **11,** pp. 281-287

19 RUETSCHI, P., ANGSTADT, R.T., and CAHAN, B.D.: *J. Electrochem. Soc.,* 1959, **106**, p. 547

20 BARAK, M., GILLIBRAND, M.I.G., and PETERS, K.: 'Oxygen overpotential on lead dioxide electrodes'. Proceedings of the second International symposium on batteries, October, 1960, p. 9, Ministry of Defence Inter-departmental Committee on Batteries, UK

21 RUETSCHI, P., and CAHAN, B.D.: 'Anodic corrosion and hydrogen and oxygen overvoltage on lead and lead antimony alloys', *J. Electrochem. Soc.,* 1957, **104**, 7, pp. 406-412

22 CRAIG, D.N., and VINAL, G.W.: *J. Res. Natl. Bur. Standards,* 1939, **22**, p. 55

23 AFIFI, S.E., EDWARDS, W.H., and HAMPSON, N.A.: 'The formation of lead dioxide electrodes by the Planté process', *Surface Technology*, 1976, **4**, pp. 173-185

24 BARAK, M., GILLIBRAND, M.I.G., and LOMAX, G.R.: 'Hydrogen overpotential with particular reference to the negative electrode in accumulators'. Proceedings of the first International symposium on batteries, October 1958, p. 5, Ministry of Defence Inter-departmental Committee on Batteries, UK

25 AGAR, J., and BOWDEN, F.P.: *Proc. Roy. Soc.*, 1938, **A169**, p. 206

26 BOCKRIS, J. O'M.: *Modern aspects of electrochemistry* (Butterworth's Scientific Publications, 1954), p. 174

26*a* BARAK, M., GILLIBRAND, M.I.G., and LOMAX, G.R.: 'Polarisation phenomena on porous electrodes'. Proceedings of the second International symposium on batteries, Paper 30, 1960, p. 6, Inter-departmental Committee on Batteries, Ministry of Supply, UK

27 THOMAS, U.B.: 'The electrical conductivity of lead dioxide', *J. Electrochem. Soc.,* 1948, **94**, pp. 42-49

28 FLEISCHMANN, M., and THIRSK, H.R.: *J. Electrochem. Soc.,* 1963, **110**, p. 688

29 BAIKIE, P.E., GILLIBRAND, M.I.G., and PETERS, K.: 'The effect of temperature and current density on the capacity of lead-acid battery plates', *Electrochim. Acta*, **1972, 17**, pp. 839-844

30 RUETSCHI, P.: 'Review of the lead-acid battery science and technology', *J. Power Sources*, 1977/78, **2**, pp. 3-24, Elsevier Sequoia SA, Lausanne, Switzerland

31 GILLIBRAND, M.I.G., and LOMAX, G.R.: 'The discharge characteristics of lead-acid battery plates', *Electrochim. Acta*, 1963, **8**, pp. 693-702

32 BODE, H., PANESAR, H.S., and VOSS, E.: *Chem. Ing. Techn.,* 1969, **41**, pp. 878-879

33 PETERS, K., HARRISON, A.I., and DURRANT, W.H.: 'Charge acceptance of the lead cell at various charging rates and temperatures'. *In* COLLINS, D.H. (Ed.) *Power sources 2* (Pergamon Press, 1970), pp. 1-16

34 RUETSCHI, P., SKLARCHUK, J., and ANGSTADT, R.T.: 'Stability and

reactivity of lead oxides'. *In* COLLINS, D.H. (Ed.) *Batteries* (Pergamon Press, 1963), pp. 89-103

35 BURBANK, J.: 'Cycling anodic coatings on pure and antimonial lead in H_2SO_4'. *In* COLLINS, D.H. (Ed.) *Power sources 3* (Oriel Press Ltd., England, 1971), pp. 13-34

36 GILLIBRAND, M.I.G., and LOMAX, G.R.: 'Hydrogen over-potential on porous lead electrodes', *Trans. Faraday. Soc.*, 1959, **55**, pp. 643-653

37 SIDGWICK, N.V.: *The electronic theory of valency* (Clarendon Press, 1927), p. 310

38 HOARE, J.P.: *The electrochemistry of oxygen* (Interscience Publishers, New York, 1968), p. 423

39 CARR, J.P., and HAMPSON, N.A.: *'The lead dioxide electrode', Chem. Rev.*, 1972, **72**, 6, 6, pp. 679-703

40 KATZ, T., and LE FAIVRE, R.: *Bull. Soc. Chim. France*, 1949, **16**, D124

41 POHL, J.P., and RICKERT, H.: 'On the electrochemical behaviour of the lead dioxide electrode'. *In* COLLINS, D.H. (Ed.) *Power sources 5* (Academic Press Inc., London, 1975), pp. 15-22

42 BARAK, M.: British Patent 839979, 1956 (The Chloride Electrical Storage Co., Ltd.)

43 FREY, D.A., and WEAVER, H.E.: *J. Electrochem. Soc.*, 1960, 107, p. 930

44 RUETSCHI, P., and ANGSTADT, R.T.: 'Self-discharge reactions in lead-acid batteries', *ibid*., 1958, **105**, 10, pp. 555-563

45 KAMAYAMA, N., and FUKUMOTO, T.J.: *J. Chem. Soc. Ind. Japan,* 1946, 49, p. 155

46 ZASLAVSKI, A.I., KONDRASHOV, Y.D., and TOLKACHEV: *Dokl. Akad. Nauk, SSSR,* 1950, 75, p. 559

47 BODE, H., and VOSS, E.: *Zeitschr. Elektrochem.*, 1956, **60**, p. 1053

48 BURBANK, J.: *J. Electrochem. Soc.,* 1959, **106**, p. 369

49 BURBANK, J.: 'Anodic oxidation of $PbSO_4$ etc.'. *Power sources 1966* (Pergamon Press, London, 1966), pp. 147-161

50 RUETSCHI, P., and CAHAN, B.D.: *J. Electrochem. Soc.,* 1958, **60**, p. 369

51 VOSS, E., and FREUNDLICH, J.: 'Discharge capacities of alpha and beta PbO_2 electrodes'. *In* COLLINS, D.H. (Ed.) *Batteries* (Pergamon Press, Oxford, 1963), pp. 73-87

52 BAGSHAW, N.E., and WILSON, K.P.: 'Microscopic examination of active materials from lead-acid cells', *Electrochim. Acta*, 1965, 10, pp. 867-873

53 HATTORI, S., YAMAURA, M., KOHNO, M., OHTANI, Y., YAMANE, M., and NAKASHIMA, H.: 'Periodic observation of the same pin-point of battery plates with the S.E.M.'. *In* COLLINS, D.H. (Ed.) *Power sources 5* (Academic Press Inc., 1975), pp. 139-153

54 VETTER, K.J.: 'Bedeutung der Loslichkeit von Elektrodenmaterialen fur die Kinetik poroser Elektroden', *Chem. -Ing. Tech.,* 1973, **45**, 4, pp. 231-236

55 FEITKNECHT, W., and GAUMANN, A.: *J. Chem Phys.,* 1952, **49,** C135

56 THIRSK, H.R., and WYNNE JONES, W.F.K.: *ibid.,* 1952, **49,** C131

57 KABANOV, B.N., LEIKIS, D.I., and KREPAKOVA, E.I.: *Dokl. Akad. Nauk. SSSR,* 1954, **98**, p. 989

58 SIMON, A.C., and CAULDER, S.M.: 'Recent developments in the NRL-ILZRO investigation of the porous $PbO_2/PbSO_4$ electrode'. *In* COLLINS, D.H. (Ed.) *Power sources 5* (Academic Press Inc., 1975), pp. 109-122

59 DAWSON, J.L., RANA, M.E., MUNASIRI, B., and McWHINNIE, J.M.:

'Morphological and kinetic studies on lead dioxide electrodes'. *In* THOMPSON, J. (Ed.) *Power sources 7* (Academic Press Inc., 1979), pp. 1-15

60 BERNDT, D., and VOSS, E.: 'The voltage characteristics of a lead-acid cell during charge and discharge'. *In* COLLINS, D.H. (Ed.) *Batteries 2* (Pergamon Press Ltd., Oxford, 1965), pp. 17-27

61 WADE, E.J.: *Secondary batteries. their construction and use* (The Electrical Printing and Publishing Co., Ltd., London, 1902), p. 492

62 FABER, P.: 'The use of titanium in lead-acid batteries'. *In* COLLINS, D.H. (Ed.) *Power sources 4* (Oriel Press Ltd., England, 1973), pp. 525-540

63 BAGSHAW, N.E., BROMELOW' K.P., and EATON, J.: 'The effect of grid conductivity on the performance of tall lead-acid cells'. *In* COLLINS, D.H. (Ed.) *Power sources 6* (Academic Press Inc., 1977), pp. 1-9

64 BRINKMANN, J.: 'Performance of double-decker cells', *ibid.*, pp. 9-14

65 Hazelett Storage Battery Co.: British Patent 185148, 1921

66 HELMS, J.H., COYNER, J.H., and HILL, C.W.: 'The Delco-Remy freedom battery'. US Soc. Automotive Engrs., Report No. 770325, 1977, pp. 1-6

67 RUETSCHI, P., and ANGSTADT, R.T.: *J. Electrochem. Soc.,* 1958, **105**, p. 555

68 CRENNELL, J.T., and MILLIGAN, A.G.: 'The use of antimonial lead for accumulator grids; a cause of self-discharge of negative plates', *Trans. Faraday Soc.,* 1931, , **27**, p. 103

69 VINAL, G.W., CRAIG, D.N., and SYNDER, C.L.: 'Composition of grids for positive plates as a factor influencing the sulphation of negative plates', *J. Res. Natl. Bur. Standards,* 1933, **10**, p. 795

70 DAWSON, J.L., GILLIBRAND, M.I.G., and WILKINSON, J.: 'The chemical role of antimony in the lead-acid battery'. *In* COLLINS, D.H. (Ed.) *Power sources 3* (Oriel Press Ltd., England, 1971), pp. 1-11

71 HOLLAND, R.: 'The evolution of stibine from lead-acid batteries'. Proceedings of the International symposium on batteries, 1958, p. 10, Signals Res. and Development Establishment, Ministry of Supply, UK

72 PERKINS, J., and EDWARDS, G.R.: 'Review of the micro-structural control in lead alloys for storage battery applications', *J. Mater. Sci.,* 1975, **10**, pp. 136-150

73 MAO, G.W., and LARSON, J.G.: 'Effect of arsenic additions on the characteristics of antimony lead battery alloys', *Metallurgia*, 1968, **76**, pp. 236-245

74 WHITE, J.W.R., and ROGERS, P.: 'A practical castability test for battery grid alloys'. Lead 74, Proceedings of the 5th International conference on lead, 1974, pp. 67-77, Lead Development Association, London, UK

75 TORRALBA, M.: 'Present trends in lead alloys for the manufacture of battery grids', *J. Power Sources*, 1976/77, 1, pp. 301-310

76 HEUBNER, U., and SANDIG, H.: 'Lead alloys for grids of SLI and sealed batteries'. Proceedings of the fourth International conference on lead 1971, pp. 29-36, Lead Development Association, London, UK

77 *Lead-acid batteries - a reference and data book* (compiled by the Indian Lead/Zinc Information Centre, Elsevier Sequoia S.A. Lausanne, Switzerland 1977), p. 143

78 DASOYAN, M.A.: 'Der Einfluss des Struktur des Bleis auf seine Korrosion in Swefelsaure', *Dokl. Akad. Nauk, SSSR,* 1956, **107**, pp. 863-866

79 MAO, G.W., and RAO, P.: *Brit. Corrosion J.*, 1971, **6**, p. 122
80 WATERHOUSE, H., and WILLOWS, R.: British Patent 622 512, 1949
81 NOLAN, K.W., HIRSCH, J.L., and POPE, D.M.: 'Seven-year trials with railway batteries using low antimony grids'. Lead 74, Proceedings of the fifth International conference on lead, 1974, pp. 79-87, Lead Development Association London, UK
82 BORCHERS, H., NIJHAWAN, N.S.C., and SCHARFENBERGER, W.: 'Entwicklung Antimonarmer Bleilegierungen fur Starterbatteriegitter' *Metall.*, 1974, **28**, pp. 863-867
83 BORCHERS, H., and NIJHAWAN, S.C.: 'Uber das Aushartungsverhalten von Antimonarmen Bleilegierungen', *ibid.*, 1975, **29**, pp. 465-471
84 Compagnie Europeene d'Accumulateurs: German Patent 512049, 1974
85 SUNDBERG, E.: 'New developments in tubular lead-acid batteries'. Lead 65, Proceedings of the second International conference on lead, 1965, pp. 227-234, Lead Development Association, London, UK
86 SUNDBERG, E.: 'Lead technology in battery making - what next?'. Lead 74, Proceedings of the fifth International conference on lead, 1974, pp. 105-112, Lead Development Association, London, UK
87 HEUBNER, U., and UEBERSCHAER, A.: 'Castability of lead-antimony alloys for battery grids, with particular reference to nucleation techniques', *ibid.*, pp. 59-66
88 ROBERTS, D.H., RATCLIFFE, N.A., and HUGHES, J.E.: *Powder Metallurgy,* 1962, **10**, p. 132
89 BAGSAHW, N.E., and HUGHES, T.A.: 'The use of dispersion-strengthened lead in positive grids in the lead-acid battery'. *In* COLLINS, D.H. (Ed.) *Batteries 2* (Pergamon Press Ltd., Oxford, 1965), pp. 1-15
90 BIRD, T.L., DUGDALE, I., and GRAVER, G.G.: 'Construction and service testing of cells made with grids of dispersion strengthened lead'. Lead 68. Proceedings of the third International conference on lead, 1968, pp. 221-225, Lead Development Association, London, UK
91 Chloride Electrical Storage Co., Ltd., and BARAK, M.: British Patent 689003, 1951
92 COTTON, J.B., and DUGDALE, I.: 'A survey of the possible uses of titanium in batteries'. *In* COLLINS' D.H. (Ed.) *Batteries* (Pergamon Press 1963), pp. 297-307
92*a* SLADIN, H.A., and The Tudor Accumulator Co.: British Patent 691 712, 1951
92*b* ROBSON, R.M., and The D.P. Battery Co.: British Patent 719 598, 1952
93 BOOTH, F.M.: 'The development of a low-loss plastic grid lead-acid cell'. Proceedings of the first International symposium on batteries, 1958, p. 6, Signals Research and Development Establishment, Ministry of Supply, UK
93*a* FABER, P.: 'A new metal/plastic compound electrode for traction batteries'. *In* THOMPSON, J. (Ed.) *Power sources 7* (Academic Press Inc., London, 1979), pp. 79-103
94 THOMAS, R.: 'The developmnt of a low-loss light-weight communications battery'. Lead 68, Proceedings of the third International conference on lead, 1968, pp. 169-185 Lead Development Association, London, UK
95 SCHUMACHER, E.E., and BOUTON, G.M.: *Metals and Alloys,* 1930, **1**, p. 405
96 HARING, H.E., and THOMAS, U.B.: *Trans. Electrochem. Soc.,* 1935,

68, p. 293

97 ROSE, M.V., and YOUNG, J.A.: 'Lead-calcium (- tin) alloys properties and prospects'. Lead 74, 1974, pp. 37-52. Lead Development Association, London, UK

98 WALSH, S. and The Chloride Electrical Storage Company Ltd: British Patent 712 798, 1952

99 BOUTON, G.M., and PHIPPS, G.S.: Paper presented at the 92nd General Meeting of the Electrochemical Society, Boston, October, 1947

100 'The St. Joe Ternary Bloom Test', 1976, Technical Leaflet from St Joe Minerals Corporation, New York City, USA

101 WEINLEIN, C.E., PIERSON, J.R., and MARSHALL, D.: 'A new grid alloy for high performance maintenance-free lead acid batteries'. *In* THOMPSON, J. (Ed.) *Power sources 7* (Academic press Inc., London, 1979), pp. 67-77

102 BAGSHAW, N.E.: 'Lead-strontium alloys for battery grids', *J. Power Sources*, 1978, **2**, 4, pp. 337-350

103 HAWORTH, D.J.., PETERS, K., and THROW, J.S.: 'Gassing rates of automotive batteries'. *In* COLLINS, D.H. (Ed.) *Power sources 5* (Academic Press Inc., London, 1975), pp. 123-137

104 COUCH, W.M.C., CRENNELL, J.T., MILLIGAN, A.G., WATERHOUSE, H., and MacLAREN, H.: British Patent 613308, 1946

105 BAGSHAW, N.E.: 'Lead-antimony-cadmium alloys for battery grids'. Lead 68, Proceedings of the third International conference on lead, 1968, pp. 209-225, Lead Development Association, London, UK

106 ABDUL AZIM, A.A., EL-SOBKI, K.M., and KEDR, A.A.: 'The effect of some alloying elements on the corrosion resistance of lead antimony alloys: Pt. 1. Cadmium', *Corrosion Science*, 1976, **16**, 4, pp. 209-218

107 SABATINO, A.: 'Maintenance-free batteries: ideas for the 80's' *In* 'Lead and zinc into the 80's'. Proceedings of an International conference, 1977, pp. 5-11, Lead/Zinc Development Association, London, UK

108 WILLIAMS, W.B.: 'The development of grid casting machines and oxide mills'. Lead 65, Proceedings of the second International conference on lead, 1965, pp. 211-218 Lead Development Association, London, UK

109 SHIMADZU, G.: *Japanese lead-acid battery industry and the present status of technology* (US Office of the Electrochemical Society of Japan, Ohio), chap. 2

110 BRACHET, M.: 'Six years operation of a high lead content litharge reactor'. Lead 74, proceedings of the fifth International conference on lead, 1974

111 ALLEN, T.: *Particle size measurement* (Chapman & Hall Ltd., London, 1975, 2nd edn.), p. 454

112 BRUNAUER, S., EMMETT, P.H., and TELLER, E.: *J. Amer. Chem. Soc.*, 1938, **60**, p. 309

113 BODE, H., and VOSS, E.: *Electrochim. Acta,* 1959, **1**, p. 318

114 BARNES, S.C., and MATHIESON, R.T.: 'The potential-pH diagram of lead in the presence of sulphate ions and some of its implications in lead-acid battery studies'. *In* COLLINS, D.H. (Ed.) *Batteries 2* (Pergamon Press Ltd. Oxford, 1965), pp. 41-45

115 LANDER, J.J.: *J. Electrochem. Soc.*, 1949, **95**, pp. 174-186

116 MRGUDICH, J.N.: *Trans. Electrochem. Soc.,* 1942, **81**, pp. 165-173

117 Chloride Electrical Storage Co., Ltd., and BARAK, M.: British Patent

638 967, 1948
118 WILLIHNGANZ, E.: *J. Electrochem. Soc.,* 1947, **92**, pp. 281-284
119 BURBANK, J.: 'Morphology of PbO_2 in the positive plates of lead acid cells', *ibid.,* 1964, **111**, 7, pp. 765-768
120 BURBANK, J.: 'The role of antimony in positive plate behaviour in the lead-acid cells', *ibid.*, 1964, **111**, 10, pp. 1112-1116
120*a* SHARPE, T.F.: 'The behaviour of lead alloys as PbO_2 electrodes', *ibid.*, 1977, **124**, 2, pp. 168-173
120*b* RUETSCHI, P.: 'Ion selectivity and diffusion potentials in corrosion layers', *J. Electrochem. Soc.,* 1973, **120**, 3, pp. 331-336
120*c* HAMPSON, N.A., KELLEY, S., PETERS, K., and WHYATT, P.: 'Morphological examination of the effect of antimony on the electrochemistry of lead', *J. Appl. Electrochem.*, 1979
121 ARMSTRONG, J., DUGDALE, I., and McCUSKER, W.J.: 'Phase changes during the manufacture of lead-acid battery plates'. *In* COLLINS, D.H. (Ed.) *Power sources 1966* (Pergamon Press Ltd., Oxford, 1967), pp. 163-177
122 PIERSON, J.R.: 'A study of some of the crystallographic and microscopic aspects of the curing of positive lead-acid battery plates'. *In* COLLINS, D.H. (Ed.) *Power sources 2* (Pergamon Press Ltd., Oxford, 1970), pp. 103-119
123*a* HUMPHREYS, M.E.D., TAYLOR, R., and BARNES, S.C.: 'The curing of lead-acid battery plates', *ibid.*, pp. 55-67
123 HATTORI, S., YAMAURA, M., KONO, M., YAMANE, M., NAKASHIMA, H., and YAMASHITA, D.: Yuasa Battery Co., Ltd., Japan, 'Antimony-free grids for deep discharge', ILZRO Project LE-276, Progress Report No. 2, 1978
124 PETERS, K., Chloride Technical Ltd.: Private communication, 1978
125 BIAGETTI, R.V., and WEEKS, M.C.: 'Tetrabasic lead sulphate as a paste material for positive plates', *Bell Syst. Tech. J.,* 1970, **49**, 7, pp. 1305-1319
126 YARNELL, C.F.: 'Tetrabasic lead sulphate as a negative paste material in the lead-acid battery'. Extended abstract 48, *Society of Electrochem.* Soc. Meeting, Oct. 1971
127 Boliden Batteri A.B. (Sweden): British Patent 875 974, 1960
128 Chloride Electrical Storage Co., Ltd., and BARAK, M.: British Patent 863 048, 1958
129 Chloride Batteries Ltd., and HALL, C.J.: British Patent 871 348
130 Electric Storage Battery Co., and HOWELL' W.H.: British Patent 808 640, 1956
131 BODE, H.: *Lead-acid batteries.* Translated by BRODD, R.J., and KORDESCH, K.V. (John Wiley & Sons Inc., USA, 1977), p. 278
132 HALSALL, V.M., and PIERSON, J.R.: 'Plate processing - the heart of the lead-acid battery'. Proceedings of the 84th Annual Convention of Battery Council International, June 6-8, 1972, pp. 47-59, BCI, Burlinghame, USA
133 HALSALL, V.M., and SABATINO, A.: 'Plastic cases up-grade lead-acid batteries', *J. Soc. Automotive Engrs.,* 1969, pp. 53-56
134 BARAK, M.: 'Batteries and fuel cells', *IEE Reviews*, 1970, **117**, pp. 1561-1582
135 KOZAWA, A., and TAKAGAKI, T.: *Japanese lead-acid battery industry and the present status of technology* (US Office of the Electrochemical

Society of Japan, Ohio, 1977), p. 57
136 SMITH, G.: *Storage batteries* (Sir Isaac Pitman & Sons Ltd., London, 1964), p. 211
137 Lucas Electrical Ltd.: Alternators (Lucas Technical Series Publn. PLT 6174, Birmingham, UK, 1977), p. 31
138 *Battery truck book* (Lead Development Association, London, 1974), p. 44
139 *Battery-powered electric work vehicles* (Lead Industries Association, Inc., New York, 1975), p. 15
140 *EVA Handbook 1978* (Electric Vehicle Association of Great Britain Ltd., 1978)
141 *Electric Vehicle News,* Westport, Conn., USA
142 *Electric Vehicle Developments,* Institution of Electrical Engineers, London, UK
143 BARAK, M.: 'Fuel economy through battery power', *Elect. Rev.,* 1975, **196**, 17, pp. 535-539
144 GALLOT, J.: 'Latest developments in sponsored test-programme for electric vehicles in France'. Proceedings of International Conference on Electric Vehicle Development (Peter Peregrinus Ltd., 1977), pp. 79-84
145 'Battery-powered buses in Japan today'. Lead 74, proceedings of fifth International conference on lead, 1976, pp. 9-22, Lead Development Association, London, UK
146 MULLER' H.G.: 'Road vehicles with combined, at least partly, electrical driving systems and energy supplies'. Proceedings of International conference on Electric Vehicle Development. (Peter Peregrinus Ltd., 1977), pp. 48-52
147 JACOB, H.: 'Battery-driven railcars'. Lead 65, proceedings of the second International conference on lead, (Pergamon Press Ltd., Oxford, 1967), pp. 183-192
148 Chloride Electrical Storage Co., Ltd., and BARAK, M., PETERS, K., and JACKSON, F.: British Patent 947 796, 1961
149 BUSHROD, C.J.: 'Tubular lead-acid batteries in motive power service'. Lead 68, proceedings of the third International conference on lead (Pergamon Press Ltd., Oxford 1969), pp. 187-190
150 'Lead-acid storage batteries'. Type review 10 100e, Varta Batterie AG., p. 36, Hannover, W. Germany
151 *Understanding motive power batteries.* (Chloride Industrial Batteries Ltd., Manchester, 1972), publn. M685, p. 28
152 TAKAGAKI, T., ANDO, K., and YONESU, K.: 'Improving lead battery performance'. Lead 74, proceedings of the fifth International conference on lead, 1976, pp. 113-122, Lead Development Association, London
153 SUNDBERG, E, E.: 'New developments in tubular lead-acid batteries'. Lead 65, proceedings of the second International conference on lead (Pergamon Press Ltd., Oxford, 1967) pp.. 227-234
154 KOZAWA, A., and TAKAGAKI, T.: *Japanese lead-acid battery industry* (US Office of the Electrochemical Society of Japan, 1977), chap. 2, p.124
155 *Spegel automatic charger for motive power batteries.* Leaflet SP/3M 278 Chloride Legg Ltd., Wolverhampton, UK, 1978, p. 5
156 AUSDERAU, A.: 'Stationary lead batteries in Switzerland'. Lead 71, proceedings of the fourth Intenrational conference on lead, 1971, pp. 12-

14, Lead Development Association, London, UK

157 EDWARDS, P.J., and BAXTER, B.W.: 'Use of Planté cells by the British Post Office', *ibid.*, pp. 1-7

158 KOONTZ, D.E., FEDER, D.O., BABUSCI, L.D., and LUER, H.J.: 'Reserve batteries for Bell system's use. Design of the new cell', *Bell Syst. Tech. J.*, 1970, **49**, 7, pp. 1253 *et seq*

159 SZPER, A.J.: 'Lead-acid batteries, built on the supported active material principle'. Proceedings of the International symposium on batteries, 1958, paper x, p. 10, Signals Research and Development Establishment, Ministry of Supply, Christchurch, Hants., UK

160 SMYTH, J.R., MALLOY, J., and FERRELL, D.T.: 'Maintenance-free lead storage batteries'. Lead 65, proceedings of the International conference on lead (Pergamon Press Ltd., Oxford, 1967), pp. 193-197

161 EBERTS, K., and JACHE, O.: 'Miniature sealed lead-acid batteries and their automatic charging', *ibid.*, pp. 199-210

161*a* EBERTS, K.:'Specific properties of small closed lead accumulators, using an immobile electrolyte'. *In* COLLINS, D.H. (Ed.) *Power sources 2* (Pergamon Press Ltd., Oxford, 1970), pp. 69-92

162 WALLINDER, J.A.: *Lead-acid batteries: the use of catalators.* (Lead Development Association, London, 1965), pp. 12-16

163 DYSON, J.I., and SUNDBERG, E.: 'Aspects of catalytic re-combination: approach of sealed operation of lead-acid cells'. *In* COLLINS, D.H. (Ed.) *Power sources 4* (Oriel Press, UK, 1973), pp. 505-523

164 WATANABE, M., and YONESU, K.: 'Development and present status of the catalyst plug for the stationary lead-acid battery in Japan'. Technical Leaflet, 1978, Japan Storage Battery Co., Ltd., Kyoto, Japan

165 RUETSCHI, P., and OCKERMAN, J.B.: *J. Electrochem. Soc.,* 1969, **116**, p. 1222

166 Bureau Technique Gautrat; British Patent 677 770, 1952

167 NEUMANN, G., and GOTTESMANN, U.: US Patent 2 571 927, 1951

168 JEANNIN, R.A.: US Patent 2 646 455, 1953

169 FALK, U., and SALKIND, A.J.: *Alkaline storage batteries* (John Wiley & Sons Inc., New York, 1969), pp. 192-197

170 MAHATO, B.K., WEISSMAN, E.Y., and LAIRD, E.C.: *J. Electrochem. Soc.,* 1974, **121**, p. 13

171 MAHATO, B.K., and LAIRD, E.C.: 'Gas recombination lead-acid batteries'. *In* COLLINS, D.H. (Ed.) *Power sources 5* (Academic Press Inc. (London) Ltd., 1975), pp. 23-41

172 ATKIN, J., BONNATERRE, R., and LAURENT, J.F.: 'Sealed nickel-cadmium and lead-acid batteries: comparison of functioning mechanisms'. *In* COLLINS, D.H. (Ed.) *Power sources 6* (Academic Press Inc. (London) Ltd., 1977), pp. 91-101

173 McCLELLAND, D.H.: US Patent 3 862 461, 1976

174 BULLOCK, K.R., and McCLELLAND, D.H.: *J. Electrochem. Soc.,* 1976, **123**, p. 327

175 'Energy cells'. Technical Leaflet 20M3/77, 1977, Gates Energy Products, Denver, USA

176 'Chloride Cyclon range of rechargeable sealed lead-acid batteries', Technical Leaflet 6100/1, 1977, Chloride Power Ltd., London, UK

177 'Battery application manual', Booklet 20M6/77, 1977. Gates Energy

Products, Denver, USA, p. 74
178 HAMEL, R.: 'NiCd and lead-acid D-size cells meet in the rechargeable area', *Product Engineering*, Nov. 1975
179 HARRISON, M.: 'Sealed recombining lead-acid batteries', *Electron*, January 1978
180 HARRISON, A.I., and PETERS, K.: 'Batteries for cordless power equipment'. *In* COLLINS, D.H. (Ed.) *Power sources 3* (Oriel Press Ltd., Newcastle upon Tyne, 1971), pp. 211-225
181 HAMMEL, R.: 'Charging sealed lead-acid batteries'. 27th Annual proceedings power sources conference, USA, June 1976
182 ROBINSON, R.G., and HEINSON, J.T.: 'The design and development of lead-acid aircraft batteries'. *In* COLLINS, D.H. (Ed.) *Batteries 2* (Pergamon Press Ltd., Oxford, 1965), pp. 401-418
183 BAGSHAW, N.E., BROMELOW, K.P., and KIRKPATRICK, J.: 'Modern lead-acid battery designs for aircraft'. *In* COLLINS, D.H. (Ed.) *Power sources 6* (Academic Press Inc. (London) Ltd., 1977), pp. 25-34
184 EARWICKER, G.A.: 'Aircraft batteries and their behaviour on constant-potential charge', *Proc. IEE*, 1956, **103A**, Suppl. No. 1, pp. 180-191

Chapter 5

Alkaline storage batteries

U. Falk

5.1 Introduction

Alkaline storage batteries may be defined as electrically rechargeable batteries using an alkaline electrolyte generally consisting of a solution of potassium hydroxide.

The advantages of an alkaline electrolyte instead of an acid in a storage battery were first perceived by the Swedish inventor Waldemar Jungner in the early 1890s. He realised that using an alkaline electrolyte would make it possible to charge and discharge electrodes under a simple transport of oxygen or hydroxyl ion from one electrode to the other without changing the composition or bulk density of the electrolyte. This would mean that a smaller amount of electrolyte could be used, that the risks for freezing of the electrolyte would be diminished, and, furthermore, that metals could be employed which would be completely inert in the electrolyte. Jungner stated that accumulators based on alkaline electrolytes would be light and mechanically strong and that they would have low stand losses on open circuit.

T.A. Edison worked along similar lines in the US and around the turn of the century Jungner and Edison started manufacture of the first alkaline storage batteries. In both cases the nickel/iron system was chosen. Edison continued with this system for many decades, whereas Jungner soon went over to nickel/cadmium. Further details of the historical background of alkaline storage batteries are given in Chapter 1 and by Falk and Salkind.[1] To the family of alkaline accumulators belong: nickel/cadmium batteries of various designs, nickel/iron batteries, silver/zinc and silver/cadmium batteries and nickel/zinc batteries. All these battery types will be dealt with in some detail on the following pages. A few other systems might also be considered as members of the family. Alkaline manganese/zinc, mercury/zinc, and mercury/cadmium, although normally primary batteries, have also been tried in rechargeable versions. These systems have, however, a limited reversibility, and as they are of little practical importance they will not be discussed further in this chapter. The nickel/hydrogen system, which is highly rechargeable and which has an alkaline electrolyte, is more related to hybrid systems such as metal/air cells, which are not dealt with in this volume.

The worldwide manufacture of alkaline storage batteries is estimated by the author to approximately £400 million for 1978. In Table 5.1, which shows the production of alkaline batteries in different regions, the batteries have arbitrarily been grouped in 'industrial batteries' and 'consumer batteries'. The industrial category contains pocket type nickel/cadmium batteries, vented sintered plate nickel/cadmium batteries and the main part of silver/zinc, silver/cadmium and nickel/zinc batteries. The consumer category contains all sealed nickel/cadmium batteries. Speciality batteries made for military use have not been considered in the table.

About half of the total world production comes from Eastern Europe. The USSR is by far the biggest producer of industrial alkaline batteries in the world. In the USSR the nickel/iron system has been specifically chosen for a number of important uses such as propulsion of industrial trucks and mining locomotives, train lighting, and diesel engine cranking.[2] Also, it should be noted that alkaline batteries are used much more for industrial purposes in Western Europe than in

Table 5.1 *Estimated world alkaline battery production 1978*

Region	Production in £ million		
	industrial batteries	consumer batteries	total
Western Europe	70	25	95
Eastern Europe	180	10	190
USA, Canada	20	40	60
Latin America	7	-	7
Japan	12	28	40
Asia (except Japan), Australia, Africa	1	2	3
	290	105	395

Table 5.2 *Battery production 1978 in some important countries: estimated values in £ millions*

Country	Nickel/cadmium pocket type	Nickel/cadmium vented sintered plate	Nickel/cadmium sealed	Nickel/iron	Silver/zinc Silver/cadmium Nickel/zinc	Total
United Kingdom	12	-	6	-	-*	18
France	10	5	11	3	1	30
Germany (BRD)	16	2	8	1	1	28
Sweden	17	-	-	-	-	17
U S A	5	11	40	-	4	60
Japan	6	4	28	-	2	40
	66	22	93	4	8	193

*Some production

America. This has an historical background and is connected with the relatively late introduction of pocket type nickel/cadmium batteries to the USA. On the other hand, the manufacture of alkaline consumer batteries is more extensive in the USA than in any other country. Japan has a similar general situation to the USA: the Japanese production of consumer batteries doubles that of industrial batteries. Table 5.2 shows the estimated 1978 production of different battery systems in certain important countries.

Some basic technical data for the alkaline systems mentioned are given in Table 5.3. In the table the theoretical energy density data listed are entirely based on the EMF values and on the simplified cell reactions given. Also, for the silver/zinc and silver/cadmium systems energy density and EMF values are given both for the reactions divalent silver oxide → metallic silver and monovalent silver oxide → metallic silver. For silver cells starting from the divalent stage an average of the two EMF values has been used to calculate the energy density.

The solid materials employed in alkaline cells are either inert materials, usually in the form of elements, or electrochemically active materials in the form of metals or compounds, in the latter case metal oxides or hydroxides. As indicated in Table 5.3, certain materials are used in more than one system. In Table 5.4 some physical and electrochemical data are given for elements and compounds commonly used in alkaline batteries. Beside these conductive or active materials, plastics are also employed in alkaline batteries for a variety of purposes, e.g. in cell containers and lids, in gaskets, in separators and in vents.

Another component of great importance is the electrolyte. As previously mentioned, the electrolyte of alkaline cells generally consists of a solution of potassium hydroxide, although sodium hydroxide is used in a few cases. For obvious reasons the electrolyte should have high electrolytic conductance, low freezing point, low reactivity to the electrode materials on open circuit and low cost. Potassium hydroxide solutions of suitable concentrations meet these requirements better than any other alkaline solutions. In Table 5.5 some important properties of potassium hydroxide solutions are listed. In alkaline cells the electrolyte concentration normally varies from 19% to 45% depending on the electrochemical system. The effects of additives and impurities in the electrolyte on the electrochemical behaviour will be separately discussed for each system.

Table 5.3 *Alkaline battery systems*

System	Symbols	EMF	Theoretical energy density	Cell reactions used for energy density calculations
		V	Wh/kg	
Nickel/cadmium	Ni/Cd	1·290	209	$2\,NiOOH + Cd + 2H_2O \rightleftharpoons 2\,Ni(OH)_2 + Cd(OH)_2$
Nickel/iron	Ni/Fe	1·370	267	$2\,NiOOH + Fe + 2H_2O \rightleftharpoons 2\,Ni(OH)_2 + Fe(OH)_2$
Silver/zinc	Ag/Zn	1·856	434	$Ag_2O_2 + 2\,Zn + 2H_2O \rightleftharpoons 2\,Ag + 2\,Zn(OH)_2$
		1·602	273	$Ag_2O + Zn + H_2O \rightleftharpoons 2\,Ag + Zn(OH)_2$
Silver/cadmium	Ag/Cd	1·380	267	$Ag_2O_2 + 2\,Cd + 2H_2O \rightleftharpoons 2\,Ag + 2\,Cd(OH)_2$
		1·155	171	$Ag_2O + Cd + H_2O \rightleftharpoons 2\,Ag + Cd(OH)_2$
Nickel/zinc	Ni/Zn	1·735	326	$2\,NiOOH + Zn + 2H_2O \rightleftharpoons 2\,Ni(OH)_2 + Zn(OH)_2$

Table 5.4 *Elements and compounds used in alkaline storage batteries*

Material	Symbol	Molecular weight	Density	Electrical resistivity	Ampere-hours per gram	Valence change
			g/cm^3	$\mu\Omega$ cm		
Cadmium	Cd	112·41	8·64	7·63	0·477	2
Cadmium hydroxide	$Cd(OH)_2$	146·41	4·79	-	0·366	2
Cadmium oxide	CdO	128·40	8·15, 6·95	-	0·417	2
Iron	Fe	55·85	7·86	~10	0·960	2
Iron (II) hydroxide	$Fe(OH)_2$	89·86	3·4	-	0·597	2
Iron oxide	Fe_3O_4	231·54	5·18	-	0·696	2
Zinc	Zn	65·38	7·14	6·0	0·820	2
Zinc hydroxide	$Zn(OH)_2$	99·38	3·05	-	0·539	2
Zinc oxide	ZnO	81·37	5·61	10^{10}-10^{14}	0·659	2
Nickel	Ni	58·71	8·90	7·0	0·457	1
Nickel (II) hydroxide	$Ni(OH)_2$	92·72	4·15	-	0·289	1
Nickel (III) hydroxide	β-NiOOH	91·71	4·6	-	0·292	1
Silver	Ag	107·87	10·5	1·62	0·248	1
					0·497	2
Silver (I) oxide	Ag_2O	231·74	7·14	-	0·232	1
Silver (II) oxide	Ag_2O_2	247·74	7·44	-	0·432	2
Graphite	C	12·01	2·25	120	-	-

Table 5.5 *Properties of potassium hydroxide solutions*

KOH concentration							
Weight	Molality	Molarity at 20°C	Density d_4^{20}	g/l at 20°C	Specific conductance at 20°C	Freezing point	Viscosity at 20°C
%					Ω^{-1} cm^{-1}	°C	centipoise
4	0·7427	0·7372	1·034	41·36	0·15	-2	1·08
8	1·550	1·527	1·071	85·68	0·28	-6	1·18
12	2·431	2·373	1·109	133·1	0·38	-11	1·29
16	3·395	3·271	1·147	183·5	0·47	-17	1·44
20	4·456	4·229	1·186	237·2	0·54	-24	1·62
24	5·629	5·246	1·226	294·2	0·60	-35	1·86
28	6·932	6·326	1·267	354·8	0·62	-49	2·17
32	8·388	7·469	1·309	418·9	0·60	-61	2·57
36	10·03	8·678	1·352	486·7	0·56	-46	3·11
40	11·88	9·911	1·396	558·4	0·52	-35	3·88
44	14·00	11·31	1·442	634·5	-	-29	5·11
48	16·45	12·73	1·488	714·2	-	-4	6·73

5.2 Nickel/cadmium pocket type batteries

The pocket type is the oldest of the various nickel/cadmium designs. This battery has been manufactured on a commercial scale since 1909 on the basis of Jungner's fundamental investigations around the turn of the century and on work by A. Estelle and others on design and manufacturing processes. At the present time pocket type nickel/cadmium batteries are produced in the United Kingdom, Sweden, West Germany, France and several other European countries as well as in the USA, Latin America and Japan.

5.2.1 Reaction mechanisms

The basic reaction mechanisms are the same for all types of nickel/cadmium cells. The most common overall cell reaction is the following:

$$2\,NiOOH + Cd + 2H_2O \underset{\text{charge}}{\overset{\text{discharge}}{\rightleftarrows}} 2\,Ni(OH)_2 + Cd(OH)_2 \tag{5.1}$$

This equation indicates that, on discharge, trivalent nickel hydroxide reacts with metallic cadmium and water to form divalent nickel hydroxide and cadmium hydroxide. On charge the opposite reactions take place. Although this equation gives a very simplified picture of what happens in the nickel/cadmium cell during operation, it represents a convenient way to illustrate the main reactions.

A few words should be said about the reactions at the separate electrodes. At the nickel hydroxide electrode the reactions can be represented by the following equation:

$$NiOOH + H_2O + e^- \underset{\text{charge}}{\overset{\text{discharge}}{\rightleftarrows}} Ni(OH)_2 + OH^- \tag{5.2}$$

Trivalent nickel hydroxide is reduced during discharge to divalent nickel hydroxide under consumption of water and formation of hydroxyl ion.

These seemingly simple reactions are, however, complicated by a series of phenomena:

(*a*) there is no simple transformation of an exactly trivalent nickel hydroxide to an exactly divalent one, since the states of oxidation are not precisely fixed to +3 and +2

(*b*) there are at least two modifications of both the charged material (β - NiOOH and γ - NiOOH) and of the discharged material

(α - $Ni(OH)_2$ and β - $Ni(OH)_2$), each being formed under specific circumstances

(*ç*) the oxidised nickel hydroxides may adsorb or trap some potassium hydroxide which is partially desorbed during discharge

(*d*) in both states of charge the hydroxides contain some trapped water, the exact amount of which is not precisely known

(*e*) there is a notable hysteresis between charge and discharge curves.

The mechanism and the rate-controlling step of the charge and discharge reactions of the nickel hydroxide electrode is still a matter of considerable controversy. Although it is recognised that, under some conditions, mass transfer is the rate-controlling step and that the proton is the mobile species, there are also indications that other mechanisms may be dominant under other circumstances.

An excellent review of the extensive nickel electrode literature up to 1972 has been made by Briggs.[3] During the last few years further studies of imporance have been made to elucidate various aspects of the nickel hydroxide electrode.[4-8]

The reactions at the cadmium electrode are somewhat less complicated than those at the nickel electrode. The overall reactions is generally recognised as

$$Cd + 2OH^- \underset{\text{charge}}{\overset{\text{discharge}}{\rightleftarrows}} Cd(OH)_2 + 2e^- \tag{5.3}$$

Metallic cadmium is oxidised on discharge reacting with hydroxyl ions to form cadmium hydroxide. Soluble intermediates play a role in the charge-discharge reactions, the majority species being $Cd(OH)_4{}^{2-}$ or $Cd(OH)_3{}^-$.

Although there are different opinions regarding the details of the reactions, the main mechanism during discharge can be described as follows. First there is a dissolution of active cadmium material as soluble complex ions such as $Cd(OH)_4{}^{2-}$ or $Cd(OH)_3{}^-$:

$$Cd + 3OH^- \rightarrow Cd(OH)_3{}^- + 2e^- \tag{5.4}$$

These complex ions may be precipitated within the electrode as cadmium hydroxide

$$Cd(OH)_3{}^- \rightarrow Cd(OH)_2 + OH^- \tag{5.5}$$

The dissolution–precipitation results in a reduction of the area of the active cadmium metal. The polarisation within the electrode increases until such a potential is reached that the direct growth of a layer of cadmium hydroxide or possibly cadmium oxide on the cadmium metal can take place. This thin layer is believed to be the direct cause of passivation of the electrode.

The reaction product, cadmium hydroxide, exists in two modifications, namely, $\beta - Cd(OH)_2$ and $\gamma - Cd(OH)_2$. The β - form is the normal product of the discharge, but the γ - form may also be present to some extent, especially at low temperatures.

A very good review of the cadmium electrode literature up to 1972 has been presented by Armstrong *et al.*[9] This is recommended for those wanting a deeper discussion. Since 1973, a number of investigations have been carried out which further discuss various aspects of the reactions at the cadmium electrode.[10-15]

5.2.2 Manufacturing processes

Active materials: The electrochemically active materials are manufactured according to processes which are well-known in their main features but often of a proprietary nature when it comes to details of methods and equipment.

The positive active material consists mainly of nickel hydroxide mixed with graphite and some additives such as cobalt or barium compounds for improved capacity and life. The positive active material is generally produced in the following manner.

A nickel sulphate solution, often containing a relatively small amount of cobalt sulphate, is prepared. The nickel solution is mixed with a sodium hydroxide solution and a voluminous precipitate of nickel hydroxide is formed. The nickel hydroxide is separated from the mother liquor by filter pressing or some other kind of separation technique. The precipitate is then often dried and after that washed virtually free of sodium and sulphate ions. The nickel hydroxide is then again dried, ground and mixed with graphite for improved conductivity. The graphite is usually added in two forms: powder graphite and flake graphite. There are many deviations among manufacturers from this simplified scheme. As an example, some manufacturers prefer to add the graphite at the precipitation step.

Typically, the positive active material has the following composition:

nickel hydroxide	80%
cobalt hydroxide	2%
graphite	18%

The negative active material consists of cadmium hydroxide or cadmium oxide mixed with compounds of iron and sometimes nickel. Today this material is generally prepared by a dry-mixing process. Cadmium hydroxide or oxide is mixed with iron sponge or iron oxides, or in some cases with nickel powder, in a suitable blender. The iron and nickel materials are added to prevent agglomeration and crystal growth of cadmium which otherwise will take place during battery operation. Sometimes graphite or paraffin is added for improved mechanical properties at the subsequent briquetting operation.

The negative active material may also be manufactured by electrolytic methods where cadmium and iron (or nickel) are simultaneously co-precipitated from an aqueous sulphate solution. A typical composition of the negative active material may be:

cadmium hydroxide	78%
iron	18%
nickel	1%
graphite	3%

Being the heart of the battery, the active materials must meet stringent specifications with regard to purity, crystal structure and particle size distribution. A large active surface area is necessary for satisfactory electrochemical behaviour.

Manufacture of electrodes: Normally, both positive and negative electrodes are built up of flat 'pockets' made of perforated steel strips incapsulating the active materials.

Steel ribbon with a thickness of about 0·1 mm and a width of 15-30 mm is perforated by an array of fine hardened steel needles which punch small circular holes in the ribbon. This can be done either by punching from one side, or, as in the novel 'double perforation', from both sides simultaneously. It is of essential importance for the high rate performance of the electrode that the superficial porosity of the ribbon is as high as possible and from this point of view the needle perforation, and especially the double perforation, is to be preferred.

Another perforating technique uses profiled roller dies. This technique is more rapid and less noisy than the needle perforation. On the other hand, only about 15% porosity or specific hole area can be obtained with rollers as compared to 23-30% with needles.

The perforated strips for positive electrodes are always nickel plated to prevent anodic oxidation of the iron which may lead to 'iron poisoning' of the positive active material. The active material is either pressed to 125-200 mm long briquettes which are fed into the preshaped perforated steel strip, as shown in Fig. 5.1, or it is continuously fed into the ribbon as a powder. The briquetting technique is

Fig. 5.1 *Briquette of active material being enclosed in perforated steel strips*

slower but will give a more precise amount of active material per unit of length. Different briquette weights are used to produce electrodes of different thicknesses. Most manufacturers use two plate thicknesses, but some use three for high rate, medium rate and low rate cell types. The upper and lower steel strips are folded together by means of rollers and are cut to the specified length (5 to 10 m). A number of these folded strips containing the active material are arranged to interlock with each other forming a long electrode sheet which is then cut to pieces of proper size. These electrode pieces are equipped with steel frames for mechanical stability and for current takeoff. This operation can be carried out manually, but during the last few years automatic machines have been designed which provide the electrode pieces with side frames and lugs and weld the top pocket to the lug for better conductivity. The electrodes are pressed or rolled to secure proper contact between frame and pockets.

Assembly of electrode groups and elements: Electrodes are assembled to electrode groups either by bolting or by welding. Bolted and welded electrode groups are shown in Fig. 5.2. Positive and negative electrode groups are intermeshed to form an element. Plates of opposite polarity must be separated from each other and this is often done by inserting plastic pins between the plates and insulators of U-bent plastic at the plate edges. Another method of separation is to use perforated plastic sheets or plastic 'ladders' between the electrodes. The distance between the plates thus secured by the separator may vary from less than 1 mm for high rate cells to 3 mm for low rate cells. The insertion of the separators has traditionally been carried out manually, but automatic machines have been developed for this operation during the 1970s.

Cell containers, vents: Nickel-plated steel containers have been used for pocket type cells for many years. Such containers are made by bending and welding sheet steel. Lids and bases are punched out from sheet steel. The lids are attached to the container by gas welding. After insertion of the element and an electrochemical forming operation the bases are welded on.

Fig. 5.2 ***Pocket type electrode groups: welded design to the left, bolted to the right***

During the last decade plastic cell containers have been increasingly used. The chemical and mechanical requirements for plastic materials to be used in nickel/cadmium cells are very severe. The strongly alkaline environment, often in combination with rather high temperatures, and the long life of the batteries call for materials which are extremely resistant and which do not change their properties with time. Polyethylene and high impact polystyrene were the first materials to be tried. Polystyrene is still widely used but polypropylene or copolymers of propylene and ethylene seem to be the ones best suited for the purpose at the present time.

Some important advantages of plastic containers over steel containers are that they require no protection against corrosion and that they allow visual control of the electrolyte levels. Furthermore, they are lighter and they can be more closely packed in a battery. The drawbacks are mainly that the plastic containers are more sensitive to high temperatures and that they require somewhat more space than steel containers.

Plastic containers and lids are made by injection moulding. The traditional method is to mould one container and one lid for each cell and, after insertion of the cell element, cement the lid to the container. The most recent design is, however, to use an open polypropylene box which is sealed along the flat side edge by welding on of another container or an end cover. This will give a strong battery block where the possibilities of stress cracking of the joints are eliminated. These alternatives are shown in Fig. 5.3. Another modern design is the monoblock, where two to six cells are housed in the same plastic container unit which is sealed by one common lid.

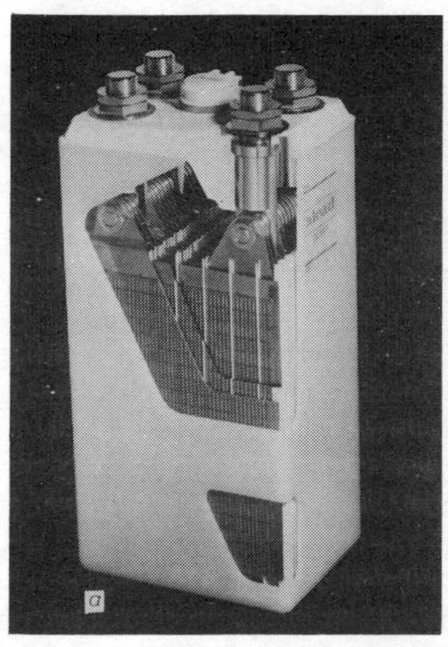

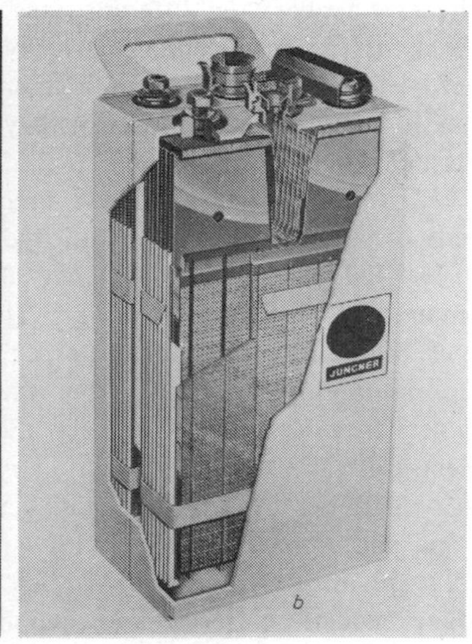

Fig. 5.3 *Nickel/cadmium pocket type cells in plastic containers*
a **Normal cemented design**
b **welded polypropylene design**

Many types of cell vents have been designed and used in the course of the years. Steel cells usually have a neck welded on to the opening in the lid. The neck may be provided with a spring-loaded flip top to which a plastic cone is sometimes attached. The cone seals the filler opening when the flip top is closed. Smaller cell types often have steel screw plugs provided with openings sealed by a rubber ring.

For cells with plastic containers, plastic screw or bayonet vents are generally preferred. During the last few years there has been a trend towards the use of flame arresting vents to prevent a possible outside explosion to be transferred to the interior of the cell. This kind of vent can be seen in Fig. 5.3*b*.

Formation: Before the final assembly of the element in the cell container, the element is given a 'formation' treatment. This is done to remove impurities from the active materials and to activate the masses by a charge-discharge procedure which will increase lattice defects and surface area. Also, sludge escaping from the electrodes during the first few cycles is removed at the formation.

The formation processes may vary in detail among manufacturers but the principles are the same. The elements are soaked in an alkaline electrolyte and are then charged and discharged for a specified number of cycles. Renewal of the formation electrolyte and washing of the elements between the different steps may be included in the operations. Elements for steel cells are usually formed in their steel containers before the base is welded on. Elements for plastics cells may be formed free in large formation tanks or they may be formed assembled in their plastic container.

Electrolyte: The electrolyte used in nickel/cadmium pocket type cells is a solution of potassium hydroxide with a density of 1·18 to 1·23 g/ml. The electrolyte may also contain lithium hydroxide in amounts varying from 15–50 g/1. This addition is made to improve the cycle life of the positive plates, especially at elevated temperatures.

Assembly of batteries: When the cells are mounted in plastic containers there is no need to keep them separated from each other for electrical reasons. Cells for stationary applications are often assembled in to batteries by putting the single cells close together on a rack or stand and connecting them with intercell connectors. Sometimes the cells are assembled in to batteries by using wooden or steel crates. Many modern batteries are assembled by connecting a number of plastic monoblocks. These monoblocks may be put close together. However, in many cases they are mounted spaced out to facilitate cooling.

Cells in steel containers are often assembled in wooden battery crates. Large stationary batteries are usually mounted on racks or stands. The stationary battery may be assembled from battery units in wooden crates which are put on the racks and connected by intertray connectors. However, the normal way is to assemble the battery from individual cells which rest on insulating materials. The cells are kept apart by intercell connectors. Pocket plate cells are normally manufactured to be vented or open, so that gases evolved during charging can escape and water can be added during operation. However, there is also a sealed type of pocket plate cell and this will be briefly discussed in the section dealing with sealed sintered type nickel/cadmium batteries.

5.2.3 Performance characteristics

Nickel/cadmium pocket type cells are available in low rate, medium rate and high rate types. They cover a capacity range from 5 Ah to more than 1200 Ah. There are different designs manufactured to meet the requirements of special applications. Prominent manufacturers of nickel/cadmium pocket type batteries, together with main types of cells supplied, are listed in Table 5.6. Properties of importance for pocket type batteries, as for all secondary batteries, are

(*a*) energy density
(*b*) discharge properties at various rates and temperatures
(*c*) charge characteristics
(*d*) charge retention
(*e*) life
(*f*) mechanical stability.

Energy density: The energy density of a battery is the amount of energy that is available per unit of battery weight or volume. The energy density depends not only on the theoretical electrochemistry of the system (see Table 5.3) but also on a series of practical factors, such as the utilisation of the active materials involved, the internal cell design and the cell arrangement in the battery. Pocket type cells have typical energy densities of 20 (±7) Wh/kg and 40 (±15) Wh/dm^3. Corresponding values for complete batteries are 19 (±8) Wh/kg and 32 (±12) Wh/dm^3. All these figures are based on the nominal capacity of the cells and the average discharge voltage at the normal discharge rate.

Generally speaking, the energy density increases as the high rate properties decrease. Also, batteries assembled from plastic cells have a considerably higher energy density than those assembled from steel cells in wooden crates.

Discharge properties: Discharge rate and temperature are important parameters for the discharge characteristics of all electrochemical cells. Increased rate and low temperature result in lower cell voltage and decreased capacity during discharge. The nickel/cadmium cell is, however, much less influenced by these parameters than some other systems, such as the lead-acid battery. This is, especially with regard to the influence of the rate, a fundamental difference which is connected with the different parts that the electrolytes play in the alkaline and in the acid system. Nickel/cadmium pocket type cells can be effectively discharged at high rates without losing very much of the rated capacity and they can be operated at both high and low temperatures.

Discharge curves at 25°C for high rate and medium rate cells at various constant rates are shown in Fig. 5.4. As can be seen in the upper diagram, high rate pocket type cells can deliver as much as 60% of their capacity even at a discharge current as high as 5 x C, where C is the

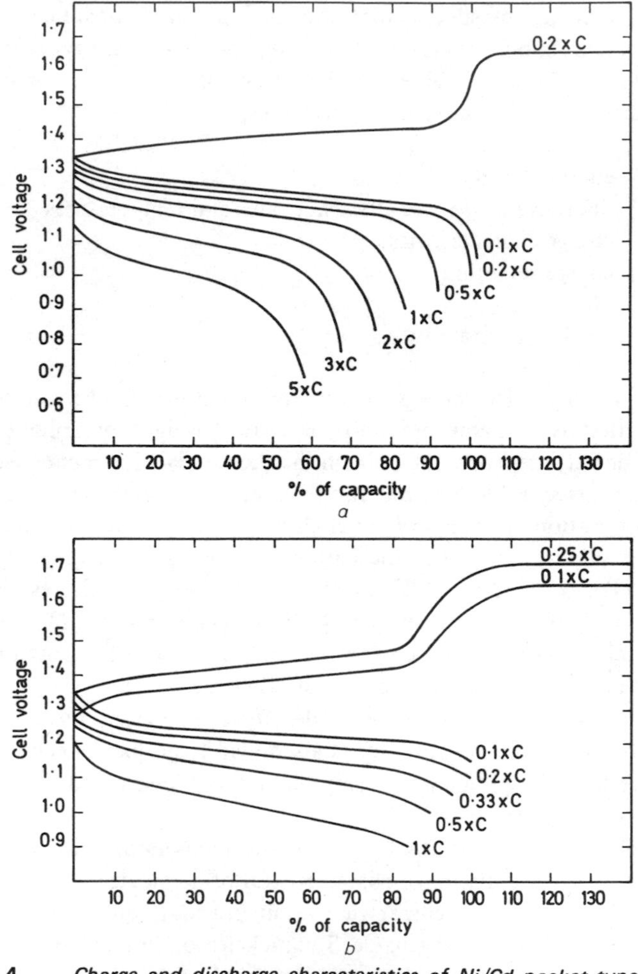

Fig. 5.4 *Charge and discharge characteristics of Ni/Cd pocket type cells at various constant rates: temperature 25°C*

a High rate cells type H manufactured by NIFE JUNGNER AB

b Medium rate cells type RV and RVP manufactured by Chloride Alcad Ltd.

Table 5.6 (cont.) *Manufacturers of nickel/cadmium pocket type cells and basic ranges supplied*

Manufacturer	Trademark	Cells manufactured				Notes
		Designation	Capacity	Container material	Performance type	
			Ah			
Chloride Alcad Limited, UK	ALCAD	EP	5-285	plastic	low rate	
		RV	65-1040	steel	medium rate	double plate design
		RVP	16-295	plastic	medium rate	double plate design
		DLS	40-900	steel	high rate	
		DLP	11-210	plastic	high rate	
NIFE JUNGNER AB, Sweden	JUNGNER (outside the UK also NIFE)	L	7·5-475	plastic	low rate	block design
		M	8·5-375	plastic	medium rate	block design
		H	8·5-375	plastic	high rate	block design
		LFB	66-363	plastic	low rate	monoblock design
		MFB	100-275	plastic	medium rate	monoblock design
		HFB	80-180	plastic	high rate	monoblock design
		KAP	10-415	plastic	low rate	
		KA	90-1245	steel	low rate	
		MDP	13-330	plastic	medium rate	
		MD	75-1180	steel	medium rate	
		HIP	8·5-235	plastic	high rate	
		HI	66-570	steel	high rate	
		MEP,MDE	18-41	plastic	medium rate	low maintenance types
VARTA Batterie AG, Germany	VARTA	D, DTN	5-22	plastic	medium rate	double cells
		T	75-1250	steel	medium rate	
		TP, TNP	6-140	plastic	medium rate	

		TSA, TSM	65-650	steel	high rate	
		TSP	7·5-125	plastic	high rate	
		DTN	4·5-11	plastic	medium rate	double cells
		BT/BTS	10-60	plastic	medium rate	monoblock design
SAFT-Société des Accumulateurs Fixes et de Traction, France	SAFT	KPL	80-360	steel	low rate	
		KPMP	10-320	plastic	medium rate	
		KPM	32-520	steel	medium rate	
		KPHP	14-250	plastic	high rate	
		KPH	30-310	steel	high rate	
Friemann & Wolf GmbH, Germany	FRIWO	T	75-520	steel	medium rate	
		TP	10-140	plastic	medium rate	
		HK	85-585	steel	medium rate	
		HKP	11-158	plastic	medium rate	
		TS	65-300	steel	high rate	
		TSP	7·5-125	plastic	high rate	
		DTN	4·5-7	plastic/steel	medium rate	double cells
McGraw-Edison Company, Edison Battery Division, USA	EDISON Americad	CED	16-250	plastic	low rate	
		MED	10-231	plastic	medium rate	
		HED	10-180	plastic	high rate	
		ED	80-400	plastic	medium rate	cliplock assembly
Japan Storage Battery Company, Japan	GS	BP	4-1000	plastic	low rate	
		DP	8-700	plastic	medium rate	
Yuasa Battery Co., Ltd., Japan	YUASA	QKC	10-800	plastic	low rate	
		QSC	10-900	plastic	medium rate;	
		QHC	20-250	plastic	high rate	
Honda Denki Co., Ltd., Japan	Honda	KAP	10-1000	plastic	low rate	
		MDP	10-700	plastic	medium rate	
		HIP	10-400	plastic	high rate	
		VHP	20-500	plastic	high rate	

numerical value of the capacity in ampere-hours (for instance, 500 A on a 100 Ah cell). During this discharge the average voltage is still above 1·0V. Medium rate cells in the lower diagram give 85% of the nominal capacity when discharged at the *C* rate. Occasional overdischarge or reversal of pocket type cells is not detrimental.

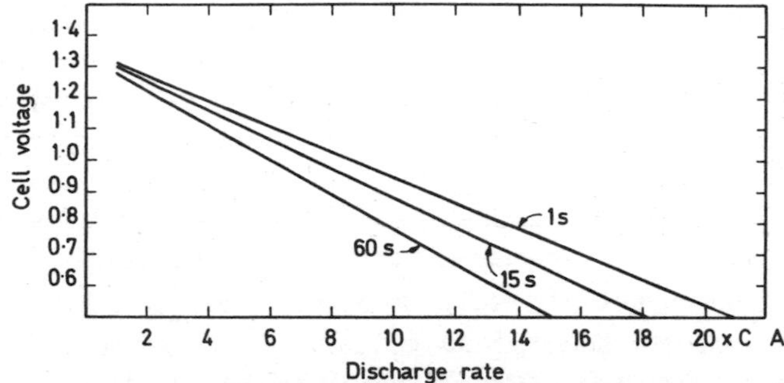

Fig. 5.5 *Voltage-current curves for fully charged Ni/Cd pocket type cell at 25°C*
High rate cell type H 310 manufactured by NIFE JUNGNER AB. C rated capacity

In Fig. 5.5 typical voltage/current curves for high rate cells are shown. This kind of diagram is generally used only for very high rate discharges and the curves are often called starter curves. The high rate cell in this diagram will deliver as much current as 20 x *C* A during 1s to a final voltage of 0·6V. The corresponding value for medium rate cells would be 6–9 x *C* A. The discharge curves shown in Figs. 5.4 and 5.5 are typical curves, which are strictly valid only for the cell types and makes specified in the captions.

From what has been stated already it is obvious that pocket type cells generally have a low internal resistance. Fully charged high, medium and low rate 100 Ah pocket cells have typical internal resistances of approximately 0·4, 1 and 2 mΩ, respectively. For a certain series of cells the internal resistance is largely inversely proportional to the size of the cell, i.e. a 200 Ah cell has only half the internal resistance of a 100 Ah cell of the same general design. The internal resistance increases with decreasing temperature and decreasing state of charge of the cell.

Pocket plate cells with normal electrolyte density can be used at temperatures down to −25°C. If a more concentrated electrolyte is provided, it is possible to use the cells at temperatures as low as −50°C. At −30°C such cells give 60-70% of their nominal capacity and at −40°C they give 50-60%. Should complete freezing of a pocket cell occur, this is not harmful. After warming-up the cell will function normally.

The decrease in discharge voltage experienced at lower temperatures is not related to a decrease of the electromotive force (EMF). On the contrary, the temperature coefficient of the electromotive force (dE/dT) of the nickel/cadmium system is slightly negative, which means that the EMF actually increases somewhat with decreasing temperature. The decrease in discharge voltage and capacity at low temperatures is instead connected with changes in the electrolyte, e.g. decreased conductivity and increased viscosity and also with some changes in the electrode processes, especially at the cadmium electrode. Pocket cells can also be used at elevated temperatures. For extended periods of use 45-50°C is generally considered as the maximum operating temperature.

Charge characteristics: Nickel/cadmium pocket type cells have favourable charge characteristics and can be charged according to either of the two basic systems: constant current or constant potential (or combinations of these).

Constant current charging is one of the standard procedures, although it is not very frequently used in modern applications. Charge characteristics obtained at such charging are shown in Fig. 5.4. The charge current often recommended is $0{\cdot}2 \times C$ A, (i.e. the 5 h rate) for 7 h for a fully discharged battery. The voltage increases slowly to about 1·5V in 4·5 h after which it rapidly increases to approximately 1·7V, where it remains until the charging is finished. The rapid voltage increase derives from the cadmium electrode which has a pronounced voltage step when approaching the fully charge state. The nickel hydroxide electrode on the other hand lacks such a step and the voltage increases steadily during charge.

When charging at normal ambient temperatures, gassing is insignificant until the voltage step is reached. After this point the major part of the current is used for the evolution of hydrogen and oxygen gas. Overcharging will lead to increased decomposition of water and accordingly to more maintenance as the intervals between water additions will be shorter. For this reason overcharging should be avoided, although it is not harmful to the pocket type cell. Charging can be carried out in the temperature range −50 to +45°C, but at the very low temperatures, as well as above 40°C, the charging efficiency suffers.

Constant potential charging involves the applications of a suitable constant voltage d.c. source to the battery and letting the charge current vary in accordance with the battery demand. A modified constant potential charging with current limitation is frequently used today. The current is often limited to $0{\cdot}4 \times C$ A or less and the constant potential used is in the range of 1·50 to 1·65V per cell at normal ambient. The charging time may vary from 5 to more than 25 h, depending on cell type and current limitation value.

The temperature during constant potential charging is of importance. At elevated temperatures there is an increased input of energy for a given potential and a greater decomposition of water. At low temperatures, on the other hand, the input of energy decreases and may be too low.

In many standby and emergency applications it is necessary to keep the battery in a high state of charge. In such cases it is usual to connect the battery in parallel with the ordinary current source and the load. The battery is continuously charged or 'floated' at 1·40-1·45V per cell. This kind of charging may be combined with a supplementary charge after each discharge or at fixed intervals.

The charge efficiency may be defined as the ratio of the output of a battery to the input required to restore the capacity. The electrochemical efficiency of the pocket type cell is approximately 72% when going from the discharged to the fully charged state. The corresponding energy efficiency is about 60%.

Charge retention: All batteries on open circuit lose some of their charge with time, owing to various internal electrochemical or chemical reactions. Such reactions also exist in nickel/cadmium pocket type cells but they are very slow, and, accordingly, the capacity losses are small. After a 6 month rest period at 25°C the available capacity is 65 to 85%, depending on make and cell type. The self-discharge reactions are rather temperature dependent. At temperatures lower than −20°C there is virtually no self-discharge, whereas at 45°C the rate of capacity loss is about three times higher than at 25°C.

Life: Nickel-cadmium pocket type batteries are well known for their excellent reliability and very long life. The mechanically strong design and the absence of corrosive attack of the electrolyte on electrodes and other cell components are contributing factors and so is the ability to stand electrical abuse such as overcharging and reversal and to stand long time storage in any state of charge.

It is difficult, however, to give definite figures on the number of charge and discharge cycles that pocket type batteries can deliver and on the total lifetime in years. This is due to the fact that battery life may vary considerably depending on the operation conditions. Factors such as operating temperature, discharge depth and charging regime influence the life. For a long life in general it is essential to keep the battery temperature below 35°C. For a good cycle life it is important that the cells are filled with an electrolyte to which lithium hydroxide has been added.

Under normal conditions the cycle life of a pocket type battery is more than 2000 cycles and the total lifetime may be between 8 and 25 years; e.g. batteries for train lighting normally have a life of 10-15 years, batteries for diesel engine cranking about 15 years and stationary standby batteries a life of 15-25 years.

Mechanical stability: Generally speaking, nickel/cadmium pocket type cells and batteries are mechanically very rugged, owing to the fact that they are built up from steel and high quality plastic components. The electrode groups are either carefully bolted, or, in more recent designs, welded together, and the cell containers are made of sheet steel or high impact types of plastic. Steel cells are assembled in wooden cases in iron-banded battery crates or in steel cradles.

Pocket type cells and batteries can stand all the stresses that may occur in battery operation in the form of vibrations, shocks, and rough handling in general. As there is no corrosive attack of the electrolyte on any of the components in the cell, there is no risk for decreased strength during the life-time of the battery or for 'sudden death' owing to corroded lugs or pole bolts.

Pocket type batteries are also resistant to extreme temperatures. The mechanical strength is little influenced by low temperatures and the batteries can stand elevated temperatures up to 70°C or more without mechanical deterioration. This is especially true for cells in polypropylene or steel containers. Cells in plastic containers can also operate in saline or corrosive environments without problems.

Properties in summary: Nickel/cadmium pocket type batteries have a good but not outstanding energy density. Their capacity is rather independent of the discharge rate and they can be effectively used at high discharge rates and in a wide temperature range. Their charge properties are favourable. They have a very good charge retention and they can be stored in any condition for long periods of time without deterioration. Pocket type cells have a very long and reliable life in most applications. They are very rugged and can stand both mechanical and electrical maltreatment. Little maintenance is required beyond occasional addition of water. Nickel/cadmium pocket type batteries have the lowest cost of all alkaline storage batteries.

5.2.4 Applications

Nickel/cadmium pocket type batteries are used in a large variety of applications where their favourable electrical properties, excellent reliability, and low maintenance requirements are of importance. Most of these applications are of an industrial nature.

One of the largest fields for pocket type batteries is railway service. In this sector, pocket batteries are used for train lighting, air conditioning, diesel engine cranking, rail cars and signalling. Batteries for train lighting applications are shown in Fig. 5.6. These batteries may have cells with steel containers, but the modern trend is to go over to plastic monoblock batteries. Such monoblocks have a very strong plastic casing, they are corrosion free, they are often equipped with flame-arresting vents and they are provided with a so-called 'dead top' Monoblock batteries of the high rate type are used for diesel engine cranking in locomotives. Pocket type batteries are also used for railway signalling at railroad crossings and for operating railway gates. This is normally an application for low rate or medium rate cells.

A related application field comprises trams, subways and trolley buses. Modern vehicles of this kind normally have considerable control equipment. Sometimes the battery takes part in the power supply under normal service conditions by covering short peak loads. In other cases the battery is used exclusively for emergency breaking or for feeding the control equipment in case of primary power failure. The batteries used are normally of medium or high rate types in plastic containers.

Pocket type batteries are used on board ships for feeding vital equipment in case of power failure. The most important purposes are emergency lighting, radio communication, starting emergency engines, fire alarm and normal lighting.

Stationary uses is a large application field. In these applications the battery normally has the function of a reserve or emergency power source and usually operates only in case of power failure. Some of the most important of these stationary uses should be mentioned briefly.

Switchgear in electric power and transformer stations very often have considerable auxiliary electrical equipment for control and protection. Pocket plate batteries of high or medium rate type are used here since an important part of the function of the batteries is to take care of short peak loads occurring, for example, when closing and tripping breakers.

In power stations, and in industry, there are processes which must not stop even in case of power failure, e.g. pumps, motor operated valves, fans etc. The task of the battery is to take over the load immediately when the power line fails and maintain the service until the normal power supply is restored. The complete power supply unit will contain battery, charger and voltage regulator. The battery is normally

Fig. 5.6 *Ni/Cd pocket type batteries for train lighting applications*
a Steel cells in wooden crates.
b Plastic monoblock batteries

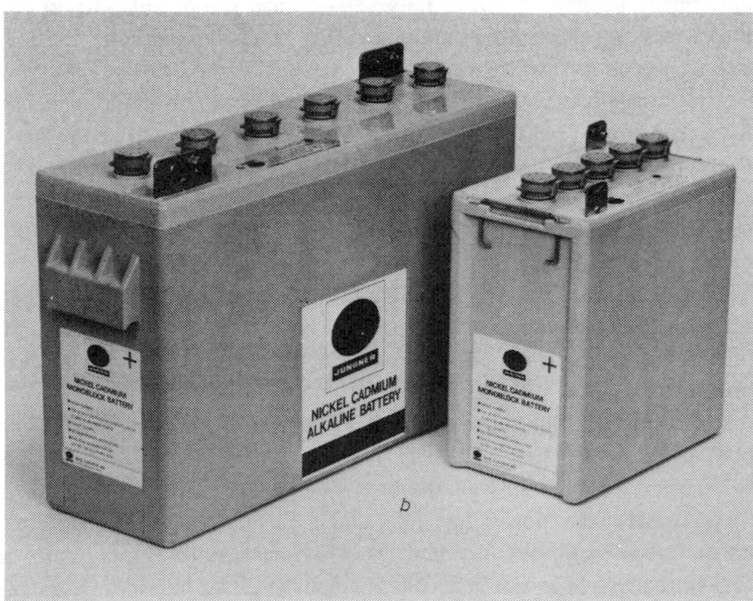

of high or medium rate type with cells in plastic containers.

In some countries there are state or other regulations requiring emergency lighting. This is needed in the first place in public buildings such as hospitals, theatres, department stores, restaurants and schools. Emergency lighting is also required in some outdoor public places. A requirement of importance for the batteries used in emergency lighting is a minimum of maintenance. Normally low rate and medium rate pocket type cells are used in this application. Special types have been developed intended for very long maintenance periods of up to 10 years.

In some cases the power required for emergency service and lighting is too big to use batteries. In this case an emergency generator is used that is driven by a diesel engine. This engine must start automatically in case of power failure. High rate pocket type batteries are used in this engine starting application.

Nickel/cadmium batteries of the pocket type are also used for telecommunication of various kinds, and are used in telephone exchanges and amplifications stations. Especially in small telephone exchanges, where the service conditions are poor, the nickel/cadmium battery offers better resistance than lead/acid batteries against poor maintenance. Pocket type batteries are also used for microwave stations.

Inverter service is an interesting application for nickel/cadmium batteries. An inverter is a device transforming DC into AC. There is an increasing demand for inverters used as emergency power sources for electronic equipment such as computers or microwave stations. The battery is used as a power source in the case of primary power supply failure.

The last field of applications to be mentioned here is that of portable applications. Some portable devices require currents that are too high to be supplied by ordinary dry batteries. In such cases nickel/cadmium batteries are often used as power sources. Examples of applications of this kind are handlamps, searchlights, signal lamps, telecommunication sets and portable instruments. In bigger devices pocket plate batteries are used, especially the twin cell type, in plastic containers. In smaller devices sealed nickel/cadmium cells are used.

5.3 Nickel/iron batteries

Nickel/iron batteries have been produced on a commercial scale since around 1900. The two main designs have been the tubular positive type and the flat pocket plate type. Cells with sintered type negatives are also manufactured.

The tubular positive or Edison nickel/iron cell was popular, especially for traction purposes, up to the 1960s. During the last decade or so it has, however, gradually lost much of its importance. It is still manufactured in West Germany and in France. The flat plate type, which is less expensive, is produced on a very large scale in the USSR.

There is, however, a revived interest in nickel/iron, as it seems to be one of the few systems which may be developed into a high energy density battery for electric vehicles. Research and development work in this direction is going on in several places in Europe, USA and Japan.[16–19]

5.3.1 Reaction mechanisms

A common but simplified way to illustrate the reactions in the nickel/iron cell is the following:

$$2\,NiOOH + Fe + 2H_2O \underset{\text{charge}}{\overset{\text{discharge}}{\rightleftarrows}} 2\,Ni(OH)_2 + Fe(OH)_2 \qquad (5.6)$$

The reactions at the nickel electrode were briefly discussed when dealing with the nickel/cadmium pocket type system (Section 5.2.1). The reaction at the iron electrode is generally recognised as

$$Fe + 2OH^- \underset{\text{charge}}{\overset{\text{discharge}}{\rightleftarrows}} Fe(OH)_2 + 2\,e^- \qquad (5.7)$$

Metallic iron reacts on discharge with hydroxyl ions to form divalent iron hydroxide. As in the cadmium electrode, soluble intermediates are involved in the reactions, probably $HFeO_2^-$ or $Fe(OH)_3^-$.[20]

The discharge of the iron electrode can be carried further to a second voltage plateau[21]

$$3\,Fe(OH)_2 + 2\,OH^- \underset{\text{charge}}{\overset{\text{discharge}}{\rightleftarrows}} Fe_3O_4 + 4\,H_2O + 2\,e^- \qquad (5.8)$$

The formation of FE_3O_4, β-FeOOH or δ-FeOOH in this second step as well as other aspects of the reactions at the iron electrode have been discussed in the literature during the 1970s.[22–25]

5.3.2 Manufacturing processes

Active materials: The positive active material for flat plate nickel/iron cells is made in the same way as that for nickel/cadmium pocket type cells. The active material for tubular positive plates is a nickel hydroxide which is prepared in a similar way but it has a more highly crystallised structure. This nickel hydroxide is not mixed with graphite but is arranged in alternate layers with nickel flake in the tubes.

The negative active material is a pure mixture of metallic iron and iron oxide (Fe_3O_4). It is normally prepared by a process in which pure iron slugs are dissolved, recrystallised, roasted, reduced and partially reoxidised. However, in the USSR iron ore is the starting material, which is concentrated, reduced, purified and finally mixed with iron powder and some additives.

Manufacture of electrodes: Flat plate positive and negative electrodes for nickel/iron cells are manufactured in the same way as their nickel/cadmium pocket type counterparts. The tubular positive type is made by clamping or welding tubes containing the active material into a nickel/plated sheet steel frame. The tubes are made of spirally wound roller perforated and nickel-plated steel ribbon. They are reinforced with steel rings after filling with alternative layers of nickel hydroxide and nickel flake.

The sintered type of negative electrode, which is manufactured by SAFT in France, is produced by mixing the active iron material with fine copper particles and then pressing and sintering this mixture into a steel frame. The iron electrode is calculated to give a higher capacity than the nickel electrode in order to avoid operation at the second voltage level of the iron electrode which is considered harmful.

Assembly of electrode groups and elements: Electrodes are assembled to electrode groups by bolting or welding and the groups are intermeshed to form elements. The insulation between plate groups of different polarity is either secured by inserting perforated sheets of plastic between the plates or by winding plastic strings around the plates. The former method is used for tubular positive cells, the latter for flat pocket type cells. Tubular positive and sintered negative plate groups separated by PVC sheets can be seen in Fig. 5.7.

Cell containers, vents: All tubular positive cells and most flat plate cells are assembled in nickel-plated steel containers as described for nickel/cadmium pocket type cells. These containers are provided with an outer coating of hard rubber or epoxy for insulation and they are furthermore equipped with corner or edge reinforcements which also provide the necessary free space between cells for sufficient ventilation. Some flat plate cells are assembled in containers of nylon or polypropylene. The vents are generally of a simple flip-top type for easy topping-up with water. They are made of steel or plastics.

Formation: Nickel/iron cells are normally given a formation treatment to remove impurities and activate the masses. This procedure is carried out in the completed cells which are filled with the formation electrolyte and are subjected to a number of charge-discharge cycles.

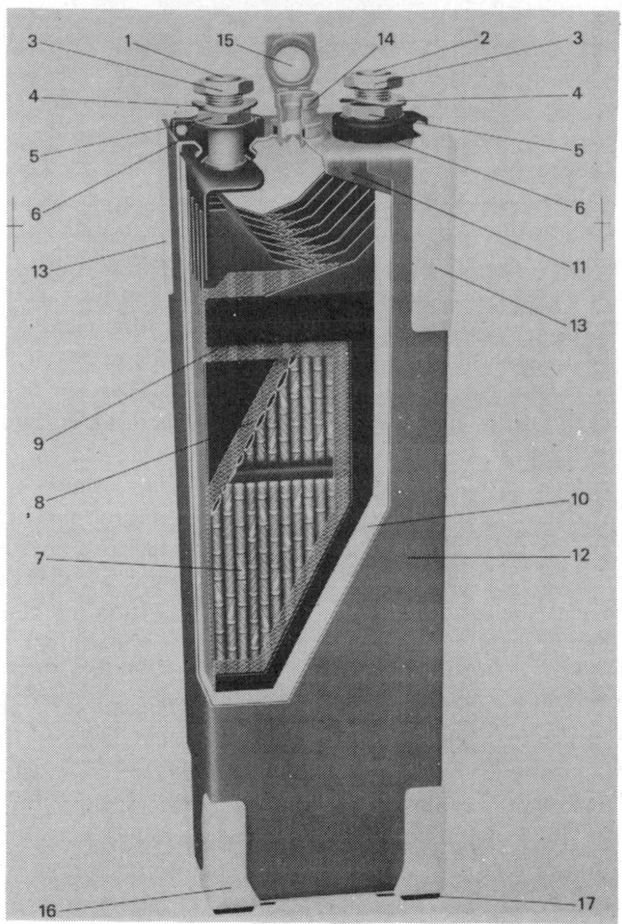

Fig. 5.7 *Nickel/iron cell in steel container with tubular positive and sintered negative plate groups*

1 Positive polebolt
2 Negative polebolt
3 Connector nuts
4 Spring washers
5 Pole sealing nuts
6 Polyamide pole insulator
7 Positive tubes
8 PVC separator
9 Sintered negative plate
10 Internal PVC insulation
11 Steel cell container
12 Epoxy coating
13 Polyamide corner reinforcements
14 Polyamide vent
15 Vent cap
16 Polypropylene corner reinforcements
17 Elastic support

Electrolyte: The electrolyte in nickel/iron cells is usually a mixture of potassium hydroxide and lithium hydroxide solutions. The lithium addition is mainly made because of its stabilising effect on the capacity of the nickel electrode during cycling. Lithium, however, also has a beneficial influence on the iron electrode. Typically, the composition of the electrolyte is 240 g/l of potassium hydroxide and 50 g/l of lithium hydroxide corresponding to an electrolyte density of 1·23 g/ml. However, densities down to 1·17 g/ml are used. In the USSR it is common to use sodium hydroxide in combination with a smaller addition of lithium hydroxide.

Assembly of batteries: Cells in steel containers are provided with a coating of plastic as described previously. They are assembled in crates of hard wood or in steel boxes and connected by means of intercell connectors and jumpers. They are separated from each other by plastic insulators. Cells in plastic containers may be close-packed in an outer crate.

5.3.3 Performance characteristics

Nickel/iron cells are available in sizes from 8 to 1150 Ah. Manufacturers of nickel/iron batteries are listed in Table 5.7 as well as main types of cells supplied.

Energy density: Nickel/iron cells are made either with normal space for electrolyte or with an extra large space. This and other design factors influence the energy density of the system. Normal figures at present are 28 ±7 Wh/kg and 60 ±15 Wh/dm^3 for cells. For complete batteries typical values are 22 Wh/kg and 40 Wh/dm^3. As before, the figures are based on the nominal capacity of the cells and the average discharge voltage at normal discharge rate.

Present research and development work on high energy density nickel/iron batteries indicates that figures as high as 50 Wh/kg and 120 Wh/dm^3 for batteries are possible.

Discharge properties: Nickel/iron cells, particularly those with tubular positive electrodes, are intended for low or moderate rates of discharge. They are typically working between the 8 and 1 h discharge rates.

Discharge curves at 20°C for a modern range of nickel/iron cells at various constant rates are shown in Fig. 5.8. The influence of the discharge rate on the cell voltage is rather pronounced, reflecting the relatively high internal resistance (R_i) of the system. For a charged 100 Ah tubular positive cell the value of the internal resistance is

Table 5.7 *Manufacturers of nickel/iron cells and basic ranges supplied*

Manufacturer	Trademark	Cells manufactured: Designation	Capacity	Container material	Element design	Notes
			Ah			
SAFT-Société des Accumulateurs Fixes et de Traction, France	SAFT	RP, RT, RU	200-960	Steel	tubular positives,	cells epoxy treated
		MP, MT, MU	240-1040	Steel	sintered copper-	
		ZP, ZT, ZU	144-1008	Steel	iron negatives	cells hard rubber coated
VARTA Batterie AG, W. Germany	VARTA	A, C	75-675	Steel	tubular positives,	
		E, LE	225-1050	Steel	pocket negatives. A and C types also made with cadmium pocket negatives.	LE-type hard rubber coated
Association Istochnik, Leningrad and other USSR producers		TNZh	300-1150	Steel	pocket positives and negatives	for motive power
		(TNZhK)	300-1000	Steel	pocket positives, pressed powder negatives	for motive power
		NZh	22-100	Steel	pocket positives and negatives	for general purposes
		FNZh	8	Steel	pocket positives and negatives	for portable lamps

approximately 4-5 mΩ. This value will increase with decreasing state of charge and decreasing temperature. Flat pocket type cells have a lower internal resistance. The nickel/iron system is more sensitive to low temperatures than nickel/cadmium. The inefficiency at low temperatures is mainly connected with a temporary passivity of the iron electrode. The critical temperature is dependent on the discharge rate. At the 8 h rate 50% of the capacity may be obtained at −18°C, whereas at the 2 h rate only 25% of the capacity is available at 0°C.

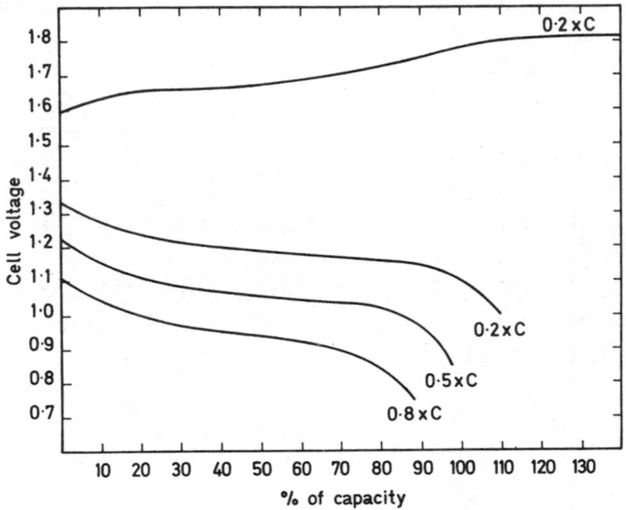

Fig. 5.8 *Charge and discharge characteristics of Ni/Fe cells at various constant rates*
Temperature 20°C. Cells type R with tubular positives and sintered copper-iron negatives manufactured by SAFT

Nickel/iron batteries can be used at elevated temperatures, although 45°C should not be exceeded for long periods of time. As there may be an appreciable rise in temperature on discharge due to the R_i losses in the cells, care must be taken to provide reasonable ventilation. Also, it is recommended to limit the discharge to 80% of the capacity to minimise the temperature rise.

Charge characteristics: Nickel/iron cells can be charged with constant current or at constant potential. The standard procedure is to use constant current (or somewhat tapering current). An often recommended charge rate is 0·2 x *C* A for 7h. Fig. 5.8 shows the charge characteristic of a nickel/iron cell charged in this manner. This curve differs from that of a pocket type nickel/cadmium cell in that it starts at a higher voltage and that the voltage step is much less marked. These differences are related to the iron electrode at which hydrogen gas will be envolved from the very beginning of the charge.

Constant rates higher than 0•2 x *C* may be used but care must be taken not to overheat the battery. Charge rates lower than 0•1 x *C* are not recommended.

Floating charge at constant potential with the battery in parallel with the charger and the load has the drawback that the potential must be chosen rather high (1•50-1•70V per cell) in order to reach the fully charged state of the battery, and, as a consequence, rest current and water consumption will also be high.

The charge efficiency defined as the ratio of the output of the battery to the input required to restore the capacity is similar to that of pocket type nickel/cadmium batteries. Thus, the electrochemical efficiency is the same, approximately 72%, whereas the energy efficiency is lower, 50-55%.

Charge retention: Nickel/iron cells lose 1-1•5% per day of their charge on open circuit stand. This means that virtually no capacity remains in the cells after a 3 month storage period. This relatively high self-discharge is caused by the iron electrode which corrodes in the electrolyte according to

$$Fe + 2H_2O \rightarrow Fe(OH)_2 + H_2 \qquad (5.9)$$

This reaction is temperature dependent.[26]

Life: One of th most important advantages of nickel/iron batteries is their very long life, even in demanding services. They are mechanically very robust and they are able to withstand both electrical and physical abuse. They can also be stored for long periods of time, preferably in the discharged and short-circuited condition, without any permanent deterioration of the electrical properties.

The total life-time of nickel/iron batteries is 7-25 years, depending on the service. The cycle life is better than for any other type of battery, namely, 2000-4000 deep cycles. This is particularly true for cells with tubular positive electrodes.

Mechanical stability: The mechanical stability of nickel/iron batteries is excellent. The reasons are the same as those given for nickel/cadmium pocket type batteries. Nickel/iron batteries are often used in applications where shocks and vibrations are very severe and where temperatures occasionally can be high. They have proved to withstand such conditions for very long periods without deterioration.

Properties in summary: Nickel/iron batteries have, at present, a somewhat higher energy density than pocket type nickel/cadmium batteries. They have, however, inherent possibilities to reach considerably higher levels. These batteries have a relatively high internal resistance and are mainly used for low and medium rate applications. Their capacity decreases rapidly with decreasing temperature. Charge retention properties are similar to those of lead/acid batteries but are poorer than for most alkaline systems. Nickel/iron batteries have an excellent life both in years and in cycles. They are mechanically very robust and will withstand severe electrical or mechanical abuse. The cost is relatively low, but the tubular positive type is more expensive than pocket type nickel/iron or nickel/cadmium.

5.3.4 Applications

Nickel/iron batteries are normally used in such applications where their very good cycle life, their favourable energy density and their excellent mechanical properties are of importance. Owing to their inherent limitations they are not used for high rate or low temperature service or in duties where a low self-discharge rate is required.

A very large application field for nickel/iron batteries has traditionally been motive power: fork lift trucks, mine locomotives and shuttle cars, railway switching locomotives and motorised hand trucks. In this area industrial type lead/acid batteries have captured a considerable part of the business both in Europe and in the USA due to their combination of a good energy density, a life of approximately 1500 cycles and a low price. Nickel/iron batteries are, however, still used in many of these applications. In the USSR it is the only system used for this kind of service. Fig. 5.9 shows a mine locomotive nickel/iron battery assembled in a steel box.

A possible future motive power application may be in electric cars, vans or buses, provided the correct combination of high energy density, long cycle life and low cost will result from the current research and development work.

Another field where nickel/iron batteries are used is for lighting and air conditioning of railway cars. In Western Europe and USA the use of nickel/iron batteries in trains is rather limited. Again, in the USSR nickel/iron is the predominating system for these applications. Nickel/iron cells are also used in mine lamps and other kinds of portable lamps.

Fig. 5.9 *Mine locomotive nickel/iron battery assembled in steel box*

5.4 Nickel/cadmium sintered plate batteries

Although the fundamental patent on sintered plate cells was applied for as early as in 1928, no production of such cells started until the Second World War, when the Germans manufactured the first sintered type batteries for military aircraft and rockets. After the war sintered cells were introduced in several countries and found applications mostly in the military field because of their outstanding high rate and low temperature properties. Sealed varieties of the sintered plate cell requiring no maintenance were mainly developed during the 1950s. Sintered plate batteries are at the present time manufactured in the United Kingdom, France, West Germany, and USSR, as well as in the United States, Japan and Israel.

5.4.1 Reaction mechanisms

The fundamental charge and discharge reactions of the nickel/cadmium system are discussed in Section 5.2.1. These reactions are the same for all kinds of nickel/cadmium cells.

A considerable part of the total sintered plate cell production comes as sealed cells and in these cells some special reactions have been arranged to take place in order to provide for the recombination of gas evolved during charging.

All sealed cells have a negative electrode with an excess of uncharged active material. On charge the positive electrode will thus first become charged in the normal manner and will evolve oxygen gas before the Negative electrode has reached the fully charged state. The oxygen evolved will diffuse to the cadmium electrode where it is reduced with the formation of cadmium hydroxide. In this manner the negative electrode is discharged by the oxygen at the same rate as it is charged by the charging current. The cadmium electrode will, therefore, never reach the fully charged state, and, consequently, will not evolve hydrogen. This is important as there are no simple recombination mechanisms for hydrogen.

The oxygen recombination reactions during charging could be represented as

$$O_2 + H_2O + 2e^- \rightarrow OH^- + HO_2^- \tag{5.10}$$

$$HO_2^- \rightarrow OH^- + 1/2\,O_2 \tag{5.11}$$

$$Cd + 2OH^- \rightarrow Cd(OH)_2 + 2e^- \tag{5.12}$$

$$Cd + 1/2\,O_2 + H_2O \rightarrow Cd(OH)_2 \tag{5.13}$$

Oxygen is transformed to hydroxyl ion via perhydroxyl ion, as shown in ractions 5.10 and 5.11. The hydroxyl ions react with metallic cadmium with formation of the normal discharge product, cadmium hydroxide. Eqn. 5.13 is the overall reaction for the recombination.

Eqn. 5.13 is discussed above as being electrochemical in nature. This reaction could, however, also describe a direct chemical oxidation of cadmium. Probably both kinds of reactions take place simultaenously in most sealed cells. Some sealed cells are provided with an auxiliary electrode containing a noble metal catalyst to facilitate the oxygen recombination.

5.4.2 *Manufacturing processes*

Sintered plaques: In sintered plate cells the porous sintered nickel plaque serves the same purpose as the perforated pocket of the pocket type electrode. It retains the active material and it acts as a conductor for the electric current.

Sintered plaques can be produced either according to the loose powder process or to the slurry method. The latter method is generally preferred today. In the slurry method a strip of nickel wire gauze or perforated and nickel-plated steel is continuously carried through a container filled with a nickel slurry. This slurry is normally prepared by mixing a low density carbonyl-nickel powder with a viscous aqueous solution of a cellulose derivative. The strip takes up a certain amount of slurry and is then passed through a vertical drying oven and from there through a vertical sintering furnace where sintering takes place at 800-1000°C in a protective atmosphere. The sintered strip is collected on a roller or is cut to pieces of suitable size, so-called master plaques, intended for more than one electrode. The thickness of the sinter is 0·5-1·0 mm depending on the type of cell for which it is intended.

Important features of the sintered plaque are high porosity (80-87%) and large surface area (0·25-0·50 m^2/g) in combination with good mechanical properties and high electrical conductivity.

The master plaques may be compressed or coined at the edges for increased strength and at the portions where tabs are to be attached. Plaques with perforated steel sheet grids need no coining, since the tabs are welded directly to an uncovered portion of the grid edge.

Impregnation of electrodes: The sintered plaque structure is filled with electrochemically active materials by some kind of impregnation process. These processes differ considerably in detail and can be more or less mechanised but they still have certain features in common. The plaque material in the form of coils or master plaques is submerged in an aqueous solution of a nickel or cadmium salt, usually a nitrate solution. Vacuum is often used to facilitate the impregnation. The material is then subjected to an electrochemical, chemical, or thermal process by which finely divided active materials are precipitated in the pores of the nickel sinter. For positive electrodes the active material will be in the form of nickel hydroxide and for negative electrodes in the form of cadmium, cadmium hydroxide or cadmium oxide. The impregnation cycle is normally completed by a washing operation and finally a drying step. As one impregnation cycle is not sufficient in most cases to introduce the desired amount of active material into the electrodes the cycle of operations is usually repeated 1-5 times depending on the type of process used. Typically, the positive electrodes are impregnated to give a 5 h capacity of 0·35-0·45 Ah/cm^3 and the negative 0·40-0·50 Ah/cm^3. These values are considerably higher than those for pocket type electrodes.

The positive impregnation solution has sometimes an addition of cobalt nitrate, the cobalt content being up to 10% of that of nickel. The cobalt will maintain capacity during cycling.

When impregnating electrodes for sealed cells some manufacturers introduce a certain amount of so-called antipolar material, i.e. cadmium hydroxide in the positive plates and nickel hydroxide in the negatives. This is done to minimise the risk of gas evolution in the cells on unintentional reversal of polarity. Impregnation methods are under continuous study and several new or modified processes have been suggested in recent years.[27–29]

Electrode formation: After impregnation the electrode material may be subjected to a formation treatment in order to remove harmful impurities such as nitrates and carbonates as well as loose particles and to activate the nickel and cadmium materials. Master plates or material in rolls are arranged to formation elements which are cycled one or more times in a potassium hydroxide electrolyte and then washed and

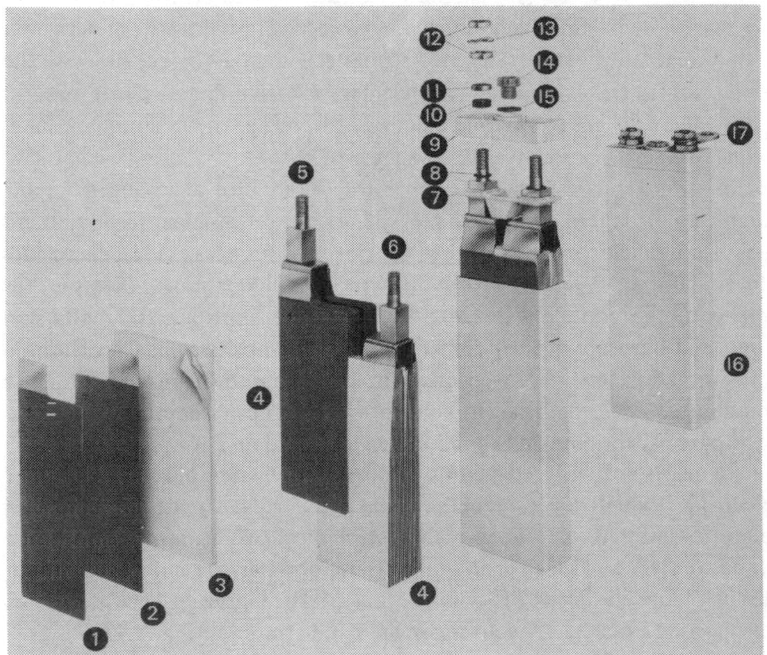

Fig. 5.10 *Vented Ni/Cd sintered plate cell in plastic container*

1 Positive plate
2 Negative plate
3 Separator
4 Electrode group
5 Positive polebolt and busbar
6 Negative polebolt and busbar
7 Twist protection device
8 Pole gasket
9 Lid
10 Pole sealing gasket
11 Steel gland ring
12 Pole nuts
13 Spring washer
14 Vent plug
15 Vent gasket
16 Cell container
17 Intercell connector

dried. At the drying operation care must be taken to avoid carbon dioxide pick up in the active materials, especially in such cases where no cell formation will be carried out at a later stage. Excessive carbonate in the cell electrolyte will impair the performance of the cadmium electrode.

Assembly of elements: The electrode material is cut to final plate sizes and one or more tabs are welded on at the coined areas or at the uncovered steel sheet edge.

Electrodes of different polarities and plastic sheet separators are intermeshed to form elements of various shapes. In vented cells this operation is similar to that described when dealing with pocket plate cells. The elements can be welded or bolted. Fig. 5.10 shows the construction of a vented sintered plate element and cell.

Sealed cells are made in three basic configurations: rectangular, cylindrical and button. The rectangular cell element is assembled in the same way as the vented cell. The element for cylindrical cells is made in the shape of a coil. Only one electrode of each polarity is used. A sandwich containing a positive plate, a layer of separator material, a negative plate, and a second layer of separator is fed into a coiling machine. Manually operated machines are still used, but automatic equipment has been developed in recent years. For obvious reasons the electrodes to be used in cylindrical cells are relatively long. As an example, the plates in the common D-size cell (4 Ah) are approximately 400 mm long and have a width of about 50 mm. The positive tabs protrude in one direction and the negative tabs in the opposite direction. Some modern cell designs have no conventional tabs but operate with special components for connecting the plates to the lid and container.

Elements for sintered plate button cells are assembled from circular electrodes which are punched out in pairs from a master plate. This plate is sintered on nickel-plated steel sheet which is provided with unperforated portions for the tabs. These portions are not covered with sinter. Punched pieces of separator material are inserted between plates of opposite polarity. The arrangement is illustrated in Fig. 5.11.

Separators: The separator is a component of vital importance to all sintered plate cells. The thin sintered electrodes require an equally thin separator. Plastic pins and the like are not suitable. Practically all separators for sintered plate cells are in the form of thin, porous, plastic sheets.

The requirements of the separator are severe. It should have a low electrolytic resistance and be as thin as possible to give high energy density of the cell. The pores should be very small to prevent 'treeing' of active material. The separator must be chemically resistant to the cell

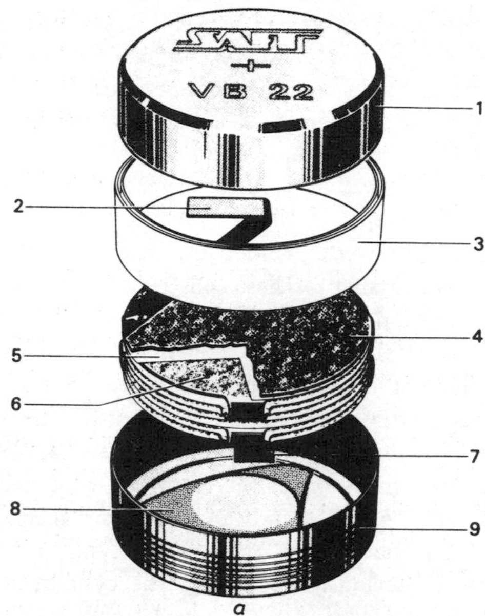

Fig. 5.11 *Exploded view of Ni/Cd sintered plate button cell*

1 Positive cup
2 Positive connector (welded to cup 1)
3 Insulating ring
4 Positive plate
5 Separator
6 Negative plate
7 Negative connector
8 Contact spring
9 Negative cup

components over a wide temperature range and also be mechanically resistant and flexible.

For use in sealed cells there are additional demands. The separator must have a structure which provides voids for the transport of oxygen gas but also electrolyte-retaining parts for ionic conduction. It must further have the properties of absorbing and retaining a sufficient amount of electrolyte.

Separators for vented cells are normally of the multilayer type. They may consist of one layer of cellophane between two layers of thin polyamide cloth or of one layer of cellophane and one or more layers of non-woven polyamide material. The semipermeable cellophane acts as a barrier against 'treeing' but also against oxygen gas on charge. Although cellophane is the most frequently used barrier material, other kinds of membranes are being tried, e.g. irradiated polyethylene (Permion) and microporous polypropylene (Celgard).[33]

Separators for sealed cells consist of one or more layers of the same or different non-woven or woven fabrics. In cylindrical cells separators composed of one or two layers of non-woven polyamide are very common. In cells intended for high temperature operation, non-woven polypropylene is used, which has a better chemical resistance, but, generally, poorer electrolyte absorption properties. Prismatic and button cells often have two or three layers of different fabrics. The total thickness of the separator in sintered plate cells is 0·2-0·3 mm.

Cell containers and vents: Vented sintered plate cells are assembled in plastic or steel containers. These cells are often employed in applications where the requirements are very severe, e.g. in aircraft. When plastic is used as container material, polyamide is preferred, owing to its excellent mechanical strength and resistance to high temperatures. However, copolymers of styrene and acrylonitrite are also used. Steel containers are made in the same manner as for pocket plate cells.

Sealed cells are almost exclusively provided with steel containers as these are completely gas-tight and can resist internal cell pressure better than plastic containers. The rectangular types are made in the conventional way previously described. The cylindrical container is a deep-drawn nickel-plated can with a wall thickness around 0·5 mm. The button cell container is made similarly.

For cylindrical and button cells the most common cover design is a nickel-plated steel lid with a plastic grommet around the rim. In button cells the lid can also be in the shape of a cup. In these cells the negative tabs are welded to the can and the positive to the lid.

Vented cells are equipped with vent arrangements of various designs. Most cells have a plastic vent, which may be of a simple flip-top type or a more advanced pressure relief valve.

Cylindrical cells normally have some kind of safety device in the lid for protection against high internal gas pressure, which might take place as a result of excessive reversal of cells or charging at rates so high that the recombination reactions cannot take care of the oxygen evolved. There are two basic types: safety vents which burst at a given pressure and remain open, and resealing valves which open at a relatively high pressure (10-20 kg/cm^2) but close again when the pressure decreases. Button cells have no safety devices of these kinds.

Formation of cells (electrolyte): Sintered plate cells are sometimes subjected to a formation treatment after assembly. This treatment consists of a few charge-discharge cycles, often in combination with renewal of the electrolyte. For sealed cells this must be done prior to the sealing operation.

The final electrolyte used in sintered plate cells is almost exclusively a potassium hydroxide solution with or without an addition of lithium

Fig. 5.12 *Cut-away views of cylindrical sealed Ni/Cd cells*
a Sintered plate type
1 Positive connector
2 Cell lid
3 Postive pole
4 Safety vent
5 Positive plate
6 Separator
7 Negative plate
8 Cell container
9 Negative connector

b Pressed powder type

hydroxide. Sodium hydroxide is used on a limited scale for high temperature sealed cells. Vented cells operate with KOH electrolyte densities of 1·20-1·30 g/ml. In sealed cells common densities are 1·25-1·35 g/ml.

Filling of electrolyte in sealed cells is generally carried out by means of an automatic pipetting machine which adds a measured amount of electrolyte to each cell. This amount is considerably less then that used in vented cells; the sealed cells work in an 'electrolyte-starved' condition. This is done to facilitate the migration of oxygen gas through the separator to the negative electrode and to provide the three-phase boundries (metal-liquid-gas) necessary for the recombination reactions.

Final cell assembly: For rectangular cell types the final container-lid welding or cementing operations are made in the way previously described. The final operations for cylindrical and button type sealed cells are often carried out in automatic equipment where the lid and its sealing gasket is inserted and the can crimped over the rim of the lid forming a very tight seal. The cells are electrically tested and sometimes classified with regard to capacity. A cut-away view of a sintered plate cylindrical cell is shown in Fig. 5.12*a*.

Assembly of batteries: Vented cells in plastic containers are often close-packed and completely encapsulated in an outer steel box. Typical examples are batteries for aeroplanes and helicopters (Fig. 5.13). Larger cells in plastic containers (up to 1000 Ah) for stationary use are mounted on racks or stands.

Fig. 5.13 *Different types of aircraft batteries with vented Ni/Cd sintered plate cells*

Vented cells in steel containers may be assembled in wooden battery crates in the same way as pocket plate cells. They are sometimes coated with an insulating plastic film and assembled in steel cases. Here, also, they are spaced from one another and supported in the cases by steel bosses fitting into insulating plastic wedges.

Sealed cells are assembled to batteries in a variety of ways. Rectangular cells are insulated with plastic sheets and are connected in steel or plastic boxes. Cylindrical cells may be close-packed, every second cell turned upside down, and kept together with a tape or a shrink-sleeve. There are also many so-called modular arrangements in which the cells are mounted in series in special plastic cases or frames.

Button cells are often connected to batteries by stacking and welding a suitable number of cells. The stack is jacketed with plastic tubing to provide insulation and improve mechanical stability.

Cells with pressed powder electrodes: In recent years an increasing number of nickel-cadmium cell designs has appeared where one or both of the sintered electrodes are replaced by some type of pressed powder or pasted electrode. The most common design is the cylindrical sealed cell where the sintered negative has been replaced by a pressed powder cadmium electrode. There are, however, also cylindrical sealed cells with both positive and negative electrodes of the pressed powder type.

The thought behind this approach is mainly that of decreasing the high manufacturing costs associated with the sintered electrodes without significantly impairing the electrical properties.

Pressed powder electrodes in cylindrical sealed cells have a grid of woven nickel screen or expanded nickel metal. Onto a strip of this material a mixture of the active metal hydroxides or oxides and a conductive material, normally nickel powder, is pressed, rolled or pasted. The active material may be bonded with teflon, polyolefins or

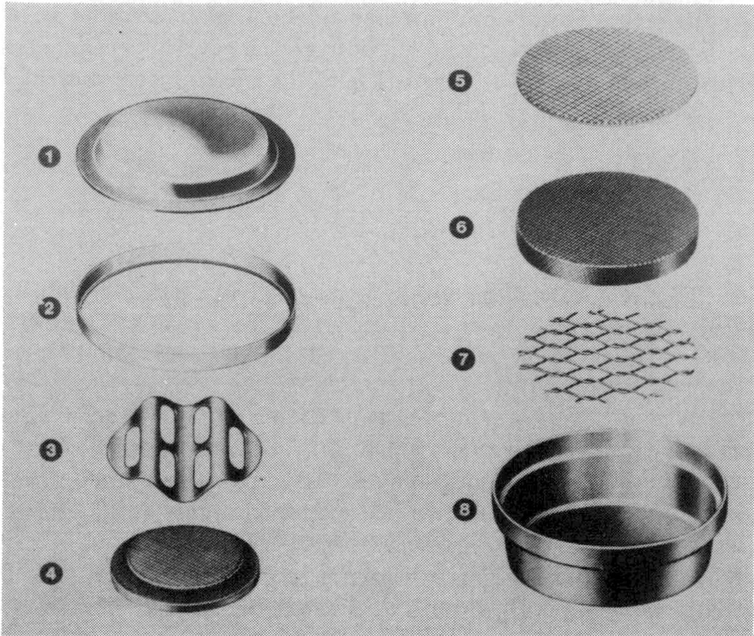

Fig. 5.14 *Exploded view of a Ni/Cd pressed powder button cell*

1 Cell lid
2 Insulating and necessary sealing gasket
3 Contact spring
4 Negative electrode
5 Separator
6 Positive electrode
7 Contact grid
8 Cell container

other suitable plastics. These electrodes have virtually the same dimensions as ordinary sintered plates for cylindrical cells and subsequent manufacturing operations are very similar to those of normal sintered plate cells. A cut-away view of a pressed powder cylindrical cell is shown in Fig. 5.12*b*.

A large number of button cells are made with positive and negative pressed powder electrodes. These electrodes are generally relatively thick and are encapsulated in pure nickel wire gauze. This kind of electrode is related to the pocket plate. The plates are separated by a non-woven plastic material. The positive electrode is connected to the can and the negative electrode is connected to the lid via contact springs. Cell and battery assembly operations are similar to those previously described for sintered plate button cells. An exploded view of a button cell is shown in Fig. 5.14. It should be mentioned in this context that there is also a more genuine sealed pocket plate cell, namely the rectangular type, manufactured in sizes up to 27·5 Ah. Here the electrodes are of the traditional pocket design with the exception that in the negative active material iron has been replaced by nickel powder in order to avoid evolution of hydrogen gas. The separator is a relatively thick polyamide fabric. These cells are assembled in welded steel containers and are provided with safety vents.

5.4.3 Performance characteristics

Nickel/cadmium sintered plate cells are available in a variety of designs and sizes. Vented cells cover a capacity range from 1 Ah to 1000 Ah. Sealed cells are available as rectangular cells from 0·8 Ah to 400 Ah, as cylindrical cells from 0·09 Ah to 10 Ah and as button cells from 0·04 Ah to 1·75 Ah. Some prominent manufacturers of sintered plate cells and the main types of cells they supply are listed in Table 5.8.

Energy density: The sintered type of nickel/cadmium cell has as, on average, a higher energy density than the pocket type. This is mainly explained by the fact that sintered electrodes can store more energy per unit of volume and weight than pocket electrodes.

There are energy density variations among the different designs of sintered plate cells, as shown in Table 5.9. Cylindrical sealed cells and batteries have generally the highest energy density of all sintered plate designs and button types have the lowest. There is, however, a considerable spread in energy density even among cells and batteries of basically the same design.

Special sealed nickel/cadmium cells for space use with extra high energy density have recently been developed in the USA.[30] 51 Wh/kg has been achieved for 20 Ah rectangular cells and an advanced design capable of delivering 58 Wh/kg is being studied.

Discharge properties: The outstanding feature of sintered plate cells is their extremely good discharge performance; the capacity of these cells is less affected by high discharge rates and low operating temperatures than any other rechargeable cell system. This is explained by the combined effect of the advantageous discharge properties of the nickel/cadmium system *per se* and the design of the sintered electrodes. The sintered plate has a highly conductive matrix with about 80% of its volume made up of very small pores. The active materials fill approximately half of the pore volume and the resulting structure exposes a very large surface area of active materials in intimate contact with the sinter to the electrolyte. There is also a very short path through the active material to the sintered material. Because of these factors, ohmic resistance and polarisation are minimised, resulting in high and nearly constant discharge voltage.

At present, the vast majority of sintered plate cells are designed with thin electrodes (0·5 - 1 mm) and thin separators and accordingly the inherent discharge properties of the system can be fully utilised.

Discharge curves at 20°C for typical vented sintered plate cells at various constant rates are shown in Fig. 5.15. From this figure it is evident that as much as 75% of the nominal capacity can be obtained at the 10 x C rate (equivalent to 1000 A on a 100 Ah cell). Even at the extreme constant rate of 20 x C about half of the capacity is still available.

The influence of temperature on the high rate performance of vented cells is demonstrated in Fig. 5.16. This voltage/current diagram shows that even at -30°C about 12 x C A can be obtained from the cell during approximately 1 s to a voltage of 0·6V. Vented cells can be used in the temperature range of -50°C to +50°C provided the cells are filled with electrolyte of suitable density.

Sealed sintered plate cells have also very satisfactory discharge properties, but they are often not quite as good as those of vented cells. This difference is mainly due to the restricted amounts of electrolyte used in the sealed cells. In Fig. 5.17 typical discharge curves for some cylindrical and rectangular types of sealed cells are presented and Fig. 5.18 shows the discharge characteristics of button cells, sintered as well as pressed powder types. A comparison between these two figures reveals that the sintered button cell will deliver somewhat less capacity than cylindrical and rectangular cells at high rates, and, furthermore,

Table 5.8 *Manufacturers of nickel/cadmium sintered plate cells and basic ranges supplied*

Manufacturer	Trademark	Cells manufactured				Notes
		Designation	Capacity Ah	Container material	Type	
Chloride Alcad Limited, England	ALCAD	W	1·2-6·0	steel	sealed,cylindrical	also rectangular pocket plate sealed cells type PS, 7·25 - 27·5
		SE	0·8-14	steel	sealed,rectangular	
The Ever Ready Company (Gt. Britain) Limited, Special Battery Division, England	EVER READY	NCC	0·1-10·0	steel	sealed,cylindrical	
		NCB	0·09-1·75	steel	sealed, button	
SAFT-Société des Accumulateurs Fixes et de Traction, France	SAFT	VOK	6·2-80	steel, plastic	vented	
		VPK	6·0-36	plastic	vented	thinner plates than VOK
		VPMS	14036	plastic	vented	for military aircraft
		GPX	35-200	steel	vented	
		VR	0·10-10	steel	sealed,cylindrical	
		VX	0·20-1·6	steel	sealed,cylindrical	rapid charge types
		VO	3·8-9·0	steel	sealed,rectangular	
		VB	0·040-0·60	steel	sealed,button	
VARTA Batteri AG, Germany (BRD)	VARTA	F	23-300	steel	vented	
		FP	4·0-40	plastic	vented	
		RS	0·10-7·0	steel	sealed,cylindrical	also cylindrical and button pressed powder cells type D, DK, and DKZ, 0·01-1·0 Ah

		SD	1•6-15	steel	sealed,rectangular	also rectangular pocket plate sealed cell type D, 23 Ah
General Electric Company Battery Business Department, USA	GE	43BO	6•0-50	plastic	vented	
		GC	0•10-3•5	steel	sealed,cylindrical	
		XGC,XKC	0•10-5•6	steel	sealed,cylindrical	goldtop types for high temperature
		XFC	0•10-3•5	steel	sealed,cylindrical	power up-15 fast charge type
		GR	4•0	steel	sealed,rectangular	also oval cell type GO, 4•0 Ah
Union Carbide Corporation Battery Products Division, USA	EVEREADY	CH	0•15-4•0	steel	sealed,cylindrical	all cylindrical cells with pressed powder negatives
		CF	0•15-4•0	steel	sealed,cylindrical	CF type for fast charge: also button pressed powder cells type B and BH, 0•02-1•0 Ah
Sanyo Electric Co Ltd., Japan	CADNICA	N	0•10-10	steel	sealed,cylindrical	
		NF	0•25-1•1	steel	sealed,cylindrical	quick charge type
		NH	1•2-3•5	steel	sealed,cylindrical	high temperature type
The Furukawa Battery Co. Ltd., Japan	Furukawa, Column	MA,CA	3•6-34	plastic	vented	for aircraft
		AHS	1•0-500	plastic	vented	
		S	0•09-10	steel	sealed,cylindrical	
		S 101 F	0•45	steel	sealed,cylindrical	quick charge type
		-	3-12	steel	sealed,rectangular	for space use

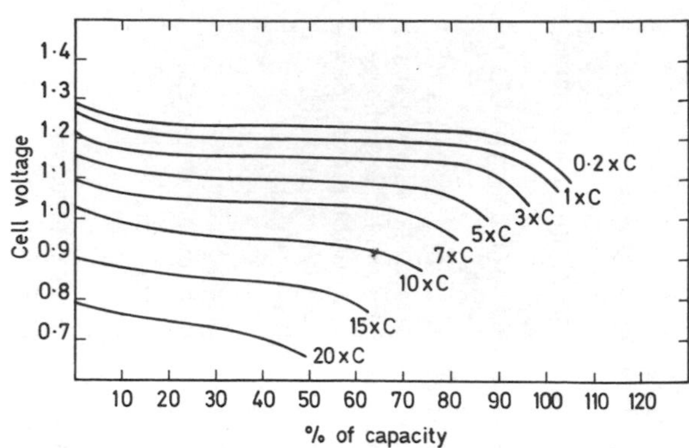

Fig. 5.15 *Discharge characteristics of vented Ni/Cd sintered plate cells at various constant rates*
Temperature 20°C. Cells type GPX manufactured by SAFT

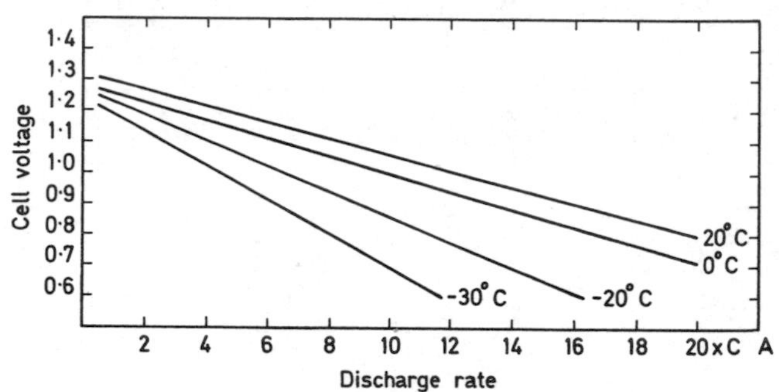

Fig. 5.16 ***Voltage-current curves for fully charged vented Ni/Cd sintered plate cells at various temperatures.***
Discharge time ~1s. Cells type FP manufactured by Varta

that the pressed powder button has much poorer discharge properties than any of the sintered types.

Sealed sintered plate cells can usually operate in the temperature range −40 to +50°C. When discharged at the normal 5 h rate these cells will typically give 95% of the nominal capacity at 0°C, 80% at −20°C and 60% at −30°C. At higher discharge currents the derating effect of temperature will be more pronounced.

The internal resistance of sintered plate cells is very low. It is roughly inversely related to the capacity of the cell and normally ranges from 0·02 to 0·04 $\Omega \times C$. As an example, a 10 Ah cell will have an internal resistance of 2-4 mΩ. Pressed powder button cells will have approximately 5 times higher internal resistance.

Charge characteristics: Vented sintered plate cells have favourable charge properties similar to those of pocket type cells. They can be charged by constant current, stepped constant current, constant potential or modified constant potential methods in a wide temperature range (−55 to +75°C).

At constant current charging the recommended procedure is often 0·2 x *C* A for 6 - 7 h or 0·1 x *C* A for 12 - 14 h for a fully discharged battery. Constant potential charging methods with current limitation are customarily used for vented type batteries, the charge potential being 1·40-1·55V per cell. This method permits a relatively fast charging in combination with low overcharging current. Without current limitation it is possible to charge these batteries very rapidly, e.g. at 1·50V per cell 70-80% of the capacity may be available after 15 min of charging.

In many applications the vented type battery is permanently floated in parallel with an ordinary constant potential source and the load. Commonly used float voltage levels are 1·35-1·42V per cell at normal ambient temperatures.

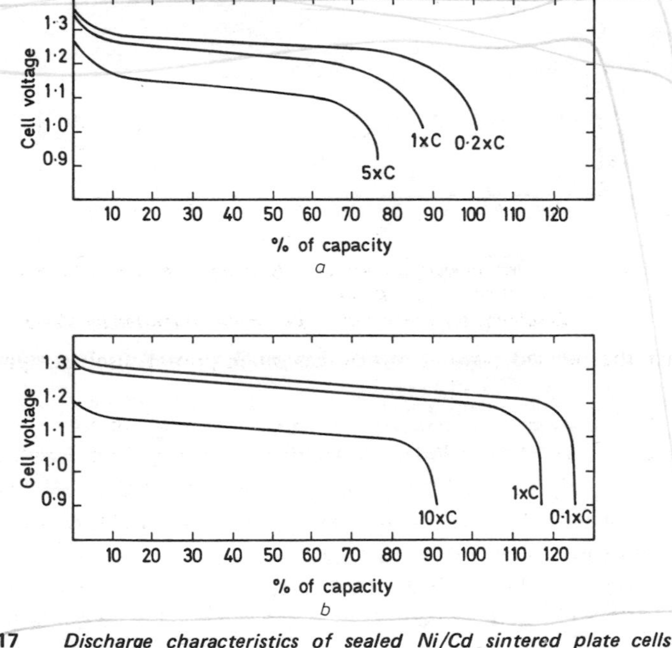

Fig. 5.17 *Discharge characteristics of sealed Ni/Cd sintered plate cells at various constant rates*
Temperature 20-25°C
a Cylindrical cells type NCC manufactured by the Ever Ready Company
b Rectangular cells type VO-4 manufactured by General Electric Company

The charging and overcharging of sealed cells is complicated by additional factors related to the gas evolution-recombination reactions in the cells. For instance, constant potential methods are not recommended as they may lead to so-called thermal runaway, a condition where the heat generated at the end of the charge results in an uncontrolled increase of the charge current, due to the fall in cell voltage as the temperature rises.

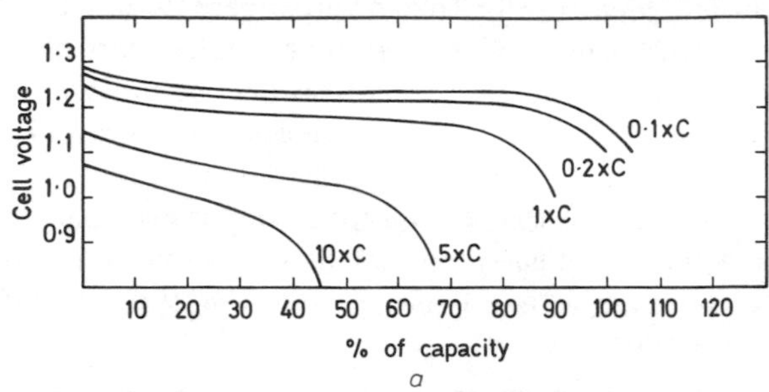

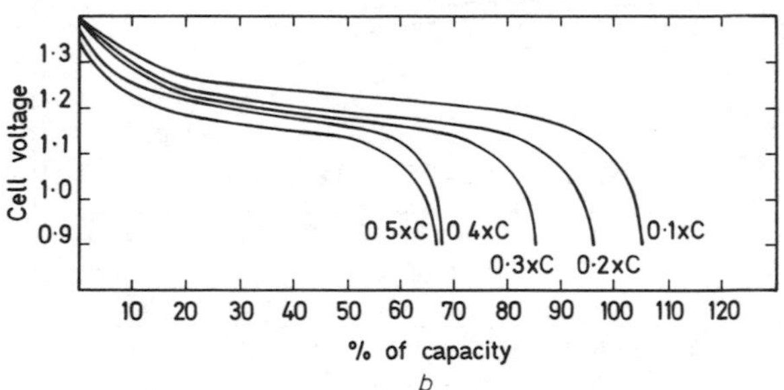

Fig. 5.18 *Discharge characteristics of sealed Ni/Cd button cells at various constant rates*
Temperature 20°C
a Sintered plate cells type VB manufactured by SAFT
b Pressed powder cells type B 225 manufactured by Union Carbide Corporation

Sealed cells are customarily charged by constant current methods and most manufacturers recommend as the normal procedure a charge current of 0·1 x *C* A, or the 10 h rate, for 14-16 h. Although normal room temperature is preferred, most sealed cells can be charged at the 0·1 x *C* rate in a temperature range of 0° to 45°C. Below this range the gas recombination reactions are slower and excessive gas pressures may build up. Above 45°C, on the other hand, the charge acceptance is poor.

Constant current charge curves of sealed cells differ significantly from those of vented cells, as illustrated by Fig. 5.19. Since the negative electrode in the sealed cell will never become fully charged, the hydrogen evolution voltage step will not appear and the end-of-charge voltage will stabilise at a lower level. Most sealed cells can be overcharged for long periods of time at the 0•1 x *C* rate and some standard types even at higher rates.

Sealed cells are often used in applications where they supply emergency power in the event of mains failure. In such applications the cells are kept in a high state of charge by permanent constant current charging. The current must be high enough to bring the cells back to the charged condition in a reasonable time. Depending on the depth and frequency of the discharges the charge currents may vary from 0•005 x *C* to 0•05 x *C* A. If the recharge must be obtained in a few hours a two rate method can be used involving a fast charge at 0•1-0•3 x *C* followed by a trickle charge at 0•01 x *C* A.

In recent years efforts have been made to shorten the relatively

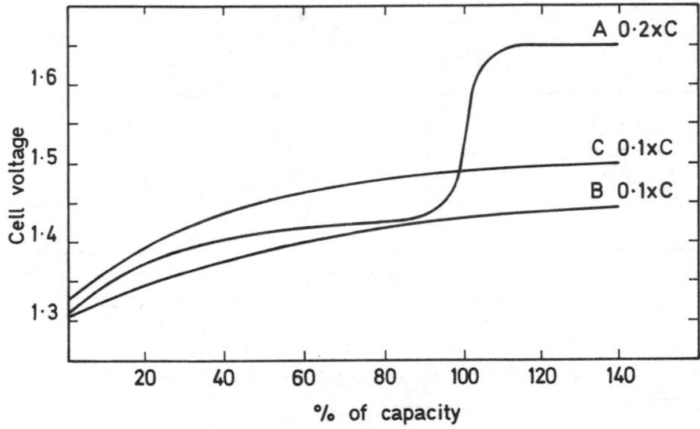

Fig. 5.19 *Charge characteristics of sintered place and pressed powder cells*
Temperature 20°C
a Vented sintered plate cell
b Sealed cylindrical sintered plate cell
c Sealed button pressed powder cell

long normal charging time. By improving the design of electrodes and separator some manufacturers have been able to produce cells capable of accepting charge and overcharge rates as high as 0•3 x *C* A. Work has also been carried out to utilise the sintered plate cell's inherent capability of efficiently accepting very high charge rates provided these rates are not continued into overcharge. Various fast-charge systems are now available based on different means for terminating the charge before internal gas pressure or cell temperature reaches too high values. The cutoff control can be based on sensing of the internal pressure, the cell temperature, the cell voltage or of combinations of these. Furthermore, auxiliary electrodes and coulometers (current integrating devices) are sometimes used for the control of rapid charging. By several fast-charge methods sealed cells can easily be charged in 1 h and by some in even 15 min. The different possibilities of charging sealed cells are dealt with at length in some of the recent trade literature.[31–32]

The electrochemical efficiency of sintered plate cells is 72-83% for vented types and 63-72% for sealed types. The corresponding energy efficiency is 62-75% and 56-65%, respectively.

Charge retention: Sintered plate nickel/cadmium cells display good charge retention properties. This is especially true for vented cells where the remaining capacity after 6 months of storage at 25°C still is 70%. For sealed cells the self-discharge rate is higher, as shown in Table 5.10.

Table 5.9 *Typical energy density data for nickel/cadmium sintered plate cells and batteries*

Cell type	Cells		Batteries	
	Wh/kg	Wh/dm^3	Wh/kg	Wh/dm^3
Vented, rectangular	25 (±10)	60 (±30)	20	45
Sealed, cylindrical	30 (±7)	90 (±20)	25	65
Sealed, rectangular	25 (±3)	60 (±10)	20	50
Sealed, button	20 (±3)	60 (±10)	15	45

Data based on nominal capacity and average discharge voltage at the normal discharge rate. Normal deviation for different cells within brackets.

Table 5.10 *Charge retention of sintered plate and pressed powder nickel/cadmium cells: typical data*

Type of cell	Available capacity after storage			
	1 month 25°C	6 months 25°C	1 month 50°C	6 months -20°C
Vented, sintered plate	90	70	<10	>90
Sealed, sintered plate	65	<10	<10	80
Sealed, pressed powder	80	60	50	90

As with all types of electrochemical power sources, the charge retention properties of sintered plate cells are influenced by the temperature. The self-discharge rate is insignificant below -20°C, whereas it increases rapidly at temperatures above normal ambient.

It is important to note that, even if the nickel/cadmium cell has lost all of its charge through self-discharge, it can easily be restored by normal recharging without any harmful effects on the electrical properties of the cell.

Life: Generally speaking, sintered plate cells and batteries have a long life. They are mechanically rugged and they are undamaged by low temperatures down to freezing of the electrolyte. They can be stored for long periods of time in any state of charge without significant deterioration of the electrical properties.

In common with other batteries, sintered plate cells are affected by the service conditions and the life-time is influenced by operating temperature, amount of overcharge, discharge depth and reversals.

For vented cells the cycle life expectancy is 1000-2000 relatively deep cycles to a remaining capacity of 50-60% of the rated value. The total life-time may vary from 2-3 years in certain military aircraft uses to 5-10 years in stationary standby applications. The rather short life-time in military aircrafts has been related to problems with the gas barrier function of the cellophane separator material often used in vented cells.[33] More resistant materials are being tried with promising results.

The cycle life of sealed cells should be longer than 392 cycles down to 60% of the rated capacity according to recommendations by the International Electrical Commission (IEC). This figure is normally easily reached and surpassed by sintered plate cells. Some manufacturers state that their sealed cells will yield at least 80% of rated capacity for 500 cycles. Others mention 1000 cycles at a depth of discharge of 50%. Moderate operating temperature, shallow discharges and minimum overcharge will help to prolong the cycle life.

The total life-time of sealed sintered plate cells may vary between a couple of years and perhaps 10 years, depending on the service. In continuous overcharge applications at least 4 years of operation should be expected provided the temperature is moderate and the overcharge current not too high. Pressed powder sealed cells have a life of 300-400 deep cycles.

Mechanical stability: Sintered plate cells and batteries display excellent resistance to vibration, shock and physical damage in general. The sintered plates are strong and all internal connections are welded, or, in some vented cells, carefully bolted. The cell containers are customarily made of steel or of high impact plastics. The elements are often compressed when inserted into the cell containers for a tight fit. The cells can be assembled into batteries which are very rugged. The mechanical strength of sintered plate cells and batteries is little influenced by low and high temperatures. There are special designs which can withstand continuous operation at as high temperature as 65°C.

Vented-cell batteries for aircraft applications are generally designed to meet the requirements of US military specifications, such as MIL-B-26220, with regard to shock and vibration.

Sealed cells, which are ordinarily used in commercial applications, can easily stand the stresses normally occurring. Cylindrical cells subjected to long-term vibration at 10 *g* over a rather wide frequency range have shown no deterioration in performance.

Properties in summary: Nickel/cadmium sintered plate cells and batteries have a good energy density, higher than that of pocket types. The outstanding feature of sintered plate batteries is, however, their extremely good discharge properties. Capacity and cell voltage are less affected by high discharge rates and low temperatures than those of any other type of rechargeable battery. This is true in particular for the vented types.

Sintered plate batteries have favourable charge properties, although there are certain limitations for sealed batteries. Charge retention is good for vented types, whereas sealed types have poorer charge retention properties, similar to those of lead/acid batteries. Both types can be stored for long periods of time without deterioration. Life expectancy for vented cells is 1000-2000 relatively deep cycles and for sealed cells 500-1000 cycles. The total life-time is 2-10 years, depending on the service.

Sintered plate batteries have a very good resistance to mechanical and electrical abuse. Little maintenance is required for vented batteries and sealed batteries require no maintenance whatsoever. The sealed types exhibit additional advantages connected with the sealed design: no evolution of gases, no risk of spilling of electrolyte and freedom to use the battery in any position. An important disadvantage of sintered plate batteries is their cost per unit of energy, which is several times higher than that of nickel/cadmium pocket type batteries.

5.4.4 *Applications*

Nickel/cadmium sintered plate batteries are employed in a great number of applications ranging from military to industrial and consumer uses.

Vented sintered plate batteries have always been valued on the military side because of their excellent discharge properties at high rates and low temperatures, their mechanical ruggedness and their relatively high energy density. These batteries are used in a large number of military aeroplanes and helicopters for starting main engines or auxliary turbines, for emergency power supply and for other on-board services. Other military applications are ground power supplies, and, to some extent, missiles.

Vented batteries are also widely used in commercial airliners and in sporting aircrafts for the aforementioned purposes. A selection of aircraft batteries is shown in Fig. 5.13. Another field where the exceptionally good discharge performance of the vented sintered plate battery is utilised is that of starting diesel locomotives and shunters, gas turbine trains and other heavy vehicles. However, in this field sintered plate batteries face competition from the less expensive nickel/cadmium pocket type batteries and also from lead/acid batteries.

A few other high power applications for vented batteries in the industrial area are braking of rapid transit trains and trams, extra high power emergency supplies and magnet cranes. It should also be mentioned that in Japan high capacity (up to 1000 Ah) sintered plate cells are used in batteries operating as stationary emergency power supplies for computer systems and for chemical plant process control. These are cells in transparent cell containers mounted on racks or stands. One of the reasons for the choice of sintered plate batteries in these applications is the high energy per unit of volume and floor space. Again, this is a field where there is strong competition from other less expensive systems, primarily nickel/cadmium pocket type.

Sealed nickel/cadmium batteries are used in a variety of applications. The combination of good electrical and mechanical properties and the special features of the sealed, maintenance-free design makes these batteries attractive both for consumer and industrial uses.

The most important fields of application for sealed cells can be grouped as follows:

(*a*) applications where the battery is the only source of energy in the equipment
(*b*) applications where the battery has an emergency function
(*c*) applications where the battery is working in parallel with another power source
(*d*) applications where the battery is used for starting purposes.

To the first group of applications belong portable instruments, cine cameras, photographic flash lamps, communications equipment, calculators, cordless power tools, portable radio and TV, razors, toothbrushes, cordless household appliances and toys. For most of these applications cylindrical cells are chosen, although rectangular cells are used for communications equipment, flash lamps, portable radio and TV.

Button cells may be used in calculators, flash lamps, toys, razors and also in hearing aids, hand torches and electric clocks. The batteries in this applications group are either charged by a separate charger or charged automatically in the equipment.

In the second group of applications the battery is normally kept on continuous charge in the apparatus and is only supplying the equipment with power during mains failure. Sealed batteries are well suited for this kind of operation as they can stand long-time overcharging requiring no maintenance. To this group of applications belong emergency lighting units, fire alarm equipment, burglar alarm devices and data system equipment. Cylindrical and rectangular cells are ordinarily used for these applications.

In group number three the battery is operating in parallel with the ordinary power source. In these auxiliary supply applications the sealed battery must often deliver current peaks during normal operation. Examples of this type of application are direction indicators for lightweight motorcycles and some kinds of telephones.

Finally, group number four deals with starting applications. Sealed sintered plate cells are well suited for starting of small engines as they can deliver very high currents for short periods of time. When the engine is running a generator will charge the battery. To this group of applications belong lawn mowers, lightweight motorcycles and chainsaws. Cylindrical cells are normally used for these applications.

5.5 Silver/zinc and silver/cadmium batteries

Both silver/zinc and silver/cadmium are rather old systems. Secondary silver/cadmium batteries were first built in 1900 by W. Jungner, and H. André started his important investigations on the silver/zinc system in the late 1920s. Silver/zinc storage batteries have been manufactured commercially since the 1940s and silver/cadmium storage batteries since the 1950s. The outstanding feature of these batteries is an exceptionally high energy density. Silver/zinc batteries are, at present, manufactured in the United States, France, Germany (BRD), the United Kingdom, Hungary, USSR and Japan. Silver/cadmium batteries are produced in the United States and in Germany (BRD).

5.5.1 Reaction mechanisms

The overall cell reaction in the silver/zinc system can be written as

$$Ag_2O_2 + 2\,Zn + 2H_2O \underset{\text{charge}}{\overset{\text{discharge}}{\rightleftarrows}} 2Ag + 2\,Zn(OH)_2 \quad (5.14)$$

The actual reaction from the divalent silver oxide state progresses through the monovalent state and the cell reaction can be written in two steps:

$$Ag_2O_2 + Zn + H_2O \underset{\text{charge}}{\overset{\text{discharge}}{\rightleftarrows}} Ag_2O + Zn(OH)_2 \quad (5.15)$$

$$Ag_2O + Zn + H_2O \underset{\text{charge}}{\overset{\text{discharge}}{\rightleftarrows}} 2\,Ag + Zn(OH)_2 \quad (5.16)$$

At the silver electrode the corresponding reactions can be represented by the following equations:

$$Ag_2O_2 + H_2O + 2\,e^- \underset{\text{charge}}{\overset{\text{discharge}}{\rightleftarrows}} Ag_2O + 2OH^- \quad (5.17)$$

$$Ag_2O + H_2O + 2\,e^- \underset{\text{charge}}{\overset{\text{discharge}}{\rightleftarrows}} 2\,Ag + 2OH^- \quad (5.18)$$

Divalent silver oxide is first reduced to monovalent oxide, which is further reduced to metallic silver at a lower potential in both cases under formation of hydroxyl ions. The reactions at the silver electrode are, however, more complicated than indicated by eqns. 5.17 and 5.18 and the chemistry and electrochemistry of this electrode have been subject to many studies in recent years. Some of this work is presented in a book on silver/zinc batteries edited by Fleischer and Lander.[34] Other important contributions are collected in the reference list.[35–46]

The reactions at the zinc electrode can be summarised as

$$Zn + 2OH^- \underset{\text{charge}}{\overset{\text{discharge}}{\rightleftarrows}} Zn\,(OH)_2 + 2\,e^- \quad (5.19)$$

However, the discharge process is believed to consist of three main steps. First, zinc is oxidised

$$Zn + 2OH^- \rightarrow Zn\,(OH)_2 + 2\,e^- \quad (5.20)$$

Then the oxidised compound dissolves forming zincate ions

$$Zn(OH)_2 + 2\,OH^- \rightarrow Zn(OH)_4{}^{2-} \quad (5.21)$$

When the electrolyte can no longer dissolve the solid zinc compounds, a passivating film, probably consisting of ZnO, is formed on the electrode and the further oxidation of zinc ceases. During charge, reactions 5.21 and 5.20 are reversed. The solubility of the zinc compounds in the electrolyte is one of the factors responsible for the problems encountered with dendritic growth of zinc on charge and with so-called shape change of the zinc electrode.

The zinc electrode, being inherently inexpensive and of high energy density, has been subject to a large number of studies with regard to electrochemical behaviour, morphology, chemical composition of reaction products etc. Some of these investigations are published in the book by Fleischer and Lander.[47] Other recent work on zinc electrodes can be found among the references.[48–62]

The reactions in the silver/cadmium system can be illustrated by the following simplified equation:

$$Ag_2O_2 + 2\,Cd + 2\,H_2O \underset{\text{charge}}{\overset{\text{discharge}}{\rightleftarrows}} 2\,Ag + 2\,Cd(OH)_2 \qquad (5.22)$$

In analogy with silver/zinc, this reaction can be written in two steps:

$$Ag_2O_2 + Cd + H_2O \underset{\text{charge}}{\overset{\text{discharge}}{\rightleftarrows}} Ag_2O + Cd(OH)_2 \qquad (5.23)$$

$$Ag_2O + Cd + H_2O \underset{\text{charge}}{\overset{\text{discharge}}{\rightleftarrows}} 2\,Ag + Cd(OH)_2 \qquad (5.24)$$

The reactions at the separate electrodes have been dealt with above and in the section on nickel/cadmium batteries.

5.5.2 *Manufacturing processes*

The manufacture of silver alkaline batteries will only be briefly discussed in this text. More detailed discussion are given in books by Falk and Salkind[63] and by Fleischer and Lander.[64]

Silver electrodes: The positive electrodes are manufactured as sintered silver plates or as chemically prepared silver oxide plates. Sintered type electrodes may be made by spreading a fine silver powder onto a silver grid and heating to fusing of the silver particles. They may also be prepared by mixing a silver powder and a thermoplastic, rolling the mix to a sheet, heating the sheet so that the silver particles sinter

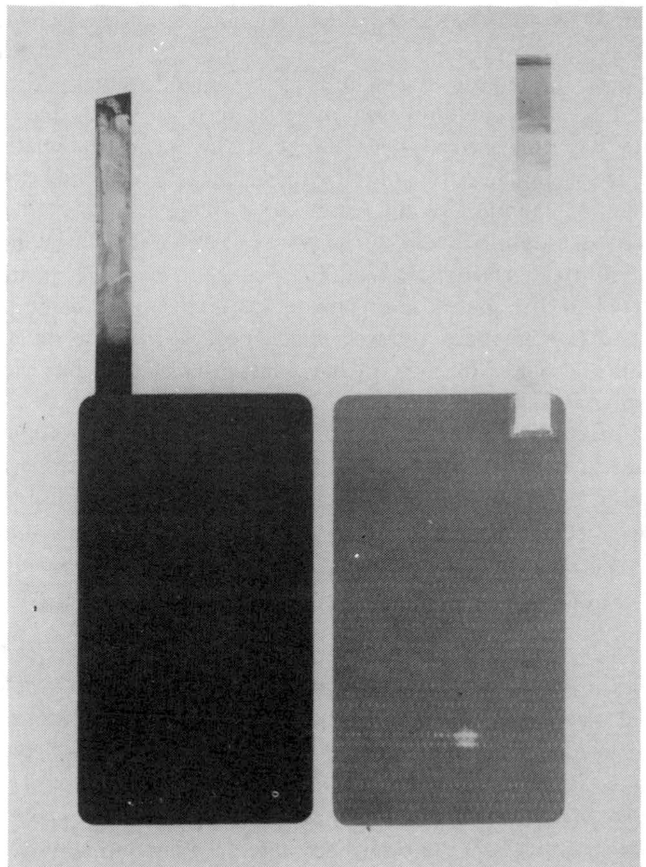

Fig. 5.20 *Sintered silver electrode to the left, zinc electrode to the right*

together while the plastic is burned off, and finally pressing the sintered sheet onto a silver grid. In both cases porous silver structures are obtained (porosity approximately 65%) with a good bond to the grid. A sintered silver electrode is shown in Fig. 5.20.

Silver oxide plates are prepared by pasting or pressing an Ag_2O_2 mass onto a grid followed by drying. This kind of plate has the advantage of being dry-charged. Silver oxide plates are more frequently used in primary cells then in secondary.

Zinc electrodes: These electrodes can be manufactured in a variety of ways. At the present time zinc electrodes to be used in secondary silver/zinc cells are often prepared by pressed powder techniques.

Zinc oxide and mercuric oxide (1-2%) powders are blended. The addition of mercuric oxide is made to minimise selfdischarge by hydrogen evolution. Sometimes teflon or other dry powder binders are also added. The mixture is put in a cavity mould together with a silver grid and a piece of non-woven material such as Viscon paper. The mixture is evenly spread over the grid, the non-woven material is folded about the plate, and the electrode is pressed to desired thickness (0·6 mm or more). The zinc electrode is then normally formed in a 5% potassium hydroxide solution, washed with water and dried. The porosity of the finished plate is 60-70%.

Instead of the pressed dry powder method, a zinc oxide pasting technique is sometimes used. Electrodeposition of zinc on a silver substrate is also used for high discharge rate electrodes. A zinc electrode is shown in Fig. 5.20.

For maximum cycle life of the cell the zinc electrodes are made with an excess of capacity as compared to the silver electrodes. They may also be made slightly larger so that they overlap the silver electrodes at all edges, and, furthermore, they may be 'contoured' to have less material in the centre and more at the edges. These measures are taken to counteract the so-called shape change taking place during cycling.

Cadmium electrodes: The negative electrodes used in silver/cadmium batteries may be of the pressed powder type, the pasted type or the sintered plate type. The manufacture of such electrodes has been discussed in the section dealing with sintered plate nickel/cadmium batteries.

Separators: Although separators are essential components in all electrochemical power sources, they are of paramount importance in silver alkaline cells and in silver/zinc cells particularly. The solubility of silver oxide and zinc oxide in the electrolyte requires separator materials of low permeability. The strongly oxidising properties of the charged silver electrode call for materials which are very resistant to chemical deterioration. Because of the high energy and power densities desired in silver cells the separators must be thin and have a low electrolytic resistance.

The practical solution is the use of multiple layer separators. These may consist of a layer of inert fabric nearest to the silver electrode, then several wraps of thin membranes to stop diffusion of silver species and minimise zinc dendrite growth, and finally, nearest the zinc, a wrap of non-woven fabric to absorb electrolyte. This last layer is often applied when preparing the zinc electrode as previously described. Separators for silver/cadmium cells are similarly designed.

The membranes used are often of regenerated cellulose, e.g. cello-

Fig. 5.21 *Selection of vented silver/zinc cells*

phane or fibrous sausage casing. However, improved membranes consisting of irradiated polyethylene or inorganic materials[65–67] have also been tried.

Assembly of cells and batteries: Normally, the electrodes in silver cells are of rectangular shape. The complete plates are wrapped with the separator material. Often, two negatives are wrapped together in one membrane sheet and are then U-folded. Positive plates are intermeshed with the negatives to form an element. The current-collecting leads extending from each plate are fed into the cell terminals and secured by crimping, soldering or welding. The dry element is inserted into a cell container of polysulfone, polyphenylene oxide, or some other strong plastic material, and the lid is secured by cementing. Rectangular cells are normally vented and provided with a simple valve in the lid. They may, however, also be made sealed.

The cells are filled with the proper amount of electrolyte consisting of a 30-45% potassium hydroxide solution. For silver/zinc cells, especially of the low rate type, this electrolyte may be partially pre-saturated with zinc oxide to minimise the solubility of the zinc electrode. Fig. 5.21 shows a selection of vented silver/zinc cells of different sizes.

Sealed silver/zinc cells are also manufactured in the button configuration. The assembly operations for such cells are similar to those described for nickel/cadmium button type cells. Silver alkaline cells are assembled into battery cases of stainless steel, glass fibre-reinforced plastic, or anodised magnesium. These cases are often made to fit a particular application and may vary considerably in design and shape. Button cells are connected to batteries by stacking and welding. The stack is normally jacketed with plastic tubing.

5.5.3 Performance characteristics

Silver/zinc cells are available in sizes from 0·05 to 600 Ah and silver/cadmium cells in sizes from 0·1 to 300 Ah. Manufacturers of silver alkaline cells and the main types supplied are listed in Table 5.11.

Energy density: The most outstanding feature of silver alkaline cells is their superior energy density. The reasons are the high energy content of the silver electrode and the compact assembly of the electrodes in the cell container. In silver/zinc cells the favourable potential of the zinc electrode is an additional factor contributing to the extremely high energy density of this system. Normal values for silver/zinc cells are 100 ± 30 Wh/kg and 180 ± 80 Wh/dm^3. Corresponding figures for silver/cadmium cells are 60 ± 15 Wh/kg and 120 ± 40 Wh/dm^3.

As always, assembled batteries have lower energy density values than cells. Because of the large variety of battery designs in use, it is difficult to give general figures, but data for batteries are normally 10·40% lower than those for cells.

Discharge properties: Silver alkaline cells are available both as high rate and as low rate types, differing in plate thickness and separator design. Typical discharge curves for silver/zinc and silver/cadmium cells are shown in Fig. 5.22. At low discharge rates there are two voltage levels present in the discharge curve corresponding to the two reactions at the silver electrode previously described (see eqns. 5.17 and 5.18). At discharge rates of 2 x *C* or more, the first voltage step will disappear.

At the lower plateau the voltage is constant. This is related to the fact that the conductivity of the silver electrode increases during discharge as more metallic material is formed. This will counteract the effects of polarisation and thus stabilise the voltage.

The discharge properties of silver cells are adversely affected by

Table 5.11 *Manufacturers of silver/zinc and silver/cadmium cells and basic ranges supplied*

Manufacturer	Trademark	System	Cells manufactured Designation	Capacity	Performance	Notes
				Ah	type	
Yardney Electric Corporation, USA	Yardney	silver/zinc	LR	0·5-525	low rate	also in modular design
	Silvercel	silver/zinc	HR	0·1-200	high rate	
	Yardney Silcad	silver/cadmium	YS	0·1-300	low rate	also in modular design
SAFT-Societe des Accumulateurs Fixes et de Traction, France	SAFT-SOGEA	silver/zinc	AMD,ALD, H	1·5-120	low rate	
		silver/zinc	AGD, RA, RB, RC, LC, LA	1-120	high rate	
Eagle-Picher Industries Inc., USA	EAGLE-PICHER, EP	silver/zinc	-	0·5-600	high and low rate	
		silver/cadmium	-	0·5-175	high and low rate	
Medicharge Ltd., England	Medicharge	silver/zinc	B	0·050-0·45	low rate	button cells
Silberkraft Leichtakkumulatoren GmbH, Germany	SILBERKRAFT	silver/zinc	S	0·5-120	low rate	
		silver/zinc	SHV	0·5-120	high rate	
		silver/cadmium	-	0·1-300	low rate	
Matsushita Electric Industrial Co., Japan	Panasonic	silver/zinc	SZL	7-150	low rate	
		silver/zinc	SZH	2-110	high rate	

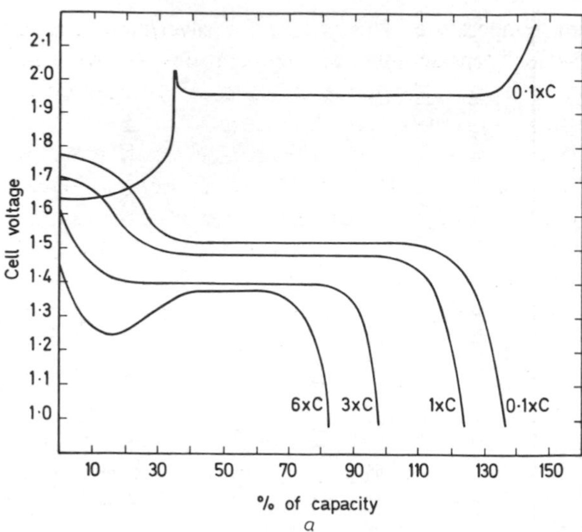

a

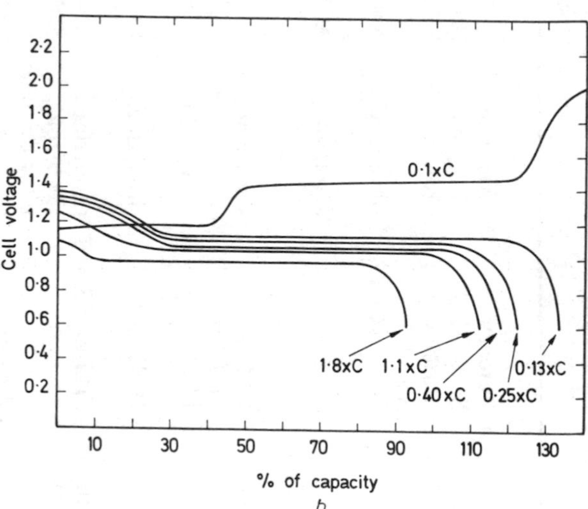

b

Fig. 5.22 *Charge and discharge characteristics of silver cells at various constant rates*
Temperature 25°C
a High rate silver/zinc cells manufactured by Silberkraft - Leichtakkumulatoren GmbH
b Silver/cadmium cells manufactured by Yardney Electric Corporation

decreasing temperature. This is true for silver/zinc cells in particular, which suffer considerably at temperatures below $-20°C$. Silver/cadmium cells are somewhat better and can be used down to $-30°C$. The maximum operational temperature is 70°C for both types.

The internal resistance of fully charged silver/zinc cells may be as low as 0·3 mΩ for a 100 Ah high rate unit. Silver/cadmium cells normally have an internal resistance which is several times higher than this value.

Charge characteristics: Silver alkaline cells are normally charged with constant current and an often recommended charge rate is $0{\cdot}1 \times C$. Typical charge characteristics of silver/zinc and silver/cadmium cells charged in this manner are shown in Figs. 5.22*a* and *b*. The two-plateau shape caused by the silver electrode is clearly visible. The charge efficiency is very good, about 90% for both systems. The energy efficiency is 70-75%. Overcharging of silver alkaline cells should be avoided. The charging of a silver/zinc cell should be interrupted when a voltage of 2·0V is reached, and a silver/cadmium cell should stop being charged when 1·6V is reached.

Charge retention: Silver alkaline cells have favourable charge retention properties. They may have as much as 80-95% of their capacity available after a rest period of one year at room temperature. However, there is a considerable spread among makes and types and less favourable charge retention values are also experienced, especially with vented types. The charge retention is affected by the temperature in the normal manner: higher temperature leads to poorer charge retention.

Life: Silver/zinc cells and batteries have a very limited life. Low rate types may deliver approximately 100 deep discharge cycles, whereas high rate types give only 1-50 cycles depending on the discharge rate and other conditions of use. The total lifetime in the wet condition is 1-2 years for low rate types and 1/2 - 1 year for high rate types. The dry storage life is up to 5 years.

Silver/cadmium batteries have better life properties than silver/zinc batteries. Approximately 500 deep cycles may be obtained and the cycle life increases with decreasing depth of discharge. The wet life is up to 3 years and the dry storage life is longer than 3 years.

As previously mentioned, much work is going on in the field of improved separator materials for silver cells.[65–67] Testing has indicated that improved cycles life may well result from the introduction of such materials, but at the present time no cells or batteries are commercially available using these new separators.

Mechanical stability: Silver cells and batteries have an excellent resistance to mechanical stress. As they are often used in military and space applications they are designed to meet stringent military requirements with regard to vibration, shock and acceleration. Accordingly, components such as cell containers and battery trays are made from very strong plastic or metallic materials and the assembly operations are carefully controlled.

Properties in summary: Silver alkaline cells have a very high energy content per unit of weight and volume. They have also, especially the silver/zinc types, excellent high rate discharge properties. The charge retention is very good and the mechanical stability favourable. On the other hand, they are not well suited for operation at very low temperatures. A serious limitation for silver/zinc batteries is the very poor life. Another disadvantage of silver alkaline batteries is their high cost, which has limited the use of these batteries to specialised areas where cost is of minor importance.

5.5.4 Applications

Silver alkaline batteries are used in applications where low weight and volume and favourable discharge properties are of importance and where the limitations with regard to cycle life, wet life and high cost can be tolerated. Such applications are mainly governmental and may be divided into four areas of use: underwater, ground, atmosphere and space.

Silver/zinc batteries have been used for a long time for the propulsion of torpedoes. They have also been used for mines, protective devices and various kinds of buoys. More recently this has expanded to underwater test vehicles, underwater searchlights and to submersible vessels such as rescue vehicles and various exploratory submarines. The ground applications are mainly military. Silver batteries are here used for communications equipment, portable radar sets and night vision equipment.

Atmosphere applications include specialised aircraft and helicopters and also target drones and stratosphere balloons. Aircraft and helicopters use vented silver/zinc batteries for emergency and sometimes also supplemental power aboard. However, in most military aircraft nickel/cadmium batteries furnish the DC power needed. Various ground-to-air and air-to-air missiles which are controlled by silver/zinc batteries can also be included in aerial applications. Most of these are primary batteries, but secondaries are used in some cases.

Both silver/zinc and silver/cadmium batteries are used in spacecraft for control and telemetry purposes. Here they work in parallel with solar panels. In these applications not only energy density is of importance but also, at least in some cases, the non-magnetic properties of the silver systems. Batteries for use in space craft are often hermetically sealed.

Although silver alkaline batteries are mainly used in military and space applications, they also find some use in industrial and consumer applications. Among those the following should be mentioned particularly: flash guns, cordless power tools, medical electronics, photographic equipment, portable transceivers and model ships and aircraft. Silver/cadmium batteries are also used in portable TV receivers and in hedge trimmers. Such batteries may be assembled in monoblock type plastic containers.

5.6 Nickel/zinc batteries

The first experiments with the nickel/zinc system were made around 1900 and nickel/zinc batteries were built and tried, to a limited extent, for train propulsion in Ireland in the 1930s. The interest for this sytem was revived around 1960 by the Russians, and during the last decade quite a lot of research and development work has been devoted to nickel/zinc batteries in several countries.

The reason for the present interest in the nickel/zinc system is that this couple has the inherent possibility of a high energy and power density in combination with a moderate production cost. This combination could make the system feasible as a power source for electric vehicles and other applications. Nickel/zinc batteries are, at the present time, manufactured on a limited scale in the USA, in the USSR and in Japan.

5.6.1 Reaction mechanisms

A simple overall reaction for the nickel/zinc cell may be written as

$$2\,NiOOH + Zn + 2H_2O \underset{\text{charge}}{\overset{\text{discharge}}{\rightleftarrows}} 2\,Ni(OH)_2 + Zn(OH)_2 \qquad (5.25)$$

On discharge, trivalent nickel hydroxide reacts with metallic zinc and water to form divalent nickel hydroxide and zinc hydroxide.

Regarding the reactions at the separate electrodes, see Sections 5.2.1 and 5.5.1, where the reaction mechanisms for the nickel hydroxide and zinc electrodes are discussed in some detail.

The nickel/zinc system is sometimes designed to be sealed. In such cells oxygen gas evolved during charging may be recombined at the zinc electrode in a similar manner as that described for sealed nickel/cadmium cells in Section 5.4.1. Oxygen could also be reduced by means of a separate auxiliary electrode connected to the zinc electrode.

5.6.2 *Manufacturing processes*

As the production of nickel/zinc batteries is still very limited and mainly of a pilot plant nature, and as the types manufactured are more or less custom-made, it is difficult to discuss production procedures for these batteries. However, some general information will be given in the following text.

Manufacture of electrodes: The positive electrodes may be of standard sintered or pocket type, but in general it is known that such electrodes are too costly or have too low an energy density. Accordingly, present efforts are directed towards pressed powder, plastic bonded or other improved types of nickel electrodes with an energy density exceeding that of the pocket electrode and at a cost considerably lower than that of the sintered nickel electrode.[68–70]

The zinc electrodes are normally manufactured in the same way as for silver/zinc cells but in some cases with special additives. A completely new approach has recently been made in Sweden. The zinc is precipitated on simple steel grids during each charge of the cell. The negative electrodes are designed to be vibrated during the charging operation which results in a smooth zinc deposit, thus avoiding the life limiting problems of zinc dendrite growth and shape change. The zinc electrodes are also designed to limit the cell capacity and all zinc is dissolved during a complete discharge. This zinc electrode construction has not yet been incorporated in commercially available nickel/zinc batteries.

Separators: The separator is the crucial component in nickel/zinc cells. It must, among other things, prevent zinc dendrite penetration and also have the capability of resisting swelling and warping of the electrodes. Separators used in nickel/zinc cells are generally of the multilayer type and are often from the same materials tried in the silver/zinc system. They include various cellulosics and non-woven materials as well as membranes of irradiated polyethylene, polyvinylalcohol and inorganics. Certain flexible separators have been developed specifically for nickel/zinc cells.[73–74]

A special case is the separator design for nickel/zinc cells with vibra-

Fig. 5.23 *350 Ah experimental nickel/zinc cell with vibrating negatives*
a Negative steel grid
b Positive nickel electrode

ting negatives. Here, a simple open plastic net is considered sufficient, as the zinc dendrite and shape change problems are eliminated.

Assembly of cells and batteries: Nickel/zinc cells are usually built in a rectangular shape, although cylindrical cells are also manufactured. Rectangular cells are assembled in plastic or steel containers in much the same manner as silver/zinc cells. Cylindrical cells are assembled in steel cans using long nickel and zinc electrodes which are coiled together with strips of separator material. The technique is very similar to that used for sealed cylindrical nickel/cadmium cells.

Batteries are assembled from these cells in conventional ways. An exception is the nickel/zinc type with vibrating zinc electrodes. Special measures must be taken to provide the vibration during charging. For this purpose, a mechanical arrangement including electric motors, rotating axles, and excentre devices is partially built into and partially attached to the battery.

The electrolyte used in nickel/zinc batteries is normally a 25-35% potassium hydroxide solution. For improved nickel electrode cycle life, sodium hydroxide with lithium hydroxide addition has also been tried.[70] The electrolyte may be saturated with zinc oxide to minimise dissolution of zinc during discharge. Fig. 5.23 shows a nickel/zinc cell with vibrating negatives.

5.6.3 *Performance characteristics*

Nickel/zinc cells are available in sizes from 0•45 to 850 Ah. Manufacturers of nickel/zinc cells and types of cells supplied are listed in Table 5.12.

Energy density: An interesting feature of the nickel/zinc system is the high content of energy per unit weight and volume. Typical data for cells are 55 ± 20 Wh/kg and 100 ± 40 Wh/dm^3. Complete batteries will have 10-20% lower values. Present research and development work indicates that 80-90 Wh/kg may be reached with nickel/zinc batteries.

Discharge properties: Nickel/zinc cells have favourable discharge characteristics. The cell voltage is high (nominally 1•60-1•65V) and the cells can be effectively discharged at high rates and in a relatively wide temperature range.

As indicated in Fig. 5.24, the discharge curves are rather flat, and even a considerable increase in current does not result in any pronounced polarisation or loss of capacity. The operating temperature range is −40 to +50°C. Sealed nickel/zinc cells generally display somewhat poorer discharge characteristics than vented cells.

Charge characteristics: Nickel/zinc cells can be charged using constant current or constant potential methods. A special problem with the charging of nickel/zinc is that although the nickel electrode will require a certain amount of overcharge to become fully charged, such overcharging is detrimental to the life of the zinc electrode. A charging technique is often applied where a constant current of 0•2-0•25 x C is used up to a preset cut-off voltage of 2•0V per cell in order to limit the overcharge to an acceptable level. A constant current charging curve is shown in Fig. 5.24. Float charging of nickel/zinc cells is also possible. The float voltage should be 1•9V at room temperature.

Charge retention: Vented nickel/zinc cells have approximately 70% of their capacity available after a rest period of 1 month at room

Table 5.12 *Manufactures of nickel/zinc cells and basic ranges supplied*

Manufacturer	Trademark	Designation	Cells manufactured Capacity Ah	Cells manufactured Type	Notes
Eagle-Picher Industries Inc., USA	EAGLE-PICHER EP	NZC	2·0-850	vented, rectangular	custom designed
USSR battery producers	-	NTs	125	vented rectangular	for electric vehicles
Tokyo Shibaura Electric Co., Japan	Toshiba	AA, C	0·45, 1·5	sealed, cylindrical	for portable appliances

A number of companies are in the pilot plant manufacturing stage and others produce special prototype nickel/zinc batteries for evaluation. Among these may be mentioned: ESB, Gould, General Motors, Energy Research, and Yardney in the USA and Furukawa in Japan.

temperature. For sealed cells the corresponding figure is 50%. The charge retention is thus relatively poor for the nickel/zinc system.

Life: The main drawback to presently available nickel/zinc batteries is their limited cycle life capability. They suffer from the usual problems associated with cycling of the zinc electrode: shape change and zinc dendrite formation. Typically, the life of a 'conventional' nickel/zinc battery is 200 relatively deep cycles.

However, much research and development work is being carried out to improve this situation. New separator materials have been studied with promising results. One of these is an inorganic/organic flexible separator developed by NASA's Lewis Research Center,[73] which has extended the life of nickel/zinc experimental cells to 800 cycles at 50% depth of discharge.

Another approach is the novel cell design using vibrating zinc electrodes. With this design it has been demonstrated that full scale battery cells may last more than 1200 deep cycles without failure and with a moderate loss of capacity.[71,72]

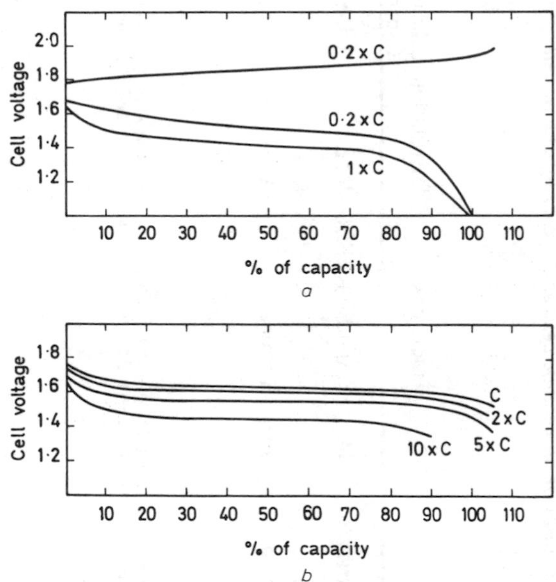

Fig. 5.24 *Charge and discharge characteristics of Ni/Zn cells at various constant rates*
Temperature 20°C
a Sealed cylindrical cell type C manufactured by Tokyo Shibaura Electric Company
b Vented rectangular cells type NZS manufactured by Eagle-Picher Industries Incorporated

Mechanical stability: Nickel/zinc cells and batteries display very good resistance to all sorts of mechanical abuse. In this respect they have properties similar to nickel/cadmium and silver alkaline batteries.

Properties in summary: Nickel/zinc cells and batteries have a high energy density, comparable to that of silver/cadmium cells. They have favourable discharge properties and exhibit small polarisation at high rate discharge. The operating temperature range is -40 to $+50°C$. The mechanical stability of nickel/zinc batteries is very satisfactory. A further advantage is the inherent possibility of a low production cost of the system. The main drawback is the limited cycle life of presently available nickel/zinc batteries. Further disadvantages are the relatively poor charge retention and the sensitivity to overcharging.

5.6.4 Applications

At the present time nickel/zinc batteries are used in a relatively limited number of applications. Sealed cells are used in portable tools and appliances where the high energy density is valued, e.g. in cine cameras, electric shavers, tape recorders, electric drills and searchlights. They are also used in tranceivers and calculators. In these applications the operation time normally varies from 1/2 to 5 h.

Bigger vented batteries, like the USSR NTs - 125 type, are used for the propulsion of certain electric vehicles. The application of nickel/zinc batteries in the traction field, fork lift trucks, electric cars, golf carts etc., may grow considerably in the future, provided the deep-discharge capability of the system can approach at least 500 cycles while maintaining a high energy density and a reasonable cost. As previously mentioned, extensive research and development efforts are being made to this end in the USA, in Japan, and in other countries.

The use of large nickel/zinc batteries for industrial standby, emergency power and diesel engine cranking has also been discussed. However, for such long life, high reliability applications it would seem that considerable long-term improvements would have to be made in the nickel/zinc system before use in these areas could become a reality.

On the military side there is, for instance, an interest to use large nickel/zinc batteries for powering ground electronics in certain weapon systems. In this case the shorter life and higher total cost of the nickel/zinc system as compared to the nickel/cadmium system might be compensated by the advantages of the lower weight and bulk of the nickel/zinc system.

References

1 FALK, S.U., and SALKIND, A.J.: *Alkaline storage batteries* (John Wiley & Sons, 1969), pp. 1-41

2 GERASIMOV, A.G.: 'Status and development prospects of chemical current sources'. Paper presented at the World electrotechnical congress, Moscow, 21st-25th June, 1977

3 BRIGGS, G.W.D.: *Electrochemistry – Vol. 4* (The Chemical Society, 1974), pp. 33-54

4 CHEREPKOVA, I.A. KAS'YAN, V.A., SYSOEVA,, V.V., MILYUTIN, N.N., and ROTINYAN, A.L.: 'Mechanism of reduction reactions at the nickel oxide electrode', *Elektrokhimiya,* 1975, **11**, 3, pp. 443-447

5 KUDRYAVTSEV, Yu. D., KUKOZ, F.I., and FESENKO, L.N.: 'An experimental study concerning the current distribution in a porous nickel electrode during polarization with alternating current', *ibid*, 1975, **11**, 3, pp. 378-382

5a NOVAKOVSKII, A.M., and UFLYAND, N.Yu.: 'Composition and thermodynamic characteristics of higher-valence nickel oxides', *Zashchita Metallov*, 1977, **13**, No. 1, pp. 22-28

6 TYSYACHNYI, V.P., KSENZHEK, O.S., and POTOTSKAYA, L.M.: 'Charging and discharge of nickel oxide films under potentiostatic conditions', *Elektrokhimija.,* 1975, **11**, 6, pp. 980-983

7 KAS'YAN, V.A., SYSOEVA, V.V., MILYUTIN, N.N., and ROTINYAN, A.L.: 'Mechanism of the processes occurring during discharge of nickel oxide electrodes', *ibid*, 1975, **11**, 9, pp. 1427-1429

8 ORD, J.L.: 'An optical study of the deposition and conversion of nickel hydroxide films', *Surface Science*, 1976, **56**, pp. 413-424

9 ARMSTRONG, R.D., EDMONSON, K., and WEST, G.D.: *Electrochemistry – Vol. 4* (The Chemical Society, 1974), pp. 18-32

10 ICHIKAWA, F., and SATO, T.: 'Amphoteric character of cadmium hydroxide and its solubility in alkaline solutions', *J. Inorganic and Nuclear Chemistry*, 1973, **45**, pp. 2592-2594

11 SIVARAMAIAH, G., VASUDEVA RAO, P.V., and UDUPA, H.V.K.: 'Direct electrochemical reduction of cadmium oxide in alkaline electrolyte', *Transactions of Saest*, 1974, **9**, pp. 167-169

12 BARNARD, R., LEE, J.A., RAFINSKI, H., and TYE, F.L.: 'Investigation into the mechanism of capacity loss by porous electrodes during cycling' *In* COLLINS, D.H. (Ed.) *Power Sources 5, 1974* (Academic Press, 1975), pp. 183-209

13 BARNARD, R., CRICKMORE, G.T., LEE, J.A., and TYE, F.L.: 'A cause of "stepped" discharge curves in nickel-cadmium cells.' Paper presented at the 10th International symposium, Brighton, September 1976

14 WILL, F.G., and HESS, H.J.: 'Morphology and capacity of a cadmium electrode', *J. Electrochem. Soc.,* 1973, **120**, pp. 1-11

15 CASEY, E.J., and GARDNER, C.L.: 'Anodic passivation by "CdO" studied by ESR', *ibid.*, 1975, **122**, pp. 851-854

16 McCORMICK, R.J., and RYAN, W.E.: 'Method of fabricating iron

electrodes for alkaline storage batteries', US Patent 3, 525, 640, Aug 25, 1970

17 BIRGE, J., BROWN, J.T., FEDUSKA, W., HARDMAN, C.C., POLLAK, W., ROSEY, R., and SEIDEL, J.: 'Performance characteristics of a new iron-nickel cell and battery for electric vehicles'. *In* COLLINS, D.H. (Ed.) *Power sources 6, 1976* (Academic Press, 1977), pp. 111-128

18 FUKUDA, M., IWAKI, T., and MITSUMATA, T.: 'Process for preparing a sintered iron negative plate for an alkaline storage battery', US Patent 3, 847, 603, Nov. 12, 1974

19 LINDSTROM, O.: 'Method of making a porous electrode for electrochemical cells', US Patent 3, 802, 878, Apr. 9. 1974

20 ARMSTRONG, R.D., and BAURHOO, I.: 'Solution soluble species in the operation of the iron electrode in alkaline solution', *J. Electroanal. Chem.*, 1972, **34**, pp. 41-46

21 SALKIND, A.J., VENUTO, G.J., and FALK, S.U.: 'The reaction at the iron alkaline electrode', *J. Electrochem. Soc.*, 1964, **111**, pp. 493-495

22 ASAKURA, S., and KEN NOBE: 'Kinetics of anodic processes on iron in alkaline solution', *ibid.*, 1971, **118**, pp. 536-541

23 KLEINSORGEN, K., NESS, P., and VOSS, E.: 'Elektrische Eigenschaften einer kunststoffverfestigten Eisenelektrode', *Metalloberfläche*, 1971, **25**, pp. 49-53

24 HAMPSON, N.A., LATHAM, R.J., and OLIVER, A.N.: 'Cathodic polarization of pure iron in concentrated alkaline solution', *J. Appl. Electrochem.*, 1973, **3**, pp. 61-64

24*a* HAMPSON, N.A., LATHAM, R.J., MARSHALL, A., and GILES, R.D.: 'Some aspects of the electrochemical behaviour of the iron electrode in alkaline solution', *Eletrochimica Acta*, 1974, **19**, pp. 397-402

25 GERONOV, Y., TOMOV, T., and GEORGIEV, S.: 'Mössbauer spectroscopy investigation of the iron electrode during cycling in alkaline solution', *J. Appl. Electrochem.*, 1975, **5**, pp. 351-358

26 ÖJEFORS, L.: 'Self-discharge of the alkaline iron electrode', *Electrochimica Acta*, 1976, **21**, pp. 263-266

27 SATHYANARAYANA, S., SRIDHARAN, L.N., and GOPIKANTH, M.L.: 'New process for the incorporation of the active mass into sintered, electrodes of the nickel-cadmium battery', *In* COLLINS, D.H. (Ed.) *Power sources 6, 1976* (Academic Press, 1977), pp. 201-213

28 KROGER, H.H.: 'Nickel hydroxide battery electrode development'. Technical report AFAPL - TR - 72 - 35, Air Force Aero Propulsion Laboratory, July, 1972.

29 SEIGER, H.N., and PUGLISI, V.J.: 'Nickel oxide electrode development'. Proceedings of the 27th power sources symposium, October, 1976, pp. 115-120

30 EAGLE-PICHER INDUSTRIES INC: 'High energy density sealed nickel-cadmium systems'. Company report, 12 December 1975

31 GENERAL ELECTRIC COMPANY: *Nickel-cadmium battery, application engineering handbook*, 1975, pp.. 4-1 - 4-40, 5-1 - 5-28

32 SAFT STORAGE BATTERY DIVISION: 'SAFT sintered plate sealed cells VR VB VX', 1974, pp. 3-11

33 BISHOP, W.S., and LANDER, J.J.: 'Low maintenance battery systems for aircraft applications'. *In* COLLINS, D.H. (Ed.) '*Power sources 6,*

aqueous KOH solution - II. Passivation experiments using linear sweep voltametry', *Electrochimica Acta*, 1972, **17**, pp. 387-394

52 POPOVA, T.I., and KABANOV, B.N.: 'Zinc passivation in alkaline solution', *Egypt. J. Chem.*, 1973, Special Issue 'Tourky', pp. 179-188

53 DIRKSE, T.P., and HAMPSON, N.A.: 'The Zn (II)/Zn exchange reaction in KOH solution - I. Exchange current density measurements using the galvanostatic method', *Electrochimica Acta*, 1972, **17**, pp. 135-141

54 BOCKRIS, J.O'M., NAGY, Z., and DRAZIC, D.: 'On the morphology of zinc electrodeposition from alkaline solutions', *J. Electrochem. Soc.*, 1973, **120**, pp. 30-41

55 JUSTINIJANOVIĆ, I.N., and DESPIĆ, A.R.: 'Some observations on the properties of zinc electrodeposited from alkaline zincate solutions', *Electrochimica Acta*, 1973, **18**, pp.. 709-717

56 DIGGLE, J.W., FREDERICKS, R.J., and REIMSCHUESSEL, A.C.: 'Crystallographic and morphological studies of electrolytic zinc dendrites, grown from alkaline zincate solutions', *J. Mat. Sci.*, 1973, **8**, pp. 79-87

57 LEE, T.S.: 'Hydrogen overpotential on zinc containing small amounts of impurities in concentrated alkaline solution', *J. Electrochem. Soc.*, 1973, **120**, pp. 707-709

58 SHAMS EL DIN, A.M., ABD EL WAHAB, F.M., and ABD EL HALEEM, S.M.: 'Effect of electrolyte concentration on the passivation of Zn in alkaline solutions', *Werkstoffe und Korrosion*, 1973, **24**, pp. 389-394

59 ARMSTRONG, R.D., and BELL, M.F.: 'The active dissolution of zinc in alkaline solution', *Electroanalytical Chemistry and Interfacial Electrochemistry*, 1974, **55**, pp. 201-211

60 TURNER, J., and HUTCHINSON, P.F.: 'Improvements to the porous zinc anode under galvanostatic and pulse discharge conditions', *In* COLLINS, D.H. (Ed.) *Power sources 6, 1976* (Academic Press, 1977), pp. 335-359

61 BRESSAN, J., and WIART, R.: 'Use of impedance measurements for the control of the dendritic growth of zinc electrodeposits', *J. Appl. Electrochem.*, 1977, **7**, pp. 505-510

62 BOBKER, R.V.: *Zinc-in-alkali batteries* (Soc. of Electrochemistry, University of Southampton, 1973)

63 FALK, S.U., and SALKIND, A.J.: *Alkaline storage batteries* (John Wiley & Sons, 1969), pp. 155-179

64 FLEISCHER, A., and LANDER, J.J.: *Zinc-silver oxide batteries* (John Wiley & Sons, 1971), pp. 183-319

65 JONVILLE, P., FRESNEL, J.M., and BEZAUDUN, J.: 'A new type of inorganic separator for alkaline silver batteries'. *In* COLLINS, D.H. (Ed.) *Power sources 5, 1974* (Academic Press, 1975), pp. 233-259

66 SHEIBLEY, D.W.: 'Improved inorganic-organic separators for silver/zinc batteries', Abstract no. 24, The Electrochemical Society, 149th meeting, Washington, D.C., May 2-7, 1976

67 CHIREAU, R.F., and BERCHIELLI, A.S.: 'Inorganic separators for alkaline batteries', Abstract no. 53, The Electrochemical Society, 150th meeting, Las Vegas, Nevada, Oct. 17-22, 1976

68 MIYAKE, Y., and KOZAWA, A.: *Rechargeable batteries in Japan* (JEC Press Inc., 1977), pp. 459-470

1976 (Academic Press, 1977), pp. 103-110
34 FLEISCHER, A., and LANDER, J.J.: *Zinc-silver oxide batteries* (John Wiley & Sons, 1971), pp. 99-153
35 FRANSSEN, H., HÄUSLER, E., and BÖHNSTEDT, W.: 'Rasterelektronenmikroskopische Untersuchungen an Silberoxidelektroden', *Metalloberfläche Angewandte Elektrochemie*, 1974, **28**, Heft 4, pp. 140-142
36 HAMPSON, N.A., LEE, J.B., and MORLEY, J.R.: 'The electrochemistry of oxides of silver - a short review', *Electrochimica Acta*, 1971, **16**, pp. 637-642
37 WALES, C.P.: 'Effects of KOH concentration on morphology and electrical characteristics of sintered silver electrodes'. *In* COLLINS, D.H. (Ed.) *Power sources 4, 1972* (Oriel Press, 1973), pp. 163-183
38 TILAK, B.V., PERKINS, R.S., KOZLOWSKA, H.A., and CONWAY, B.E.: 'Impedance and formation characteristics of electrolytically generated silver oxides - I Formation and reduction of surface oxides and the role of dissolution processes', *Electrochimica Acta*, 1972, **17**, pp. 1447-1469
39 HOAR, T.P., and DYER, C.K.: 'The silver/silver-oxide electrode - I. Development of electrode by slow ac cycling', *ibid.*, 1972, **17**, pp. 1563-1584
40 SATO, N., and SHIMIZU, Y.: 'Anodic oxide on silver in alkaline solution', *ibid.*, 1973, **18**, pp. 567-570
41 AMBROSE, J., and BARRADAS' R.G.: 'The electrochemical formation of Ag_2O in KOH electrolyte', *ibid.*, 1974, **19**, pp. 781-786
42 DYER, C.K., and HOAR, T.P.: 'The silver/silver-oxide electrode - II Influence of a superimposed sinusoidal potential on the anodising of silver', *ibid.*, 1975, **20**, pp. 161-171
43 DYER, C.K., and HOAR, T.P.: 'The silver/silver-oxide electrode - III Anodising of silver with asymmetric sinusoidal current', *ibid.*, 1975, **20**, pp. 173-177
44 SASAKI, H., and TOSHIMA' A.: 'Studies on the silver/silver-oxide electrode', *ibid.*, 1975, **20**, pp. 201-207
45 GIBBS, D.B., RAO, B., GRIFFIN, R.A., and DIGNAM, M.J.: 'Anodic behaviour of silver in alkaline solutions', *J. Electrochem. Soc.*, 1976, **122**, pp. 1167-1174
46 TURNER, J.: 'Electrolytic studies on the system $Ag/Ag_2O/AgO$ in alkaline chloride solutions', *J. Appl. Electrochem.*, 1977, **7**, pp. 369-378
47 FLEISCHER, A., and LANDER, J.J.: *Zinc-silver oxide batteries* (John Wiley & Sons, 1971), pp. 19-95
48 NAGY, Z., and BOCKRIS, J.O'M.: 'On the electrochemistry of porous zinc electrodes in alkaline solutions', *J. Electrochem. Soc.*, 1972, **119**, pp. 1129-1136
49 GREGORY, D.P., JONES, P.C., and REDFEARN, D.P.: 'The corrosion of zinc anodes in aqueous alkaline electrolytes', *ibid.*, 1972, **119**, pp. 1288-1292
50 BOCKRIS, J.O'M., NAGY, Z., and DAMJANOVIC, A.: 'On the deposition and dissolution of zinc in alkaline solutions', *ibid.*, 1972, **119**, pp. 285-295
51 DIRKSE, T.P., and HAMPSON, N.A.: 'The anodic behaviour of zinc in

69 EAGLE-PICHER INDUSTRIES INC.: 'Nickel battery systems for electric vehicles'. Company report, April 1976

70 YARDNEY ELECTRIC DIVISION: 'Design and cost study, zinc/nickel oxide battery for electric vehicle propulsion', Final report no. 2033-76 prepared for Argonne National Laboratory, Contract No. 31-109-38-3543, October 1976

71 KRUSENSTIERNA, O. von, and REGER, M.: 'A high energy nickel/zinc battery for electric vehicles'. Society of Automotive Engineers, Report no. 770384, February 28 - March 4, 1977

72 KRUSENSTIERNA, O. von: 'High-energy long life zinc battery for electric vehicles'. *In* COLLINS, D.H. (Ed.) *Power sources 6, 1976* (Academic Press, 1977), pp. 303-319

73 SHEIBLEY, D.W.: 'Separators for nickel/zinc batteries', abstract no. 25, The Electrochemical Society, 149th meeting, Washington, D.C., May 2-7, 1976

74 CHARKEY, A.: 'Advances in component technology for nickel/zinc cells'. Proceedings of 11th Intersociety Energy Conversion Engineering Conference, State Line, Nev., USA, 12-17 Sept. 1976, pp. 452-456

Chapter 6

High temperature batteries

J.L. Sudworth

6.1 Introduction

The use of aqueous electrolytes limits the number of materials which can be considered as electrodes for galvanic cells and effectively rules out the use of high energy couples. If operating temperatures higher than ambient are permissible, molten salt electrolytes and solid electrolytes can be used, allowing a much wider choice of galvanic couples to be considered.

There are a number of applications requiring batteries which are large and intensively used and thus able to maintain their operating temperature without external heating. The most important are load levelling in the electricity supply industry and vehicle propulsion. High temperature batteries are being developed for both applications.

It is not known whether these batteries can be repeatedly cooled to room temperature and reheated to the operating temperature without damage, and most proposals for the use of high temperature batteries assume that the battery will always be kept hot. This leads to two operational constraints: first, when not in use for extended periods of time (the actual length of time depends on the operating temperature, the freezing point of the active materials or electrolyte and the heat transfer properties of the thermal insulation used), an external power supply must be provided to heat the battery; secondly, if cells have to be changed in a battery, the operation must be done while the battery is hot. These constraints are not insuperable obstacles to the introduction of high temperature batteries, but they do mean that these batteries must offer significant advantages over ambient temperature systems. Briefly, these advantages are a high specific energy and power and a potentially low cost.

The cells which will be described here are of two types: molten salt electrolyte cells and solid electrolyte cells. In the former type, lithium is the negative electrode and in the latter type, sodium. The molten salt electrolyte cells described in this chapter are: lithium/chlorine; lithium-aluminium/carbon-tellurium tetrachloride; lithium sulphur; lithium-aluminium/iron-sulphide; and lithium-silicon/iron sulphide. Lithium-aluminium/iron-sulphide cells have received most attention in the literature and corresponding emphasis will be placed on them here.

The solid electrolyte cells described are: sodium/sulphur with beta alumina electrolyte; sodium/sulphur with glass electrolyte, and sodium/antimony trichloride with beta alumina electrolyte. The sodium/sulphur cell with beta alumina electrolyte is being actively developed in many centres, whereas the other two systems are being developed only in the laboratories where they were first studied; consequently, most of the section on solid electrolyte batteries will be devoted to the sodium/sulphur cell with beta alumina electrolyte.

6.2. Thermal management

The operating temperature of these batteries lies in the range 200°-650°C and obviously the cells must be contained in a thermally insulated container. The thermal characteristics of a battery are determined by the thermodynamics of the cell reaction and by the duty cycle. During operation a cell may absorb or evolve heat, depending on whether the cell voltage is above or below the thermoneutral potential; the cell voltage at which the passage of current will result in neither the absorption nor evolution of heat. It is given by

$$E_t = -\frac{\Delta H}{nF} \tag{6.1}$$

When a cell is operating reversibly, the heat absorbed or evolved is the difference between the enthalpy of the reaction and the free energy of the reaction, which, from eqn. 2.12, is $T\Delta S$. Thus we can write

$$Q = \frac{T\Delta S}{n} \tag{6.2}$$

where Q is heat absorbed or evolved from the cell per equivalent. From eqns. 2.12 and 2.13

$$Q = \frac{\Delta H}{n} + FE_{MF} \tag{6.3}$$

Using eqn. 6.1, Q can be expressed in terms of the open-circuit potential and the thermoneutral potential

$$Q = F(E_{MF} - E_t) \tag{6.4}$$

If the cell reaction is proceeding at 100% current efficiency the rate of heat generation P_t is given by

$$P_t = I(E - E_t) \tag{6.5}$$

when E is the cell potential at current I.

It is convenient to express the rate of heat generation in terms of the voltage efficiency of the cell η, E_{MF} and E_t

$$I = \frac{E_{MF} - E_{MF}\eta}{R} \tag{6.6}$$

Thus the rate of heat generation is given by

$$P_t = \frac{E_{MF}(1 - \eta)}{R}(E_{MF}\eta - E_t) \tag{6.7}$$

This simple relationship will apply provided that the activities of the reactants and products remain constant, which for many high temperature batteries will be the case over most of the discharge range. This is illustrated in Fig. 6.1 for a sodium sulphur cell operating in the region of constant sulphur activity, i.e. up to 60% depth of discharge. When the battery is operating isothermally, P_t will be the rate of heat flow in or out of the battery. If the battery is operating non-isothermally, a knowledge of the specific heat of the battery would be necessary to calculate the rate of heat flow.

Thus, provided the thermoneutral potential of a cell is known and the assumptions mentioned above are valid, the rate of heat evolution or absorption can be calculated for any cell power level. This information can then be used to calculate the thickness of thermal insulation required to maintain the battery temperature within specified limits during a given duty cycle.

For most discharge rates the thickness of insulation required will be much less than that required to keep the battery hot during an extended period on open circuit. One way of dealing with this problem is to use a thickness of insulation sufficient to keep the battery temperature within the required range during discharge and to maintain the battery temperature during periods when no power is required by electrical heaters powered by the battery. This method has two disadvantages; it degrades the battery energy density and produces a high skin temperature for the battery box.

An alternative approach is to provide thermal insulation of sufficient thickness to maintain the battery temperature during a prolonged open-circuit stand and cool the battery during discharge using either natural convection or forced air circulation. Obviously these problems of thermal management can be minimised by designing a battery with as wide a range of operating temperature as possible.

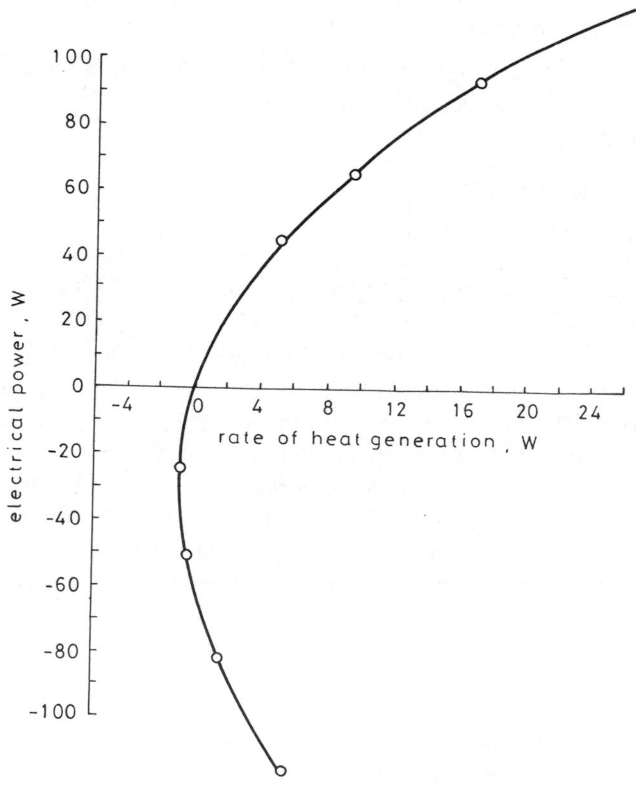

Fig. 6.1 *Rate of heat generation during charge and discharge of a sodium sulphur cell in the composition range S-$Na_2S_{5\cdot15}$ at 350°C*
E_{MF} = 2·078V, E_t = 2·20V, cell resistance 4·6 mΩ

The type of insulation used can greatly affect the specific energy, particularly the volumetric specific energy, and there are strong incentives to reduce the insulation thickness. This can be achieved by the use of so-called super insulation, which, for the same thermal transfer coefficient, can be up to an order of magnitude thinner than conventional thermal insulation. This super insulation consists of many layers of metal foil, usually aluminium, separated by fibrous material and contained in an evacuated metal container. This type of insulation is most suitable for cylindrical batteries. For prismatic (e.g. square cross-section) batteries the size of the super insulating panels is limited by the strength of the metal walls; if the panels are too large, atmospheric pressure forces the two walls of the evacuated container together, compressing the foil and increasing the thermal transfer rate. Other problems with super insulation are the provision of connections to the battery through the evacuated container and the need to allow access to the battery for replacement of failed cells.

6.3 Lithium/chlorine cells

The lithium/chlorine couple has an E_{MF} of 3·46V and a theoretical specific energy of 2200 Wh/kg at 614°C, the melting point of the most commonly used electrolyte, lithium chloride. The operating temperature is 650°C. The lithium metal is contained in a metal fibre wick and the chlorine gas electrode is made of porous carbon.

The high operating temperature, and the fact that the product of the cell reaction is the electrolyte, combine to produce very high power densities; a current density of 5 A/cm^2 at a cell voltage of 3V is typical for this cell.[1] There is a problem associated with the presence in the chlorine of inert gases which build up in the porous electrode, causing a diffusion limitation. This effect can be overcome by arranging for an excess flow of gas to purge out the impurities.

The problems which are encountered with the design of long life, rechargeable cells include the following: the severe corrosion problems arising from the use of chlorine gas at these temperatures; the solubility of lithium metal in the electrolyte causing self discharge; the difficulty of ensuring that the lithium and chlorine produced on charge do not react chemically and that the chlorine can be compressed and liquified; and the reactivity of molten lithium, which attacks most known insulators.

A more recent development[2] is the use of a mixed alkali halide electrolyte (19 m/o LiF, 66 m/o LiCl, 15 m/o KCl) which permits the cell to be operated at 450°C, at which temperature the E_{MF} is 3·61V. The lower operating temperature has led to increased lifetime, and cells have been operated for periods of up to 668 h, 210 cycles. The material problems, however, are still severe, and the chlorine electrode gradually floods with electrolyte.

The cell design is shown in Fig. 6.2. The lithium electrode is made from stainless steel or nickel wire screen. Both materials are wetted by lithium which is retained by the structure. Vertical passages in the matrix allow the free flow of electrolyte, which at certain states of charge contains solid lithium fluoride, precipitated as the lithium chloride concentration falls.

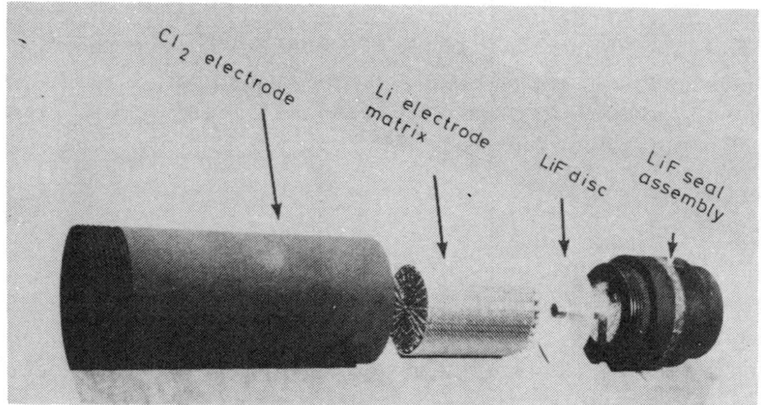

Fig. 6.2 *Lithium/chlorine cell with LiF, LicL, KCl molten salt electrolyte* [Courtesy the Electrochemical Society Inc.]

The chlorine electrode is porous graphite as in earlier versions of this cell, joined to the insulator, solid lithium fluoride, by hot pressing. The insulating seal is obtained by hot pressing lithium fluoride between the parts of the cell.

When assembling the cell, it is necessary to be scrupulous in excluding traces of oxygen and water vapour, and an extensive de-gassing and reducing procedure is followed. The lowest practicable operating temperature is 450°C, as below this temperature lithium metal is released from the matrix into the electrolyte, where it can react with chlorine gas. Solubility or dispersion of lithium into the melt is the main cause of self discharge even at 450°C. It is postulated that the dispersed lithium reacts with the graphite electrode to form Li_2C_2, which induces the wetting of the electrode by the normally non-wetting electrolyte, resulting in gradual penetration of the chlorine electrode by electrolyte. The characteristics of some of the cells tested

Table 6.1 *Experimental lithium/chlorine cell characteristics*

Cell number	Weight g	Volume cm^3	Li electrode Diameter mm	Electrolyte weight g	Energy Wh	Specific energy Wh/kg	Specific power W/kg
18	260	197	25	75	33	131	69
20	310	197	32	111	47	169	56
21	330	201	32	125	83	250	230
					92	280	100
22	360	209	33	125	128	350	47

Operating temperature 450°C for all cells except number 21 (425°C) are given in Table 6.1. Because of the problems mentioned this cell is not being actively developed.

6.4. Lithium-aluminium chlorine cells

The difficulties associated with the use of elemental lithium, and chlorine gas, led R.A. Rightmire and a group of workers at Standard Oil Ohio to develop what became known as the SOHIO battery.[3,4,5] By alloying the lithium with aluminium, a solid electrode was produced, which, although having a potential 300 mV less with respect to chlorine than elemental lithium, had many practical advantages. The problems associated with gaseous chlorine were eliminated by chemisorbing the chlorine on active carbon. The electrolyte used was LiCl/KCl eutetic, which melts at 352°C, and the operating temperature was in the range 352° - 500°C. This cell has an open-circuit voltage at top of charge of 3·2V.

6.4.1 The positive electrode

The electrodes are produced by calcining mixtures of active carbon fillers with organic binders such as pitch. The active filler is a carbon material which has been heated in an inert atmosphere between 500°C and 1000°C, producing a micro-crystalline material with a surface area in the range 600-1000 m^2/g. Of carbons derived from naturally occurring materials, coconut char gives the highest capacity.[4]

The electrode is then conditioned with a proprietary process which converts this material to a high surface area polymer of carbon, chlorine and alkali metals, lithium or potassium[5] Electrodes prepared in this way store, at best, 0·35 Wh/cm^3 (0·39 Wh/g). To improve the capacity of the positive electrode a number of additives were investigated.[6,7,8] The one which was finally adopted was tellurium tetrachloride, which is chemically bonded to the edges of the carbon microcrystallites. The tellurium chloride presumably undergoes the reaction shown in eqn. 6.8

$$4\,Li + TeCl_4 \rightarrow 4LiCl + Te \quad E^0_{625} = 3{\cdot}336\ V \tag{6.8}$$

The use of lithium-aluminium alloy anodes reduces the cell E_{MF} by 0·3V to 3·006V. The open-circuit voltage of 3·2V could indicate the presence of $TeCl_2$ or free chlorine. No evidence has been found for the formation of lithium telluride. Cathodes tested in half cells had capacities of 0·42 Wh/cm³, but the cathode in the best cell constructed had a capacity of only 0·3 Wh/cm³.

6.4.2 *The negative electrode*

As already mentioned, one of the problems of the lithium electrode is the solubility of lithium in its salts. Rightmire *et al.*[5] developed a solid lithium aluminium anode capable of charging and discharging at peak current densities of 2 A/cm². Owing to the low activity of lithium in this alloy, the solubility in the electrolyte is reduced a hundredfold over that of liquid lithium.

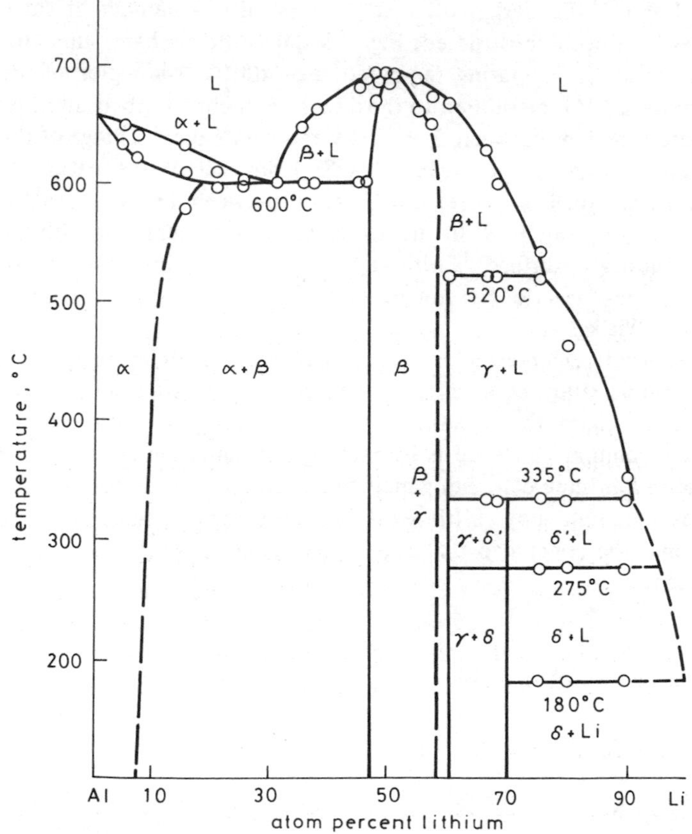

Fig. 6.3 *Phase diagram of the lithium - aluminium alloy system*
α = Al, β = LiAl, γ = Li_3Al_2, δ = δ^1 = LiqAl +

The phase diagram for Li-Al is shown in Fig. 6.3. The safe limits for operation at 500°C are 0–50 a/o lithium. This gives a negative electrode capacity of 1·3 Ah/cm^3. The electrodes can be cast from the alloy and then pre-conditioned by scraping the surface in an inert atmosphere followed by a few cycles in a cell at a low current density.[9,10] Short circuiting of the cell owing to the swelling of the negative electrode is prevented by pressing the electrode against a boron nitride separator in a sandwich type construction.

6.4.3 Cell design and performance

Fig. 6.4 shows the design of a cell from which a number of small batteries have been constructed. Fig. 6.5 shows the discharge curve for such a cell at the operating temperature of 400°C. The region of the curve above 2·9V is attributed to the discharge of chemisorbed chlorine; the 'plateau region' between 2·9 - 2·5V represents the discharge of the tellurium chloride as per reaction 6.8; below 2·5V the nature of discharge reaction is obscure, the 'plateau' between 1·7 and 1·4V is ascribed by Metcalfe *et al.* to the alkali metal ions interacting with the carbon substrate surface,[11] although what this means in terms of electrochemical reactions is not made clear. The discharge capacity to 1V was 62 Wh/kg.

More recently, workers at ESB Inc.[12] have reported on an engineering and economic study of a battery designed for fork-lift truck applications. They claim that the most suitable method of preparing the lithium-aluminium electrode is by making an aluminium electrode the cathode in a lithium containing melt and electrochemically forming the lithium-aluminium alloy. Little work has been reported since that time and it must be concluded that this battery is no longer being actively developed.

6.5 Lithium/sulphur cells

The cell reaction is

$$2\,Li + S \rightarrow Li_2S \quad \Delta G^\circ_{298} = 434 \text{ kJ/mole} \tag{6.9}$$

$$\Delta H^\circ_{298} = 446 \text{ kJ/mole}$$

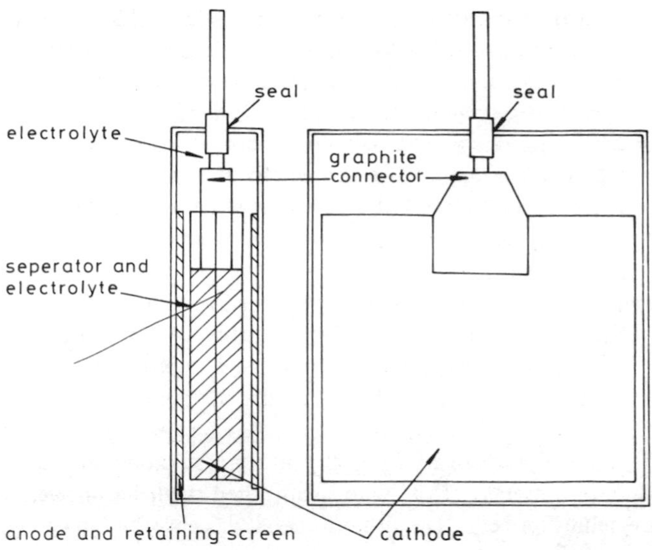

Fig. 6.4 *Schematic drawing of a lithium = aluminium/carbon-tellurium tetrachloride cell*

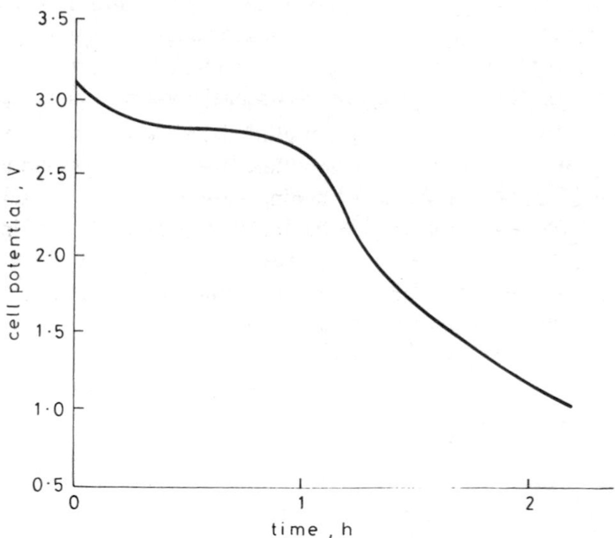

Fig. 6.5 *Discharge curve of a lithium-aluminium/carbon - tellurium tetrachloride cell at 400°C*
Discharge current 6A, electrode area 150 cm^2

This gives a theoretical specific energy of 2624 Wh/kg and a cell E_{MF} of 2·25V at 298K. The phase diagram for the lithium/sulphur system has still not been unequivocally established. Two groups of workers[13,14] report a two phase region with liquid sulphur and molten Li_2S_2. Other workers[15,16] report a two phase region with sulphur and molten Li_2S_3. There is, however, substantial agreement on the monotectic temperature in the two-liquid region of the phase diagram. This is reported to be 369·5°C,[14] 362 ± 3°C[15] and 364·8°C[16] and defines the lower limits of the operating temperature.

A number of electrolytes have been used in this cell, but the one most commonly used is the eutectic mixture of LiCl/KCl. This has a melting point of 352°C and a decomposition potential of about 3·4V. Cells containing this electrolyte are usually operated between 420° and 460°C.

The original investigation of the lithium/sulphur cell was due to Cairns and Shimotake, [17–19] who also studied the lithium/selenium and lithium/tellurium cells. The lithium metal was absorbed in spongy iron, the electrolyte was immobilised in magnesium oxide and the sulphur was absorbed in metal foam or carbon felt. These cells, which were unsealed and had to be operated in helium atmosphere glove boxes, could sustain a current of 4 A/cm^2 at a cell voltage of 1V and thus very high power densities were possible. Selenium and tellurium were not seriously considered as positive electrodes, presumably because of their toxicity and relative scarcity, and were not developed further.

Although the lithium/sulphur cell has potentially a very high specific energy and power, it also has some serious drawbacks as a secondary cell. Thus sulphur has a very high vapour pressure at the cell operating temperature [24·6kPa (0·243 atm) at 400°C]. Another problem is the difficulty of recharging the lithium electrode; filaments of lithium metal tend to form, bridging the separator and short circuiting the cell. The reactivity of molten lithium is also a formidable problem severely limiting the choice of materials for use in the cell.

One way of reducing the severity of these problems is to reduce the activity of the two reactants. This, of course, reduces the cell E_{MF}, but a reduction of the lithium activity by one order of magnitude only results in a reduction of 140 mV in the cell E_{MF} at 425°C. Thus, considerable reductions in the activities of the reactants are possible while still maintaining a reasonable cell voltage. The specific energy, however, may be considerably reduced, depending on the method used to alter the activities.

As we saw in Section 6.4, the activity of the lithium can be reduced by alloying it with aluminium. This also has the advantage of producing a solid negative electrode which can be recharged without the formation of lithium filaments. Silicon can be used instead of aluminium to form an Li/Si alloy.[20,21] The sulphur activity can be reduced by combining it with a metal to form a sulphide such as iron sulphide, FeS, or pyrites, FeS_2.[22,23]

6.6 Lithium alloy/metal sulphide cells

6.6.1 Lithium-aluminium/iron sulphide cells

6.6.1.1 Li-Al/FeS$_2$ cells: The cell reactions and corresponding potentials become[24]

$$4\,Li + 3\,FeS_2 \rightarrow Li_4Fe_2S_5 + FeS \quad 1{\cdot}8V \tag{6.10}$$

$$2\,Li + Li_4Fe_2S_5 + FeS \rightarrow 3\,Li_2FeS_2 \quad 1{\cdot}6V \tag{6.11}$$

$$6\,Li + 3\,Li_2FeS_2 \rightarrow 6\,Li_2S + 3\,Fe \quad 1{\cdot}3V \tag{6.12}$$

The theoretical specific energy is 638 Wh/kg. The negative electrode can be fabricated by loading powdered Li/Al alloy into iron Retimet, a porous iron structure which serves as a current collector. Alternatively, lithium metal can be deposited into an electrode consisting of a pressed aluminium powder plaque containing a few percent of stainless steel wire which acts as a current collector. The lithium aluminium alloy is formed by making the plaque the cathode in an LiCl/KCl eutectic at about 400°C (the anode is lithium metal absorbed on a felt metal).

The positive electrode is iron pyrites, FeS_2 and can be fabricated by loading powdered FeS_2 into a porous current collector. Vitreous carbon foam has been used as the current collector.[25] Addition of about 25 Wt% CoS_2 has been found to improve the performance of this electrode.[25]

Loading powdered FeS_2 into a vitreous carbon foam is not an economic method of fabricating these electrodes. An alternative construction, which has been evaluated at Atomics International[26] is to use a molybdenum current collector housing equipped with horizontal shelves at 12·5 mm spacing. Moybdenum is used because it is the only metal which is not corroded in this cell environment.

The positive electrode is wrapped in zirconia cloth to retain the fine particles and in a boron nitride cloth or felt for electrical insulation. The cell is assembled with the positive electrode sandwiched between two negative electrodes and with an additional sheet of zirconia cloth between the electrodes to prevent shorting by the finely divided Li-Al alloy formed during cell operation. After assembly the cell is heated to 450°C and filled with molten LiCl/KCl eutectic. A typical cell design is shown in Fig. 6.6.

One problem encountered in the operation of the FeS_2 electrode is the tendency of the active material to swell and be extruded from the electrode cavities.[26] The forces involved are often great enough to fracture a rigid separator and the use of flexible separators is probably essential.

6.6.1.2 Li-Al/FeS cells: The main drawback of the FeS_2 electrode is the need to use a costly moybdenum current collector. This is not necessary if FeS is used as the electrode.

The cell reaction in this cell appears to be[24]

$$\text{First discharge: } 2\,Li + FeS \rightarrow Li_2S + Fe \qquad (6.13)$$

$$\text{Recharge: } \begin{cases} 8\,Li_2S + 8\,Fe + 2\,KCl \rightarrow K_2Fe_7S_8 + Fe + 2LiCl + 14Li & (6.14) \\ K_2\,Fe_7\,S_8 + Fe + 2\,LiCl \rightarrow 8\,FeS + 2\,KCl + 2\,Li & (6.15) \end{cases}$$

The theoretical specific energy of the Li-Al/FeS cell is 450 Wh/kg and the average cell E_{MF} at 400°C is 1·30V.

The capacity of the FeS electrode is reported to be less than 100% of theoretical, e.g. 80% for electrodes 4·8 mm thick.[26] This is thought to be due to the formation of *J* phase ($K_2Fe_7S_8$ or possibly $LiK_6Fe_{24}S_{26}Cl$), which reportedly surrounds the particles of FeS and hinders discharge.[26] Addition of 20 mol% Cu_2S to the FeS electrode increases active material utilisation but it has been observed that copper migrates to the electrode face and into the separator, eventually causing cell shorting.[26]

The electrode structure used to contain the FeS is fabricated from iron,[27] steel or, preferably, nickel,[26] which is more corrosion resistant. These structures are much less costly and easier to fabricate than the molybdenum structures required for FeS_2 electrodes and this advantage helps to offset the relatively low specific energy of the Li-Al/FeS

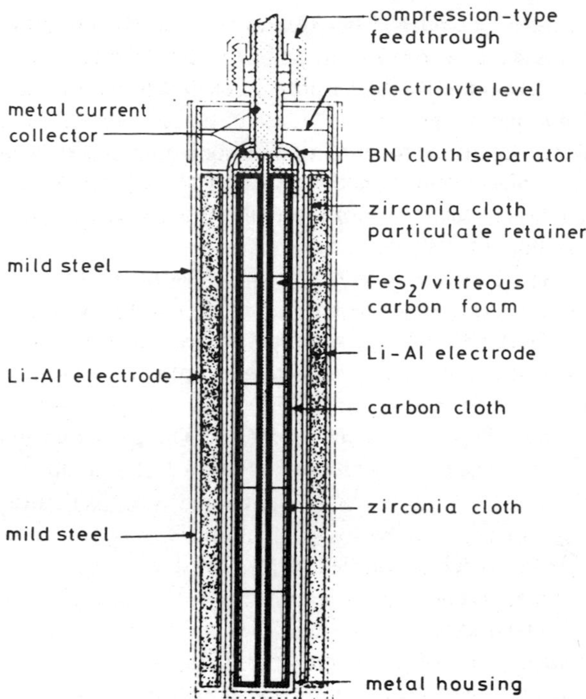

Fig. 6.6 *Sealed lithium - aluminium/iron sulphide cell*
[Courtesy Academic Press Inc. (London), *Power sources 6*, 1977, D.H. Collins (Ed.), from paper 47, E.C. Gay *et al.*]

couple. The FeS electrode, however, does suffer from one of the problems of the FeS_2 electrode; swelling during discharge. In this electrode this is attributed to the formation of *J* phase.

6.6.1.3 Cells assembled in the discharged state: The cells already described were assembled in the charged state, using Li/Al alloy and iron sulphides. These cells suffer from a number of serious drawbacks, namely, swelling of the positive electrode during the first discharge, low utilisation of the positive active material, and the evolution of large amounts of gas during starting and early cycling. Handling Li/Al alloy is also inconvenient. These problems can be largely overcome by assembling cells in the discharged state.[28,29]

For Li-Al/FeS cells the positive electrode is fabricated by mixing together iron powder, Li_2S powder (prepared by sintering Li_2S at 1200°C and grinding the resulting hard cake), and 20-35% by weight

LiCl/KCl eutectic powder. The mixture is poured over an electrode plate having a steel mesh substrate and vertical ridges to prevent slumping of the active material. The plate is then hot pressed at 380°C and 2800 kPa (400 p.s.i).

The negative electrode is fabricated from aluminium mesh; a sheet of stainless steel placed between the layers of aluminium mesh acts as a current collector. The separator configuration and the method of cell assembly are as previously described.

Figs. 6.7 and 6.8 show typical charge/discharge curves for FeS_2 and FeS cells, respectively. The capacity as a function of cycle life for the Li-Al/FeS cell is shown in Fig. 6.9. The swelling observed in cells assembled in the charged state is much reduced in this type of cell. This is obviously due to the positive active material being in its most expanded state when the cell is assembled. The gas evolution is also much reduced, presumably due to the use of aluminium instead of Li-Al alloy which is easily contaminated with oxygen, nitrogen or moisture.

The Li-Al/FeS_2 cells described so far have had two plateaux in the discharge curve. Cells operating only on the upper plateau have a number of advantages, apart from the higher average voltage and simpler voltage control requirements. These are potentially good performance at high discharge rates and a high energy efficiency. The specific energy is, of course, lower than that of two plateaux cells (514 Wh/kg).

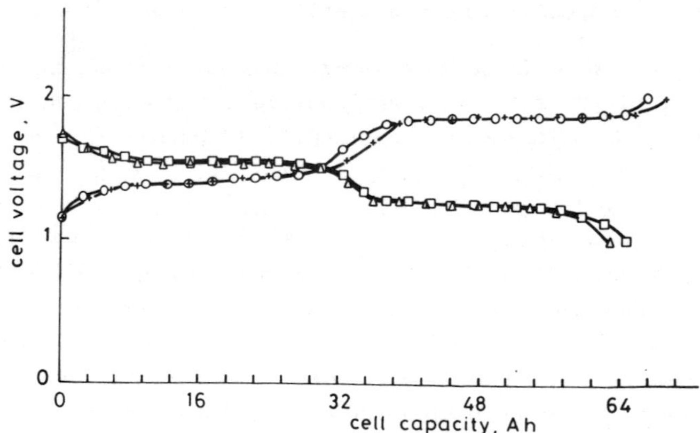

Fig. 6.7 *Charge/discharge curves for a lithium-aluminium/iron disulphide cell*

□ 5A discharge △ 7·5A discharge
○ 5A charge + 7·5A charge

[Courtesy Academic Press Inc. (London), *Power sources 6*, 1977, D.H. Collins (Ed.), from paper 46, W.J. Walsh and H. Shimotake]

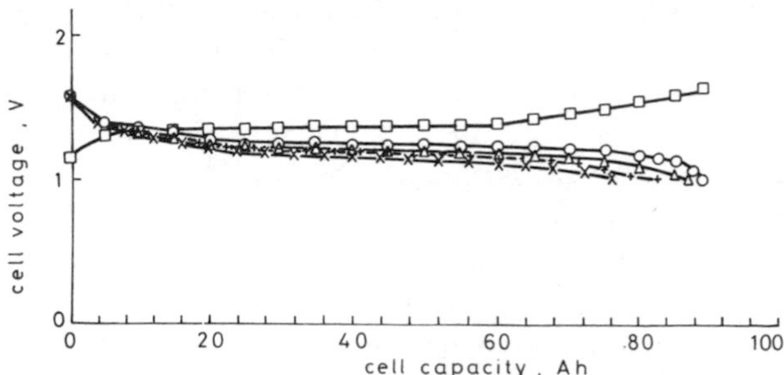

Fig. 6.8 *Charge/discharge curves for a lithium-aluminium/iron sulphide (FeS) cell*

□ 5A charge
○ 5A discharge
x 20A discharge
Δ 10A discharge
+ 15A discharge

[Courtesy Academic Press Inc. (London) Ltd., *Power sources 6*, 1977, D.H. Collins (Ed.), from paper 46, W.J. Walsh and H. Shimotake]

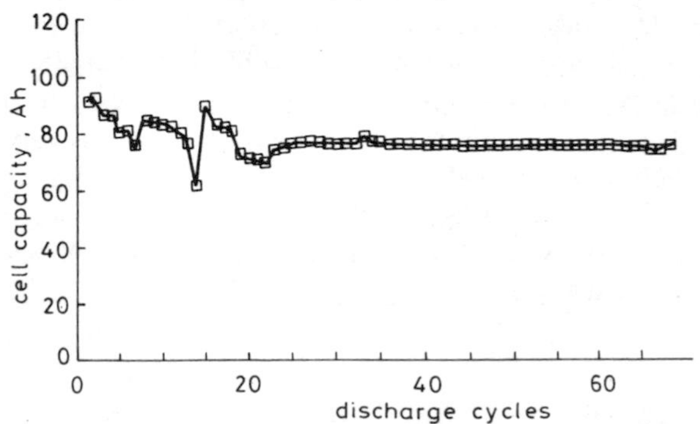

Fig. 6.9 *Capacity as a function of cycle life for a lithium-aluminium/iron sulphide (FeS) cell*

LiAl/FeS-uncharged, positive electrode area = 278 cm^2, negative electrode area - 323 cm^2, theoretical capacity of positive electrode = 121 Ah, theoretical capacity of negative electrode = 5 Ah, cell temperature = 450°C

6.6.1.4 Behaviour of Li-Al/FeS_2 cells on overcharge: The ultimate reaction on overcharge is decomposition of the electrolyte, which occurs at 3·4V. This, of course, must be prevented, as it would rapidly lead to destruction of the cell by liberated chlorine gas. Deleterious reactions can, however, occur at lower charging voltages and these have been described by Tomczuk, Holling and Steunenberg.[30] They concluded that ferrous chloride was formed at potentials in excess of 2·65V against Li, and that the reactions leading to the formation of

$FeCl_2$ were irreversible. They also concluded that occasional overcharge may not be deleterious but that consistent overcharge will probably result in a marked decline in cell performance. A battery charger which overcomes this overcharge problem has been designed and tested at Argonne National Laboratory.[31] The majority of the charging is accomplished at the full battery voltage, with the current decreasing at a preset rate throughout the charge. When the current reaches a certain value, the charging process is completed at constant voltage by an equalising charge through small auxiliary leads connected to each cell.

6.6.2 Li-Si/FeS_2 cells

The use of Li-Al alloys for the negative electrode, while overcoming many of the problems encountered in the use of pure lithium, leads to a significant penalty in terms of cell voltage and electrode capacity. McCoy *et al.*[32] and Hall *et al.*[33] have reported that superior performance can be obtained from the use of Li-Si alloys. The alloys formed and their electrode potentials with respect to lithium are shown in Fig. 6.10. The fully charged electrode is Li_5Si and the average voltage of all plateaux covered in the complete discharge of Li_5Si is 230 mV positive to Li compared with 303 mV for Li-Al alloys, and the electrode capacity is 2·12 Ah/gm compared with 0·8 Ah/gm for Li-Al alloys, giving a specific energy of 977 Wh/kg.

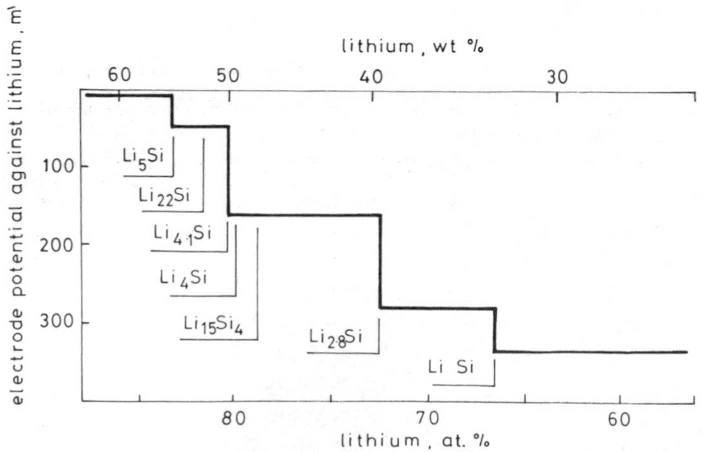

Fig. 6.10 *Electrode potential of lithium silicon alloys against the lithium electrode*

Cells containing Li-Si negative electrodes have shown good capacity retention over a long period of time,[34] i.e. no capacity loss after 1000 h and 40% loss after 11 000 h. However, accelerated tests at higher temperatures (550°C) have shown that the stainless steel honeycomb used to contain the alloy is embrittled by silicon transfer. Titanium and molybdenum appear resistant to attack by silicon at 550°C and the former is preferred for economic reasons.

6.6.3 Status of lithium alloy/iron sulphide cells

The number of different lithium alloy/iron sulphide cells under development at the present time is perhaps an indication of the problems that have been met in trying to develop this type of cell. Considerable progress has been made in improving the longevity of cells, but this has been achieved by accepting a lower specific energy. There still remains the problem of the high cost of materials such as boron nitride cloth and molybdenum. Alternatives to these materials need to be identified.

6.7 Sodium/sulphur cell

6.7.1 Introduction

This is a solid electrolyte cell which utilises liquid sodium and sulphur as the negative and positive electrodes, respectively. Two types of electrolyte are used: a polycrystalline ceramic, known as beta alumina, or a conducting borate glass. In both cases sodium is the mobile ion and the operating temperatures are normally 350° and 300°C, respectively,

Sodium and sulphur react electrochemically to form sodium polysulphides. The phase diagram for sodium sulphide/sulphur determines many of the cell characteristics and is shown in Fig. 6.11.

It is preferable to keep to the liquidus region of the phase diagram during cell operation and this puts a lower limit to the operating temperature of 285°C. The depth of discharge is normally limited to the stoichiometry Na_2S_3 and in what follows the depth of discharge, utilisation of active material and charge acceptance will be referred to this composition, as indicated in Fig. 6.12, which is a plot of the cell open-circuit voltage against sulphur electrode composition after Gupta

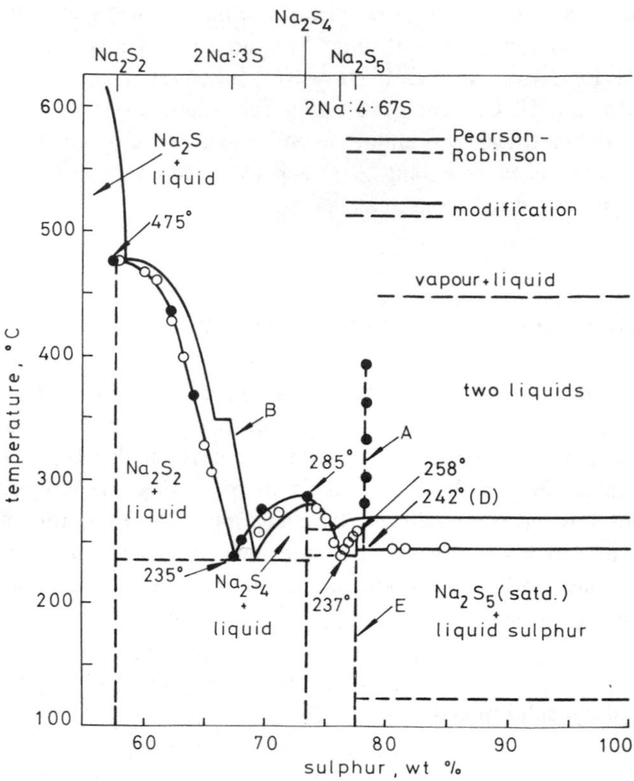

Fig. 6.11 *NaS-S phase diagram according to Gupta and Tischer (1972)* [Courtesy Electrochemical Society Inc.]

and Tischer. By adopting this convention it is possible to obtain a depth of discharge in excess of 100%.

The cell reaction is written as

$$2\,Na + 3S \rightarrow Na_2S_3 \tag{6.16}$$

The thermodynamic data pertaining to sodium polysulphides are given in Table 6.2. These can be used to calculate the heat produced on discharge or charge using the relationships given in Section 6.2. The physicochemical properties of the sodium polysulphides; conductivity, viscosity, density and surface tension, have been determined by Cleaver *et al.*[35-37] These are summarised in Table 6.3 for one temperature, i.e. 350°C. The data for elemental sulphur are also included for comparison. Sulphur, of course, is an insulator (10^{-7} Ω cm at 350°C)

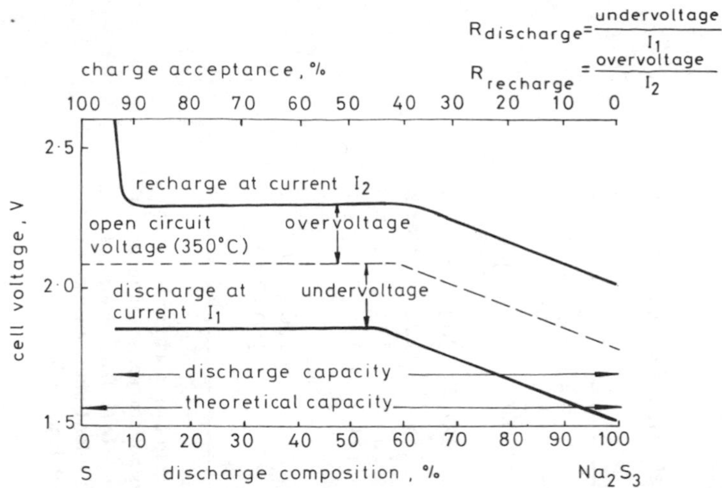

Fig. 6.12 ***Open-circuit potential against composition of the sulphur electrode in a sodium/sulphur cell***
[Courtesy the Electrochemical Society Inc.]

and its viscosity is very temperature dependent, reaching a maximum of 930 poises at 185°C,[38] although just above its melting point (119°C) the viscosity is as low as 6 centipoises. The boiling point of sulphur is 444°C and it has an appreciable vapour pressure at 350°C. The electrochemistry of the sodium sulphur cell is almost wholly concerned with the sulphur electrode, the ionisation and deposition of sodium metal at the solid electrolyte surface being a fast reversible process.

6.7.2 The sodium/sulphur cell with beta alumina electrolyte

6.7.2.1 The solid electrolyte: This cell which was first described by Kummer and Weber[39] utilises polycrystalline beta alumina as the solid electrolyte. This material was first observed as an impurity in alpha alumina at the turn of the century. At that time it was thought to be another form of alumina, hence the name, but later determinations of its composition gave the formula $Na_2O.11Al_2O_3$. The crystallographic structure of the material was first determined by Bragg,[40] and it proved to be a layer structure; a block of aluminium and oxygen ions in a spinel formation alternates with a layer of sodium ions. It is the two dimensional sodium layer which provides the sodium ion conduction.

Table 6.2 *Thermodynamic data for sodium polysulphides calculated from open circuit measurements*

Polysulphide composition (molten)	$-\bar{G}$ kJ 280°C	300°C	330°C	360°C	390°C	$-\bar{H}$ kJ	$-\bar{S}$ j/K
Na_2S_5	199	199	199	198	198	200	17 ± 4
Na_2S_4	189	189	188	188	188	202	16 ± 8
$Na_2S_{3.7}$	186	185	184	184	183	194	21 ± 6
$Na_2S_{3.1}$	176	175	174	172	171	199	42 ± 10
Na_2S_3		173					
$Na_2S_{2.6}$		163 (440°C)				200	103
$Na_2S_{2.4}$		160 (440°C)				200	103

$\bar{G}$, $\bar{H}$ and $\bar{S}$ are partial molar quantities, the differential energies and entropies of adding Na to a large excess of polysulphide melt so that the polysulphide composition does not change; in accordance with the reaction $2Na + (x - 1)\ Na_2S_x = x\ Na_2\ S_{x-1}$. Cell $E_{MF} = -\bar{G}/F$. Data abstracted from 'Thermodynamic and physical properties of molten sodium polysulphides from open-circuit voltage measurements'. N.K. Gupta and R.P. Tischer, *J. Electrochem. Soc.*, **119**, 8, p. 1033.

Table 6.3 *Physicochemical properties of sodium polysulphides and sulphur at 350° C*

	MPt		Density		Viscosity		Conductivity		Surface tension
	°C		g/cm³		cP		Ω^{-1} cm^{-1}		Mn/m
S	115·3	S	1·66	S	500	S	10^{-7}	S	46
Na_2S_5	258	$Na_2S_{4\cdot8}$	1·86	$Na_2S_{5\cdot2}$	18·6	$Na_2S_{5\cdot2}$	0·39	$Na_2S_{5\cdot2}$	114
Na_2S_4	285	$Na_2S_{3\cdot7}$	1·91	Na_2S_4	18·0	Na_2S_4	0·48	$Na_2S_{4\cdot1}$	127
Na_2S_3	235	Na_2S_3	1·87	$Na_2S_{3\cdot1}$	18·5	Na_2S_3	0·69	Na_2S_3	172

A second form of beta alumina, beta″ alumina, was first indentified by Yamaguchi,[41] who also determined its crystallographic structure. This was determined independently by Bettman and Peters,[42] who confirmed Yamaguchi's data. The structure of beta″ alumina, $R\bar{3}m$, is similar to that of beta alumina, $P6_3/mmc$, but there are three spinel blocks per unit cell compared with two in beta alumina. Both structures are shown in Fig. 6.13. The material is ascribed the composition $Na_2O.5.33Al_2O_3$[43] but in this form it has a highly defect structure and is unstable above 1550°C[43], thus the pure polycrystalline material has not been sintered to the fully dense state, although this might be possible if a very reactive powder was used. If, however, it is doped with lithia or magnesia (or both), the structure is stabilised and a high density polycrystalline ceramic can be prepared. The resistivity of this material is lower than that of beta alumina and it is the preferred electrolyte for sodium sulphur cells.

6.7.2.1.1 Properties of beta″ alumina: The polycrystalline material used as the electrolyte in cells has a density of about 3·20 g/cm³ compared with the theoretical density of 3·27 g/cm³, and an expansion coefficient of 7×10^{-6} K^{-1}. The resistivity of the material at 350°C varies according to the ratio of beta: beta″ alumina as shown in Fig. 6.14. It is also dependent on grain size and preparation route, hence the scatter in the results. The resistivity of beta alumina tubes is usually anisotropic, the radial resistivity being between 1·2 and 2·0 times the axial resistivity. The strength of this material is about 200 MN/m² and is also dependent on grain size. The aim in producing electrolyte is to obtain the maximum conversion to beta″ alumina while minimising the amount of grain growth. These two aims are conflicting in that both conversion of beta alumina to beta″ alumina and grain growth are time and temperature dependent. By using the correct sintering conditions and doping level it is possible to keep the grain size below 50 μm while obtaining $>$ 85% conversion to beta″ alumina.

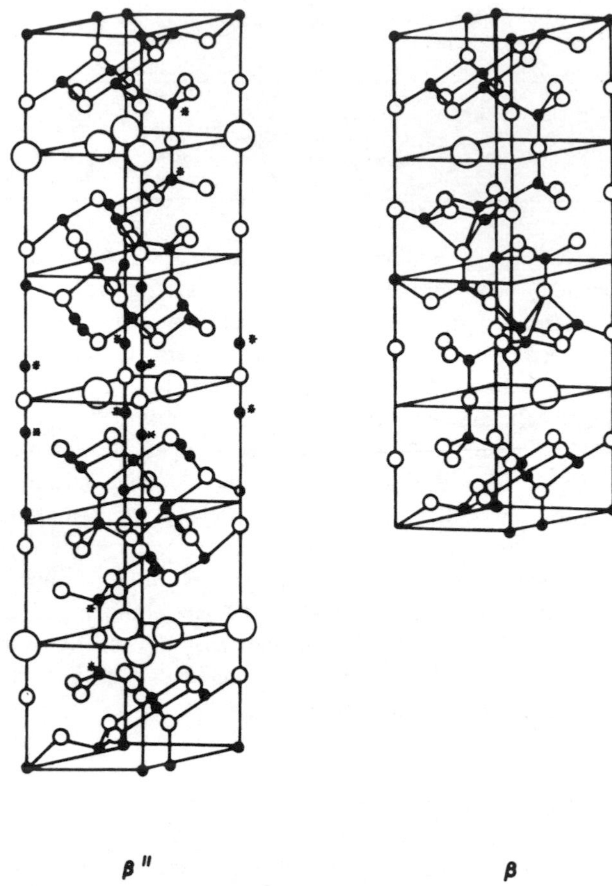

Fig. 6.13 *Crystal structures of beta and beta'' alumina*
○ Na oO ●AL ●* 0·83 Al

Although the production of a dense, low resistivity material is a prerequisite for good initial performance in a sodium sulphur cell, these properties do not guarantee an acceptable cycle life. When sodium sulphur cells are subjected to charge/discharge cycling, it has often been observed that the cell develops an internal short circuit. On investigation this proves to be due to penetration of the ceramic by filaments of sodium metal.[44] The generally accepted explanation of this phenomenon[45,46] is that sodium is deposited preferentially in the flaws or microcracks at the ceramic surface, where the sodium is generated on charge, because of current focusing. The Poiseuille pressure resulting

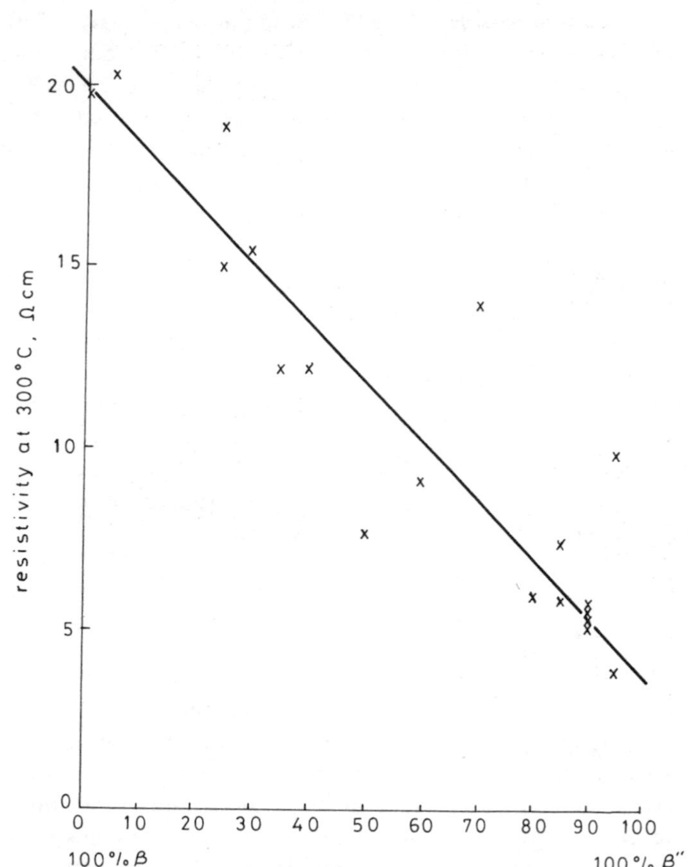

Fig. 6.14 *Resistivity of beta alumina as a function of β/β'' ratio*

from the flow of sodium metal is sufficient to propagate the crack if the current density is above a critical value.

Although there is some evidence to support this theory[46] it is not consistent with certain experimental observations. Thus materials of very similar physical properties, i.e. surface finish, grain size, beta'' alumina content, strength, but containing different amounts of lithia or magnesia, have been shown to have significantly different lives both in sodium sulphur cells and in sodium pump tests[47] (in these tests sodium forms the anode and cathode allowing high current densities to be used, and the Faradaic efficiency to be determined by measuring the volume of sodium transported per Faraday).[48,49]

Certain compositions (eg. 0·70% Li_2O, 8·9% Na_2O, 90·4% Al_2O_3)

have been shown to have very long lives in sodium pump tests at 350°C (up to 2000 Ah/cm^2 at current densities of 0·5 A/cm^2, 600 Ah/cm^2 at 2·5 A/cm^2)[47] and in sodium sulphur cells (up to 2000 cycles at current densities of 150 mA/cm^2 on discharge and 50 mA/cm^2 on charge). Formation of sodium dendrites in this material is rarely observed, a gradual darkening of the ceramic does occur[49] but does not appear to affect the performance of the electrolyte.

It is very likely that phases other than beta″ alumina, eg. sodium aluminate, will segregate at grain boundaries and the extent to which these are formed will depend on the level of dopants.[47] If these second phases are preferentially attacked or dissolved by sodium this might explain why materials of similar physical properties but different chemical composition have widely different resistances to breakdown. An alternative explanation of the failure of beta alumina ascribes breakdown of the material to the electric field which can be generated across the grain boundaries if the resistivity of the material at the grain boundary is high.[50]

6.7.2.1.2 Preparation of beta alumina ceramics: There are a number of ways of preparing beta alumina, but perhaps the most elegant method is the reactive sintering process developed by Duncan.[51] For production of the composition mentioned, alpha alumina powder, sodium hydroxide, lithium hydroxide and water in the correct proportions are milled together to produce a slurry with a 50% solids content which is spray dried to a free flowing powder. This powder can be isostatically pressed at 240 MPa (35 000 p.s.i.) into tubes 37 mm outside diameter, 360 mm long, which are then placed in magnesia crucibles and sintered at a maximum temperature of 1620°C. The resulting tubes, 33 mm outside diameter, are cut to 300 mm length using a water cooled diamond saw. The time temperature profile used during the sintering stage has to be closely controlled if a homogeneous fine grained structure is to be produced. The material is heated to about 1500°C, cooled to about 1300°C and finally sintered at 1620°C.[52] An alternative method of sintering has been developed by Wynn Jones and Miles,[53] in which the green tube is passed through a hot zone in a tubular furnace, sintering being completed in minutes rather than the several hours required for the batch process. This rapidity of sintering is offset to a certain extent by the need to pre-sinter the green tubes and to anneal after sintering. Consequently, there is little to choose between the two sintering methods. Gordon and Miller[54] have developed a pass through sintering method which eliminates the need for separate prefiring and post annealing stages, but this requires the use of a calcined powder.[55] Instead of isostatic pressing, electrophoresis can be used to form the green tube,[50,56,57] but in this case it is more convenient to start from beta alumina powder rather than a mixture of oxides.

Beta alumina has been made in disc form (up to 80 mm diameter and 2 mm thick) and tubular form (up to 600 mm in length with an outside diameter of 33 mm and in shorter lengths up to 50 mm outside diameter). The geometry of the beta alumina electrolyte largely determines the cell design. Discs[44] have been used but most designs have utilised tubes. Tubular cells can have the sodium inside the tube (central sodium cells) or outside the tube (central sulphur cells), and both designs are currently receiving attention. The central sulphur design requires a tube of at least 20 mm diameter, and for this reason all the early work was confined to central sodium cells, which usually had electrolyte tubes 10-15 mm outside diameter. Only when ceramic technology advanced sufficiently to allow the production of large diameter tubes was some consideration given to the central sulphur design.

6.7.2.2 Flat plate cells: The flat plate design uses a disc of beta alumina glass-sealed to an alpha alumina ring which can form the load bearing member in a compression seal. The cells can be stacked in series to form a bipolar battery with no external intercell connections. The conducting bipole, which is simply a disc, can be fabricated from a wide choice of materials. Thus vitreous carbon is readily available in disc form, and, since it is impermeable and intrinsically stable to sodium polysulphide melts and to sodium, it is an ideal conducting bipole.

For a given sulphur electrode thickness, the capacity of the cell is proportional to the square of the diameter, and for a practical battery a disc of at least 8 cm diameter would be required to give a reasonable specific energy.[44] Unfortunately, it has proved very difficult to seal such large diameter cells using a compression seal, because of the very tight tolerances on the flatness of the alpha alumina ring required to prevent fracture when the load is applied. This sealing problem has halted the development of the flat plate cell, but if a seal can be devised which does not require the alpha alumina ring to be mechanically loaded, the flat plate bipolar battery might be worthy of further consideration.

6.7.2.3 Central sodium tube cells: Numerous designs of central sodium cells have appeared in the literature. These can be subdivided into two main types; single electrolyte tube cells and multitube cells. The multitube cell is difficult to assemble and most practical cells have been of the single tube type, an example of which is shown in Fig. 6.15.

Referring to Fig. 6.15, it can be seen that the beta alumina tube is sealed to an alpha alumina collar which serves as an insulating load bearing member in a compression seal. Other types of seals which have been developed include glass/metal[64] and aluminium diffusion bonds (also known as thermocompression seals).[58] Almost all of these utilise an alpha alumina collar sealed to the beta alumina tube with a glass.

The sodium required for the reaction is normally stored in the reservoir at the top of the cell, the electrolyte tube remaining full of sodium throughout the discharge. Cells can be constructed in which all the sodium is stored in the electrolyte tube, a wick being used to keep the surface of the electrolyte covered with sodium metal.

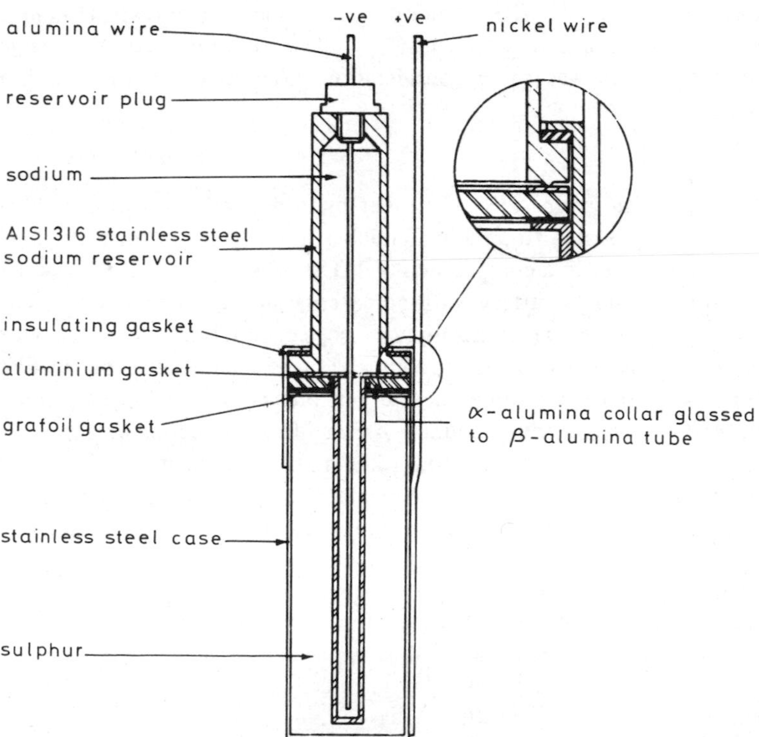

Fig. 6.15 *Central sodium electrode tubular sodium/sulphur cell*

The sulphur electrode is a carbon or graphite felt impregnated with sulphur. The resilient felt maintains good electrical contact with the cell case which is also the current collector. During discharge there is an increase in volume in the sulphur electrode due to the formation of sodium polysulphides. It is necessary, therefore, to leave approximately 50% voidage in the sulphur electrode at room temperature to ensure that the gas pressure at the end of discharge is not excessive. Evacuation of the electrode results in somewhat smaller voidage being required. The pressure in the sodium electrode decreases during discharge as the sodium is transported through the beta alumina tube.

The beta alumina tube diameter is usually 15-30 mm and the length 160-400 mm. The specific energy of the cell shown in Fig.. 6.15, (capacity 38 Ah) was 200 Wh/kg initially, but after extended electrical cycling this dropped to 80 Wh/kg due to corrosion of the stainless steel case.[59]

6.7.2.4 Central sulphur tube cells: The corrosion of the cell case by sodium polysulphides is a problem which has proved difficult to overcome. The material from which the cell case is made must not only be corrosion resistant but must also be capable of being fabricated into a fairly complex shape. This combination of properties can be avoided if the sulphur electrode is located inside the electrolyte tube; the cell case can then be made of mild steel, and the current collecting pole in the sulphur electrode is of cylindrical geometry and easily fabricated from a wide range of materials. A diagram of such a cell is shown in Fig. 6.16.[60] In this particular design the compression seal is retained but the sodium level is maintained by gas pressure above the sodium which is contained in an aluminium reservoir at the base of the cell. The current collecting pole is graphite, which for an electrolyte tube length of 160 mm has adequate conductivity. The electrolyte tube is 33 mm outside diameter, 30 mm inside diameter, which, with a 17 mm outside diameter pole results in a sulphur electrode thickness of 6·5 mm and a capacity of 38 Ah. Cells of this design have maintained greater than 80% capacity for up to 1000 deep cycles, indicating that corrosion of cell case is the major cause of loss of capacity in central sodium cells.

The change to the central sulphur design is not without attendant penalties, and for the discharge rates of interest for traction and load levelling, the specific energy is significantly less than that of the central sodium design, as can be seen from Fig. 6.17, which shows the calculated specific energy for the two types of cell. In the case of the central

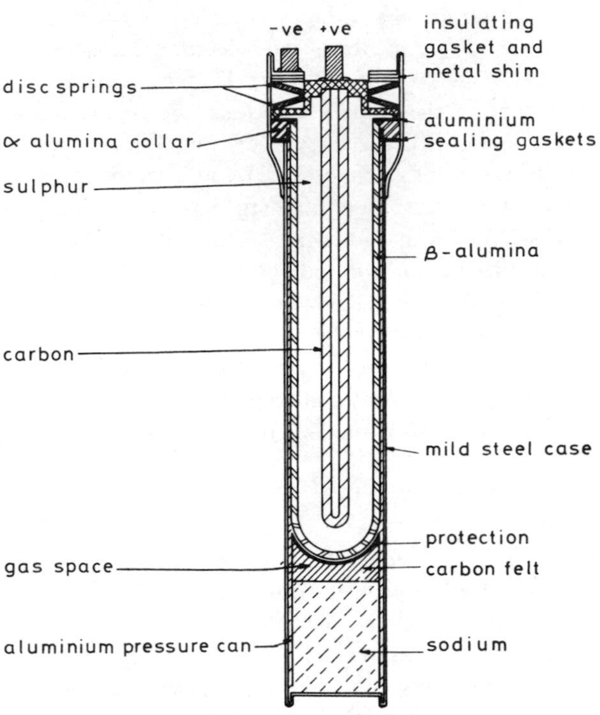

sodium cell it is assumed that all the sodium is stored within the electrolyte tube; such cells have been operated successfully but only in limited numbers. If the sodium is contained in a separate reservoir the specific energy is somewhat lower.

6.7.2.5 Cell materials: In both the designs described, the main materials problem is the corrosion of the positive current collector. Three classes of material have been used as the current collector: metals, non-metals and composites.

Metals: All metal sulphides have a negative free energy of formation and only those metals which form a stable, coherent coating of sulphide can be considered for use in the cell. Only three metals have been found to be stable: molybdenum, chromium and aluminium.[61,62]

Molybdenum forms a thin coating of MoS_2, which is electronically

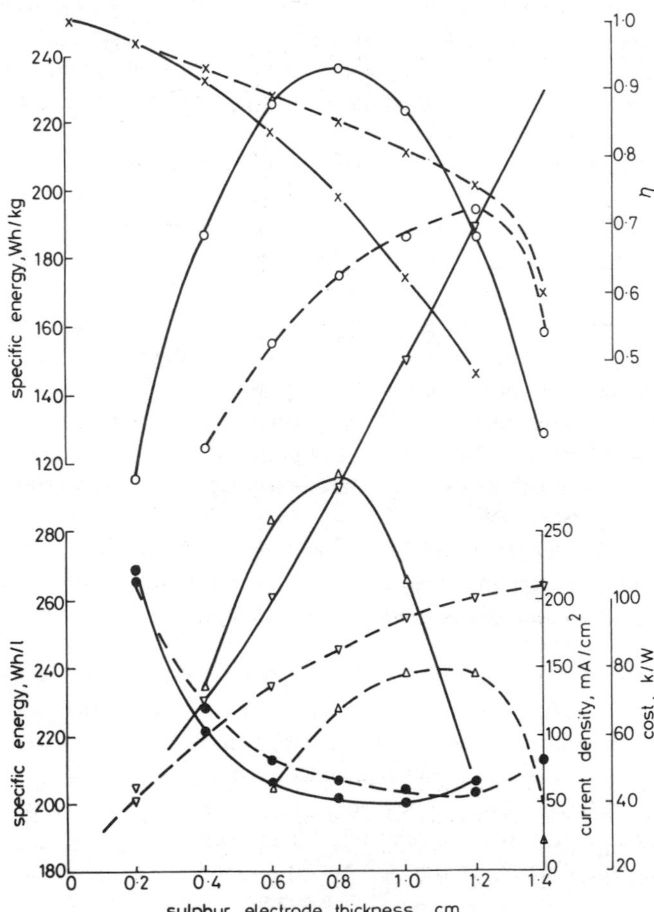

Fig. 6.17 *Specific energy and cost of central sodium and central sulphur electrode tubular sodium sulphur cells as a function of sulphur electrode thickness*

Discharge rate = 10 hours, electrolyte tube length = 300 mm, electrolyte tube diameter = 30 mm

— central sodium cells
— — — central sulphur cells
○ gravimetric specific energy
Δ volumetric specific energy
x voltage efficiency
∇ current density
● cost

conducting and insoluble in sodium polysulphides. It is a metal which is very difficult to fabricate and weld, and is reported to give rise to high cell resistances, particularly on charge.[63] It has been used as a liner in central sodium cells[64] and as a solid pole in central sulphur cells. Chromium can only be used as a coating on another metal and is discussed later under composites.

Aluminium forms a thin stable coating of Al_2S_3, but this is electronically insulating and aluminium alone cannot be used as the current collector. It is, however, used as a component of the seal in the compression seals and in the diffusion bond, or thermo compression seals, where its passivity to sulphur can be used to advantage.

Many alloys have been evaluated, and, although some such as the inconel and austenitic stainless steel group of alloys show very low corrosion rates in out-of-cell tests, they corrode rapidly in the cell environment. This difference in behaviour is probably due to the greater range of sodium sulphide compositions in the cell compared with the out-of-cell tests which are restricted to one polysulphide composition, usually Na_2S_4.

It appears, then, that no metals or alloys are suitable for use alone as the current collector in the sulphur electrode, although some could be used as coatings and others as substrates in composite electrodes.

Non-metals: If non-metals are to be used as the current collector material they should be mechanically strong, have a conductivity similar to metals and be inert to sulphur and sodium polysulphides. Ideally, any compounds used should be incapable of being reduced by sodium. The one material which comes close to meeting these requirements is carbon, but even this has a relatively high resistivity (10^{-3} Ω cm) and low strength. Vitreous carbon has been proposed as a cell case material for central sodium cells[65] and porous graphite rods have been used as current collectors in central sulphur cells.[60] The latter are acceptable provided the length of the rod is less than 200 mm and the diameter is at least 15 mm.

There are many compounds inert to sodium polysulphides and having electronic conductivity, but, with the exception of transition metal sulphides, which are often metallic conductors, the conductivity of these materials is not high enough for them to be considered as a substitute for metals. Most of the transition metal sulphides can be reduced electrochemically by sodium (cf lithium/metal sulphide cells) and are unlikely to be acceptable as current collectors.

Composites: Two types of composites have been considered: coated metals, and metal coated non-metals.

Coated metals: Some pure metals, and many alloys, form conducting oxides, and since these are almost invariably more stable than the corresponding sulphides it might be expected that oxide coated metals and alloys would be suitable as current collector materials. This does not seem to be borne out by the experimental evidence. Thus, stainless steels, which form coherent, stable and conducting oxides, are corroded rapidly in the cell environment. More protection is afforded if thicker oxide films are present, and nickel based superalloys (Inconel 600, 601, 625, 671 and 718, Incoloy 800, 825 and 840) which have been oxidised in air at elevated temperatures show a greater resistance to corrosion by polysulphides than the untreated alloys.[66] Unfortunately, the electrical resistance of the thicker coating is unacceptably high.

As already discussed, molybdenum and chromium appear to be resistant to corrosion by polysulphides and both metals have been used as coatings. A variety of techniques have been used: electroplating, plasma spraying, and vacuum deposition all result in a coating which is not strongly bonded to the substrate; reactive coating methods, such as chromising, give rise to coatings which are bonded to the substrate by chemical reaction. The latter type are expected to be more tolerant of external factors, such as thermal cycling and damage during handling, but the substrate is limited to those metals which form reaction coatings. Chromised mild steel is claimed to be resistant to corrosion in the cell environment for periods of at least three months,[50,61] but this coating, like all coatings on a readily corrodible substrate, is vulnerable to localised attack. Should a defect in the coating, or localised electrochemical attack resulting from the cell geometry, lead to penetration at just one point, it is very likely that the sulphide formed on the substrate will lead to spalling of the coating over large areas of the current collector. For this reason some workers have preferred to use aluminimum as the substrate.[67] This passivates in polysulphide melts forming a very thin insulating layer which is much less likely to lead to spalling of any coating. Although in many ways it is an ideal material for use as a substrate, having low resistivity (12×10^{-6} Ω cm at 350°C), low cost, low density (2·7 gm/cm^3), and being easily fabricated into complex shapes, it does have two disadvantages; its low melting point (666°C) and a high coefficient of thermal expansion ($23 \times 10^{-6} K^{-1}$). Its low melting point seems to rule it out as a container material in central sodium cells, since the chance of a cell rupturing would be greatly increased. It could still be considered for the central sulphur cell if a coating could be found with good adherence over the temperature range ambient–400°C. Molybdenum coated aluminium has been investigated but the mismatch in coefficients of thermal expansion

($\gamma = 5{\cdot}9 \times 10^{-6}\ K^{-1}$ for Mo) is probably too great.[63]

The materials used as coatings on metals need not be very good conductors, thus a 10 μm coating of a material having a resistivity of 100 Ω cm would only contribute 0·1 Ω cm^2 to the cell resistance.

Metal-coated non-metals: If the non-metal is used as the cell case or the current collector it must have adequate mechanical strength, which means a thickness of at least 1 mm. To limit its contribution to the cell resistance to 0·1 Ω cm^2 would require a resistivity for the material of about 1 Ω cm. Several groups of materials with conductivities of this magnitude are known. Examples of these with their room temperature resistivities (in Ωcm) are carbon (10^{-3}), SiC(10^{-1}), TiC(10^{-4}), Nb or Ta doped TiO_2(10^{-1}), $(La_{(1-x)}Ca_x)$ CrO_3 (10^{-1} for x = 0·15). One which has received considerable attention is rutile (TiO_2) doped with either Nb_2O_5 or Ta_2O_5.[68]

Tantalum doped rutile has been prepared in the form of impermeable ceramic tubes having a resistivity of approximately 0·1 Ω cm at 350°C.[68] It is stable in sodium polysulphide melts, and has a high enough mechanical strength for it to be considered as a cell case material in central sodium cells. It can be coated with nickel using an electroplating technique and the bond between the coating and the substrate will withstand thermal cycling.

A cell with a ceramic case would be rather susceptible to mechanical shock and this material may be more appropriate as the current collector in a central sulphur cell. It would need to be in the form of a tube with an internal metal coating to give acceptable longitudinal conductivity, and this would be more difficult than coating the outside of a cell case. If these fabrication difficulties can be overcome, doped rutile offers the possibility of a chemically stable, mechanically strong, and low cost current collector.

Carbon is not readily available in impermeable form and is mechanically weak and difficult to join to other cell components. Silicon carbide is stronger but is difficult to fabricate as an impermeable material.

6.7.2.6 The positive electrode: Sulphur is an insulator, has a high viscosity, and, as can be seen from the phase diagram (Fig. 6.11) it is immiscible with sodium pentasulphide. This combination of properties determines the design and performance of the sulphur electrode. The sodium polysulphides are good ionic conductors, have a relatively low viscosity and are more dense than sulphur. The melting points of the sulphides determine the lower operating temperature of the cell; the upper temperature limit is determined by the vapour pressure of sulphur.

Sodium sulphur cells are almost invariably assembled in the fully charged state and at the beginning of discharge the electrode consists of pure sulphur absorbed in a conducting matrix, usually carbon or graphite felt. Initial reduction of sulphur occurs only at the solid electrolyte/carbon matrix interface, but as the sodium polysulphides are formed, the reaction quickly spreads through the whole of the electrode. Provided that the current density is not too high or the operating temperature too low, the formation of solid Na_2S_2 can be avoided if the overall composition of the electrode is limited to Na_2S_3. As inspection of the phase diagram shows, discharge much beyond this point will result in precipitation of solids which can adversely affect the performance of the electrode.

On recharge, the lower sulphides are first oxidised to higher sulphides, but when the composition Na_2S_5 is reached, elemental sulphur is produced. Depending on the electrode structure and the current density or charging voltage, the cell may polarise at this point or continue to charge almost to pure sulphur. This ability to charge well into the two phase region is essential if a high specific energy is to be obtained. This and a stable cell resistance are the two main objectives in the development of sodium sulphur cells.

The polarisation of the sulphur electrode, which terminates the charging of the cell, is ascribed to the formation of a film of sulphur on the surface of the solid electrolyte, effectively blocking the transport of sodium ions.

Modifications to the electrode to defer the formation of this blocking film of sulphur have been proposed. For electrodes greater than 10 mm thick, incorporation of channels in the carbon felt matrix to improve convective flow of sulphur has been shown to be beneficial.[69] For thinner electrodes this is less appropriate and modification of the matrix to ensure removal of sulphur from the vicinity of the solid electrolyte by capillary action is more effective. One way of achieving this is to incorporate a layer of alpha alumina fibres (which are preferentially wetted by sodium polysulphides) between the carbon felt and the solid electrolyte surfaces.[70] Other electrode modifications include use of a carbon mat having gradated electrical resistance to bias the electrode reaction away from the solid electrolyte;[71] addition of tetracyanoethylene to the sulphur, to form a charge transfer complex which prevents the sulphur layer from blocking off the solid electrolyte[72]; and the creation of a physical gap between the solid electrolyte and most of the felt matrix.[73] The most effective method appears to be the inclusion of an alpha alumina layer adjacent to the solid electrolyte. The charge acceptance of such a cell is shown in Fig. 6.18.

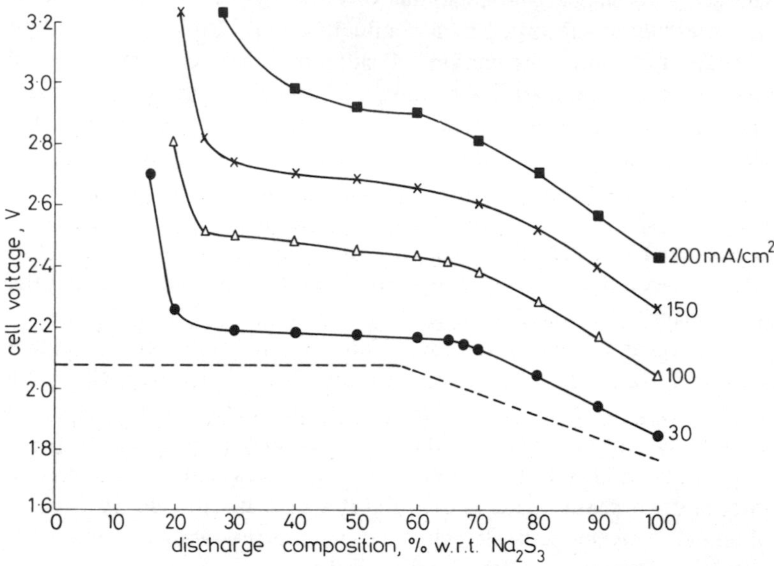

Fig. 6.18 *Charge characteristics for a central sulphur electrode tubular sodium/sulphur cell with a layer of alpha alumina fibres adjacent to the solid electrolyte surface*
Cell capacity = 38 Ah, active electrolyte area = 100 cm^2

Although high charge acceptance can be achieved using the methods discussed, it is not always possible to maintain this for the life of the cell. A decline in charge acceptance of up to 20% over the first 20 cycles is not uncommon, with a more gradual decline thereafter. In cells with metal current collectors this is usually due to interaction of corrosion products with the sulphur electrode. In cells with non-metallic current collectors the effect is less marked and sometimes absent altogether, and, clearly, more subtle effects are at work. With graphite felt electrodes it has been observed that the felt structure is destroyed by prolonged cycling, whereas carbon felts show much less damage.[74] There is, however, little difference in the charge acceptance of electrodes fabricated from either of these materials even after hundreds of cycles, and it seems that mechanical damage of the felt is not an important factor. It has been suggested that heterogeneous reaction within the electrode results in progressive passivation of the carbon felt with solid sulphides,[75] but there is no evidence to support this. Whatever the causes of this decline in charge acceptance, it is not an intrinsic property of the sulphur electrode, since cells have accumulated hundreds of cycles with no detectable loss of capacity after the first 20 cycles, as shown in Fig. 6.19.

6.7.2.7 The negative electrode: Ionisation of sodium at the solid electrolyte surface is a simple fast reaction and presents no problems provided that the beta alumina surface is wetted by the sodium metal and that the surface area in contact with the sodium metal remains constant. Operation of the cell at 350°C or above is usually sufficient to guarantee the former, and the latter can be achieved by a variety of methods.

In central sodium cells it is usual to provide a reservoir on top of the cell to contain the sodium required in the cell reaction. This means that excess sodium, equivalent to the volume of the beta alumina tube, must be incorporated in the cell. The provision of excess sodium can be avoided by providing a wick inside the beta alumina tube. Materials such as stainless steel felts, perforated stainless steel foil and iron foil have been used.[76] Depending on the dimensions of the beta alumina tube, the reservoir may be dispensed with completely.

In central sulphur cells the same principles apply, but an additional

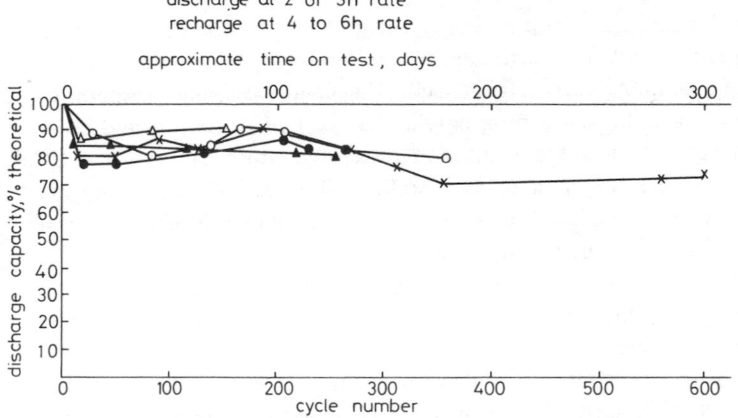

Fig. 6.19 *Capacity as a function of a cycle number for central sulphur electrode tubular sodium/sulphur cells with carbon poles*
Cell capacity = 10 Ah, active electrolyte area = 33 cm^2, electrolyte tube diameter = 33mm, discharge rate = 2 or 3 h, charge rate 4 to 6 h

method has been used, namely, the pressure can or pressurised sodium reservoir.[60] A metal can, usually aluminium, is filled with sodium, except for a gas space at the closed end defined by a piece of carbon felt. On insertion into the cell, sealing and evacuation of the sodium compartment, and heating to the operating temperature, the gas expands and forces the molten sodium to the top of the cell. The change in sodium level during the discharge and charge is then confined to the pressure can.

6.7.2.8 Cell safety: With two highly reactive liquids separated only by a thin membrane of ceramic at 350°C, the possibility exists of a very vigorous chemical reaction.

Owing to limitations in ceramic technology, most of the early cells utilised small diameter (10-15 mm) electrolyte tubes, and when these fractured it was found that high melting polysulphides formed and attenuated the reaction. When larger diameter tubes became available it was observed that in some cases the chemical reaction was not attenuated and breaching of the cell case resulted. In an extensive series of experiments Hames and Tilley[77] established three safety principles, which, if adhered to, almost eliminate the possibility of the cell case breaching following fracture of the ceramic electrolyte. These were as follows: the volume of sodium immediately available for reaction, i.e. adjacent to the electrolyte tube, should be minimised; the flow of the remainder of the sodium from the reservoir should be restricted; and, finally, the inside of the cell case should be coated or lined with a material resistant to corrosion by molten polysulphides at the temperatures obtaining in the failed cell. The central sulphur design with its narrow annulus containing only a small volume of sodium and in which can easily be inserted a lining of flexible graphite, and its tight fitting pressure can which acts as a sodium flow restrictor, lends itself to incorporation of the three safety principles. Hundreds of 38 Ah capacity cells have been deliberately failed without breaching of the cell case, and cells of up to 100 Ah capacity containing 30 mm outside diameter × 300 mm long ceramic tubes have also survived deliberate fracture of the electrolyte.

In all the safety tests described, the sulphur electrode was fully packed with carbon felt (density 0·15 g/cm^3). There is some evidence to show that lower felt densities result in a higher risk of cell breaching.[57,60] Thus, of about 1000 cells of the type shown in Fig. 6.16, only one case of cell breaching on failure was recorded. In this cell a low density felt had been used in the sulphur electrode. This may not be significant, but sulphur electrodes containing felt-free channels to aid convection should be rigorously tested to ensure that they do not result in breaching of the cell case in the event of electrolyte fracture.

6.7.2.9 Present status of sodium sulphur cells with beta alumina electrolyte: The life of the ceramic electrolyte has been increased considerably and is probably not the life limiting factor in most cell designs. The problem of corrosion of the cell case or current collector has proved more intractable, but some materials have now been identified which remain uncorroded for hundreds of cycles. The related problem of low charge acceptance has been overcome by the development of structured electrodes which can be repeatedly charged to greater than 80% of theoretical capacity.

Because of their liquid electrodes, sodium sulphur cells have always raised questions of safety, particularly in transport applications, but cells have now been developed which are safe even under the most severe failure conditions.

The major remaining problem is to combine all the above characteristics in a cell design which will give a high energy density, low cost and acceptable reliability. The central sulphur electrode design is closest to achieving this, and will almost certainly be used for the first commercial prototype batteries. The central sodium electrode cell which has a potentially higher specific energy might well be preferred for the second generation of sodium sulphur cells.

6.7.3. The sodium sulphur cell with glass electrolyte

6.7.3.1 The solid electrolyte: The use of glass as the electrolyte in the sodium sulphur cell was first described by Levine[78] in 1968, around the same time that the Ford Motor Company disclosed their work on sodium sulphur cells with beta alumina electrolyte. Since that time Levine and his co-workers at the Dow Chemical Company have developed cells up to 40 Ah capacity.[79-81] The glass used as the electrolyte in these cells is a sodium borate glass. It contains small amounts of additives (sodium halides) to prevent crystallisation and to improve conductivity, and, although its resistivity is low compared with other glasses at 2×10^4 Ω cm at 300°C, it is still several orders of magnitude greater than that of beta alumina. This problem has been overcome by

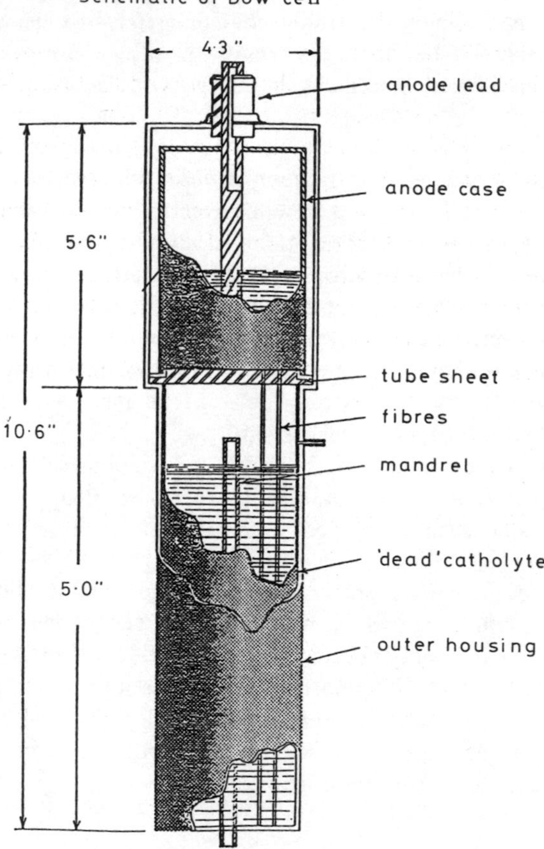

Fig. 6.20 *Sodium/sulphur cell with glass electrolyte*
[Courtesy the Electrochemical Society Inc.]

an ingenious cell design which uses thousands of hollow glass fibres as the electrolyte tubes.[78] These fibres, about 100 mm long, 50 μm inside diameter and 70 μm outside diameter, are placed in parallel, as shown in Fig. 6.20. The fibres are sealed at the bottom end and open at the top. They are held together by a glass 'tubesheet' or header. The open ends of the fibres communicate with the sodium reservoir and are filled with sodium metal. The sulphur electrode consists of aluminium foil interleaved between adjacent rows of fibres, the whole assembly being immersed in sulphur. Because of the high surface area of the electrodes, even at relatively high discharge rates the current density is very low (about 2 mA/cm^2) on the glass electrolyte. The interelectrode spacing can be as low as 40 μm and this combined with the very low current density minimises concentration polarisation in the sulphur electrode.

The glass tubes are spun from molten glass by a proprietary process, chopped into 10 cm lengths, sealed at one end by fusing and attached to aluminium strips on a polythene sheet. The fibres are then automatically transferred to the aluminium foil at the correct spacing, aluminium strips somewhat thicker than the fibres are used to space the foil at its bottom edge, the foil is wound onto a mandrel and a slurry of glass applied at one end. After this the roll of foil and tubes is heated in an oven to fuse the glass powder and form the tube sheet. An aluminium reservoir for the sodium is attached to the tube sheet and an aluminium lead welded to the bottom of the roll of foil. The assembly is then inserted in a stainless steel tube and the cell filled with sodium and sulphur.

6.7.3.2 The negative electrode: Impurities in the sodium have been the cause of very short cell lifetimes, especially on deep discharge cycling. Calcium and $NaOH/Na_2O$ are the main culprits. During discharge, sodium flows down the capillaries, and, as sodium ions, through the glass walls. The calcium and sodium oxides concentrate inside the capillary. The calcium deposits in the glass surface as Ca^{++}, causes strains and breaks the glass fibre. The oxides cause gross glass corrosion leading to fibre breakage.

If the cell is not cycled deeply, these impurities do not concentrate and lifetimes of a few months are obtained. Deep cycling, however, results in lives of a few days or weeks.

Use of very pure sodium results in an increased cell lifetime but failure can still occur due to oxide corrosion. The water contents of the glass parts of the cell slowly leach out during cell operation and react with sodium to form new oxide. This can be prevented by placing an active metal scavenger, e.g. Ti, Zr, in the sodium to remove any oxide that is formed.

Sodium in which the impurity content has been reduced to a few parts per million of calcium and oxide, wets the glass extremely well, and the glass tubesheet, which separates the sodium chamber from the sulphur chamber, must be completely leaktight.

6.7.3.3 The positive electrode: The electrode reactions take place at the surface of the aluminium foil. Since aluminium forms an insulating layer of aluminium sulphide it is usual to coat the foil with an inert material. Moybdenum can be vapour deposited onto the aluminium and at the current densities used this is apparently quite stable. Alternatively, a coating of carbon can be used. This is applied to the aluminium by dipping the foil in an aqueous suspension of graphite while abrading the surface of the metal to disrupt the tenacious film of aluminium oxide. It is necessary to use fairly pure aluminium, as the magnesium found in some aluminium alloys can dissolve in the melt and degrade the glass electrolyte. Because of the low current densities used and the small electrode spacing, the cell can be cycled over 90% of its theoretical capacity. One cell has operated continuously over this range for 3800 cycles (5½ months).

6.7.3.4 Cell materials: The cell case is stainless steel which appears to be sufficiently corrosion resistant. This corrosion resistance of stainless steel may be related to the different operating condition in these glass electrolyte cells. Thus the operating temperature of 300°C is lower than is normally used for beta alumina electrolyte cells; the internal resistance is lower and consequently the voltage limits between which the cell operates are less (typically 100 mV polarisation on charge and discharge), and, probably most important, the current flow across the stainless steel/polysulphide interface is negligible.

The most critical problem is the interaction between the hollow glass fibres and the glass tubesheet. The glass of the tubesheet must melt at a lower temperature than that of the fibres and must have a similar coefficient of thermal expansion, $100 - 125 \times 10^{-7}\ K^{-1}$. To make a good seal between the fibres and the tubesheet it is necessary to control very closely the particle size of the glass powder used and to rigorously exclude moisture during the milling and sieving operations. Because of the low melting point of the glass tubesheet it is essential to keep the operating temperature of the cell close to 300°C. Cell failure is usually the result of fibre breakage either by the corrosion mechanism discussed above or by mechanical effects such as defective sealed ends. Failure of the fibre just below the tubesheet is probably caused by the weakening of the fibre during fusion of the tubesheet. There is some evidence that fibre damage may arise because the solid sodium sulphide formed when one fibre breaks can rupture the adjacent tubes, thus causing cumulative damage and eventual failure of the whole cell. For a similar reason it is necessary to avoid freezing the sodium sulphides, which imposes a lower temperature limit of 285°C on the cell operating temperature. It can be seen that very good temperature control and accurate control of the depth of discharge will be necessary for this battery.

6.7.3.5 Outstanding problems: The immediate problem with this cell is the limited lifetime due to breakdown of the hollow glass fibres. To increase this lifetime, improvements in the technology of sealing the tubesheet to the glass fibres are required. Most cells constructed to date have a capacity of 6 Ah, although a few 40 Ah cells have been tested. Scaling up involves building up the tubesheet, handling and assembling the bundle and making a leaktight seal between the fibres and the tubesheet.

As with the sodium sulphur cell with beta alumina electrolyte, safety is a very important consideration. The volume of sodium in the fibres is very small so the main hazard is fracture of the tubesheet allowing the sodium in the reservoir to react with the sulphur. The presence of aluminium foil is an additional factor in this cell; if the temperature exceeds 660°C, the aluminium melts and reacts with the sulphur liberating more heat. The cell needs to be designed so that bulk mixing of sodium and sulphur is prevented.

The advantages of the cell are its potentially low cost, and, arising from the very low operating current density, the ability to withstand considerable overload on both charge and discharge.

6.8 Sodium/antimony trichloride cell

6.8.1 Introduction

This cell utilises beta alumina as the solid electrolyte, and liquid sodium as the negative electrode. The positive electrode is antimony trichloride ($SbCl_3$) dissolved in sodium chloroaluminate ($NaAlCl_4$).[82-84] The operating temperature of the cell is 210°C and the overall cell reaction is

$$3Na + SbCl_3 \rightarrow 3NaCl + Sb \qquad (6.17)$$

The theoretical specific energy of this reaction is 752 Wh/kg. Although $E^0{}_{473}$ for reaction 6.17 is 2·90V, the actual open-circuit voltage of the cell varies slightly from this, as can be seen from Fig. 6.21. The increase in open-circuit voltage at compositions near the top of charge could possibly be due to the presence of $SbCl_5$ or to the formation of complexes such as $AlCl_4^-$. A typical constant current charge/discharge curve is shown in Fig. 6.22. Cell design is determined by the geometry of the electrolyte. Early cells were based on beta alumina discs, but, in spite of the low operating temperature, no stable seal has been found for this design. To overcome the sealing problem a central positive

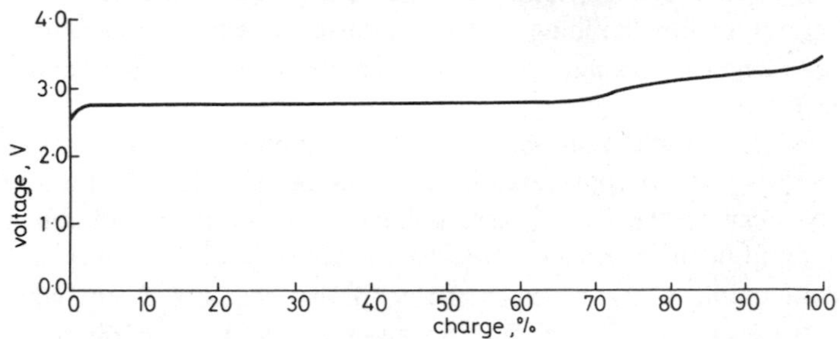

Fig. 6.21 *Open-circuit voltage as a function of positive electrode composition for the sodium/antimony trichloride cell*

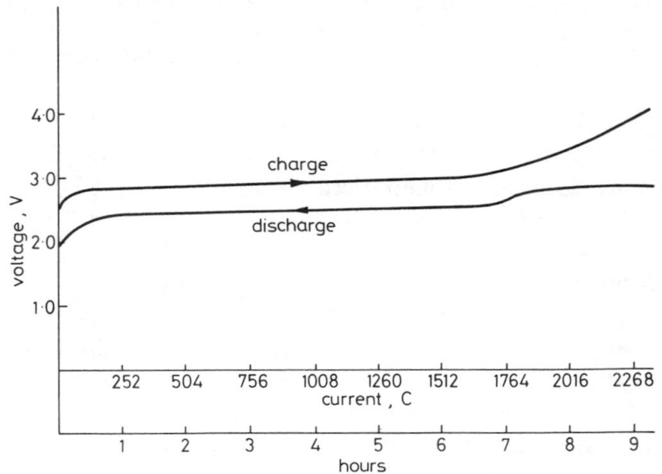

Fig. 6.22 *Typical constant current charge/discharge curves for the sodium/antimony trichloride cell*
Cycle 10 discharge: 2393 coulombs, 2·45V, 850 J/g, active mix utilisation 82%, 1·6 Wh, charge 3·07V, 80% turnaround efficiency, 22mA/cm^2 beta alumina

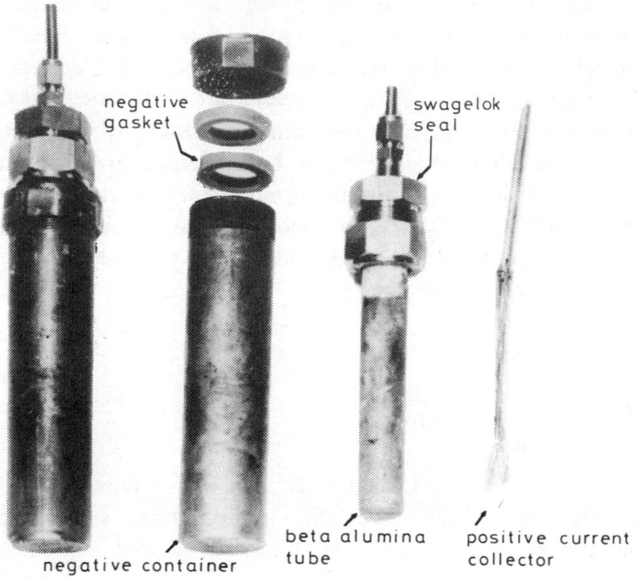

Fig. 6.23 *Tubular design sodium/antimony trichloride cell*

tubular cell design has been developed in which the seals are not in contact with the reactants. This is shown in Fig. 6.23. At 210°C beta alumina electrolyte has a resisitivity of 20-30 Ω cm and beta ″alumina has a resistivity of 10-15 Ω cm. Two compositions which have been used are 8·0 wt % Na_2O, 2·0 wt % MgO, 90·0 wt % Al_2O_3 (95-100% beta alumina) and 8·7 wt % Na_2O, 0·7 wt % Li_2O, 90·6 wt Al_2O_3 (75-100% beta″ alumina).

6.8.2 *The negative electrode*

Because of the low temperature of operation (compared, that is, to sodium sulphur cells) the beta alumina is not wetted by sodium. With disc cells this has been overcome by pasting the surface of the beta alumina with a thin film of sodium metal. This is obviously impracticable for the tubular design and the same effect can be obtained by heating the beta alumina tube to 600°C in contact with liquid sodium. The beta alumina is irreversibly wetted by sodium at this temperature and the film of sodium remains in place even after cooling to room temperature. This is an improvement on the pasting technique but still appears impractical for producing large numbers of cells.

It has been shown that beta alumina can be wetted with sodium at temperatures below 200°C if it is sufficiently pre dried. Heating the electrolyte to 1000°C for a few minutes achieves the desired result, but even in an inert atmosphere the beta alumina reverts to its original state in a few hours. A proprietary protective coating has been developed, which, it is claimed, protects the beta alumina from moisture for several hours after drying and which can be easily removed prior to cell assembly.

6.8.3 *The positive electrode*

Cells are assembled in the fully charged condition and the positive electrode mix contains antimony trichloride, sodium chloride, aluminium chloride, carbon black and small amounts of sulphur. For maximum conductivity and energy content the composition shown in Table 6.4 is preferred.

Table 6.4 *Sodium/antimony trichloride cell preferred positive electrode compositions*

	Fully charged	
	Wt%	Mole %
Carbon*	7·7	-
$SbCl_3$	29·4	17·3
$AlCl_3$	47·8	48·1
NaCl	15·1	24·6
	100·0	100·0

*Vulcan XC72-R or Shawinigen black

Sulphur (2 wt %) is usually added to give a higher average discharge voltage. This improvement is thought to be related to the elimination of high melting phases (sodium chloride and antimony) with poor ionic conductance which form at certain stages of cell charge or discharge and are deposited on the beta alumina. Addition of Sb_2S_3 or $SbSe_3$ has a similar effect.

The boiling point of $SbCl_3$ is 219°C, and so it might be expected that the molten salt mixture would have a high vapour pressure at 210°C. Melts corresponding in composition to the fully charged state have a vapour pressure of only about 666 Pa (5 Torr) at 200°C indicating complex formation. Excessive pressure in the positive electrode should not therefore be a problem.

6.8.4 Cell design

A principal advantage of operating at a relatively low temperature is the ability to use elastomeric materials for the cell seals. Silicone rubber is used to seal the sodium electrode. Immersion tests followed by weight loss measurements have shown it to be only slowly attacked by sodium over a period of 300 days (an average of 0·036 mg/cm^2 per day).

PTFE is used as the seal for the positive electrode. It is attacked by the melt and for this reason disc type cells, where the melt is in contact with the gaskets, can only be run for extended periods by renewing the gaskets at frequent intervals. The problems are less acute with tubular cells where the seal is only exposed to the vapour, but gradual carbonisation still occurs. A proprietary composite seal with PTFE as one of the elements has been developed and has been shown to be resistant to direct contact with the positive mix for at least a year.

The other major materials problem of the cell is the current collector in the positive electrode. Tungsten and molybdenum are the only metals or alloys which show any corrosion resistance in the melts of similar composition to that in Table 6.4, but both these metals are subject to embrittlement and contact problems in the area of welds. To overcome this problem a 'broom' electrode has been constructed from 10 cm strands of 0·5 mm diameter wire tied to a 6 mm diameter central spine. This and a brush design are shown in Fig. 6.24.

It has now been shown that molybdenum will corrode at voltages as low as 3·75V with respect to sodium. Controlling the cell voltage to this value in a battery would be very difficult and it is unlikely that molybdenum can be used as the current collector. Tungsten, however, is not corroded, even at potentials which result in chlorine evolution. It is too costly to be used as the current collector, but as a coating on other metals such as nickel it would be acceptable. A vapour deposited coating of tungsten on an 8 mm diameter nickel wire showed no sign of attack after exposure to the positive melt for one week. It has recently been shown[85] that pure nickel will not corrode in the positive melt if sulphur is omitted, and is stable even during chlorine evolution,

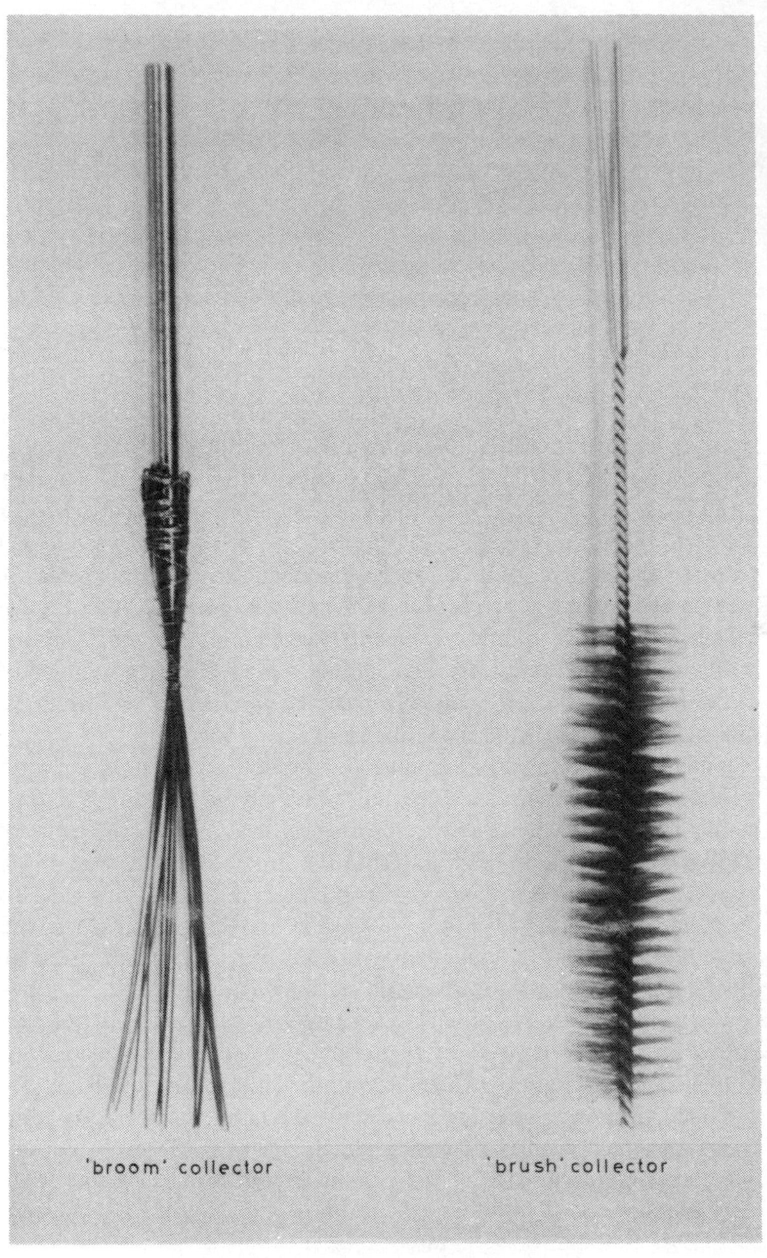

Fig. 6.24 *Designs of current collector for sodium/antimony trichloride cell*

probably due to the formation of a film of $NiCl_2$ which breaks down on discharge. Unfortunately, omission of sulphur adversely affects the performance of the electrode.

The utilisation of antimony for the broom and brush electrodes at a current density of 75 mA/cm^2 was 88% and 96%, respectively. This improvement is attributed to the increased surface area of the current collector (25 cm^2 and 82 cm^2). It is suggested that deposition of antimony on the metal surface on discharge results in displacement of the carbon of the mix away from the metal. On charge, the antimony starts to dissolve isolating from the metal current collector some of the antimony left on the carbon. This effect is less with the thinner layers formed on the higher surface area current collectors and is much reduced if a nickel current collector is used, probably because of the formation of nickel antimony alloys.

6.8.5 *Cell performance*

Most of the tubular cells have used ceramic tubes 150 mm long by 25 mm outside diameter. The relative performance of these cells as a function of current collector design is shown in Table 6.5. Each cell had a molybdenum wire current collector and a theoretical capacity of 6·84 Ah.[86] A cell having a capacity of 28 Ah has been constructed and tested. This was similar in design to previous tube cells having swagelock polymeric seals and a molybdenum brush current collector, but the electrolyte tube was 190 mm long, 44 mm outside diameter, the discharge performance is shown in Fig. 6.25. The cell failed on the 83rd cycle when sodium leaked past the silicone seal.

The projected specific energy for a 200 Wh cell varies from 113 to 95 Wh/kg for discharge rates of 13 and 3 h, respectively. These figures assume 100% utilisation of antimony and no degradation in performance with cycling.

The specific energy of a battery comprising the above cells will be about 70% of the cell specific energy, i.e. 67-79 Wh/kg. This is considerably below that projected for the sodium sulphur battery, and, despite the lower operating temperature, the materials problems with this cell, particularly the seals and current collector appear equally if not more severe. The cost of the battery seems unlikely to be less than that of the sodium sulphur battery and the only advantage the sodium antimony chloride battery appears to offer, provided the materials problem is solved, is the opportunity to use flexible seals thus reducing limits on tolerances of cell components.

Table 6.5 *Effect of positive current collector design on energy output on rate tests before and after cycling*

Design or performance parameter	Unit	Cell type			
		A	B	C	D
Collector area	cm^2	94	76	50	48
Collector weight	g	5·8	6·4	4·4	3·4
Stem diameter	cm	0·051	0·076	0·076	0·076
Time on test	h	4008	5056	4911	4935
Cycles between rate tests		130	250	260	260
Cycles on test (total)		244	450	502	499
Mean capacity first 150 cycles	Ah	5·7	5·4	4·5	4·1
Utilisation (Sb)	%	83	79	66	60
Accumulative capacity to test end∮	Ah/cm^2 β alumina	44	50	54	32
Energy output before cycling loss after cycling discharge rate*	Wh (%)	Wh (%)	Wh (%)	Wh (%)	Wh (%)

0·5 A		17·5(5)	17·7(26)	16·4(24)	16·3(43)
1·0 A		16·3(8)	17·2(34)	14·5(28)	14·2(46)
2·4 A		13·0(13)	14·9(33)	10·3(20)	10·0(51)
4·8 A		10·4(25)	-	-	-
charge rate †					
0·5A		19·8(13)	19·7(27)	16·6(20)	16·5(20)
1·0 A		19·3(15)	19·4(32)	16·1(30)	14·5(24)
2·4 A		18·7(24)	15·0(24)	7·9(–)	5·6(–)
Turnaround energy efficiency	%				
discharge rate*					
0·5 A		88	90	91	86
1·0 A		86	87	83	82
2·4 A		71	77	74	60
Failure mode		Beta cracked	Seal	VT	VT

* Charge rate held constant at 14 h

† Discharge rate held constant at 10 h

∮ Sum of charge and discharge capacity

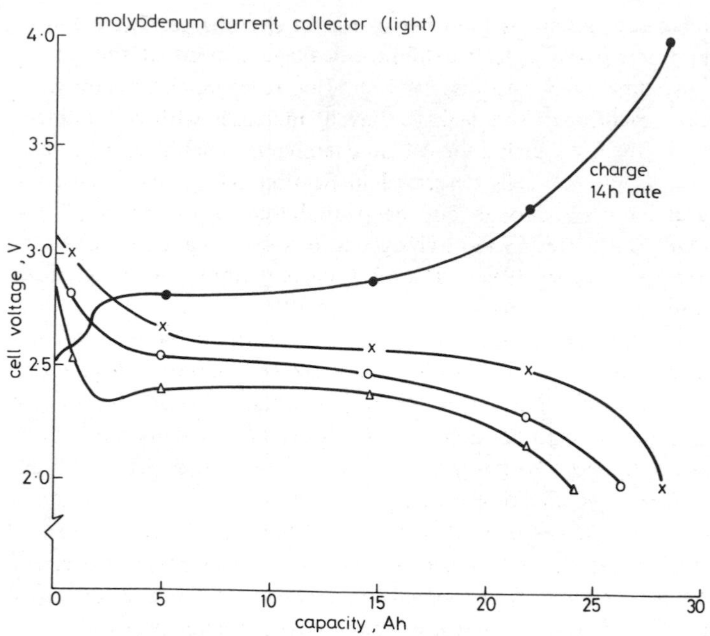

Fig. 6.25 *Charge/discharge curve for a 28 Ah sodium/antimony trichloride cell*

	Rate h	Energy Wh	Efficiency %
×	10	71	83
○	7	64	81
Δ	5	56	78

6.9 Comparison of battery systems

The performance of a cell in a battery is unlikely to be identical to that of the same cell when tested individually under well controlled conditions. Thus, in a battery there will inevitably be temperature fluctuations about the mean battery temperature; cells connected in long series chains will have current forced through them resulting in very low discharge voltages and very high charge voltages. The way in which cells are connected in series and parallel in a battery is very important and will probably be different for each type of cell.

6.9.1 Molten salt electrolyte batteries

The most important characteristic of these cells, from the point of view of battery design, is their behaviour on overcharge. The lithium-aluminium/carbon-$TeCl_4$ cell exhibits a leakage current at the top of charge, presumably due to dissolved chlorine reacting chemically with the lithium electrode. This leakage current increases with cell voltage, as shown in Fig. 6.26, giving the cell an overcharge capability. A battery comprising 12 of the cells described in Section 6.4.3. has been constructed and cycled through 100 deep discharge cycles over a period of 12 days. The battery was discharged at 6 A for 2 h and recharged to a maximum voltage of 39V with a 12 A current limit; the charge time was 90 min.

The only Li-Al/FeS_x secondary battery known to have been constructed and operated successfully was a six cell battery constructed at Argonne National Laboratory. Few performance details have been reported. As we saw in Section 6.6.1.4, these cells are damaged if they are consistently overcharged and batteries of these cells will required a very sophisticated charging system.

From the aforementioned it will be seen that there is scant information available on the performance of molten salt electrolyte batteries. Multikilowatt batteries are expected to be constructed in the very near future and it will be very interesting to see how these behave.

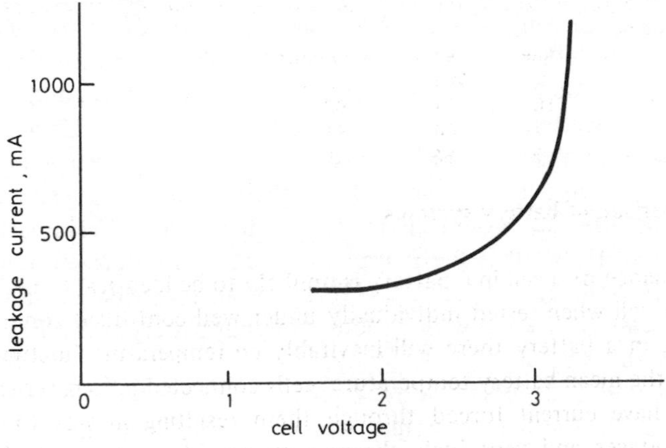

Fig. 6.26 ***Leakage current as a function of charge voltage for the lithium-aluminium/carbon - tellurium tetrachloride cell***

6.9.2 Solid electrolyte batteries

The main feature of these batteries is that their current efficiency is 100%. The only batteries constructed to date have been sodium/

sulphur batteries with beta alumina electrolyte. Several laboratories have reported results on batteries up to 10 kWh capacity[87,88] and two batteries have been installed in road vehicles.[89,90] The first battery to power a road vehicle was built by the UK Electricity Council in 1971[89] and powered a light van. Recently Yuasa Battery Company installed a battery in a small commercial vehicle.[90] The Electricity Council battery comprised about 1000 central sodium cells of 30 Ah capacity. The rated capacity was 50 kWh and the battery voltage was 100V. The specific energy of the cell was 160 Wh/kg but the weight of the containment and auxiliaries reduced the specific energy of the battery to 63 Wh/kg. The range of the 18 cwt van was between 60 and 100 miles, depending on road and driving conditions. The life of the battery was 20 cycles.

The Yuasa battery was, in many ways, similar to the Electricity Council battery. The cells were central sodium electrode type, capacity 27 Ah. The cell seal, however, was considerably more advanced, utilising a glass/metal and alpha alumina/glass/beta alumina seal. The cells were assembled in the partly discharged stage, i.e. the positive electrode was Na_2S_5. The specific energy of the cell was 180 Wh/kg^{-1}. 28 of these cells were assembled into a self contained module fitted with thermal insulation, a heating device and a cooling device. This module had an energy capacity of 1·3 kWh, a specific energy of 91 Wh/kg and a maximum cycle life of 220 cycles. Following this a complete battery was constructed having an energy capacity of 29 kWh and comprising 630 cells. The battery voltage was 60V, capacity 486 Ah. This battery was tested in combination with a mini van, from which the engine had been removed, on a dynamometer. From these studies it was estimated that the performance would be similar to that of a lead/acid battery powered vehicle, but that the range would be 2·4 times greater, implying a specific energy of at least 60 Wh/kg. Road tests were conducted with the same vehicle but with a sodium sulphur battery two thirds the size of the above battery.

The batteries described were constructed from central sodium electrode cells. The largest battery which has been constructed from central sulphur electrode cells is that described by Sudworth and Hames.[87] This was a 10 kWh battery comprising 176 cells of the type shown in Fig. 6.16. The operating voltage was 48V, capacity 288 Ah. This battery had a specific energy of 60 Wh/kg and has completed 100 electrical cycles. The major difference between this battery and other sodium sulphur batteries was that the cell, which incorporated no high cost materials such as molybdenum, was capable of being easily scaled up to 100 Ah capacity by doubling the electrolyte length and increasing the sulphur electrode thickness. Individual cells of this latter type have been constructed and tested[91]

One feature of these batteries which is very important for traction applications is their ability to withstand thermal cycles. The Japanese workers have reported that the individual cells can be cycled between 300°C and room temperature up to 30 times without damage[64] and there seems no reason to doubt that the battery can be cycled similarly. Cells of the type shown in Fig. 6.16 have been cycled between room temperature and 350°C for up to 13 times without any adverse effects and a 48 cell battery has been cycled in the same manner for 12 cycles with the loss of about 10% of the cells.

6.9.3 Future prospects

Of the molten salt electrolyte batteries, only the lithium alloy/iron sulphide system is being actively developed. This has evolved from the lithium-aluminium/carbon-$TeCl_4$ cell and the lithium/sulphur cell, and well engineered, high capacity cells are now being constructed and tested. The first generation batteries will probably be lithium-aluminium/iron sulphide batteries with a relatively low specific energy (less than 100 Wh/kg). The development of lithium alloy/iron disulphide batteries must await a solution to the problem of corrosion of the positive current collector.

The sodium sulphur cell with beta alumina electrolyte is at a more advanced stage of development than the related glass electrolyte cell or the sodium antimony trichloride cell. Indeed, several large sodium sulphur batteries have been constructed and some of these have been used to propel vehicles. The battery based on the central sulphur electrode design has now reached the commercial prototype stage and it seems likely that it will be the first high temperature battery to reach the market. A specific energy of 100 Wh/kg will probably be achieved for these first batteries, but development of the central sodium electrode cell could result in a higher specific energy for future batteries.

Acknowledgments

I would like to thank my fellow workers at British Rail for many helpful comments and suggestions. I am also indebted to my colleagues at the Admiralty Marine Technology Establishment, ESB, Rayovac Corp and Dow Chemicals for their comments on the relevant sections of my manuscript. Thanks are also due to British Railways Board for permission to use previously unpublished data on the sodium sulphur cell.

References

1 HEITBRINK, E.H. *et al.*: General Motors Company Technical Report. US Air Force Aero Propulsion Laboratory TR-67-89, August 1967

2 BRADLEY, T.G., and SHARMA, R.A.: 'Rechargeable lithium-chlorine cell with mixed ionic molten salt electrolyte'. Proceedings of the 26th Annual Power Sources Symposium, 1974, p. 60-64 (PSC Publication Committee, Red Bank, N.J.)

3 SPRAGUE, J.W. *et al.*: 'Active carbon plate cathodes for molten salt batteries', Abstracts of the Fall Meeting, Electrochemical Society, Philadelphia, Pa., October 1966

4 ANDERSON, R.A. *et al.*: 'Development of electrochemical energy storage unit for vehicle propulsion'. Final Technical Report, US Army Mobility Equipment Research and Development Center, Contract DA-44-009-AMC-1843(T), Stadard Oil Company (Ohio), December 1967

5 RIGHTMIRE, R.A. *et al.*: 'A sealed lithium-chloride, fused salt secondary battery.' Society of Automotive Engineers, Paper No. 690206, January 1969

6 British Patent 1 177 012: 'Tungsten-containing electrode', The Standard Oil Company, 7 January 1970

7 RIGHTMIRE, R.A. *et al.*: US Patent 3 462 312 'Electrical energy storage device comprising fused salt electrolyte, tantalum containing electrode and method for storing electrical energy', The Standard Oil Company, 3 January 1966

8 RIGHTMIRE, R.A. *et al.*: US Patent 3 462 313 'Electrical energy storage device comprising molten metal halide electrolyte and tungsten-containing electrode', The Standard Oil Company, 3 January 1966

9 British Patent 1 177 011: 'Electrode and electrical storage device', The Standard Oil Company, 7 January 1970

10 BENAK, J.L.: US Patent 3 501 349 'Method of treating aluminium-lithium electrode', The Standard Oil Company, 16 May 1966

11 METCALFE, J.E., CHANEY, E.J., and RIGHTMIRE, R.A.: 'Characteristics of a tellurium-lithium/aluminium battery'. Proceedings of the Intersociety Energy Conversion and Engineering Conference, Boston, Mass. August 1971, pp. 685-689

12 SCHAEFER, James C., *et al.*: 'The ESB-SOHIO Carb Tek^{R} molten salt cell'. Proceedings of the 10th Intersociety Energy Conversion and Engineering Conference, New York, August 1975, pp. 649-650

13 ROWLANDS' G.J.: Ph.D. thesis, University of Southampton, England, 1975

14 PEARSON, T.G., and ROBINSON, P.J.: *J. Chem. Soc.*, 1931, p. 413

15 SHARMA, R.A.: 'Equilibrium phases in the lithium-sulphur system', *J. Electrochem. Soc.*, 1972, **119**, 11, pp. 1439-1443

16. CUNNINGHAM, P.T., JOHNSON, S.A., and CAIRNS, E.J.: 'Phase equilibria in lithium-chalcogen systems. II. Lithium sulphur, *ibid.*, 1972, **119**, 11, pp. 1448-1450

17. CAIRNS, E.J., and SHIMOTAKE, H.: 'High-temperature batteries', *Science*, 1969, **164**, pp. 1347-1355
18 SHIMOTAKE, H., and CAIRNS, E.J.: 'Extended abstracts of the spring meeting of the electrochemical society', New York, May 1969, Abstract No. 296
19 CAIRNS, E.J. *et al.*: 'Lithium/sulphur secondary cells'. Paper presented at the 23rd meeting of the International Society of Electrochemistry, Stockholm, Sweden
20 LAI, S., and McCOY, L.R.: 'Lithium-silicon electrode', Extended abstracts of the fall meeting of the electrochemical society, Dallas, Texas, October 1975, Abstract No. 21
21 ZEITNER, Edward J., and DUNNING, John S.: 'High performance lithium/iron disulphide cells.' Proceedings of the 13th Intersociety Energy Conversion and Engineering Conference, San Diego, August, 1978, pp. 697-701
22 VISSERS, D.R., TOMCZUK, Z., and STEUNENBERG, R.M.: 'A preliminary investigation of high temperature lithium/iron sulphide secondary cells', *J. Electrochem. Soc.* 1974, **121**, 5, pp. 665-667
23 NELSON, P.A., GAY, E.C., and WALSH, W.J.: 'Performance of lithium/iron sulphide cells'. Proceedings of the 26th annual power sources symposium, 1974, pp. 65-68 (PSC Publication Committee, Red Bank, N.J.)
24 CAIRNS, E.J., and DUNNING, J.S.: 'High-termperature batteries'. Proceedings of the symposium and workshop on advanced battery research and design, March 1976. Electrochemical Soc. & Argonne National Laboratory, ANL.76-8 p. A.81-A.97
25 GAY, E.C., *et al.*: 'Electrode designs for high performance lithium-aluminium/iron sulphide cells'. *In* COLLINS, D.H. (Ed.) *Power sources 6* (Academic Press, 1977), pp.. 735-749
26 ELECTRIC POWER RESEARCH INSTITUTE, EPRI-EM-460: 'Development of lithium-metal sulphide batteries for load levelling'. Progress Report No. 3. July 1977
27 WALSH, W.J., *et* al.: 'Development of prototype lithium/sulphur cells for application to load levelling devices in utilities'. Proceedings of the 9th intersociety energy conversion and engineering conference, San Francisco, August 1974, p. 911
28 SHIMOTAKE, H., and BARTHOLOME, L.: 'Development of uncharged Li-Al/FeS cell'. Proceedings of the symposium and workshop on advanced battery research and design, Electrochemical Soc., & Argonne National Laboratory, ANL. 76-8, p B210-B218
29 WALSH, W.J., and ARNZEN, J.D.: Argonne National Laboratory, Progress Report ANL-76-9, 1975
30 TOMCZUK, Z., HOLLINS, R.E., and STEUNENBERG, R.K.: 'Overcharge studies of the FeS_2 electrode in Li/FeS_2 cells'. Proceedings of the symposium and workshop on advanced battery research and design, March 1976, Electrochemical Soc., & Argonne National Laboratory, ANL. 76-8, p. B.99-B.116
31 ARGONNE NATIONAL LABORATORY: 'High-performance batteries for off-peak energy storage and electric-vehicle propulsion'. Progress report for the period July-Sept 1976, Dec. 1976, Report ANL-76-98
32 McCOY, L.R., and LAI, S.: 'Lithium silicon electrodes for the lithium iron

sulphide battery'. Proceedings of the symposium and workshop on advanced battery research and design, Argonne National Laboratory, ANL-76-8, March 1976, p. B.167-B.175

33 HALL, John C., and McCORMICK, James T.: 'Development of initially discharged lithium-silicon/iron sulphide load-levelling cells'. Electrochemical Society Symposium on Load Levelling, October 1978, pp. 358-376

34 ELECTRIC POWER RESEARCH INSTITUTE, EPRI-EM-460: Development of lithium-metal sulphide batteries for load levelling'. Progress Report No. 3, July 1977

35 CLEAVER, B., DAVIES, A.J. and HAMES, M.D.: 'Properties of fused polysulphides-I, *Electrochimica Acta*, 1973, **18**, 10, pp. 719-726

36 CLEAVER, B., and DAVIES, A.J.: 'Properties of fused polysulphides - II, *ibid.*, 1973, **18**, 10, pp. 727-731

37 CLEAVER, B., and DAVIES, A.J.: 'Properties of fused polysulphides -III', *ibid.*, 1973, **18**, 10, pp. 733-739

38 TULLER, W.N.: *The sulphur data book* (McGraw-Hill, 1954)

39 KUMMER, J.T., and WEBER, N.: 'A sodium sulphur secondary battery'. Society of Automotive Engineers' Congress, Detroit, Janaury 1967, Paper No 670179 (also *S.A.E. Trans*, 1968, **76**, pp. 1003-1007)

40 BRAGG, W.L., GOTTFRIED, C., and WEST, J.: *Z.Krist.*, 1931, **77**, p. 255

41 YAMAGUCHI, G., and SUZUKI, K.: *Bull. Chem. Soc. Japan*, 1968, **41**, p. 93

42 BETTMAN, M., and PETERS, C.R.: 'The crystal structure of Na_2O-MgO-$5Al_2O_3$ with reference of Na_2O-$5Al_2O_3$ and other isotypal compounds', *J. Phys. Chem.*, 1969, **73**, 6, pp. 1774-1780

43 DE VRIES, R.C., and ROTH, W.L.: *J.Am.Ceramic Soc.,* **1969, 52**, p. 304

44 SUDWORTH, J.L., *et al.*: 'An analysis and laboratory assessment of two sodium sulphur cell designs'. *In* COLLINS, D.H. (Ed.) *Power sources 4* (Academic Press, 1973), pp. 1-20

45 RICHMAN, R.H., and TENNENHOUSE, G.J.: 'A model for degradation of ceramic electrolytes in Na-S batteries', *J. Am. Ceramic Soc.,* 1975, **58**, 1-2, pp. 63-67

46 ARMSTRONG, R.D., DICKINSON, T., and TURNER, J.: 'The breakdown of β-alumina ceramic electrolyte', *Electrochimica Acta*, 1974, **19**, 5, pp. 187-192

47 BUGDEN, W.G., and DUNCAN, J.H.: 'Effect of dopants on beta-alumina resistivity and reliability'. *In.* DE VRIES, K.J. (Ed.) *Science of ceramics– Vol. 9* (Nederlandse Keramische Vereniging, December 1977), pp. 348-355

48 DEMOTT, D.S., and HANCOCK, P.: 'Sodium transport in beta alumina', *Proc. Brit. Ceramic Soc.,* 1971,**19**, p. 193

49 DEMOTT, D.S., and REDFERN, B.A.W.: 'The deterioration of beta alumina during electrolysis', *J. de Physique*, Colloque C7, suppl. to No. 12, December 1976, p.C7-423-C7-427

50 ELECTRIC POWER RESEARCH INSTITUTE, EPRI EM-683: 'Development of sodium-sulphur batteries for utility application', D. Chatterji. May 1978. xiii

51 DUNCAN, J.H., and HICK, B.K.: British Patent 1331 321 'Beta-aluminium polycrystalline ceramics', British Railways Board, 26 Sept. 1973

52 BUGDEN, W.G., and DUNCAN, J.H.: 'Controlling the Duplex structure of beta alumina'. Second International meeting on Solid Electrolytes, Univer-

sity of St. Andrews, Sept. 1978, pp. 4.4.1 - 4.4.4
53 JONES, I.W. and MILES, D.J.: *Proc. Brit. Ceramic Soc.,* 1971, **19**, p. 161
54 GORDON, R.S.: 'Fabrication and characterisation of β''-alumina electrolytes and electronic-ceramic containers for the sodium-sulphur battery'. University of Utah, Report UTEC-MSE 76-286. December 1976
55 MILLER, M.L., McENTIRE, B.J., and GORDON, R.S.: 'A pre-pilot process for the fabrication of polycrystalline β'' alumina electrolyte tubing'. Paper presented to the Basic Science and Nuclear Divisions Fall Meeting (American Ceramic Society), Sept. 1977
56 FALLY, J., *et al.*: 'Study of a beta-alumina electrolyte for sodium-sulphur battery', *J. Electrochem. Soc.,* 1973, **120**, 10, pp. 1296-1298
57 ELECTRIC POWER RESEARCH INSTITUTE EPRI EM-266: 'Development of sodium-sulphur batteries for utility application', D. Chatterji, December 1976, xvi
58 DESPLANCHES, G., *et al.*: British Patent 1 499 497 'An electric cell using beta sodium alumina electrolyte, E.G.A. sodium-sulphur electric cell', Compagnie Generale d'Electricite S.A., 1st Feb. 1978
59 DELL, R.M., SUDWORTH, J.L., and JONES, I.W.: 'Sodium sulphur development in the United Kingdom'. Presented to the 11th Intersociety Energy Conversion and Engineering Conference 1976
60 SUDWORTH, J.L., TILLEY, A.R., and BIRD, J.M.: 'A potentially low-cost sodium/sulphur cells for load levelling'. Presented at the 152nd meeting of the Electrochemical Society, Atlanta. October 1977
61 McGUIRE, M.: British Rail R & D Division (Unpublished results)
62 WEINER, S.A., *et al.*: 'Research on electrodes and electrolyte for the Ford sodium sulphur battery'. Ann. Reports 1975. Nat. Science Foundation, NSF-RANN Contract No. HSF-C.805
63 MARKIN, T.L., *et al.*: 'Current collectors for sodium sulphur batteries'. Paper 48 presented to the Power Sources Symposium, Brighton 1978
64 HATTORI, S., *et al.*: 'A new design for the high-performance sodium-sulphur battery'. Paper presented to the Society of Automotive Engineers, Feb. 1977, (Paper No. 770281)
65 PETT, R.: 'Sodium-sulphur battery review: supporting research'. Presented to DoE-Ford sodium-sulphur battery development program review, 1st Nov. 1978, Marriott Hotel, Newport Beach, California
66 KINSMAN, K.R., and OEI, D.G.: 'Corrosion of nickel base superalloys in sodium tetrasulphide'. Extended Abstracts of the Fall Meeting of the Electrochemical Society, Atlanta, October 1977
67 HARTMANN, B.: 'Casing materials for sodium/sulphur cells', *J. Power Sources*, 1978, **3**, 3, pp. 227-235
68 JOHNSON, O.W., and MILLER, G.R.: 'A metal-clad electronic ceramic container and electrode system', *Am. Ceramic Soc. Bull.,* 1977, **56**, 8, p. 706
69 MINCK, R.W.: 'The performance of shaped graphite electrodes in sodium sulphur cells'. Proceedings of the Symposium and Workshop on Advanced battery research and design, March 1976. Electrochemical Soc. and Argonne National Laboratory, ANL.76-8, p.B. 199-B.209
70 FALLY, J., *et al.*: 'Some aspects of sodium sulphur cell operation', *J. Electrochem. Soc.,* 1973, **120**, 10, pp. 1292-1295
71 BREITER, M., and DUNN, B.: *J. Appl. Electrochem.* (in press)

72 WEDDIGEN, G.: 'Problems of the sulphur electrode in the sodium-sulphur system'. International Society of Electrochemistry, 29th meeting, Druzhba-Varna, Sept. 1977, Extended Abstracts No. 114, pp. 472-483

73 LE MEHAUTE, A., *et al.*: US Patent 4 037 028 'Sodium-sulphur type electric cell', Compagnie Generale d'Electricite, Paris, France, 19 July 1977

74 SOUTH, K.D.: British Rail R & D Division (unpublished results)

75 GIBSON, J.G.: 'The distribution of potential and electrochemical reaction rate in molten polysulphide electrodes', *J. Appl. Electrochem.* 1974,4, pp. 125-134

76 KNODLER, R., BAUKAL, W., and KUHN, W.: 'Investigation into the immobilisation of the sodium electrode in the sodium-sulphur battery', *J. Electrochem. Soc.* 1977, **124**, 2, pp. 236-237

77 HAMES, M.D., and TILLEY, A.R.: 'Safety of sodium/sulphur cells'. 24th International Society of Electrochemistry meeting, Marcoussis, France, May 1975

78*a* US Patent 3 476 602: 'Battery cell', Dow Chemical Co. Midland, Michigan, 4 Nov. 1969

b BROWN, W.E., *et al.*: US Patent 3 679 480 'Electrical cell assembly', Dow Chemical Company, 25th July 1972

79 LEVINE, C., and ANAND, J.: 'The sodium-sulphur battery with glass electrolyte'. Symposium on load levelling, Fall meeting of the Electrochemical Society, Atlanta, October 1977, pp. 292-331

80 LEVINE, DEAN, and WU.: 'Sodium-sulphur battery'. US Dept. of Energy 2nd annual Battery and Electrochemical Conference, 780603, May 1978

81 LEVINE, C.A.: 'Progress in the development of the hollow fiber sodium-sulphur secondary cell'. Proceedings of the 10th Intersociety Energy Conversion and Engineering Conference, New York, August 1975, pp. 621-623

82 ELECTRIC POWER RESEARCH INSTITUTE, EPRI. EM-230: 'Sodium-chloride battery development program for load levelling'. Interim report (Research Project 109), December 1975

83 ELECTRIC POWER RESEARCH INSTITUTE, EPRI.EM-751: 'Sodium-antimony trichloride battery development program for load levelling'. Interim report (Research Proect 109-3), April, 1978, xiv

84 WERTH, J., KLEIN, I., and WYLIE' R.: 'The sodium chloride battery', *Electrochem. Soc. Extended abstract*, 1976 (Reprinted from *Energy storage*, p. 198-204

85 WERTH, J., and CHREITZBERG, A. (Private communication)

86 CHREITZBERG, A.M., *et al.*: 'Performance of molten salt sodium/beta alumina/$SbCl_3$ cells', *J. Power Sources*, 1978, 3, pp. 201-214

87 SUDWORTH, J.L., and HAMES, M.D.: 'A 10 kWh sodium sulphur battery', Presented to the Electrochemical Society Fall Meeting, Pittsburgh, 1978

88 FISCHER, W., *et al.*: 'Sodium sulphur battery development in Germany – a status report'. Presented to the Electrochemical Society Fall Meeting, Pittsburgh, 1978

89 CHURCHMAN, A.T.: 'Tomorrow's battery today'. Paper presented to the Royal Society of Arts, West Midlands Region at the University of Aston, 14 March 1975

90 YUASA Battery Co.: 'Past and present research effort on development of Na/S batteries'. Press release, October 1977

91 BIRD, J.M., TILLEY, A.R., and SUDWORTH, J.L.' 'Sodium/sulphur cell designed for quantity production'. Proceedings of the 13th Intersociety Energy Conversion Engineering Conference, Nov. 1978, (pp. 685-689) Society of Automotive Engineers Reprint No 789201

Chapter 7

Room temperature cells with solid electrolytes

T. Dickinson

7.1 Introduction

Most of the reported cell systems employing solid electrolytes also have solid anodes and cathodes. The advantages of wholly solid state systems have been described by Hull;[1] the principal ones being their ruggedness, ability to operate over wide temperature ranges, ease of miniaturisation and the possibility of obtaining very long shelf- and operating-lives.

Until relatively recently, however, the foregoing advantages were seriously outweighed by the low ionic conductivities, at room termperature, of the known solid electrolytes. Two factors have mitigated this shortcoming. The first has been the discovery of new or modified electrolytes with higher conductivities. Secondly, the power requirements of many electronic devices are being continuously reduced. Cells for heart pacemakers,[2] for example, are now required to produce only 15 μA at a voltage between 5V and 3·5V, electronic wrist watches, using liquid crystal displays and tritium powered sources for back lighting, need less than 10 μA at a voltage in excess of 2·4V, and standby power sources for CMOS solid-state memory devices require[3] only a few milliamperes at 2-3V.

Wholly solid-state systems have the disadvantages[4] that the electrolyte cannot permeate the electrodes, the cell cannot accommodate gross changes in electrode volume and optimising the composition for electrode compatibility can prove troublesome. As will be apparent later, these limitations have not caused excessive difficulties.

To date, almost all the systems reported are primary cells and the majority of the electrolytes are cationic conductors with only one mobile ion. The nature of this ion, has therefore, determined the anode material, and it is consequently convenient to classify the systems according to the nature of the mobile ion.

It is highly unlikely that electrolytes, with other than monovalent ions as the mobile species, will show appreciable conductivities near room temperature. This arises because the coulombic interaction between, for example, divalent ions is likely to increase the activation energy for conduction four-fold over that for monovalent ions.[5]

In the known solid electrolytes which have adequate conductivities near room temperature, the mobile ion is one of the following; Ag^+, Cu^+, Li^+, Na^+, H^+, and F^-; cell systems have been devised around all of these. As will become apparent, the only commercially viable cells appear to be those based on lithium ion conductors. The bulk of the material reported here is therefore concerned with these systems.

It is not possible here to give a detailed account of the many cell systems which have been devised. Owens[4] has, however, comprehensively reviewed cells based on silver ion conductors, a recent review of silver, copper and lithium solid-state power sources is also available,[6] while Owens *et al.*[7] have described selected silver and lithium solid-state cells.

7.2 Silver ion conductors

7.2.1 Electrolytes

Until very recently, the most highly conducting solid electrolyte known was rubidium silver iodide ($RbAg_4I_5$). Its ionic conductivity at room temperature is $0{\cdot}27\ \Omega^{-1}\ cm^{-1}$. Many other compounds based on silver iodide have also been found to possess appreciable ionic conductivity. They fall into two main categories. In the first, relatively small amounts of large foreign cations are incorporated into the compound. The second group is composed of a range of silver-iodide/silver-oxysalt electrolytes.[6] They are reported to be much more stable toward moisture and iodine than $RbAg_4I_5$. These electrolytes possess a glass-like structure and may allow the preparation of transparent glasses with fairly high conductivities.

7.2.2 Power sources

Rubidium silver iodide has been used as the electrolyte in many cell systems.[4] Its low decomposition voltage (0·69V), however, precludes the use of strongly oxidising cathode materials. The preferred cathode material was $(CH_3)_4NI_5$, which, on discharge, probably produces $[(CH_3)_4]Ag_{13}I_{15}$, which itself has an appreciable ionic conductivity.[8]

These cells gave very flat discharge characteristics and predicted shelf-lives of 20 years. The low cell voltage, energy density and volumetric energy density along with their relatively high cost has prevented them from being commercially viable. The same is expected to be true of cells based on the silver-iodide/silver-oxysalt electrolytes. Performance data for both these types of cell have been reported[6,7] and the design of cells based on $RbAg_4I_5$ has also been described.[7]

7.3 Copper ion conductors

7.3.1 Electrolytes

A variety of complex cuprous halides have been found to possess quite high ionic conductivities.[9–12] The best conductor discovered to date is $Rb_3Cu_{12}I_5Cl_{10}$, which has an ionic conductivity[12] of 0·28 Ω^{-1} cm^{-1} at 25°C

7.3.2 Power sources

At least two of these compounds[13,14] are, unfortunately, unstable toward cathode materials, giving a cell voltage of about 0·6V. This arises because the cuprous ions are oxidised to the cupric state, generating electronic conductivity in the solid.[14] It is to be expected that this will be a general property of all these materials, although the decomposition voltage of $Rb_3Cu_{12}I_5Cl_{10}$ has been reported[12] to be 0·69V. In addition, the exchange current density for the copper/cuprous ion reaction is very low,[15] leading to marked polarisation at the anode.[16]

Although experimental cells have been constructed using a variety of cathode materials,[6] systems based on these electrolytes seem unlikely to be produced commercially.

7.4 Proton conductors

7.4.1 Electrolytes

Although the mixed acids $H_3(PW_{12}O_{40})29\,H_2O$ and $H_3(PMo_{12}O_{40})$ $30H_2O$ are reasonably good proton conductors,[17] they are corrosive toward metals and tend to lose water in air with a consequent decrease in conductivity. A much more attractive material[18] is $HUO_2\,PO_4.4H_2O$ (HUP). This compound has a conductivity of 0·004 Ω^{-1} cm^{-1} at 23° C. It is insoluble in water, stable in air and can readily be pressed into translucent discs.

Hydronium beta″ alumina $[0{\cdot}8\,H_2(H_2O)_{2{\cdot}8}\,0{\cdot}084\,MgO\,.0{\cdot}5\,Al_2O_3]$ has been reported[19] to have a conductivity of 10^{-4} Ω^{-1} cm^{-1} at 25°C.

7.4.2 Power sources

Takahashi[17] has reported a current/voltage curve for a fuel cell employing the mixed phospho-tungstic acid. The use of HUP in a solid state hydride battery[20] has been investigated.

7.5 Sodium ion conductors

7.5.1 Electrolytes

Although a substantial number of sodium ion conductors have been discovered, those with the highest conductivities at room temperature are the sodium beta aluminas and their gallium analogues. The preferred beta alumina for use at room temperature was $Na_2O\,.9Al_2O_3$ containing 1% MgO and 0·5% Y_2O_3 as stabilisers.[21] This material has a conductivity of 10^{-3} Ω^{-1} cm^{-1} at 25°C and is apparently stable toward moisture.

Single crystals of the gallium analogues of beta and beta″ alumina have conductivities at 25°C about 1·5 times larger than the aluminium based compounds.[22] Comparative results on sintered ceramics are not yet available.

7.5.2 *Power sources*

Primary cells, employing a beta alumina electrolyte and operating at or near room temperature, do not appear to be commercially available. For a period, GEC advertised cells of the type[21]

$$\text{Na(Hg)/beta}'' \text{ alumina}/Br_2 \text{ in 1 M NaBr(aq)}$$

Because severe polarisation was encountered with solid anodes, sodium amalgam was used as the anode material, and, to ensure the existence of a liquid phase in the amalgam, the cells had to be operated at temperatures above 21°C.

The published information on such cells[21] made them seem attractive for application such as heart pacemakers. Presumably, the new lithium-based systems have made these cells redundant.

7.6 Fluoride ion conductors

7.6.1 *Electrolytes*

Two groups of fluorides show reasonable ionic conductivities; those of the rare-earth metals and those of Ca, Ba, Sr and Pb. No cells have been devised using the first group, presumably because of the costly materials involved. Within the second group, $\beta\text{-}PbF_2$ has by far the highest conductivity and hence attention has been focused on it.

Since these materials are defect conductors, their conductivities can be considerably enhanced by doping with aliovalent cations. Thus, although pure $\beta\text{-}PbF_2$ has a conductivity of only $5 \times 10^{-7}\ \Omega^{-1}\ cm^{-1}$, the addition of 2m/o† KF raises this[23] to $1 \times 10^{-3}\ \Omega^{-1}\ cm^{-1}$. It has been claimed[24] that $PbSnF_4$ is the best fluoride ion conductor discovered so far.

7.6.2 *Power sources*

Thin film cells using either pure $\beta\text{-}PbF_2$ or material doped with 0•5 m/o AgF have been reported.[25,26] The first cell gave a voltage of only 0•932V at a current drain of 1 $\mu A/cm^2$, while the second produced only 0•15V at 0•5 $\mu A/cm^2$.

† m/o = mol %

Attempts[23] to use the potassium fluoride doped electrolyte in cells such as

$$Pb/\beta\text{-}PbF_2\ .KF(2m/o)\ /AgF$$

gave promising open circuit voltages (1·30V for the above cell). Unfortunately, they suffered severe polarisation under load. This was because α–PbF_2 (conductivity = $4 \times 10^{-8}\ \Omega^{-1}\ cm^{-1}$) was the anodic dissolution product. The formation of this material is at least partially due to the tendency of β–PbF_2 to convert to α–PbF_2 under pressure.

7.7 Lithium ion conductors

7.7.1 Electrolytes

The high voltage and energy density anticipated from cells employing lithium anodes has prompted considerable research into lithium ion conductors. Some of this work was directed toward devising electrolytes, stable toward molten lithium, which could be used in systems analogous to the sodium/sulphur cell. Such electrolytes normally have too low a conductivity at room temperature for them to compete with other lithium ion conductors. A few, however, although still relatively poor conductors, may offer advantages of long term stability. The electrolytes described here are grouped under the headings of lithium halides and other lithium compounds.

7.7.1.1 Lithium halides: Pure lithium iodide has the highest conductivity of the lithium halides but this amounts to only $1 \cdot 2 \times 10^{-7}\ \Omega^{-1}\ cm^{-1}$ at 25°C. Many attempts have been made to improve this conductivity by doping or other expedients. Liang and co-workers were among the pioneers in this field. The preferred material, discovered by this group, was $LiI\ .Al_2O_3$(33–45m/o) which has a conductivity[27] of $1 \times 10^{-5}\ \Omega^{-1}\ cm^{-1}$ at 25°C.

Conduction of LiI is believed to occur because of the presence of lattice vacancies. Since the concentration of vacancies is likely to be much greater near the surface of crystals than in their bulk, the addition of Al_2O_3 may enhance the conductivity by providing a large amount of internal surface area.[28] Harris[29] has argued that such an explanation would not account for the large increase in conductivity obtained. He has proposed that segregation of residual impurity to, and the formation of precipitates within, the interfacial layer would produce a region of fragmented structure and complex chemical composition which should increase the conductivity by transport of iodide ions. In accord with either of these views, it has been reported[30,31] that the conductivities of the other lithium halides can also be enhanced by addition of Al_2O_3. The very high conductivity ($3{\cdot}4 \times 10^{-4}\ \Omega^{-1}\ cm^{-1}$) claimed[31] for LiBr $.Al_2O_3$(40m/o), taken in conjunction with the information provided below suggests that moisture was present in this electrolyte.

Addition of SiO_2 can also increase the conductivities of lithium halides. With either Al_2O_3 or SiO_2 present, the conductivity is further increased by the presence of a small amount of water.[7,32] This water is evidently strongly bound within the electrolytes since they are essentially dry, can be ball-milled into fine powders and are apparently stable in contact with solid lithium.

The properties of the most highly conducting lithium halide electrolytes are summarised in Table 7.1. In all cases the electronic conductivities are negligibly small. Where the decomposition voltage is queried, the value quoted is the theoretical value for the pure lithium halide.

7.7.1.2 Other lithium compounds: Although lithium beta alumina is a reasonably good conductor,[33] sodium beta alumina with 50% of the sodium atoms replaced by lithium is substantially better.[34] The ceramic material has a conductivity of about $10^{-3}\ \Omega^{-1}\ cm^{-1}$ at 25°C, making it one of the best lithium ion conductors available to date. It is also stable toward lithium at ambient temperature and the transport number of Li^+ is very close to unity.

Table 7.1 *Properties of some lithium halide based electrolytes at 25°C*

Electrolyte	Ionic conductivity	Decomposition voltage	Reference
	$\Omega^{-1}\ cm^{-1}$	V	
LiI	$1{\cdot}2 \times 10^{-7}$	2·90	27
LiI $.Al_2O_3$(33–45m/o)	1×10^{-5}	2·90(?)	27
LiBr $.Al_2O_3$(40m/o)	$3{\cdot}4 \times 10^{-4}$	3·56(?)	31
LiCl(46·3w/o) $.Al_2O_3$(34·7w/o) H_2O(19·Ow/o)	$1{\cdot}4 \times 10^{-4}$	3·95(?)	32
LiF(56·6w/o) $.Al_2O_3$(33·6w/o) H_2O(9·8w/o)	$1{\cdot}3 \times 10^{-4}$	6·05(?)	32

† w/o = weight %

The most serious rival for this position is Li_3N. The conductivity of single crystals has been measured as about $4 \times 10^{-4}\ \Omega^{-1}\ cm^{-1}$ using Cr electrodes and $3 \times 10^{-3}\ \Omega^{-1}\ cm^{-1}$ using lithium electrodes.[35] The difference in these values has been attributed to changes in the Li^+ activity between the two conditions and a contact effect when preparing the electrodes with molten lithium. There seems good reason to believe[36] that, with careful control of the sintering conditions, ceramic pellets can be prepared with conductivities of about $6{\cdot}6 \times 10^{-4}\ \Omega^{-1}\ cm^{-1}$ at 25°C. The theoretical decomposition voltage of lithium nitride is only 0•44V so that it may not be stable in cells producing a large voltage. However, application of 3•8V for 6 weeks to the cell $Au/Li_3/Au$ failed to cause any decomposition.[37] It may be that any decomposition reactions are extremely slow.

Lithium aluminium chloride has also been investigated[38] as a solid electrolyte. Although its intrinsic conductivity is low it is possible that suitable doping may improve this value.

The properties of these materials are summarised in Table 7.2 along with those of two lithium oxysalts. These latter materials were devised for use with molten lithium and do not have very high conductivities at 25°C. Their stability toward lithium, and probably to strong oxidising agents, may, however, make them of interest. These materials are ceramics and it should be noted that their conductivities increase with their densities. An increase of about 5-fold is expected for a 100% dense sample over that of a 75% dense one.[39]

7.7.2 Power sources

For a number of years, Liang and co-workers developed cells based on lithium iodide. An early difficulty was the tendency of the cations from the cathode material (e.g. AgI) to interdiffuse with Li^+, resulting in an internal short circuit. This problem does not appear to exist with the cell system now manufactured by the Mallory Battery Company. The system employed is

$$Li/LiI,Al_2O_3(33m/o)/PbI_2(40w/o),PbS(40w/o),Pb(20w/o)$$

The electrolyte is prepared by heating a powdered mixture of LiI and Al_2O_3 at 550°C for 17 h in helium containing less than 15 parts per million H_2O and O_2. The cathode is a well blended mixture of the powdered components.

The construction of a typical cell is shown in Fig. 7.1. Although a

Table 7.2 *Properties of some lithium ion conductors at 25°C*

Electrolyte	Conductivity Ionic	Electronic	Decomposition voltage	Theoretical density	Reference
	Ω^{-1} cm^{-1}		V	%	
Li beta alumina	$1{\cdot}3 \times 10^{-4}$	negligible	very high	$\simeq 100$	33
Li/Na beta alumina	$\simeq 10^{-3}$	negligible	very high	$\simeq 100$	34
Li_3N	$6{\cdot}6 \times 10^{-4}$	$\simeq 10^{-8}$	0·44	$\simeq 100$	36
$LiAlCl_4$	$1{\cdot}2 \times 10^{-6}$	$< 10^{-8}$	?	Compressed powder	38
$LiHf_2(PO_4)_3$	$3{\cdot}8 \times 10^{-6}$		?	79	39
$Li_{0,8}Zr_{1{\cdot}8}Ta_{0{\cdot}2}(PO_4)_3$	$4{\cdot}7 \times 10^{-6}$		?	85	39

variety of sizes and shapes can be produced, the dimensions of the cell shown are given in the legend. Clearly, multi-cell stacks can readily be assembled. The cells must be assembled, of course, in a dry atmosphere.

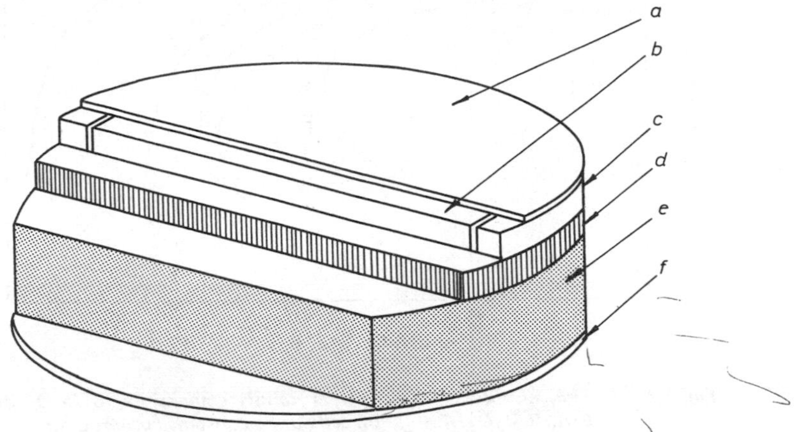

Fig. 7.1 *Sectional view of the Mallory solid state electrochemical cell*
a Steel anode current collector
b Lithium anode (area 1·48 cm^2)
c Polypropylene anode retaining ring
d Solid electrolyte (0·02 ± 0·005 cm thick, 1·81 cm^2 in area)
e Cathode
f Pb cathode current collector

The cathode, electrolyte and anode with its retaining ring are consolidated by pressing in a die at 7030 kg/cm^2. The complete assembly is hermetically packaged in a steel housing incorporating a glass to metal seal,[40] having an outside diameter of 1·73 cm.

The open-circuit voltage of the cell is 1·91 ± 0·01V at 25°C and the cathode reactions are believed to be

$$2\,Li^+ + PbI_2 + 2e = 2\,LiI + Pb$$
$$2\,LI + PbS + 2e = Li_2S + Pb$$

Performance curves for cells having stoichiometric capacities (limited by the amount of the cathode) of 33 mA h and 136 mA h have been reported.[40] Even though the cells had electrodes of the same diameter little change in performance resulted from increasing the amount of cathode material. Typical discharge curves at room temperature for the 136 mAh cell are shown in Fig. 7.2.

Mallory claim that these cells will operate over the temperature

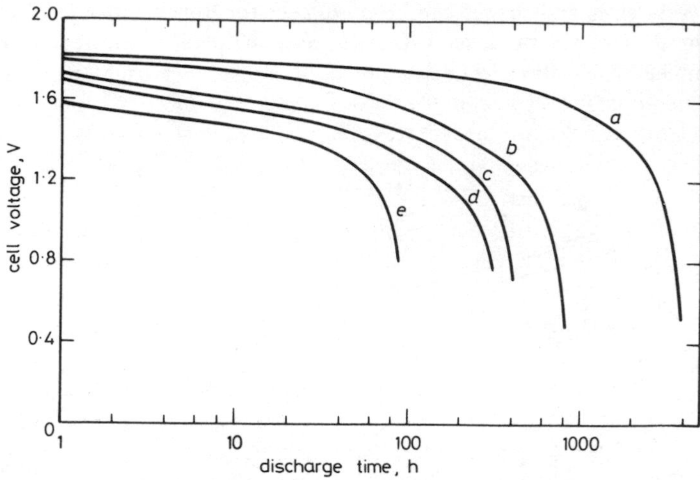

Fig. 7.2 *Typical discharge curves, at room temperature, of a 0·136 Ah Li/LiI(Al_2O_3)/PbI_2, PbS cell during constant current drain*
a 18 μA
b 36 μA
c 54 μA
d 72 μA
e 90 μA
(After Reference 33.)

range −40°C to 220°C, although a modified electrode structure is required for operation at temperatures above 150°C. At −40°C the current capability of the cells is, however, only 2 to 3% of that at room temperature. As might be expected the current capability at 100°C is 10 to 20 times that at room temperature.

These cells have been evaluated as power sources for heart pacemakers. There is some disagreement as to the life to be expected under these conditions. Results published by Liang and Barnette[40] show that when discharged at 1·3 μA/cm^2 (equivalent to about 16 μA for 8 batteries in parallel) the voltage of a 3-cell battery was initially 5·64V (at 37°C) but fell almost linearly with time reaching 5·1V after two years. Assuming this behaviour continues, an operational life of 7-8 years, to 3·5V, is anticipated. Owens *et al.*,[2] as a result of a three-year discharge under closely similar conditions, predicted that the usable life would be only 4-5 years. Until these tests have been completed, however, predictions based on extrapolations must be treated with caution.

This system has been shown to withstand storage for one year at temperatures between 25°C and 100°C without any measurable loss in capacity[40] and is anticipated to have a shelf life exceeding 20 years under normal storage conditions. With a battery containing three cells in series, a volumetric energy density of 0·49 Wh/cm^3 is obtained at a current density of about 1·3 $\mu A/cm^2$.

Applications envisaged for the system include memory circuits, oil-well drilling equipment, heart pacemakers, wrist watches and safety devices (smoke alarms, sprinkler heads etc.).

The use of a cathode consisting of As_2S_3 (70w/o), Bi (25w/o) and electrolyte (5w/o), in place of the PbI_2,PbS combination allows 95% of the theoretical capacity (0·065 Ah) to be obtained at a current density of 12 $\mu A/cm^2$ to a cut-off voltage of 1·0V. The cell has essentially the same open-circuit voltage as the PbI_2,PbS system but the latter is estimated to give only about 60% of the theoretical capacity under the same conditions. The volumetric density of the As_2S_3 system is therefore about 0·80 Wh/cm^3. No data on its shelf-life is available.[41] Various other cathode materials have been patented in connection with this cell[42] but insufficient data were reported to assess whether they are better than As_2S_3.

Cells employing electrolytes consisting of LiF, LiCl or LiBr with Al_2O_3 or SiO_2 and controlled amounts of water have also been patented.[32] With the system Li/LiI,Al_2O_3,H_2O/Me_4NI_5 a cell voltage of 2·75V was obtained and a projected energy density of 0·40 Wh/cm^3 at the 10 year discharge rate. Cathode efficiencies of 95-100% were obtained in accelerated discharge tests. It is of interest that, with $FeCl_3$ as the cathode, the cell could be recharged but its performance and cycle life are not described.

Although pure LiI has a low conductivity, it has been used as the electrolyte in a very successful cell originally devised by Schneider *et al.*[43] Cells based on the same system but constructed in differing ways are marketed by both Catalyst Research Corporation and Wilson Greatbatch Ltd. The system employed is

$$\text{Li/LiI/poly-2-vinylpyridine},6{\cdot}2I_2$$

The cathode material (abbreviated to P2VP,nI_2 is a semiconducting charge-transfer complex. Its resistivity[44] falls from an initial value of about 2000 Ω cm to about 1000 Ω cm when the composition is P2VP, 3·5I_2. The resistivity rises again as the iodine content decreases, the rise being rapid for compositions below P2VP,2I_2. This gives rise to a fairly abrupt drop in cell voltage, since the system is designed to be cathode limited.

Both manufacturers use stainless steel cases for their cells. They differ, however, in the internal arrangement. The 'Catalyst' cell has a fluoropolymer plastic envelope, which is highly resistant to iodine attack, inside the steel case. Within this is the lithium anode which envelopes the cathode material. This cell is constructed in a dry room where the relative humidity is 1%. After cleaning the lithium metal, it is formed into a cup and the depolariser poured in. A spontaneous reaction occurs to form a very thin film of LiI between the anode and the depolariser. Following solidification of the depolariser, the lithium cup is cold-welded shut and then the plastic envelope added. The leads pass through glass-to-metal hermetic seals. A sectional view of the cell and a photograph of the completed cell are shown in Fig. 7.3.

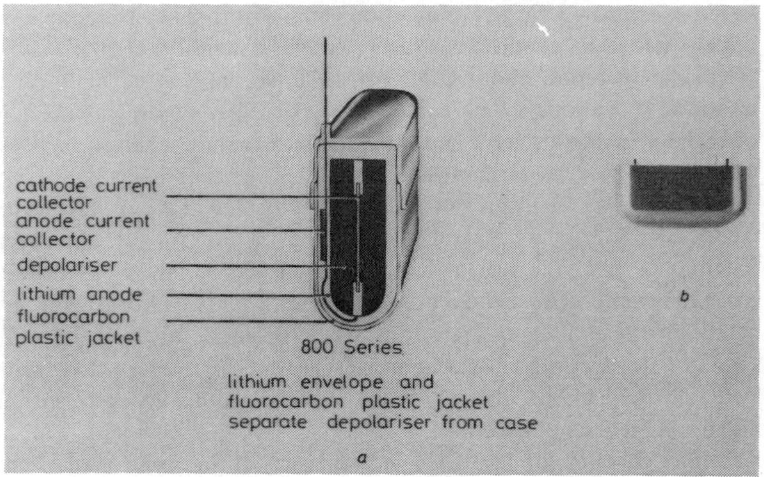

Fig. 7.3 *The Catalyst Research Corporation Li/LiI/P2VP.nI$_2$ cell*
a Sectional view
b Complete cell

Although a range of sizes are made, that shown in the photographs and to which the discharge curve refers, Fig. 7.4, has the characteristics shown in Table 7.3 where they are compared with the Wilson Greatbatch (WG) cell.

A sectional view of the WG cell is shown in Fig. 7.5. It differs from the Catalyst cell in that the lithium is surrounded by the cathode material which is in direct contact with the stainless steel case. This has the advantage that only one hermetic seal is required, but it has been reported[45] that tantalum, used as the cathode current collector in the Catalyst cell, is the only metal to wholly withstand attack by iodine.

As shown by Table 7.3, the performance characteristics of the two cells are very similar. A discharge curve for the Catalyst cell, at an average current of 15 μA, is shown in Fig. 7.4; the WG cell gives a very

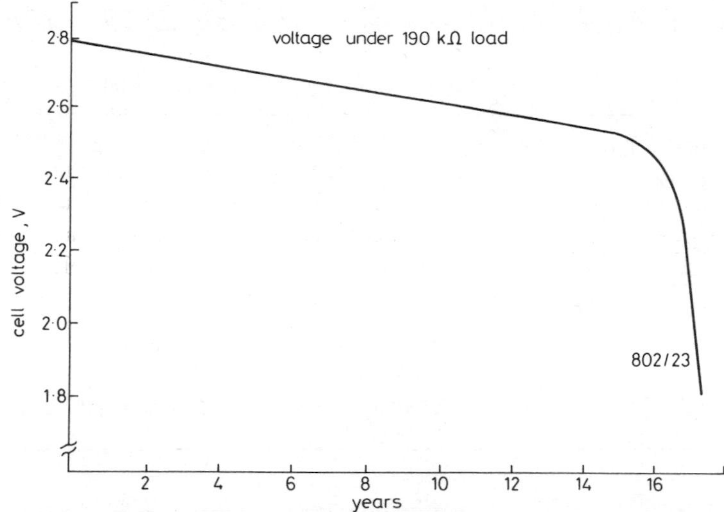

Fig. 7.4 *Projected performance of Li/LiI/P2VP .nI$_2$ type 802/23 cells based on 5 years' test data*

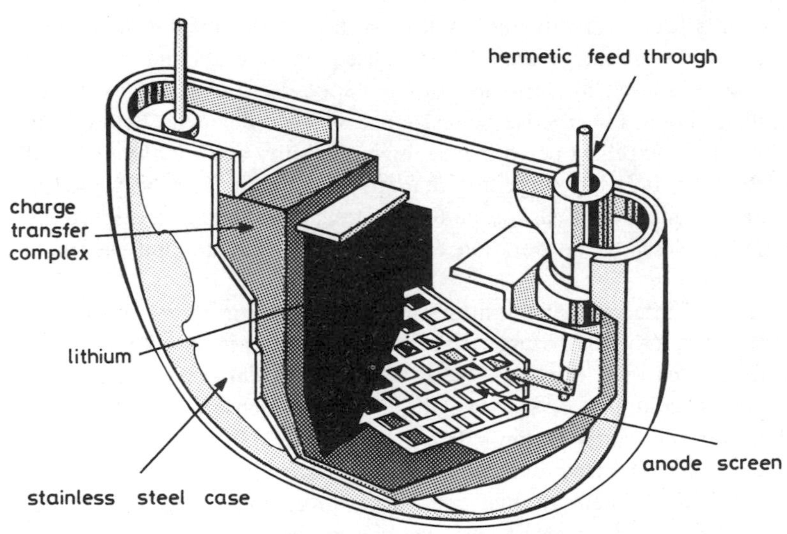

Fig. 7.5 *Sectional view of the Wilson Greatbatch Li/LiI/P2VP .nI$_2$ cell*

Table 7.3 *Performance characteristics of $Li/LiI/P2VPnI_2$ cells at 37°C*

	Catalyst model 802/23	WG model 762
Dimensions, mm	45·1 x 13·5 x 23	45 x 9 x 28
Volume, cm^3	11·2	8·6
Mass, g	30	30
Open-circuit voltage, V	2·8	2·8
Rated current, μA	50	30
Rated capacity, Ah	2·3	2·5
Energy density	0·53 Wh/cm^3	0·73 Wh/cm^3
	0·20 Wh/g	0·20 Wh/g
Self-dischrarge	< 10% in 10 years	< 10% in 10 years

similar curve. The reaction at the cathode is

$$2Li^+ + P2VP.nI_2 + 2e = 2LiI + P2VP.(n-1)I_2$$

The increase in thickness of the lithium iodide layer during discharge accounts for the almost linear fall of voltage with time during the first part of the discharge curve. The subsequent more rapid decline in voltage is caused by depletion of the depolariser, which, as explained earlier, shows a sharp rise in resistance when n decreases below 2. In view of the excellent performance, high reliability and very long lives of these cells, it is not surprising that they are being extensively used in heart pacemakers. Cells of modified design are being developed by Catalyst Research Corporation to power wrist watches and electronic calculators.

It is of interest that the lithium iodide formed in these cells has an ionic conductivity 10x larger than normal lithium iodide. Photomicrographs of partially discharged cells have shown that the lithium iodide grows in columnar crystallites, the axes of which are parallel to the direction of current flow. This might provide high surface area pathways with many defects, accounting for the high conductivity.

A similar system to that reported above, employing LiBr as the electrolyte and a bromine charge-transfer complex, has been patented.[46] It is reported[7] to have an open-circuit voltage of 3·50V and a volumetric energy density of 1·25 Wh/cm^3. Few other details of its performance are available to date.[7]

Acknowledgments

I wish to express my sincere thanks to Dr. C.C. Liang (Wilson Greatbatch Ltd.), Mr. B. MacDonald (Mallory Battery Co. Ltd.) and Dr. A.A. Schneider (Catalyst Research Corporation) for the considerable amount of information they provided on the cells manufactured by their companies.

References

1 HULL, M.N.: 'Recent progress in the development of solid electrolyte batteries', *Energy Convers.*, 1970, **10**, pp. 215-226

2 OWENS, B.B., FESTER, K., and KUDER, T.: 'Battery applications of lithium conducting solid electrolytes', Ext. Abst. Elect. Soc. Meeting, Las Vegas, Nevada, Oct. 1976

3 OWENS, B.B., and UNTEREKER, D.F.: 'Acclerated testing of long-life primary cells'. Paper 40, 11th International Power Sources Symposium, 1978

4 OWENS, B.B.: 'Solid electrolyte batteries'. *In* TOBIAS, G.W. (Ed.) *Advances in electrochemistry and electrochemical engineering – Vol. 8,* (Wiley, 1971), pp. 1-62

5 MEHAN, G.D.: 'Theoretical issues in superionic conductors'. *In* MAHAN, G.D., and ROTH, W.L. (Eds.) *Superionic conductors* (Plenum Press, 1976), pp. 128-134

6 LAZZARI, M., and SCROSATI, B.: 'A review of silver, copper and lithium solid-state power sources', *J. Power Sources*, 1977, **1**, pp. 333-358

7 OWENS, B.B., OXLEY, J.E., and SAMMELLS, A.F.: 'Applications of halogenide solid electrolytes' *In* GELLER, S. (Ed.) *Topics in applied physics – Vol. 21* (Springer-Verlag, 1977), pp. 67-104

8 OWENS, B.B.: 'A new class of high conductivity solid electrolytes; tetraalkylammonium iodide-silver iodide double salts', *J. Electrochem. Soc.*, 1970, **117**, pp. 1536-1539

9 TAKAHASHI, T., YAMAMOTO, O., and IKEDA, S.: 'Solid state ionics: high conductivity solid copper ion conductors: N-alkyl (or hydro)-hexamethylenetetramine halide-copper (I) halide double salts', *ibid.*, 1973, **120**, pp. 1431-1434

10 TAKAHASHI, T., and YAMAMOTO, O.: 'Solid state ionics: high conductivity solid copper ion conductors: N,N' dialkyl (or dihydro)-triethylenediamine dihalide-copper(I) halide doublt sales', *ibid.*, 1975, **122**, pp. 83-86

11 SAMUELS, A.F., GOUGOUTAS, J.Z., and OWENS, B.B.: MHigh conductivity solid electrolytes, double salts of substituted ammonium halides and cuprous halides', *ibid.*, 1975, **122**, pp. 1291-1296

12 TAKAHASHI, T., YAMAMOTO, O., YAMADA, S., and HAYASHI, S.: 'Solid-state ionics: high copper ion conductivity of the system CuCl-CuI-RbCl'. Extended Abstracts Second International Meeting on Solid Electrolytes, St. Andrews, Scotland, 1978, pp. 6.2.1-6.2.4

13 LAZZARI, M., PACE, R.C., and SCROSATI, B.: 'Electrochemical properties of the N–N′-dimethyl triethylenediamine dibromide-cuprous bromide solid electrolyte', *Electrochim. Acta,* 1975, **20**, pp. 331-334

14 ARMSTRONG, R.D., DICKINSON, T., and TAYLOR, K.: 'The anodic decomposition of copper(I) N-methyl hexamethylenetetramine bromide', *J. Electroanal. Chem.*, 1975, **64**, pp. 155-162

15 ARMSTRONG, R.D., DICKINSON, T., and TAYLOR, K.: 'The kinetics of the Cu/Cu^+ electrode in solid electrolyte systems', *ibid.*, 1974, **57**, pp. 157-163

16 LAZZARI, M., RAZZINI, G., and SCROSATI, B.: 'An investigation on various cathode materials in copper solid state cells', *J. Power Sources,* 1976, **1**, pp. 57-63

17 TAKAHASHI, T.: 'Some superionic conductors and their applications'. *In* MAHAN, G.D., and ROTH, W.L. (Eds.) *Superionic conductors* (Plenum Press, 1976), pp. 379-394

18 SHILTON, M.G., and HOWE, A.T.: British Patent Application 47470/76

19 FARRINGTON, G.C., and BRIANT, J.L.: *Mat. Res. Bull.* (in press)

20 CHILDS, P.E., HOWE, A.T., and SHILTON, M.G.: *J. Power Sources* (In press)

21 WILL, F., and MITOFF, S.P.: 'Primary sodium batteries with beta-alumina solid electrolyte', *J. Electrochem. Soc.*, 1975, **122**, pp. 457-461

22 CHANDRASHEKHAR, G.V., and OSTER, L.M.: 'Ionic conductivity of monocrystals of the gallium analogues of β– and β''–alumina', *ibid.*, 1977, **124**, pp. 329-332

23 KENNEDY, J.H., and MILES' R.C.: 'Ionic conductivity of doped beta-lead fluoride', *ibid.*, 1976, **123**, pp. 47-51

24 REAU, J.M., LUCAT, C., PORTIER, J., HAGENMULLER, P., and COTT, M.: *Mat. Res. Bull.* (in press)

25 KENNEDY, J.H., and HUNTER, J.C.: 'Thin-film galvanic cell $Pb/PbF_2/PbF_2,CuF_2/Cu$', *J. Electrochem. Soc.,* 1976, **123**, pp. 10-14

26 SCHOOMAN, J.: 'A solid-state galvanic cell with fluoride-conducting electrolytes', *ibid.*, 1976, **123**, pp. 1772-1775

27 LIANG, C.C.: 'Conduction characteristics of the lithium iodide-aluminium oxide solid electrolytes', *ibid.*, 1973, **120**, pp. 1289-1292

28 RAYLEIGH, D.O.: 'Solid electrolytes', *In* RANGARAJAN, S.K. (Ed.) *Topics in pure and applied electrochemistry* (S.A.E.S.T., 1975), pp. 103-121

29 HARRIS, L.B.: 'Fast ion conduction by an interfacial mechanism'. Extended Abstracts Second International Meeting on Solid Electrolytes, St. Andrews, Scotland, 1978, pp. 1.7.1.-1.7.4

30 NOMURA, E.: 'Lithium-ion conductive solid electrolytes', *Chem. Abstr.* 85: 49217v. Japanese Patent 76. 12,633

31 NOMURA, E.: 'Solid electrolyte with lithium-ion conductivity', *Chem. Abstr.* 85: 145660e. Japanese Patent 76. 23,497

32 OWENS, B.B., and HANSON, H.J.: 'Solid electrolytes for use in solid state electrochemical devices' US Patent 4,007,122 (Feb. 8th, 1977)

33 WHITTINGHAM, M.S., and HUGGINS, R.A.: *In* ROTH, R.S., and SCHNEIDER, S.J. (Eds.) *Nat. Bur. Standards. Special Publication 364* (Washington, 1972).

34 FARRINGTON, G.C., and ROTH, W.L.: 'Li^+–Na^+ beta alumina - a novel Li^+ solid electrolyte', *Electrochim. Acta,* 1977, **22**, pp. 767-772

35 ALPEN, U. von, and BELL, M.F.: 'Impedance measurements on Li_3N single crystals' (submitted to *J. Electroanal. Chem.)*
36 BOUKAMP, B.A., and HUGGINS' R.A.: 'Fast ionic conductivity in lithium nitride', *Mat. Res. Bull.,* 1978, **13**, pp. 23-32
37 ALPEN, U. von., Private communication
38 HUGGINS, R.A.: 'Recent results on lithium ion conductors', *Electrochim. Acta,* 1977, **22**, pp. 773-781
39 SHANNON, R.D., TAYLOR, B.E., ENGLISH, A.D., and BERZINS, T.: 'New Li solid electrolytes', *ibid.,* 1977, **22**, pp. 783-796
40 LIANG, C.C., and BARNETT, L.H.: 'A high energy density solid-state battery system', *J. Electrochem. Soc.,* 1976, **123**, pp. 453-564
41 LIANG, C.C., JOSHI, A.V., and BARNETT, L.H.: 'Metal salt depolarisers in solid state batteries' Power Sources Symposium Proceedings, 1976, **27**, **pp. 141-144**
42 LIANG, C.C., and BARNETTE, L.H.: Geman Patents Nos. 2,517,882 (13th Nov. 1975) and 2,517,883 (13th Nov. 1975)
43 SCHNEIDER, A.A., MOSER, J., WEBB, T., and DESMOND, J.: 'A new high energy density solid electrolyte cell with a lithium anode' Proceedings 24th Power Sources Symposium, 1970, pp. 27-30
44 SCHNEIDER, A.A., and KRAUS, F.E.: 'End-of-life characteristics of the lithium-iodide cell'. National Bureau of Standards Workshop III 'Reliability Technology for Cardiac Pacemakers', 1977
45 SCHNEIDER, A.A., GREATBATCH, W., and MEAD, R.: 'Performance characteristics of a long life pacemaker cell' *In* COLLINS, D.H. (Ed.) *Power sources 5* (Academic Press, 1975), pp. 651-659
46 GREATBATCH, W., MEAD, R.T., McLEAN, R.L., RUDOLPH, F., and FRENZ, N.W.: US Patent 3,994,747 (30th Nov. 1976)

Index

Absorbent pad,
use in mercury/zinc and silver/zinc cells, 76
Acetonitrile,
organic non-aqueous electrolye, 8
Acetylene black,
component of cathodes of Leclanche cells, 67
Acid,
hydrocholoric, electrolyte for hydrogen electrode, 25
nitric, electrolyte in Grove cell, 10
nitric, corrodant for Plante plates, 13
perchloric, electrolyte in PbO_2/Pb cell, 3
perchloric, corrodant for Plante plates, 13
chloric, corrodant for Plante plates, 13
phosphoric, additive to electrolyte of lead-acid cells, 220
sulphuric, electrical resistivity, 157
sulphuric, concentration in lead-acid cells, g/L v.S.G., 157
Sulphuric, purity standards for lead-acid cells, 157
A.C. perturbations, applications to electrochemical reactions, 44
Acopor WA, ion-exchange membrane separator for silver/zinc cells, 76
Active materials,
alkaline manganese cells, 67, 69
lead-acid batteries, 156, 157
Leclanche primary cells, 67, 69
mercury/zinc cells, 75
nickel/cadmium cells, 329, 333
nickel/iron cells, 329, 350
nickel/zinc cells, 391
silver/zinc cells, 75
sodium/sulphur cells, 435, 438
Activation polarisation, lead-acid cells, 167
Activities, reactants and products, relation to cell potential, 28
Activation energy, zincate and ions in alkaline manganese cells, 129
Additives to electrolytes,
lithium hydroxide in alkaline cells, 338
phosphoric acid in lead-acid cells, 220
Adsorption, relation to kinetics, 32
Aircraft batteries,
lead-acid, 306
sintered-plate nickel/cadmium, 378
silver/zinc, 390
Air-conditioning batteries, nickel/cadmium, 347
Air-depolarised cells and batteries, 3, 56, 72, 98, 107
Alkaline storage batteries, world production, 325

Alkaline manganese batteries, components and construction, 67
Alloys for grids for lead-acid batteries,
automotive, 199
traction flat plate, 263
tubular positive plate, 268
Alternating current, measurement of resistive effects in batteries, 47
Alternators, charging lead-acid automotive batteries, 259
Aluminium, grids for lead-acid batteries, 208
Amalgamation,
zinc in Leclanche cells, 69
zinc in alkaline manganese cells, 69
effect on passive dissolution of zinc, 131
Ammonium chloride,
electrolyte in Leclanche cells, 56, 67, 69, 73
volume of solid in Leclanche cells, 73
Amorphous, zinc hydroxide in primary alkaline cells, 129
Anodes,
definition, 5
alkaline manganese cells, 69
Anode discharge mechanisms,
alkaline manganese cells, 128
Leclanche cells, 125
silver/zinc or cadmium button cells, 76
Andre, H. silver/zinc cell, 16
Anti-polar mass, sealed nickel/cadmium sintered plate cells, 361
Appliances, cordless, 156, 289
Applications,
air depolarised batteries, 97
alkaline manganese batteries, 87
lead-acid, gelled electrolyte cells, 290
sealed cells, 289, 293
starved electrolyte cells, 290
supported electrode cells, 291
Nickel/cadmium pocket plate batteries, 347, 349
nickel/cadmium sintered plate batteries, 378
nickel/iron batteries, 357
nickel/zinc batteries, 397
silver/zinc and silver/cadmium batteries, 391
sodium/sulphur high temperature batteries, 455
Application discharge tests for primary batteries,
button cells, 87
flat cell types, 88
round and square types, 88
Arsenic, lead-acid batteries,
alloys for grids, 201, 202
effect on negative plates, 160
Arsine, production in lead-acid cells, 160
Assembly, lead-acid,
SLI batteries, 249
flat plate traction cells, 264
tubular plate cells, 270
starved electrolyte cells, 301, 303
Assembly, primary,
alkaline manganese cells, 68
button cells, 75
flat cell batteries, 72
Leclanche round cells, 68
Automotive lead-acid batteries,
annual production in USA and Japan, 153
chargers, 258
maintenance-free, 210, 214
Automotive lead-acid batteries, specification,
S.A.E., 253
B.C.I., 253
I.E.C., 254
Avrami, theory of crystal growth, 42, 43

Bacon, F.T., fuel cell, 17
Barium sulphate, expander for lead-acid negative plates, 231
Barton pot, manufacture of oxide for lead-acid batteries, 222
Basic lead sulphates,
lead-acid batteries, 217
mono, tri, tetra-basic sulphates, 229
Battery, definition, 50
multi-cell, 55
Battery storage, alkaline nickel-cadmium, 331

Battery storage lead-acid,
automotive, 153
traction or motive power, 154
stationary, 156
Battery, high temperature systems,
comparison, 451
molten salt electrolytes, 406, 452
solid electrolytes, 452
future prospects, 454
Bellcell, lead-acid telephone battery, 287
Beta alumina solid electrolyte, 420
crystal structure, 422, 423
electrical resistivity, 424
polymorphic forms, 422
processes of manufacture, 420
Beta lead dioxide,
morphology, 185
electrode potential, 185
oxygen overvoltage, Tafel lines, 182
Bitumen, seals for primary cells, 70
Blanc fixe, expander for lead-acid negative plates, 231
Bobbin, cathode for Leclanche cells, 70
Boron nitride, insulant for lithium/iron sulphire cells, 413
Box negative plate, lead-acid batteries, 193
Braking, regenerative, motive power batteries, 262
Briquette manufacture, nickel/cadmium pocket plates, 335
Bromine/zinc cells, 9, 19
Brush C., pasted lead-acid plate, 15
Bunsen, R.W., primary cell, 10
Burglar alarms, sintered plate nickel/cadmium batteries, 380
Button cells and batteries,
dimensions and shapes, 61
performance on IEC hearing test, 99
mercury/zinc and silver/zinc, 75
nickel/cadmium, pressed powder plates, 363, 367

Cadmium,
electrodes in nickel/cadmium pocket plate cells, 332
electrodes in siver/cadmium cells, 384
electrodes in sintered plate cells, 360
hydroxide, polymorphic forms, 333
Cadmium, lead-acid cells,
alloys for grids, 213
reference electrode, 179
Capacitance, interfacial electrode, 30
Capacity, primary cells,
discharge-voltage curves, 82
alkaline manganese cells, 95, 96
Leclanche cells, 91, 92
practical discharge, 79
total capacity, 78
Capacity lead-acid cells,
effect of plate thickness, 175, 178
effect of current density, 170, 172
temperature coefficient, 173
Calcium, grid alloy for lead-acid cells, 210
Carbon, active,
for chlorine electrode, 408
air depolarised cells, 71
Carbon,
expander for lead-acid negative plates, 230, 231
felt for sodium/sulphur cells, 430
cathode rod for Leclanche cells, 70
vitreous foam for ferric sulphide electrodes, 412
vitreous material for sodium/sulphur cells, 431
Carbonate, contaminant in alkaline batteries, 361, 362
Carboxymethyl cellulose in primary batteries, 69
Catalators, use in lead-acid batteries, 296
Catalysis in electrodes, 33
Cathode,
definition, 2, 5
alkaline manganese cells, 67
Cathode, Leclanche cells, 67
discharge mechanism, 113
Cellophane, separator for silver/zinc cells, 76, 385
Ceramics, preparation of beta alumina, 425

Charge acceptance,
lead-acid batteries, 180, 255
lithium/iron sulphide high temperature cells, 416
sodium/antimony trichloride cells, 443
sodium/sulphur cells, 435
Charge retention, nickel/cadmium,
pocket plate cells, 345
sintered plate cells, 376
Charge retention
nickel/iron cells, 356
nickel/zinc cells, 394
silver/zinc cells, 389
Charge transfer,
Leclanche cells, 113, 119
alkaline manganese cells, 120
cathodes in solid electrolyte (room temps.) cells, 474, 478
Chargers, lead-acid batteries,
automotive, 258
traction, one step, 279
two step, 281
Chaney, N.K., manganese oxides, solid solutions, 101
Chloride ion,
concentration in Leclanche cells, 111
impurity level in lead-acid electrolyte, 160
Chloric acid, corrodant for Plante formation, 13
Chlorine,
gas electrode, 6
zinc cells, 19
lithium high temperature cells, 406
Clark, L., standard cell, 15
Clerici, G., zinc/bromine battery, 19
Coated metals, components in sodium/sulphur cells, 432
Cobalt hydroxide, pocket plate nickel/cadmium cells, 333
Cobalt nitrate, sintered plate alkaline cells, 360
Coefficients of use of active materials,
lead-acid automotive, 175, 177
lead-acid traction, 195, 196
Computer,
assisted design of grid for lead-acid batteries, 195
controlled systems, use of lead-acid batteries, 288, 289
Concentration polarisation,
Leclanche cells, 114
lead-acid cells, 167
Conductivity,
lead and lead alloys, 199
sulphuric acid, 159
beta alumina for sodium/sulphur cells, 422
Containers, lead acid,
automotive batteries, 248
traction cells, 265
stationary cells, 284
starved electrolyte cells 303
Containers, nickel/cadmium,
pocket plate cells, 336, 337
sintered plate cells, 364, 376
Containers,
Nickel/iron cells, 351
nickel/zinc cells, 393
Containers, primary,
air-depolarised cells, 71
button cells, 75
Leclanche cells, 70
alkaline manganese cells, 68
Containers,
sodium/sulphur high temperature cells, 427, 429
solid electrolyte, room temperature cells, lithium/iodide, 473, 475
silver/zinc or cadmium cells, 386
Copper,
additive to lead-acid grid alloys, 203, 205
chloride/lithium cells, 20
electrode in Daniell cell, 7
grids for lead-acid cells, 208
halides in solid electrolyte cells, 466
impurity level in lead-acid electrolyte, 160
ion conductor in solid electrolyte cells, 466
Cordless appliances,
lead-acid batteries, 156, 289
nickel/cadmium cells, 379
silver/cadmium batteries, 391

silver/zinc batteries, 391
Corrosion,
grids in lead-acid batteries, 154, 201, 204, 211, 214
zinc electrodes in primary cells, 69, 71, 76
steel containers for alkaline cells, 336,
problems with high temperature chlorine electrodes, 406
stainless steel alloys in sodium/sulphur cells, 431
Coup de fouet, lead-acid cells, 190
Covers and vents,
lead-acid cells and batteries, 248
alkaline cells, 336, 337, 351, 364
catalytic vent-plugs, lead-acid batteries, 296
Cranking current (C.C.A.) lead-acid batteries, 254
Crystal structure,
manganese oxides in primary cells, 100, 102
lead and lead alloys, 199, 201, 206
alpha and beta lead dioxide, 185, 186
nickel hydroxide and cadmium hydroxide, 332, 333
Curing, processes for lead-acid plates, 233, 236, 238
Current density, relation to potential, the Tafel law, 34, 37, 172, 182
Cyclon cells, sealed lead-acid, 307
Cylindrical cells,
alkaline, pressed powder electrodes, 367
Leclanche and alkaline manganese, 58, 67
mercury/zinc batteries, 78
cyclon lead-acid cells, 307

Daniell, J.F., primary cell, 7
Density, apparent, lead oxide powder, 227
Diamond grid pattern, lead-acid traction batteries, 195
Diesel starting,
lead-acid batteries, 153
nickel/cadmium batteries, 347
Diffusion,
effects in electrode systems, 34
layer on electrode surfaces, 46
effects in Leclanche cells, 114
effects in alkaline manganese cells, 95
Dimensions, standards for primary cells, 53 *et seq.*
Discharge, nominal capacities of primary cells, 80, 81, 82
Discharge characteristics,
lead-aicd plates, 170, 172, 174
nickel/cadmium cells, 339, 341
nickel/iron cells, 353
sintered plate alkaline cells, 370, 372
silver/zinc cells, 387
Dispersion-strengthened lead, 206
Dissolution-precipitation mechanism,
alkaline manganese electrodes, 121
Lead-acid electrodes, 187
nickel/cadmium electrodes, 332
mercury oxide cathodes, 122
silver oxide cathodes, 123
Double layer,
Helmholtz theory of electrodes, 30, 31, 32
Gouy-Chapman theory, 30
Double sulphate theory, lead-acid cells, 165
Dry-charging process, lead-acid plates, 243
Duplex electrodes, flat Leclanche cells, 71
Dust,
production of lead oxide by ball-milling, 221
production of lead oxide by thermal processes, 222
Dynamo, charging lead-acid batteries, 14, 259

Edison Thomas Alvin, alkaline batteries, 324
Efficiency,
based on Gibbs free energy, 29
thermal convertor, 31
charge/discharge relation of alkaline cells, 345, 356, 376

Electric vehicles,
alkaline batteries, 357
lead-acid batteries, 154, 261, 262
Electric fences, air depolarised batteries, 97
Electron,
transfer processes, 2, 5, 25, 33
diffusion in manganese dioxide, 117
Electrode,
definition, 24
potential, 1, 26, 29, 162
reference, 26
volumes in primary cells, 69, 73
Electrochemical kinetics, 31
Electrochemical reactions,
alkaline cells, 331, 350, 381, 382
lead-acid cells, 163, 164
primary cells, 99, 106, 114
solid electrolyte (room temps) cells, 473, 478
Electrolytic process for cadmium/iron electrodes, 334
Electrocrystallisation, 34, 37
Electrolyte,
definition, 24
alkaline manganese primary cells, 77
alkaline storage batteries, 327, 330, 338, 353, 385
lead-acid batteries, 158, 159
Leclanche cells, 67, 73
molten salts, high temperature cells, 413
organic, for lithium cells, 20
solid, for high temperature cells, 418
Electromotive force (e.m.f.), definition, 26
Electrostatic interactions at electrode surfaces, 31
Emergency lighting, stand-by power applications,
lead-acid, 156, 284
alkaline, 347, 380
Energy density,
definition, 5
alkaline storage batteries, 328, 353, 368, 376
lead-acid batteries, automotive, 4, 250
lead-acid batteries, traction, 273, 275
nickel/zinc batteries, 394
silver/zinc or cadmium cells, 386
sodium/sulphur cells, 430
Enthalpy, relation to Gibbs free energy, 27, 29, 403
Estelle, A., pocket plate cell designs, 331
Electrolyte-starved,
alkaline cells, 365
lead-acid cells, 302
Engine starting,
lead-acid SLI batteries, 251 *et seq.*
alkaline storage batteries, 347, 379
Expanders, lead-acid negative plates, 230
Expanded metal, wrought grids for lead-acid cells, 197

Faraday,
laws of electrochemistry, 7
constant, 26
current, 35
Fast charging,
lead-acid batteries, 180, 281
sintered plate nickel/cadmium batteries, 375
Faure, C., process for lead-acid plates, 15
Fick, diffusion law, 41
Final voltages on discharge,
lead-acid batteries, 252, 273
alkaline storage batteries, 340, 335, 372, 396
silver/zinc or Cadmium cells, 388
Leclanche primary batteries, 94, 95
Flat Leclanche batteries, designation and dimensions, 54
Flat plate lead-acid cells, 263
Float service,
lead-acid batteries, 283, 284, 289
nickel/cadmium batteries, 345
Fluoride ion conductors, solid electrolyte (room temps) cells, 468
Fluorine, electrode potential, 4

Formation,
lead-acid plates and cells, 191, 243, 265, 302
nickel/cadmium pocket plate cells, 338
nickel/iron cells, 351
sintered plate nickel/cadmium cells, 361, 364
Fuel cells, 1, 10, 16, 31, 33
Fugacity, gaseous electrodes, 28

Galvani, L.,
electrochemical cell, 7, 25
potential, 26
Galvanostatic experiment, 35
Gap fillers, button cells, 77
Gas,
polarisation of electrodes, 11
recombination in sealed alkaline cells, 259
recombination in sealed lead-acid cells, 296, 298, 299
evolution at top-of-charge, lead-acid cells, 177
-tight lead-acid cells, 293
Gauntlet, lead-acid tubular positive plate, 266, 268
Gauss, field strength theorem, 32
Gelled electrolyte, lead-acid cells, 290
Gelling agents, primary cells, 69
Gibbs,
free energy change, 28, 29, 105
-Helmholtz equation, 28
Glass, borate electrolyte for high temperature cells, 420
Glass-wool, retainers for lead-acid flat-plate cells, 263
Gouy-Chapman theory of charge distribution, 30
Grain,
boundary phases in grid alloys for lead-acid cells, 211
size, effect of additions of lead and cadmium on zinc, 70
size, pure lead and lead alloys, 199, 201, 204, 206
size, wrought lead-calcium alloy, 197
Graphite, in cathodes for,
Leclanche cells, 67
alkaline manganese cells, 67
mercury/zinc cells, 74
nickel/cadmium cells, 333
silver/zinc or cadmium cells, 74
Graphite, current collector for,
chlorine electrode, 407
sulphur electrode, 428/431
Grey oxide,
basic material for lead-acid plates, 221
methods of manufacture, 222, 223
properties, 223 *et seq.*
purity standards, 228
Grids for lead-acid batteries,
anodic corrosion, 201, 211, 213
design of die-cast grids, 193, 195
lead alloys, 199, 202, 203
die-casting machines, 220
plastic materials, 209
titanium, copper, aluminium, 208
wrought lead-calcium alloy, 197
Grinding mills for grey oxide, Tudor, Hardinge, 220
Groutite, mineral manganese dioxide, 100
Grove, S (Sir), gas electrodes, 16

Hardeners for alloys for lead-acid battery grids, 200, 201
Hearing aids, button cells, IEC hearing aid test, 97
Heart pacemakers, solid electrolyte (room temps) cells, 464
Hybrid electric vehicles, 263
Hydrazine, reducing agent for manganese dioxide, 100
Hydrox fuel battery, 17
Hydroxyl ions, concentration in,
Leclanche cells, 116
alkaline manganese cells, 109, 120
Hydroxly ions, reactions in alkaline batteries, 331, 332, 350, 381
HUP, solid electrolyte for room temperature cells, 467
Hysterisis, charge/discharge curves for alkaline batteries, 332

Immobilised electrolyte, lead-acid

cells, 290, 291
Impedance, a.c. study, 44, 46
Impurity levels in electrolyte,
lead-acid batteries, 160
alkaline batteries, 327, 361, 362
Injection moulding,
plastic containers for lead-acid batteries, 249
plastic containers for alkaline cells, 337
Insulation of high temperature cells, 405, 406
Internal resistance,
nickel/cadmium cells, 302, 343
silver/zinc or cadmium cells, 389
"starved electrolyte" lead-acid cells, 302
International common samples,
materials for primary cells, 325
Inverters,
use with lead-acid batteries, 283, 288, 289
use with nickel/cadmium batteries, 349
Ion exchange,
effect of pH in Leclanche cells, 109
resin membranes for button cells, 76
Ionic atmosphere, thickness at electrode/solution interface, 30
Iron,
anode in nickel/iron cells, 350
additive to cadmium anode in nickel/cadmium cells, 334
impurity in sulphuric acid electrolyte, 160
Iron sulphide/lithium cells, 20, 412 *et seq.*
Ironclad tubular lead-acid plate, 265

Jar formation, lead-acid cells, 243
Jelly electrolyte for lead-acid cells, 290
Jungner Waldemar, development of alkaline batteries, 16, 324

Kathanode, lead-acid traction cell, 263
Kinetics,
of electrode reactions, 31
effects of polarisation in lead-acid electrodes, 166
Kummer, J.T., sodium/sulphur cells, 20, 422
Kraft paper, separator for Leclanche cells, 70

Laurent-Cely, F., chloride process for lead-acid plates, 15
Lead,
consumption in batteries in USA and Japan, 151, 154
alloys for lead-acid grids, 199, 203, 264, 268
dispersion-strengthened lead (DSL), 206
purity standards for oxide, 228
sulphate, solubility in sulphuric acid, 165
sulphate, double-sulphate theory, lead-acid cells, 165
Lead oxides,
ball-milling processes of manufacture, 221
Barton pot or Linklater process, 222
litharge or lead monoxide, properties, 183
lead dioxide, properties, 184, 185
red lead or minium, 183
Lifetime,
lead-acid, Bellcells for telephone service, 288
lead-acid Plante batteries, 285
lead-acid traction cells, 261
nickel/cadmium, pocket plate cells, 345
nickel/cadmium sintered plate cells, 377
nickel/iron cells, 356
nickel/zinc cells, 396
silver/zinc cells, and cadmium, 389
sealed primary, lead-acid and alkaline cells, 302, 304
lighting, emergency,
lead-acid batteries, 283
nickel/cadmium batteries, 347
Lignin, expander for lead-acid nega-

tive plates, 231
Ligno-sulphonates, expanders for
lead-acid negative plates, 231
Linklater pot, lead oxide production,
222
Lithium/chlorine cells,
electrolyte, 406
problems in design, 406, 407
Lithium-aluminium/chlorine cells,
electrodes, 408
performance, 409
Lithium halide solid electrolytes
(room temps), conductivity, 470
Lithium iodide, solid electrolyte cells
(room temps), 475
Lithium ion conductors, solid
electrolyte (room temps), 469
Lithium/iron sulphide cells, 412
chemical reactions, 412
electrodes, 413
discharge voltage curves, 416
overcharge characteristics, 416
status, 418
Lithium hydroxide,
use in nickel/cadmium cells, 338
use in nickel/iron cells, 353
Lithium/selenium or tellurium high
temperature cells, 411
chemical reactions, 411
electrolytes, 411
Lithium-silicon alloy for high
temperature cells, 412, 417
Lithium/sulphur high temperature
cells,
chemical reactions 409
electrolytes, 411
Local action,
lead-acid negative plates, 161, 200
nickel/cadmium cells, 345
Liquid crystal displays, batteries for
watches, calculators, 78, 98
Luggin capilliary, potential measurements, 35
Low temperature performance,
lead-acid electrodes and batteries,
173, 256, 258
nickel/cadmium cells, 343
nickel/cadmium sintered plate
cells, 369
nickel/iron cells, 355
nickel/zinc, 394
silver zinc or cadmium, 389

Maintenance, low for emergency
lighting, nickel/cadmium cells,
349
Maintenance-free batteries, lead-acid,
210
Manchester rosette Plante lead-acid
plates, 192
Manganese dioxide,
cathode in Leclanche, alkaline
manganese cells, 67
active natural ore, 91, 93
electro-deposited, 79, 84, 91
crystal structures, 101, 102, 103,
115
products of discharge 101
Manganese oxy-hydroxide, 102, 109,
116
Manganous hydroxide, alkaline
manganese cells, 121
Mass transport,
effect on performance, 5, 27, 31
in nickel hydroxide electrode,
332
Mechanical stability,
nickel/cadmium or iron batteries,
346, 356, 377
nickel/zinc batteries, 397
silver/zinc cells, 389
Mercury, use in Leclanche and alkaline manganese primary cells, 69
Mercury oxide,
solubility in potassium hydroxide,
122
additive to zinc electrodes, 383
Mercury sulphate, standard cells, 15
Metal/air cells, 18
Metal/halogen cells, 19
Metal-coated non-metals,
current collectors, sodium/
sulphur cells, 433
tantalum-dope rutile, 433
Microporous separators, lead-acid
cells, 245
Minium or red lead, properties, 183
Miners' car-lamp batteries, lead-
acid, 156
Mining locomotive batteires,

lead-acid, 154
nickel/iron, 357
Missile batteries, silver/zinc, 390
Molten salt electrolytes, 3, 408, 453
Molybdenum current collectors, sodium/antimony trichloride cells, 446
Multi-cell primary batteries, dimensions, lay-out, 64

National Bureau of Standards (USA), standards for primary cells, 51
Negative plates for lead-acid batteries,
paste composition, 229
setting/drying processes, 233
formation processes, 242
dry-charging, 244
discharge characteristics, 171
factors affecting capacity, 174
Nernst, diffusion layer, 46
Neutral air-depolarised cells, 52, 71
Nickel/cadmium alkaline cells,
pocket plates, 331
reaction mechanisms, 331
manufacturing processes 333 *et seq.*
properties, summary, 346
low costs, 346
Nickel hydroxide, reactions in alkaline cells, 331
Nickel/iron cells,
reactions, 350
manufacturing processes, 351
performance data, 353
Nickel-plating,
steel ribbon for pocket type plates, 334
steel containers for alkaline cells, 336
Nickel sulphate, material for nickel pocket plates, 333
Nickel/zinc cells,
processes and performance data, 391, 394
vibrating zinc electrode, 396
Nickel, effect on lead-acid negative plates, 161
Nitrogen oxides, effect on lead-acid plates, 159
Nitrates,
impurities in lead-acid cells, 160
nickel with cobalt, cadmium for sintered plates, 360
Nitric acid, formation of Plante lead-acid plates, 191
Nomenclature of primary cells, 52
Non-steady state, examination of electrode systems, 39
Non-metals, use in sodium/sulphur cells, 431
Nucleation,
phase changes in electrodes, 39
passivation of zinc electrodes in primary cells, 129, 132
of lead dioxide by antimony in lead-acid cells, 217

Open-circuit potentials,
primary cells, 101
vs composition of MnO_2 in primary cells, 103
primary cells, calcd. and observd., 106
alkaline storage batteries, 328
sodium/sulphur cells, 420
sodium/antimony trichloride cells, 443
lithium/iodide solid electrolyte cells, 473
Open-circuit potential vs available capacity, sealed lead-acid cells, 305
Organic expander, Lead-acid negative plates, 231
Overcharge,
effect on sintered plate nickel/cadmium cells, 375
conditions for thermal runaway, 374
Overvoltage and polarisation, 4
oxygen, lead-acid positive plates, 165

Pacemakers, heart, solid electrolyte cells, 464, 474
Paper,
Kraft separators for primary cells, 70
pulp separators for lead-acid batteries, 246

Paraffin, use in nickel/cadmium active materials, 334
Particle size, methods of measurement, 224
Passivation, zinc anode in alkaline electrolyte, 133
Pastes, lead-acid cells,
processes of manufacture, 229, 232
curing and drying, 233
porosity and surface area, 236, 237
additives to positive and negative, 230, 231
Perchloric acid, lead-acid Plante process, 13, 191
Performance, lead-acid,
automotive batteries, 253 *et seq.*
traction batteries, 273 *et seq.*
stationary batteries, 285
starved electrolyte cells, 303
nickel/cadmium pocket plate cells, 340
nickel/cadmium sintered plate cells, 372, 373
nickel/iron cells, 355
silver/zinc and cadmium cells, 388
nickel/zinc cells, 396
primary batteries, nominal capacities, 80/81
Perforated steel ribbon,
needle process of manufacture, 334
roller die process, 334
Permion, separators for button cells, 363
Persulphuric acid, formation in lead-acid cells, 180
Perturbation, AC processes in battery research, 39
Plante, H., lead-acid process, 191
high performance lead-acid cells, 283
Plastic containers, alkaline cells, 336
Platinum, catalyst in catalator vent plugs, 295, 297
Pocket plates, processes for nickel/cadmium cells, 331, 334
Polarisation,
metals by gas, 12
lead-acid cells, kinetic aspects, 166
Polarisation, lead-acid cells,
kinetic aspects, 166
concentration, 167
resistance, 169
Polarisation, Leclanche primary cells, manganese dioxide electrode, 114
Polarity, primary cells, IEC designation, 58
Polycarbon monofluoride, primary cells, 8, 20
Polypropylene cell containers,
lead-acid cells, 198
alkaline cells, 336
Polypropylene separators, alkaline cells, 364
Polymorphism,
litharge, 183, 223
lead dioxide, 185
nickel oxyhydroxide, 331
Manganese dioxide, 102
zinc hydroxide, 129
Polyamide separators for alkaline cells, 363
Polytetrafluorethylene (PTFE), seal for sodium electrodes, 448
Polyvinyl chloride, separators for lead-acid cells, 246, 247
Porosity,
Leclanche positive electrode, 118
lead-acid positive plates, 237
lead-acid negative plates, 238
lead-acid separators, 247
sintered nickel plaques, 360
methods of measurement, 227
Porvic, poly-vinyl chloride separators for lead-acid batteries, 246
Potassium hydroxide,
alkaline manganese primary cells, 77
alkaline storage batteries, 327, 339, 343, 365
Power density, definition, 5
Pressed powder electrodes, nickel/cadmium cells, 367, 374
Proton conduction, HUP solid electrolyte, room temps cells, 467
Purity,
sulphuric acid electrolyte, 160

water for topping up lead-acid cells, 161
lead for oxide production, 228
Potassium hydroxide, properties of solutions, 330
Potential, electrode, definition, 35
Potentiostatic methods of testing, 35
Power stations,
lead-acid battery applications, 156
alkaline battery applications, 349

Racks, stationary alkaline batteries, 338
Radio communications, nickel/cadmium batteries, 347
Radar, silver/zinc batteries, 390
Railway applications, signalling etc.,
air-depolarised cells, 97
pocket-plate nickel/cadmium batteries, 347
nickel/iron batteries, 357
lead-acid batteries, rail-cars, 263, 155
lead-acid batteries, general services, 156
Rectifier, property of lead dioxide, 184
Recuperation, primary cells, 82
Reference electrode,
definition, 25, 35
cadmium electrode for lead-acid cells, 179
Red lead or minium, properties and uses in lead-acid cells, 183
Redox reactions, 2, 26
Reserve capacity, lead-acid battery SAE specification, 255
Resistance, internal,
lead-acid Plante cells, 285
lead-acid tubular plate cells, 285
lead-acid starved electrolyte cells, 302
nickel/cadmium, pocket plate cells, 343
nickel/cadmium sintered plate cells, 372
Resistance, specific resistivity,
pure lead, 170
sulphuric acid, 159
Resistance,
study of effects in cells by a.c. methods, 47
inter-cell connectors, SLI lead-acid batteries, 249
Retainers, lead-acid batteries, 246, 264, 266
Reversal,
nickel/cadmium pocket plate cells, 345
nickel/cadmium sintered plate cells, 364
Ring disc electrodes,
means of exploring reaction kinetics, 37, 41
silver/zinc electrodes, 124
Rubber,
vulcanised for lead-acid containers, 248
microporous separators, lead-acid cells, 246
Rubidium silver iodide,
solid electrolyte, 21, 467
power sources 466

Safety precautions, sodium/sulphur cells, 437
Sausage casing, segmented cellulose separator, 384
Scanning electron microscope, (s.e.m.) studies of electrodes, 216, 236
Scott volumeter, apparent density of lead oxides, 227
Sealed cells,
starved electrolyte lead-acid cells, 307
Leclanche and alkaline manganese cells, 68
nickel/cadmium sintered plate cells, 358
oxygen re-combination mechanisms, 298, 359
Seals, high temperature cells, 427, 446
Self-discharge,
lead-acid negative plates, 161, 200
nickel/iron iron negative plate in,

356
silver alkaline cells, 389
lithium/chlorine high temperature cells, 406
Selenium, alloy for lead-acid battery grids, 204
Semi-conductors, use as galvanic electrodes, 24
Separators,
lead-acid cells, 245 *et seq.*
Leclanche cells, 70
button cells, 76
nickel/cadmium pocket plate cells, 335
nickel/cadmium sintered plate cells, 362, 363
nickel-iron cells, 351
nickel/zinc cells, 392
silver/cadmium or zinc cells, 384
Setting, pastes for lead-acid plates, 233 *et seq.*
Sieving, particle size of lead oxides, 225
Signalling on railways,
nickel/cadmium batteries, 347
neutral air-depolarised batteries, 97
Silicone rubber, seal for sodium electrode, 446
Silica gel, lead-acid cells, 290
Silver,
alloys for lead-acid grids, 202
ion conductors, solid electrolyte cells, 465
chloride/magnesium seawater batteries, 3
hydroxide, solubility in alkali, 123
Silver oxide cathodes, properties and reactions, 99, 123, 381, 382
Silver/zinc or cadmium cells,
cycling lifetimes, 389
charge characteristics, 389
energy densities, 386
manufacturers, ranges supplied, 387
manufacturing processes, 382, 383, 385
Silver rubidium iodide solid electrolyte, 465
reaction mechanisms, 381
Sintered iron-copper negative plate, 351
Sintered plate nickel/cadmium cells,
chemical reactions, 359
manufacturing processes, 359, 361
separators, 362
Sodium hydroxide,
electrolyte for alkaline primary cells, 77
electrolyte for button cells, 78
electrolyte for nickel/iron cells, 353
passivation of zinc in primary cells, 135
Sodium polysulphides, properties, 421, 422
Sodium/sulphur cells,
cel reactions 418, 419
sulphur electrode, O/C potential, 420
beta alumina electrolyte, 422, 423, 425
flat plate cells, 426
central sodium tube cells, 427
Sodium/sulpher cells, central sulphur tube cells, 428
capacity and cycle number, 436
charge characteristics, 435
Sodium/sulpher cells,
sodium electrode, provision of steel wick, 436
cell safety, 437
glass electrolyte, construction, 438, 439
glass electrolyte, causes of failure, 441
glass electrolyte, outstanding problems, 442
Sodium/antimony trichloride cells,
electrodes, 442, 444, 445
tubular cell design, 444
cell reactions at 210°C, 442
seals, teflon (PTFE) for positive, 446
seals, silicone rubber for negative, 443
current collectors, broom and brush, 447

charge/discharge curves, 443
performance, energy density, 450, 453

Solid electrolyte, room temps.,
rubidium silver iodide, 465
complex uranium phosphate, HUP, 467
lithium halide compounds, 470

Solid electrolyte, high temps
batteries, future prospects, 454

Solid solutions in primary cells,
magnanese oxides, 101, 111, 113

Solubility,
lead sulphate in sulphuric acid, 165
zinc oxide and hydroxide in potassium hydroxide, 128
mercury oxide in alkali, 122
silver oxide in potassium hydroxide, 123

Space applications, silver/zinc
batteries, 390

Specific gravity,
sulphuric acid for lead-acid cells, 158
potassium hydroxide for alkaline cells, 327, 338, 353

Specifications,
primary cells and batteries, 55, 56, 57
lead-acid automotive batteries, 253

Spine castings, lead-acid tubular
positive plates, 265

Square cell batteries, primary, 54, 57, 63

Stack-built,
Leclanche batteries, 344
lead-acid traction cells, 265

Standard cell potential,
definition, 28
primary cells, calculated, 105

Standardisation, US national
standards, primary batteries, 51

Starch, use in Leclanche cells, 69

Starting, lighting and ignition
(s.l.i.) batteries, lead-acid, 153

Starting batteries for diesel engines,
lead-acid, 153
nickel/cadmium, 347

Starved electrolyte cells,
lead-acid, 307
nickel/cadmium, 365

Stationary batteries, lead-acid, 156, 282

Steady state galvanostatic/potentio-
static procedures, 35

Steel, ribbon,
alkaline pocket plates, 334
wick for sodium electrode, 436
can for alkaline manganese cells, 70
containers for alkaline cells, 337
wire screen for lithium electrodes, high temps cells, 407

Stibine, production in lead-acid cells, 200

Storage batteries,
classes of lead-acid, 152
alkaline batteries, 324, 328

Strontium-lead alloy for m.f. lead-
acid battery grids, 212

Submarine batteries, lead-acid,
standard, 308
double-decker cells, 312, 313

Sulphur, effect on lead alloys for
lead-acid grids, 205

Sulphur electrode in sodium/sulphur
high temps cells 433
cause of polarisation, 434
charge-discharge characteristics, 435

Sulphuric acid,
purity levels for lead-acid batteries, 157
concentrations for different services, 158
electrical resistivity, 159

Super-saturation,
zinc in alkaline batteries, 129
lead sulphate in lead-acid cells, 191

Surface area,
B.E.T. measurement, 226
lead-acid positive and negative materials, 237, 238
sintered plaques for alkaline plates, 360

Switch operation,
lead-acid batteries, 283

nickel/cadmium batteries, 347
Symbols, conventions of use, 24

Tafel Law,
definition, 34, 37, 180
oxygen over-voltage on alpha and beta lead dioxide, 182
Teflon, binder for pressed powder electrodes, 367
Telephone batteries,
air-depolarised cells, 97
lead-acid, Plante, 286
lead-acid Bellcell, 286
Tellurium,
alloy for lead-acid grids, 265
effect on negative plates, 265
tetrachloride, lithium chlorine cells, 408
Temperature,
coefficient of capacity, lead acid plates, 173
vs capacity, starved electrolyte lead-acid cells, 304
coefficient of e.m.f. nickel/cadmium cells, 344
Terylene, lead-acid tubular plates, 266
Thermal management, high temperature cells, 403
Thermal runaway, alkaline batteries, 374
Thermodynamics, basic principles, 27
Thermoneutral potential, definition, 403
Thermal converter, 31
Tin, alloys for lead-acid grids, 202
Titanium, grids for lead-acid cells, 208
Torpedoes, silver/zinc batteries, 390
Tortuosity,
Leclanche active materials, 118
active materials, lead-acid plates, 169
separators, lead-acid cells, 247
Traction batteries,
lead-acid flat plate, 263
lead-acid tubular plate, 265
lead-acid nickel/iron batteries, 357
Transistors, alternators for charging lead-acid s.l.i. batteries, 258
Trees, dendritic growth of zinc in silver/zinc cells, 382
Tri-ethanolamine, use in alkaline manganese cells, 121
Trucks, battery-powered,
UK production, 155
nickel-zinc batteries, 397
Tubular plates,
lead-acid cells, manufacturing processes, 265 *et seq.*
nickel-iron cells, processes, 349 *et seq.*
Tubular construction, beta alumina solid electrolyte 425
Tudor, H., development of Plante process, 14
Tungsten, current collectors in molten salt cells, 446

Uninterruptible power systems (u.p.s.), lead-acid batteries, 156, 288

Vapour pressure, sulphur at 400°C, 413
Vents,
cylindrical mercury/zinc batteries, 78
nickel/cadmium cells, 337
nickel/iron cells, 351
sintered plate nickel/cadmium cells, 364
flame-arresting, nickel/cadmium train-lighting batteries, 347
Vibrating electrode, nickel/zinc batteries, 393
Volta A., pile and crown of cups, 10
Voltage, calculated and observed, primary cells, 106
Voltammetry, alkaline cells, anodic processes, 131

Watches, button batteries for liquid crystal displays, 78
Warburg impedence, relation to diffusion, 44
Water,
absorption test for lead oxides,

228
purity standards for lead-acid batteries, 161
Webber N., development of sodium/ sulphur batteries, 20, 420
Weight analysis, lead-acid batteries,
automotive, 250
traction, 278
Wick, stainless steel felt for sodium electrodes, 436
Wood,
separators for lead-acid cells, 245
saw-dust, expander for lead-acid negative paste, 231

X-rays,
examination of primary cell reactions, 99, 100, 102, 113
study of polymorphism of lead dioxide, 185

Zamboni. G., galvanic "dry pile", 11

Zinc,
negative electrode in various systems, 3, 4, 8, 55, 326
concentration profiles in solution in Leclanche cells, 126
hydroxide, solubility in alkali, 128
oxide, solubility in alkali, 128

Zinc chloride, concentration in Leclanche cells, 56, 67, 69, 73, 120

Zincate,
formation in alkaline manganese cells, 130, 134
effect on solubility of silver oxide in alkali, 123
decomposition of ions, 129